CW01189674

AIRPORT DESIGN AND OPERATION

Third Edition

AIRPORT DESIGN AND OPERATION

Third Edition

by

ANTONÍN KAZDA

University of Žilina, Žilina, Slovakia

ROBERT E. CAVES

Loughborough University, Loughborough, UK (Retired)

Emerald

United Kingdom – North America – Japan
India – Malaysia – China

Emerald Group Publishing Limited
Howard House, Wagon Lane, Bingley BD16 1WA, UK

First edition 2000
Second edition 2007
Third edition 2015

Copyright © 2015 Emerald Group Publishing Limited

Reprints and permissions service
Contact: permissions@emeraldinsight.com

No part of this book may be reproduced, stored in a retrieval system, transmitted in any form or by any means electronic, mechanical, photocopying, recording or otherwise without either the prior written permission of the publisher or a licence permitting restricted copying issued in the UK by The Copyright Licensing Agency and in the USA by The Copyright Clearance Center. Any opinions expressed in the chapters are those of the authors. Whilst Emerald makes every effort to ensure the quality and accuracy of its content, Emerald makes no representation implied or otherwise, as to the chapters' suitability and application and disclaims any warranties, express or implied, to their use.

British Library Cataloguing in Publication Data
A catalogue record for this book is available from the British Library

ISBN: 978-1-78441-870-0

ISOQAR certified Management System, awarded to Emerald for adherence to Environmental standard ISO 14001:2004.

Certificate Number 1985
ISO 14001

INVESTOR IN PEOPLE

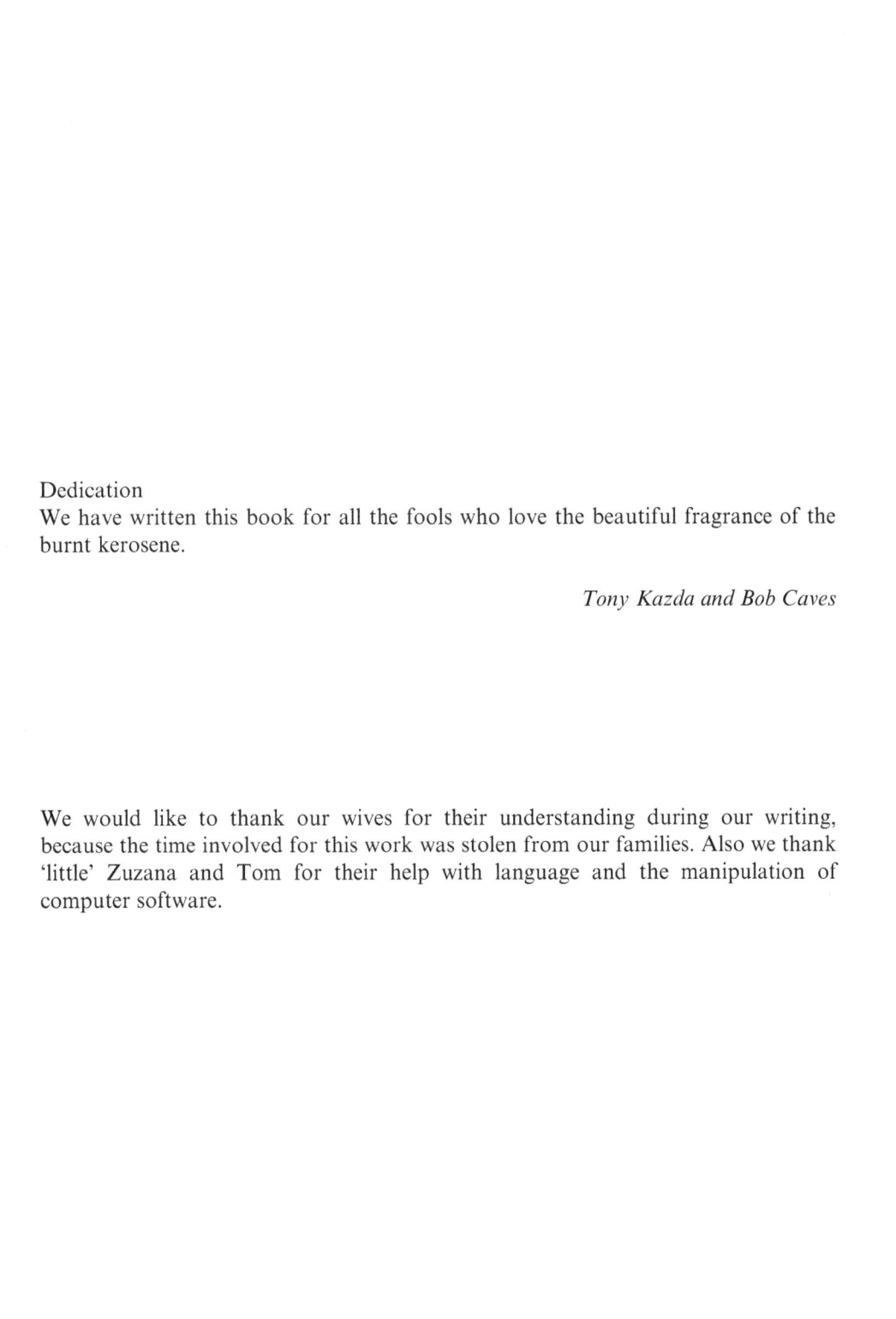

Dedication
We have written this book for all the fools who love the beautiful fragrance of the burnt kerosene.

Tony Kazda and Bob Caves

We would like to thank our wives for their understanding during our writing, because the time involved for this work was stolen from our families. Also we thank 'little' Zuzana and Tom for their help with language and the manipulation of computer software.

Contents

Abbreviations	*xv*
Preface	*xxiii*
Acknowledgements	*xxv*

1.	Air Transport and Airports	*1*
	1.1. Development of Airports	*1*
	1.2. Standards	*14*
	1.2.1. ICAO Legislation	*14*
	1.2.2. National Standards and Recommended Practices	*17*
	1.3. Airport Development Planning	*18*
2.	Predicting Traffic	*23*
	2.1. Introduction	*23*
	2.2. Types of Forecast Needed	*23*
	2.3. Methods of Analysis	*24*
	2.3.1. Informed Judgement	*25*
	2.3.2. Trend Extrapolation	*26*
	2.3.3. Econometric Models	*27*
	2.3.4. The Travel Decisions	*30*
	2.3.5. Modal Shares	*30*
	2.3.6. Discrete Choice Models	*31*
	2.3.7. Revealed and Stated Preferences	*33*
	2.3.8. Effects of Supply Decisions	*34*
	2.3.9. Uncertainty	*34*
	2.3.10. Scenario Writing	*35*
	2.4. Historic Trends in Traffic	*36*
	2.5. Factors Affecting the Trends	*37*
	2.5.1. Economic Factors	*37*
	2.5.2. Demographic Factors	*38*
	2.5.3. Supply Factors	*38*
	2.5.3.1. Cost per passenger kilometre (pkm)	*38*
	2.5.3.2. Aircraft size	*39*

viii Airport Design and Operation

	2.5.3.3. Cost of input factors	39
	2.5.3.4. Technology	40
	2.5.3.5. Management	40
	2.5.3.6. Capacity constraints	40
	2.5.3.7. Fares	41
2.5.4.	Economic Regulation	41
2.5.5.	Environmental Regulation	41
2.5.6.	Cargo	42
2.6. Conclusions		42
3. Airport Site Selection and Runway System Orientation		45
3.1. Selection of a Site for the Airport		45
3.2. Usability Factor		48
3.3. Effect of Low-Visibility Operations		51
3.4. Control of Obstacles		62
3.5. Other Factors		69
4. Runways		73
4.1. Aerodrome Reference Code		73
4.2. Runway Length		77
4.3. Declared Distances		93
4.4. Runway Width		95
	4.4.1. Runway Width Requirements	95
	4.4.2. Runway Shoulders	96
	4.4.3. Runway Turn Pads	97
4.5. Runway Slopes		98
	4.5.1. Transverse Slopes	98
	4.5.2. Longitudinal Slopes	99
5. Runway Strips and Other Areas		103
5.1. Runway Strips		103
5.2. Clearways		106
5.3. Runway End Safety Areas		106
6. Taxiways		113
6.1. Functional Criteria and Taxiway System Design		113
6.2. Rapid Exit Taxiways		115
6.3. Taxiway Separations		120
6.4. Taxiway Geometry		122
7. Aprons		127
7.1. Apron Requirements		127
7.2. Apron Sizing		128
7.3. Apron Location		130

	7.4. Apron Concepts	*131*
	7.4.1. Simple Concept	*132*
	7.4.2. Linear Concept	*132*
	7.4.3. Open Concept	*133*
	7.4.4. Pier Concept	*134*
	7.4.5. Satellite Concept	*135*
	7.4.6. Hybrid Concept	*136*
	7.5. Stand Types	*136*
	7.6. Apron Capacity	*142*
	7.7. Isolated Aircraft Parking Position	*144*
8.	Pavements	*145*
	8.1. Background	*145*
	8.2. Pavement Types	*146*
	8.2.1. Unpaved Movement Areas	*147*
	8.2.2. Pavements	*148*
	8.2.2.1. Use of hard surface pavements	*148*
	8.2.2.1.1. Subgrade	*149*
	8.2.2.1.2. Sub-base	*150*
	8.2.2.1.3. Bearing course/base course	*150*
	8.2.2.2. Flexible (asphalt) pavements	*150*
	8.2.2.3. Rigid (concrete) pavements	*152*
	8.2.2.4. Combined pavements	*158*
	8.2.2.5. Block paving	*159*
	8.3. Pavement Strength	*160*
	8.3.1. Pavements-Aircraft Loads	*160*
	8.3.2. Pavement Strength Reporting	*162*
	8.3.3. Overload Operations	*167*
	8.4. Runway Surface	*168*
	8.4.1. Runway Surface Quality Requirements	*168*
	8.4.2. Methods of Runway Surface Unevenness Assessment by the Dual Mass Method	*168*
	8.4.3. Pavement Texture	*172*
	8.4.4. Runway Braking Action	*176*
	8.5. Pavement Management System	*180*
9.	Aircraft Ground Handling	*185*
	9.1. Aircraft Handling Methods and Safety	*185*
	9.2. Aircraft Ground Handling Activities	*191*
	9.2.1. Deplaning and Boarding	*191*
	9.2.2. Supplies of Power, Air-Conditioning and Compressed Air	*195*
	9.2.3. Cargo and Baggage Loading	*195*
	9.2.4. Push Back Operations	*197*

	9.3. Collaborative Decison-Making (CDM)	*200*
	9.4. Visual Guidance Systems	*204*
10.	Aircraft Refuelling	*211*
	10.1. Background	*211*
	10.2. Fuel — Requirements	*214*
	10.2.1. Requirements for Fuel Quality	*214*
	10.2.2. Fuel Deliveries and Storage	*217*
	10.3. Fuel Distribution	*221*
	10.4. Safety of the Refuelling Operation	*225*
	10.4.1. Ecological Damage	*226*
	10.4.2. Fire Safety	*229*
	10.4.3. Fuel Farms Security	*231*
	10.5. Aircraft Fuel — Future Trends	*231*
11.	Cargo	*233*
	11.1. Introduction	*233*
	11.2. The Freight Industry's Characteristics	*237*
	11.3. Airside Design Considerations	*241*
	11.4. Terminal Design and Operating Considerations	*243*
	11.4.1. Location	*243*
	11.4.2. Design Parameters	*243*
	11.4.3. Mechanisation	*245*
	11.4.4. Terminal Functions and Operations	*246*
	11.4.5. Documentation	*248*
	11.4.6. Utilities	*250*
	11.4.7. Security	*251*
	11.5. Cargo Terminal Layout and Sizing	*251*
	11.5.1. Layout	*251*
	11.5.2. Functions and Facilities	*253*
	11.5.3. Sizing	*254*
	11.6. Landside Design and Operations	*256*
	11.7. Future Trends	*257*
	11.8. DHL Case Study	*257*
12.	Passenger Terminals	*261*
	12.1. Airport Terminal Design Principles	*261*
	12.2. Airport Terminal Layout	*266*
	12.3. Airport Terminal Concepts	*269*
	12.4. Terminal Design	*271*
	12.4.1. Design Methods	*271*
	12.4.2. Component Design	*275*
	12.5. The Handling Process	*286*

	12.5.1. Passenger Handling	*286*
	12.5.2. Baggage Handling	*295*
12.6.	Non-Aeronautical Services	*296*
12.7.	Passenger Transportation — People Movers	*298*
12.8.	Mobile and IT Technologies	*302*

13. Security *305*
- 13.1. Unlawful Acts and Air Transport *305*
- 13.2. Legal Framework of International Aviation Security *312*
- 13.3. The Airport System and Its Security *315*
- 13.4. Safeguarding of Airport Security *320*
    - 13.4.1. Security as a Service *320*
    - 13.4.2. Airport Perimeter Security and Staff Identification *321*
    - 13.4.3. Employee Security Procedures *325*
    - 13.4.4. Measures in Relation to Passengers *326*
- 13.5. Detection of Dangerous Objects *331*
    - 13.5.1. Metal Detectors *332*
    - 13.5.2. Millimetre-Wave Scanners *332*
    - 13.5.3. Backscatter Screening *333*
    - 13.5.4. X-ray Units *333*
    - 13.5.5. Gas Analysers *337*
    - 13.5.6. Vacuum Chambers *338*
    - 13.5.7. Dogs *338*
    - 13.5.8. Liquid Scanners *340*
- 13.6. Conclusion *340*

14. Landside Access *343*
- 14.1. Access and the Airport System *343*
- 14.2. Selection of the Access Modes *346*
- 14.3. Categories of Surface Transport Users *348*
- 14.4. Access and Terminal Operations *348*
- 14.5. Access Modes *350*
    - 14.5.1. Passenger Car *350*
    - 14.5.2. Taxi *353*
    - 14.5.3. Minibus *355*
    - 14.5.4. Bus *355*
    - 14.5.5. Railway Transport *357*
    - 14.5.6. Unconventional Means of Transport *362*
- 14.6. Airport Ground Access Improvements *364*

15. Visual Aids for Navigation *367*
- 15.1. Markings *367*
    - 15.1.1. Markings Requirements *367*

xii   Airport Design and Operation

15.1.2. Marking Types		369
15.1.3. Signs		372
15.2. Airport Lights (*Author*: František Bělohradský, deceased, Consultant, Prague, CZ)		373
15.2.1. Characteristics and Components of Airport Lighting Systems		373
15.2.1.1. Introduction		373
15.2.1.2. Light sources		375
15.2.1.3. Lights and fittings		376
15.2.1.4. Frangible safety masts		378
15.2.1.5. Requirements for aerodrome lights		379
15.2.2. Characteristics and Components of Airport Lighting Systems		381
15.2.2.1. Approach and runway systems		381
15.2.2.1.1. Non-instrument and instrument runways		381
15.2.2.1.2. Precision approach runway		384
15.2.2.2. Approach slope indicator systems		392
15.2.3. Heliport Lighting Systems		397
15.2.4. Lighting of Obstacles		399
15.2.5. Light Control		401
15.2.5.1. Remote control equipment		401
15.2.5.2. Single lamp control and monitoring		402
15.2.6. Lighting Systems Construction and Operation		404
15.2.6.1. Lighting systems design and installation		404
15.2.6.2. Maintenance of the lighting systems		405
15.2.7. Trends in Lighting Systems Development		408
16. Electrical Energy Supply		409
16.1. Background		409
16.2. Electrical Systems Reliability and Backup		409
16.3. Supply Systems		416
16.3.1. Parallel System		416
16.3.2. Serial System		416
16.3.2.1. Serial System — The Principle		416
16.3.2.2. Serial System — Components		417
16.4. Electrical Supply to Category I–III Lighting Systems		422
17. Radio Navigation Aids		423
17.1. Background		423
17.2. Radio Navigation Aids		424
17.2.1. Instrument Landing System (ILS)		424
17.2.2. Microwave Landing System (MLS)		429
17.2.3. Global Navigation Satellite Systems (GNSS)		430

	17.2.4. VHF Omnidirectional Radio Range (VOR)	*432*
	17.2.5. Non-directional Radio Beacon (NDB)	*433*
	17.2.6. Distance Measuring Equipment (DME)	*434*
	17.2.7. Transponder Landing System (TLS)	*435*
17.3.	Radar Systems	*435*
	17.3.1. Precision Approach Radar (PAR)	*435*
	17.3.2. Surveillance Radar Element (SRE)	*436*
	17.3.3. Surface Movement Radar (SMR)	*436*
	17.3.4. Advanced Surface Movement and Guidance Control Systems (A-SMGCS)	*437*
17.4.	Flight Inspections and Calibrations	*438*

18. Airport Winter Operation *441*
- 18.1. Snow and Aircraft Operation *441*
- 18.2. Snow Plan *442*
- 18.3. Mechanical Equipment for Snow Removal and Ice Control *446*
- 18.4. Chemicals for Runway De-Icing *453*
- 18.5. Thermal De-Icing *457*
- 18.6. Runway Surface Monitoring *459*
- 18.7. Aircraft De-Icing *460*

19. Airport Emergency Services *471*
- 19.1. Roles of the Rescue and Fire Fighting Service *471*
- 19.2. Level of Protection Required *473*
  - 19.2.1. Response Times *473*
  - 19.2.2. Aerodrome Category for Rescue and Fire Fighting *474*
  - 19.2.3. Principal Extinguishing Agents *475*
  - 19.2.4. Complementary Extinguishing Agents *478*
  - 19.2.5. The Amounts of Extinguishing Agents *479*
- 19.3. Rescue and Fire Fighting Vehicles *482*
- 19.4. Airport Fire Stations *487*
- 19.5. Emergency Training and Activity of Rescue and Fire Fighting Unit *490*
  - 19.5.1. Personnel Requirements and Training *490*
  - 19.5.2. Preparation for an Emergency Situation and Rescue and Fire Fighting Intervention Control *492*
- 19.6. Runway Foaming *494*
- 19.7. Post Emergency Operations *496*
- 19.8. Emergency Services and Environment Protection *499*
- 19.9. Final Thoughts *500*

20. Environmental Control *503*
- 20.1. Background *503*

	20.2. Noise	506
	20.2.1. Characteristics	506
	20.2.2. Descriptors Used for Aircraft Noise Rating	508
	20.2.3. Evaluation of Noise in the Vicinity of Airports	514
	20.2.4. Land Use and Compatibility Planning	517
	20.2.5. Aircraft Noise Measurement	521
	20.2.5.1. Short-Term Measurement	521
	20.2.5.2. Long-Term Noise Monitoring	522
	20.2.6. Prediction of Air Transport Noise	524
	20.2.7. Airport Noise Mitigation and Noise Abatement Procedures	530
	20.3. Control of Gaseous Emissions and Energy Conservation	533
	20.4. Protection of Water Sources	535
	20.5. Landscaping	537
	20.6. Waste Management	538
21.	Wildlife Control	541
	21.1. Introduction	541
	21.2. Bird Strike Statistics	543
	21.3. Passive Management Techniques — Habitat Modification	546
	21.4. Active Management Using Dispersal Techniques	548
	21.5. Mammals Control	552
	21.6. Ornithological Protection Zones	552
References		555
Index		561

ABBREVIATIONS

μm	Micrometres
A	Ampere
a/c	Aircraft
ABAS	Aircraft-Based Augmentation System
AC	Advisory Circular
A-CDM	Airport Collaborative Decision Making
ACARE	Advisory Council for Aeronautics Research in Europe
ACI	Airports Council International
ACLU	American Civil Liberties Union
ACMI	Aircraft, Crew, Maintenance, and Insurance
ACNSG	Aircraft Classification Number Study Group
ACRP	Airport Cooperative Research Program
AD	Aerodrome
ADF	Automatic Direction Finder
ADSG	Airport Design Study Group
AEA	Association of European Airlines
AFFF	Aqueous Film Forming Foam
AFTN	Aeronautical Fixed Telecommunication Network
AGA	Aerodromes, Air Routes and Ground Aids
AGNIS	Azimuth Guidance for Nose-in Stands
AIP	Air Information Publication
AIP	Airport Improvement Program
AMAN	Arrival Manager
AMC	Acceptable Means of Compliance
ANP	Aircraft Noise-Power
AODB	Airport Operations Database
API	Advance Passenger Information
APIS	Aircraft Parking and Information System
APN	Apron
APP	Approach
APP	Advanced Passenger Processing
APU	Auxiliary Power Unit
ARCP	Aerodrome Reference Code Panel
ASD	Addressable Switching Device
ASDA	Accelerate-stop Distance Available

A-SMGCS	Advanced Surface Movement, Guidance and Control Systems
ASPSL	Arrays of Segmented Point Source Lighting
ASTM	American Society for Testing and Materials
ATB	Automatic Ticket and Boarding
ATC	Air Traffic Control
ATFCM	Air Traffic Flow and Capacity Management
ATFM	Air Traffic Flow Management
ATM	Air Traffic Management
atm	Air Transport Movements
ATSA	Aviation and Transportation Security Act
AWB	Air Waybill
AWOP	All Weather Operation Panel
BAA	British Airports Authority
BANANA	Build Absolutely Nothing Anywhere Near Anything
BOD	Biochemical Oxygen Demand
BOLDS	Burroughs Optical Lens Docking System
C	Degrees Celsius (Centigrade)
CAA	Civil Aviation Authority
CAD	Computer Added Design
CAN	Aircraft Classification Number
CAP	Civil Aviation Publication
CAT	Category
CBP	Customs and Border Protection
CBR	Californian Bearing Ratio
CCR	Constant Current Regulators
CCS	Cargo Community System
CCTV	Close Circuit TV
cd	Candelas
CDA	Continuous Descent Approach
CDD	Charge-Coupled Devices
CDG	Charles de Gaulle Airport
CDM	Collaborative Decision Making
CFMU	Central Flow Management Unit
CFRP	Carbon Fibre Reinforced Polymer
CIP	Commercially Important Passengers
CL	Centre Line
CLOS	Command Line-of-Sight
cm	Centimetre
CMC	Cargo Movement Control
CNEL	Community Noise Equivalent Level
CNR	Composite Noise Rating
CO	Carbon Monoxide
CO_2	Carbon Dioxide
CPM	Computational Pipeline Monitoring

CRM	Collision Risk Model
CSA	Common Situational Awareness
CUSS	Common User Self Service
CUTE	Common Use Terminal Equipment
CWY	Clearway
DA	Decelerating Approaches
dB	Decibel
DC	Direct Current
DDF	Data and Documentation Flow
DDM	Difference in the Depth of Modulation
DH	Decision Height
DHS	Department of Homeland Security
DIN	Deutsches Institut für Normung (German Institute for Standardization)
DLR	Deutschen Zentrums für Luft- und Raumfahrt (The National Aeronautics and Space Research Centre of the Federal Republic of Germany)
DM	Dual Mass
DMAN	Departure Manager
DME	UHF Distance Measuring Equipment
Doc	Document
DOT	Department of Transport
E	Dynamic Modulus of Elasticity
EASA	European Aviation Safety Agency
EC	European Commission
ECAC	European Civil Aviation Conference
EDDS	Explosive Device Detection System
EDI	Electronic Data Interchange
EDMS	Emissions and Dispersion Modelling System
EDS	Explosive Detection System
EFSO	Emergency Fuel Shutoff System
EGNOS	European Geostationary Navigation Overlay Service
EIA	Environmental Impact Assessment
EMA	East Midlands Airport
EMAS	Engineered Material Arresting System
EPA	Environmental Protection Agency
EQA	Equivalent Gates
ERCD	Environmental Research and Consultancy Department
ETDS	Explosive Trace Detection Equipment
ETV	Elevating Transfer Vehicles
EU	European Union
EUR	Europe or European
EXP	World Exports
FAA	Federal Aviation Administration of the USA
FAF	Final Approach Fix

FAR	Federal Aviation Regulation
FASG	Frangible Aids Study Group
FATO	Final Approach and Take-off Area
FFFP	Film Forming Fluoroprotein Foam
FIDS	Flight Information Display System
FLIR	Forward Looking Infrared
FMCW	Frequency Modulated Continuous Wave
FOD	Foreign Object Damage
ft	Feet
ftk	Freight Tonne Kilometres
g	Gram
GBAS	Ground-Based Augmentation System
GCA	Ground Controlled Approach
GHG	Greenhouse Gas
GLONASS	Globalnaya Navigatsionnaya Sputnikovaya Sistema
GNP	Gross National Product
GNSS	Global Navigation Satellite System
GP	Glide Path
GPS	Global Positioning System
GPWS	Ground Proximity Warning System
GRI	Global Reporting Initiative
HAPI	Helicopter Approach Path Indicator
HC	Hydrocarbons
HGS	Head up Guidance System
Hi-Lo	High-Loader
HOP	Helicopter Operations Panel
HOT	Hold Over Time
HV	High Voltage
Hz	Hertz
IATA	International Air Transport Association
ICAO	International Civil Aviation Organisation
ID	Identification
IEC	International Electrotechnical Commission
IFE	In-flight Entertainment
IFR	Instrument Flight Rules
ILS	Instrument Landing System
IM	Inner Marker
IMC	Instrument Meteorological Conditions
INM	Integrated Noise Model
IR	Infra-Red
ISBN	International Standard Book Number
ISO	International Standards Organization
Istr	Intensity in a Space Beam
IT	Information Technologies

JFK	John F. Kennedy International Airport
kHz	Kilohertz
km	Kilometre
kN	Kilo-newton
Kt	Knots
L_{AE}	Sound Exposure Level
LAN	Local Area Network
lb	Pound
LBS	Location Based Services
LCC	Low Cost Carrier
LCM	Lamp Control and Monitoring
LDA	Landing Distance Available
LED	Light Emitting Diode
L_{eq}	Equivalent Continuous Sound Level
LF	Low Frequency
LGT	Light, Lighting
LLZ	Localiser
LMS	Logistics Management Systems
LOC	Localiser
LOS	Level of Service
LP	Luminescent Panel
LTO	Take-off and Landing
LV	Low Voltage
LVO	Low Visibility Operations
LVTO	Low Visibility Take-Off
m	Metre
MAGLEV	Magnetic Levitation
MANPADs	Man-Portable Air Defence Systems
MCT	Maximum Continuous Thrust
MD	McDonnell-Douglas
MEL	Minimum Equipment List
MF	Medium Frequency
MHz	Megahertz
MIS	Management Information System
MKR	Marker
MLS	Microwave Landing System
mm	Millimetre
MM	Middle Marker
MoU	Memorandum of Understanding
MPa	Megapascal
mppa	Million Passengers per Annum
MTD	Mean Texture Depth
MTOM	Maximum Take-off Mass
NADP	Noise Abatement Departure Procedures

NASP	National Aviation Security Programme
NDB	Non-Directional Radio Beacon
NEF	Noise Exposure Forecast
NFC	Near Field Communications
NIMBY	Not in My Back Yard
NM	Nautical Mile
NNI	Noise and Number Index
NOTAM	Notice to Airmen
NOx	Nitrous Oxides
NPD	Noise-Power-Distance
NPIAS	National Plan of Integrated Airport Systems
NTK	Noise Monitoring and Track Keeping
OAS	Obstacle Assessment Surfaces
OAT	Outside Air Temperature
OCP	Obstacle Clearance Panel
O-D	Origin Destination
OFZ	Obstacle Free Zone
OM	Outer Marker
OPS	Aircraft Operations
p.a.	Per annum
Pa	Pascal
PANS	Procedures for Air Navigation Services
PAPA	Parallax Aircraft Parking Aid
PAPI	Precision Approach Path Indicator
PAR	Precision Approach Radar
PAX	Passenger
PBFM	Passenger and Bag Flow Model
PCN	Pavement Classification Number
PEDS	Primary Explosive Detection System
PETN	Pentaerytrytol Tetranitrate
PIN	Personal Identification Number
pkm	Passenger Kilometres
PLASI	Pulse Light Approach Slope Indicator
PMS	Pavement Management Systems
PNdB	Perceived Noise Decibel
PNL	Perceived Noise Level
PNR	Passenger Name Record
ppm	Parts Per Million
PSZ	Public Safety Zone
ptf	Propensity to Fly
QR code	Quick Response Code
RAVC	Reduced Aerodrome Visibility Conditions
RDF	Radio Direction Finders
RDX	Cyclotrimethylentrinithramin

RER	Regional Rapid
RESA	Runway End Safety Areas
RETIL	Rapid Exit Taxiway Indicator Light
RFF	Rescue and Fire Fighting
RFFS	Rescue and Fire Fighting Service
RFFSG	Rescue and Fire Fighting Study Group
RFID	Radio Frequency Identification
RMS	Root Mean Square
rpk	Revenue Passenger Kilometres
RPM	Revolutions Per Minute
rpm	Revenue Passenger Miles
RSA	Runway Safety Area
rtk	Revenue Tonne Kilometres
RVR	Runway Visual Range
RWY	Runway
SAE	Society of Automotive Engineers
SAGA	System of Azimuth Guidance Approach
SAMs	Surface-to-Air Missiles
SARPs	Standards and Recommended Practices
SAS	Scandinavian Airline Systems
SBAS	Satellite Based Augmentation System
SBR	Standard Busy Rate
SCC	Series Circuit Coupler
SEL	Single Event (Sound) Exposure Level
SeMS	Security Management System
SITA	Société Internationale de Télécommunications Aéronautiques
SLA	Service Level Agreements
SMB	Side Marker Board
SMR	Surface Movement Radar
SMS	Safety Management System
SNAP	Significant New Alternatives Policy
SPL	Sound Pressure Level
SRE	Surveillance Radar Element
SS	Settleable solids (mg/litre) - suspended solid after one hour quiescent settlement
SSI	Sensitive Security Information
SSR	Secondary Surveillance Radar
SVR	Slant Visual Range
SWY	Stopway
TDD	Telecommunications Display Device
TDZ	Touchdown Zone
TGV	Train à Grande Vitesse
TIP	Threat Image Projection
TLS	Transponder Landing System

TNT	Trinitrotoluene
TO	Take-off
TODA	Take-off Distance Available
TOGA	Take-off and Go-Around
TORA	Take-off Run Available
TPHP	Typical Peak Hour Passengers
TRA	Task and Resource Analysis
TSA	Transportation Security Administration
TSC	Terrorist Screening Center
TWY	Taxiway
UH	Unburned Hydrocarbons
UHF	Ultra High Frequency
UK	United Kingdom
ULD	Unit Load Devices
UPS	Uninterruptible Power Supply
US	United States
USA	United States of America
USAF	US Air Force
USD	United States Dollars
V	Volt
V	Velocity
V_1	Take-off Decision Speed
V_2	Take-off Safety Speed (Applicable to Larger Multi-engine Aircraft)
VA	Volt-ampere
VAGS	Visual Alignment Guidance System
VAP	Visual Aids Panel
VAT	Value Added Tax
VASIS	Visual Approach Slope Indicator System
VDGS	Visual Docking Guidance System
VFR	Visual Flight Rules
VHF	Very High Frequency
VLA	Very Large Aircraft
VMC	Visual Meteorological Conditions
VOR	Very High Frequency Omnidirectional Radio Range
VRLA	Valve Regulated Lead Acid
W	Watt
WAAS	Wide Area Augmentation System
WCO	World Customs Organization
WECPNL	Weighted Equivalent Continuous Perceived Noise Level
Wi-Fi	Wireless Fidelity
WRS	Wide-area Reference Stations
WS	Wing Span

PREFACE

This book is titled 'Airport Design and Operation'. However, the reader will not find chapters devoted exclusively to airport design or airport operation. Airport design and airport operation are closely related and influence each other. A poor design affects the airport operation and results in increased costs. On the other hand it is difficult to design the airport infrastructure without sound knowledge of airport operations. This is emphasised throughout the book.

The book does not offer a set of simple instructions for solutions to particular problems. Every airport is unique and a simple generic solution does not exist. The book explains principles and relationships important for the design of airport facilities, for airport management and for the safe and efficient control of operations. We hope that we have been able to overcome the traditional view that an airport is only the runway and tarmac. An airport is a complex system of facilities and often the most important enterprise of a region. It is an economic generator and catalyst in its catchment area. However, this book is focused on one narrow part of the airport problem, namely design and operation, while bearing the other aspects in mind.

This third edition includes some important changes in the international regulations covering design and operations. It reflects the greater attention being given to security, safety and changes on the air transport market with respect to the impact of low cost carriers operations. The third edition contains new parts on airport long-term planning; Reduced Aerodrome Visibility Conditions operation; RESA and EMAS construction; aircraft performance; rapid exit taxiway specifications; dowelling technology in concrete pavement design; fuel storage and leak detection; impact of mobile and IT technologies on passenger terminal design and operations; developments in security risks and their impact on security SARPs; ground transport system improvements and their impact on airport attractiveness; new de-icing chemicals and procedures; changes in rescue and fire fighting and a new chapter on wildlife control. All these changes and additions strengthen the operational content of this book.

Tony Kazda and Bob Caves

Žilina, Slovakia and Loughborough, UK, January 2015

ACKNOWLEDGEMENTS

We appreciate all the help from the professionals who have contributed to the text or have given freely of their time and expertise to advise and correct our draft texts. Without their help, it would not have been possible to complete this book. We would like to express special thanks to:

Charles Allen, DHL
Zoltan Bazso, Heathrow Airport
Bill Blanchard, East Midlands airport
Petr Čiviš, AGA – Letiště, Prague
Zbyněk Hackl, Airport Systems Design Agency, spol. s r.o., Prague
Petr Hloušek, Prague Airport
Dr. Stephanie Hochreuther, BU ICS, Technical Application EMEA, Aviation, Clariant Produkte (Deutschland) GmbH
Dr. Martin Hromádka, Žilina Airport
Tomáš Hruška, Air Transport Europe
Abdul Nasser Chakra, Emirates Aviation University, Dubai
Bohdan Koverdynsky, Prague Airport
Dr. Libot Kurzweil, Prague Airport
Andrea L. Manning, Zodiac Aerospace
Matěj Mareth, Slovak Transport Authority
Darren Maynard, DHL
Richard Moxon, Cranfield University
Kai Nieruch, Lufthansa
Prof. Andrej Novák, University of Žilina
Jan Pojezny, Vestergaard Company A/S

1

AIR TRANSPORT AND AIRPORTS

Tony Kazda and Bob Caves

1.1. DEVELOPMENT OF AIRPORTS

First, consider the well-known question: 'Which came first'? In the context of this book, it does not refer to the notorious problem about a chicken and an egg but about an airport and an aircraft. In fact, the answer is clear. The aircraft came first. When aviation was in its infancy, the aviator first constructed an aircraft and *then* began to search for a suitable 'airfield', where he could test the machine. The aerodrome parameters had to be selected on the basis of performance and geometrical characteristics of the aircraft. That trend to accommodate the needs of the aircraft prevailed, with some notable exceptions like New York's La Guardia airport, until the end of the 1970s. This was despite the increasing requirements for strength of pavements, width and length of runways and other physical characteristics and equipments of aerodromes. The aerodromes always had to adapt to the needs of the aircraft.

The first aircraft was light, with a tail wheel, and the engine power was usually low. A mowed meadow with good water drainage was sufficient as an aerodrome for those aircraft. The difficulty in controlling the flight path of these aircrafts required the surrounding airspace to be free of obstacles over a relatively wide area. Since the first aircraft were very sensitive to crosswind, the principal requirement was to allow taking off and landing always to be into wind (Figure 1.1). In the majority of cases, the aerodrome used to be a square or circle without the runway being marked out. The wind direction indicator, which was so necessary in those days, still has to be installed at every aerodrome today, although its use now at big international airports is less obvious. Other visual aids that date from that period are the landing direction indicator and the boundary markers. The latter aid determined unambiguously where the field was and where the aerodrome was, this flight information for the pilot not always being evident in the terrain.

Immediately after World War I in 1919–1920, the first air carriers opened regular air services between Paris and London, Amsterdam and London, Prague and Paris, among others. However, in that period no noticeable changes occurred in the airport equipment, or in the basic operating concept, other than some simple building for the processing of passengers and hangars for working on the aircraft.

2 *Airport Design and Operation*

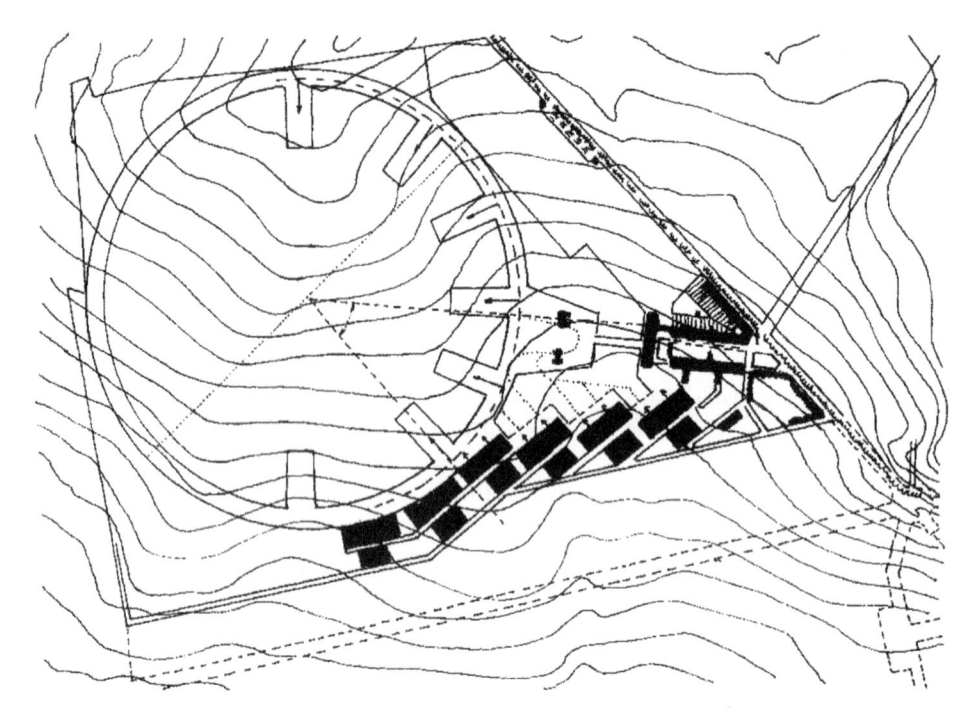

Figure 1.1: The second prize winner in a competition in 1931 for the design of Praha-Ruzyně airport development. *Source*: The Czech Airports Authority.

Even in the 1930s, the new technology of the Douglas DC-2 and DC-3, which were first put into airline service in 1934 and 1936, respectively, was not significantly different to require large changes in the physical characteristics of aerodromes, so the development of airports up to that period may be characterised as gradual. The first passengers on scheduled airlines were mostly business people or the rich and famous, but this was a small-scale activity, most of the flying being done by the military. The main change in the airfield's physical characteristics was the runway length. The multiengine aircraft required the length to increase to approximately 1,000 m.

The increasing number of aircraft and the training of the military pilots required more support facilities at airfields, such as hangars, workshops and barracks.

War does not benefit mankind but, for aviation, it has meant a rapid step change in development. After World War II, there were unusually favourable conditions for the development of civil aviation and air transport. On one hand there were damaged ground communications, while on the other hand, there were plenty of surplus former military aircraft. There was also the requirement to support the supply chains from the United States to Latin America, to Japan, to Europe under the Marshall Plan or the largest post-war air cargo operation during the Berlin Blockade (24 June 1948–12 May 1949) — the Berlin airlift. All of that activity allowed civil air transport to recover quickly and then to continue to a higher level than before World War II. The requirements for aerodromes changed dramatically in that same short period of time.

The new aircraft required paved runways, partly because they were heavier and partly because regularity of service became more important. However, they were still relatively sensitive to the crosswind, despite having nose-wheel steering. Therefore, the big international airports adopted a complicated system of between three and six runways in different directions in order to provide sufficient operational usability from the entire runway system (Figure 1.2). The large number of runways often reduced the amount of land available for further development of the airport facilities. One of the runways, most often the runway in the direction of the prevailing winds, was gradually equipped with airport visual and radio-navigation aids, thereby being regarded as the main runway. At the same time terminal facilities were constructed which, besides the services required for the processing of passengers and their baggage, provided also the first non-aeronautical services, such as restaurants, toilets and duty free shops.

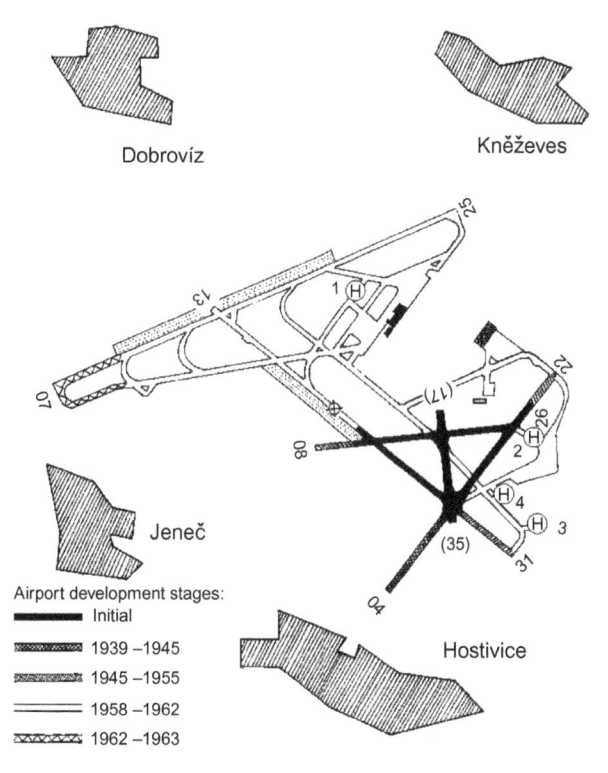

Figure 1.2: Development of Praha-Ruzyně runway system. *Source*: Čihař (1973).

The next substantial change that significantly influenced the development of airports was the introduction of aircraft with jet propulsion.

Jet aircraft required further extension of the runway, together with increases in its width and upgrading its strength. The operation of jet aircraft had an effect also upon other equipment and technical facilities of the airport. One of them was the fuel supply system. Not only did the fuel type change from gasoline to kerosene but also the volume per aircraft increased considerably, requiring reconstruction of the fuel farms and the introduction of new refuelling technologies.

The introduction of the first wide-body jet aircraft, the Boeing B 747-100 in 1970, had a large impact on the design of terminals. Before the B 747-100, the runway or apron were limiting capacity factors for some airports but, after it was introduced, the terminal building capacity became critical. The B 747-100 capacity could replace two or three existing aircraft. Thus, the number of aircraft movements was relatively reduced, and the number of passengers per movement increased. The B 747-100 required a further increase in the strength of manoeuvring areas, the enlargement of stands and other changes such as airport visual aids which resulted from greater height of the cockpit giving a different view from the cockpit during approach and landing.

The B 747-100 in fact symbolized a whole new era of wide body air transport, as well as causing the system to adapt to it. At the same time, it demonstrated that there had to be a limit to which airports could adapt fully to whatever the cutting edge of aircraft technology demanded of them. Not only was there a reaction from the international airport community; the manufacturers themselves also came to realise that if they constructed an aircraft with parameters requiring substantial changes of ground equipment, they would find it difficult to sell it in the marketplace. Futuristic studies of new aircraft in the early 1980s, with a capacity of 700–1000 seats, were not taken beyond the paper stage, partly for this reason as well as because the airlines had found it hard to sell all the capacity offered by the B 747. Following this argument, the Boeing B 777-200 was designed with folding wingtips, though this option has not yet been taken up by any airline. The Airbus A 380 was designed to fit into an 80 m box which the airport industry regarded as the maximum it could cope with economically. Although all new large hub airports like Munich, Hong Kong, Kansai, Beijing, Dubai and others have been designed to cope with the A 380, only minor changes like the location of airside signs has been necessary. The airports which were originally designed around the needs of piston-engine aircraft have had to make very substantial changes to accept it. London's Heathrow airport has lost more than 20 stands due to having to increase taxiway separations and has had to build a new pier, the total cost being £450 million. The Airbus A 380 also has a considerable impact on the design and operations of airport passenger terminals and access roads. The passenger capacity of the A 380 depends on the seat configuration chosen by an airline. The A 380-800 is certified for up to 853 passengers (538 on the main deck and 315 on the upper) in a one-class configuration though now no airline has configured A 380 with so many seats

(Figure 1.3). Airlines announced seat capacities ranging from 407 passengers (Korean Air) to over 644 (Emirates Airline) in two classes.

Figure 1.3: Airbus 380 will be used on the densest routes. *Copyright*: Airbus; *Photo*: J. Pommery.

Most recent changes to airports have not been provoked by new aircraft technology but by political and economic developments. The airport situation in Europe has changed considerably since the 1960s. The airport in the past was a 'shop-window' of the state, and together with the national flag carrier, it is also an instrument to enforce state policy. After the successful corporatisation and then the privatisation of the British Airport Authority and some other airports, many governments have gradually changed their policy towards airports, particularly in regard to subsidy.

The following important factors influenced the entire development of airports from 1975 to 2014:

1. The threat of terrorism and a fear of unlawful acts.
2. The privatisation of airports.
3. The progressive deregulation of air transport.
4. The increasing environmental impact around airports.
5. Growth of low-cost carriers.

The threat of terrorism, and in particular the bomb attack against the B 747 Pan-Am Flight 103 on 23 December 1988 near Lockerbie in Scotland and later The September 11 attacks in 2001, subsequently required expensive changes of airport terminal buildings with a consistent separation of the arriving and departing passengers and installation of technical equipment for detecting explosives. The security problems are discussed in detail in Chapter 13.

The privatisation of airports started in Great Britain in 1986, and represented a fundamental change in the manner of administering and financing the airports in Europe. It was and still is seen by most people as a success, though there are those, particularly in the United States, who believe that the emphasis on commercial viability has made it difficult to concentrate on an airport's main function of providing an effective and efficient transfer between air and ground transport. It has, though, resulted in a considerable extension and improvement of the services provided, particularly for the passengers and other visitors of the airport.

The deregulation that began in the United States in 1978 produced a revolution in the development of that industry. Up to then, air transport had been developing in an ordered fashion. Deregulation represented a free, unlimited access to the market, without any capacity and price limitations, unblocking the previously stringent regulation of the market in the United States. The percentage of the population who had never travelled by plane before reduced from 70% to 20%. However, it also brought about negative consequences for airport capacity due to the concentration of traffic at the major hubs and due to the gradual creation of extremely large airlines with the features of strong monopolies.

Therefore in Europe deregulation was approached with considerable caution, to the extent that the term 'liberalisation' has been adopted for the policy. The first measures to affect the major airlines were adopted by the states of the European Twelve in 1988, though some countries had entered into liberal bilateral agreements as early as 1984. The measures referred in particular to the determination of tariffs and the shares of route capacity. They allowed more flexibility and easier access to the market when certain requirements were fulfilled, free access for aircraft of up to 70 seats and conferment of the Fifth Air Freedom within the states of the European Community. A Third Package came into operation in 1992 which allowed any EU carrier to operate any route within Europe without control of fares, releasing the latent demand which has been exploited by the low-cost airlines.

The rate of growth of air transport worldwide since 1990 has been strong. The volume of passengers in regular air transport doubled in the period from 1990 to 2000, and in the region of the Pacific Basin it even quadrupled. The airspace in Europe became seriously congested. Airspace slots, into which a flight can be accepted by prior arrangement, became scarce. The queues of aircraft lengthened, both on the ground and in the air. The costs incurred by delayed flights reach USD hundreds of millions annually.

Besides the need of funding for reconstruction and the building of new terminals, the biggest problem for many large airports, in particular, in Europe is the lack of capacity of the runway system, leading to a requirement for the construction of new runways. This is accentuated by the development of regional transport which will continue throughout Europe, despite the EC's preference for rail travel. Regional transport serves business trips mostly or to feed long-haul flights, thereby increasing the demand for capacity of runway systems during the peak hour.

It is impossible to adopt a quick and effective solution in Europe, the construction of new capacity being hindered by the legal procedures, but also by problems involving public finance, for which projects should be submitted for public discussion in most countries. For example the opening of the new Berlin — Brandenburg airport was originally planned for 2010 but it has encountered a series of delays due to poor construction planning, management, execution and corruption. In August 2014 the former CEO estimated that the airport will open in 2018 or 2019, at the latest. In 2006, the construction cost was budgeted at €2.83 billion. By late 2012, expenditures for Berlin Brandenburg Airport totalled €4.3 billion, nearly twice the originally calculated figure. There are new runways operating at Amsterdam and Frankfurt. Also, new runways are planned at Munich, Prague and Vienna airports. A proposal for a new London airport — that is, Inner Thames Estuary (ITE) airport has been withdrawn. The UK Airports Commission evaluated a total of 58 proposals for a new runway in the London area on the basis of strategic fit, the economy, surface access, the environment, effect on people, operational viability and cost (Airports Commission, 2014). The options have, at the time of writing, been narrowed down to three, all involving Heathrow or Gatwick airports. However, the final proposal is still subject to government approval and if approved, it will have long and arduous planning inquiries to negotiate. There are some technological and managerial possibilities for obtaining better use of the existing runway capacity, such as the most recent time-based separation at Heathrow airport or making use of the different characteristics of regional transport aircraft to implement a separate system of approach and take-off, as in the United States. However, the extra capacity would be exhausted within a very few years at long-term growth rates. According to Airbus Global Market Forecast (2014–2033) air traffic will double in the next 15 years. Boeing in the Current Market Outlook (2014–2033) predicts long-term demand — till 2033 for 36,770 new aircraft of which 15,500 will replace older less efficient aircraft. The remaining 21,270 aircraft will provide the fleet growth to support expansion, particularly in emerging markets.

Both companies anticipate the fastest market growth in the Asia-Pacific (including China) region. This could be illustrated by comparing the top 15 world airports data in 2013 and 2005 by total passengers (Table 1.1). There were eight US

8 Airport Design and Operation

Table 1.1: World top 15 airports ranking by total passengers — 2013 data.

Rank	Airport/IATA code	Total PAX (million)	% change in 2012	PAX in 2005	Rank in 2005
1	Atlanta Hartsfield (ATL)	94.4	−1.1	85.9	1
2	Beijing Capital (PEK)	83.7	2.2	41.0	15
3	Heathrow (LHR)	72.4	3.3	67.9	3
4	Tokyo (HND)	68.9	3.3	63.3	4
5	Chicago O'Hare (ORD)	66.9	−0.1	76.5	2
6	Los Angeles (LAX)	66.7	4.7	61.5	5
7	Dubai (DBX)	66.4	15.2	N/A[a]	N/A[a]
8	Paris (CDG)	62.1	0.7	53.8	7
9	Dallas/Fort Worth (DFW)	60.4	3.2	59.2	6
10	Soekarno-Hatta (CGK)	59.7	3.4	N/A[b]	N/A[b]
11	Hong Kong (HKG)	59.6	6.3	40.3	16
12	Frankfurt (FRA)	59.0	0.9	52.2	8
13	Singapore Changi (SIN)	53.7	5.0	32.4	25
14	Amsterdam (AMS)	52.6	3.0	44.2	9
15	Denver (DEN)	52.6	−1.1	43.4	11

Source: Airports Council International Traffic Data.
Note: Total passengers: arriving + departing passengers + direct transit passengers counted once.
[a] Dubai appeared in the ACI statistics for the first time in the 2007, ranked 27 with 34.3 million PAX and growth 19.3%.
[b] Soekarno-Hatta appeared in the ACI statistics for the first time in the 2009, ranked 22 with 37.1 million PAX and growth 15.2%.

airports and just two airports from Asia among the top 15 airports in 2005 but five US airports and five Asian in 2013. Beijing 'leapt' from 15 to the second place in just eight years and more than doubled the number of passengers. Dubai Airport (7th in 2013) appeared among the top 30 airports for the first time in 2007, ranked 27th with 19.3% growth and 34.3 million passengers and nearly doubled its throughput within six years. Similarly Soekarno-Hatta was 10th in 2013, up from 36th in 2008.

The North American and European markets are sustainable but moderate growth is expected. Those markets retain their importance because of their size. Also, the Middle East increased its importance not only due to well-established long-haul air carriers but also due to a very good geographic position which supports their global hub strategy.

The changing structure of air transport, including not only the increasing number of small aircraft intended for direct point-to-point inter-regional transport but also the trend to liberalisation and the universally growing transport volumes, will even further increase the pressure on airport capacity. In addition, the airports must also satisfy the changing profile and new categories of passengers. They must prepare for increasing numbers of elderly people and young parents with children. New standards have made it necessary to reconstruct completely some airport terminals. All these pressures will require substantial increases in investment. Similar changes will appear in the carriage of freight.

As already mentioned, the airside characteristics have always been 'dictated' by the aircraft geometry and performance. On the other hand the landside design and in particular the passenger terminals, are influenced by the state policies and market trends. During the last two decades the Low-Cost Carrier (LCC) model has grown tremendously, pioneered by Southwest Airlines in the United States and Ryanair in Europe. The LCC model has also rapidly developed in Asia and Middle East. According to Airline Profiler, LCCs' movements in 2005 in Europe accounted for 17% of all passenger traffic, and this increased to 32% in 2013. LCCs' very efficient business model (fast turnaround times, scheduling optimisation etc.) also created pressure on the airports they serve. For example, the top three Ryanair requirements on an airport are low airport charges, fast turn-around times and single-story airport terminals. At the beginning, LCCs benefited from cheaper landing fees at previously unused secondary airports.

Some low-cost carriers prefer simple and sparse passenger terminals with a minimum of commercial facilities so as to negotiate low charges. A few airports decided to satisfy those requirements. However, it must be stressed that most low-cost carriers are extremely unstable partners and can withdraw their services within a month so that the airport is left with a facility of limited usage. Therefore, the requirement to design for high terminal flexibility is very important. It must also be stressed that airport terminal capital and operating costs are only a part of the total cost of the airport infrastructure development and its operation. In the long run, during the whole terminal life-cycle, the cost difference between a 'normal' and 'economical' terminal may not be significant.

The process of major airport development is taking progressively longer. The second Munich airport was only opened 30 years after the first plans were drawn up. The Fifth terminal at Heathrow opened in 2008, 13 years after the inquiry into it began and nearly 25 years after it was recognised as being necessary. It is therefore sensible to predict requirements perhaps 35 or even more years ahead, yet the ability to predict even 15 years ahead is questionable.

The Far East has become the most rapidly expanding region from the viewpoint of further development of airports as the economy recovers its former vitality. The airports at Beijing, Shanghai, Shenzhen, Guangzhou, Hong Kong (Chek Lap Kok) and Seoul Incheon show very strong growth, and plans for the construction of new large international airports in China and Indonesia have been announced. Due to a relatively simple system of planning and public hearing procedures, the overall planning/construction time will be considerably faster than in Europe. The same could be expected for new airports in the Middle East or Turkey.

After the opening of Munich II, Oslo Gardermoen and Athens Sparta and the major expansion of Madrid Barajas, Milan Malpensa, Manchester and Paris Charles de Gaulle (Roissy), the development of the network of new international airports in Western Europe may be considered as almost complete. Further airport capacity will be gained by reconstruction and improvement of former military bases and scarcely used 'secondary' airports flown mostly by low-cost carriers, but they are often in rather remote locations. The remaining option for increasing capacity is the further development of existing airports. In general, this will be problematic because they have not reserved sufficient land for further development or increasingly complex technological developments will have to be employed.

It is anticipated that the 'hub and spoke' system will continue to be supported in the United States, so causing further pressure on capacity. The National Plan of Integrated Airport Systems (NPIAS) for 2015–2019 identifies 3,345 public-use airports (3,331 existing and 14 proposed) eligible to receive grants under the FAA Airport Improvement Program (AIP). The FAA estimates that over the years 2015–2019, there will be a need for approximately $33.5 billion of infrastructure projects eligible for Airport Improvement Program grants. The 389 primary airports (large hubs, medium hubs, small hubs and non-hubs) account for 12% of the airports and 62% of the total development. Large hubs have the greatest estimated development needs, accounting for $8 billion (25%) of the $33.5 billion identified. The 2,939 non-primary (non-primary commercial service, general aviation and reliever airports) make up 88% of the airports and account for 38% of the total development, respectively.

Development of Heathrow Airport

The development of Heathrow airport is used as an example to illustrate how requirements for the runway system and the other infrastructure of the airport can change rapidly and unexpectedly.

The history of Heathrow airport began in 1929. Richard Fairey Great West Aerodrome, which was used mostly for experimental flights was opened on the site of the present airport. In the course of World War II the Ministry of Aviation needed to build a bigger airport in the London area with longer runways that could be used by heavy bombers and airliners. In 1942 site selection started and in 1944, it was decided to build an airport at Heathrow with the then classical arrangement of three runways forming an equilateral triangle.

The war terminated before the airport was completed. It was necessary to adapt the airport project to the needs of civil aviation. It was not a simple task. It was necessary to estimate the development of civil aviation and its requirements after a six-year stagnation. A commission of experts assessed several options for completing the construction of the runway system using the three runways under construction. Apart from others, the commission determined these requirements:

1. The runway system should allow the operation of any type of aircraft, considering a crosswind limit of 4 kts (2 m/s).

2. Two parallel runways should be constructed in each direction, with a minimum separation of 1,500 yards (1,371 m).

The resulting design was in the form of a Star of David. Originally the construction of a third runway triangle was planned to the north of the present airport, beyond the A4 trunk road. Thus a system of three parallel runways would have been available whatever the wind direction. That additional triangle was rejected in 1952. By the end of 1945 the construction of the first runway and several buildings were complete, and Heathrow airport was officially opened on 31 May 1946, passenger processing being done partly in tents.

The dynamic developments in air transport made possible by new types of aircraft required constant changes to the original project. By the end of 1947 the construction of the first runway 'triangle' was completed. Work on the second 'triangle' continued simultaneously with the construction of an access tunnel into the central area under runway No. 1. In 1950 the construction of the runway system was practically completed. In order that space may be found for the construction of terminals, apron and the remaining infrastructure in the central area, runway No. 3 and subsequently also other runways were closed very quickly, and thus only three runways have been in operation since then (see Figure 1.4). The cross runway has now also been closed to allow redevelopment of the East end of the central area, leaving only the now standard arrangement of one pair of parallel runways until and if a third parallel runway is built; this might be on the site of the proposed third triangle.

12 Airport Design and Operation

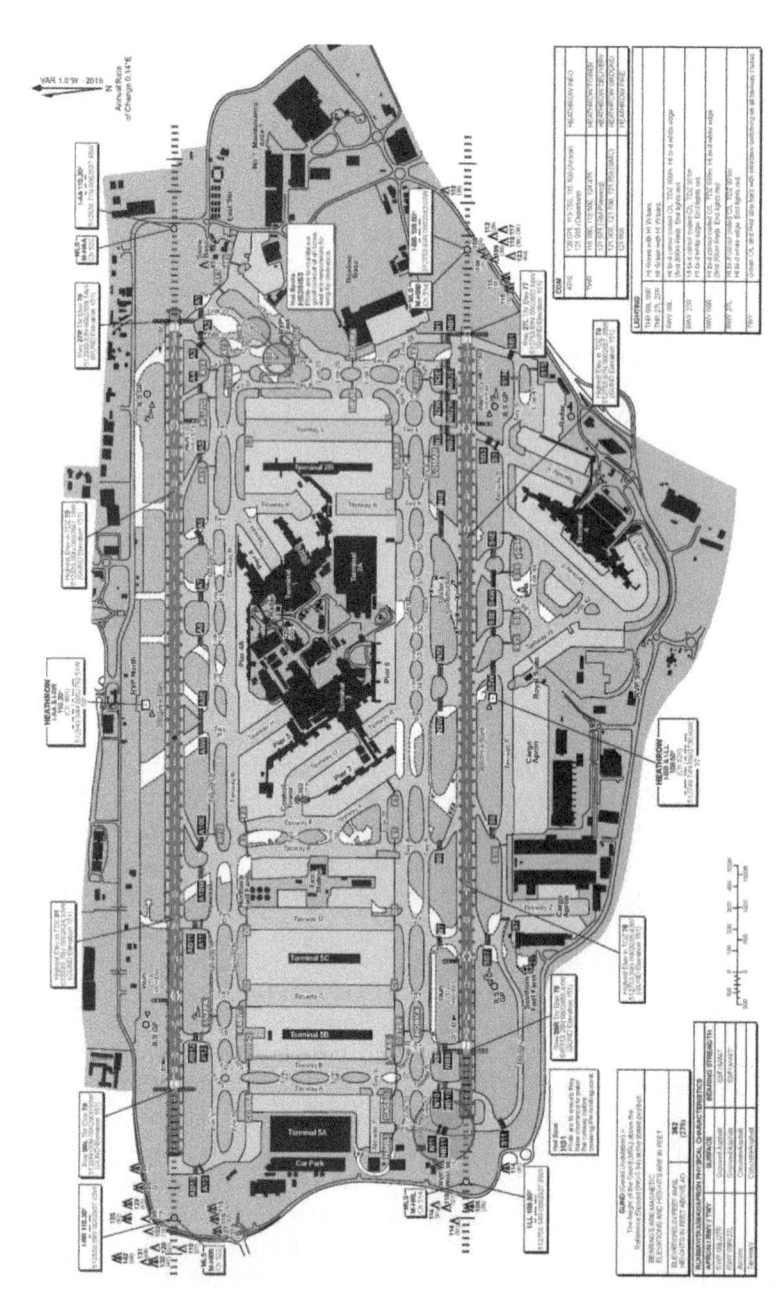

Figure 1.4: The current layout of London Heathrow. Source: NATS UK AIP Aerodrome Chart - ICAO AD 2-EGLL-2-1.

The original terminal buildings were only of a temporary nature. They were located to the north of the northern runway, and it was clear that in the future they would have to be substituted by a new complex in the middle of the runway system. The construction of the new complex began in 1950 with the control tower and terminal building designed for short routes, this becoming the present Terminal 2. It was completed and opened in 1955.

In step with the increasing demand for air transport, new terminal buildings were built. In 1962 Terminal 3 was opened, designed specifically for long-haul flights. In 1968 Terminal 1 designed for domestic airlines was opened. Terminal 4 was built in 1986 to the south of the southern runway after a protracted inquiry, and all British Airways' long-haul routes were moved into it from Terminal 3. This broke out of the central area for the first time, despite the difficulties caused by aircraft having to cross the southern runway and by passengers and bags having to be transferred between Terminal 4 and the central area.

Further increase in traffic caused the capacity of terminal buildings and stands to become the limiting factors, so Terminal 5 with its satellites was constructed in place of the Perry Oaks sewage farm, between the two main runways at the west end of the site.

The inquiry took four years and the terminal was opened in March 2008. In the same year the first commercial A380 flight arrived at Heathrow. In the autumn of 2009, after British Airways vacated Terminal 4 and moved to Terminal 5, Terminal 4 was refurbished. The old Terminal 2 was closed in November 2009 and demolished in summer 2010. In the same year Terminal 5's second satellite building was completed and in October 2010 the second phase of building Terminal 2B began. In summer 2011 the new Terminal 5C opened officially and in summer 2014 the new Terminal 2, The Queen's Terminal, was opened. Now there is a Commission working to decide on a preferred way of generating a further runway's worth of capacity in the South East of England. The options have been decided and are now out to consultation.

They include a third parallel independent runway to the northwest of the present northern runway at Heathrow. They also include a so-called extended northern runway which, with a safety area between the two runways, could be used for departures at the same time as the existing one accepts landings or could give flexible options for relieving noise under the various approach and departure paths.

14   Airport Design and Operation

1.2. STANDARDS

1.2.1. ICAO Legislation

Safety is the principal requirement in aviation. Standardisation is one of the means to achieve it. In the case of airports, it is standardisation of facilities, ground equipment and procedures. The only justification for differences is to match the types of aircraft that may be expected to use the airports. It is, of course, necessary for the standards to be appropriate and to be agreed by the aviation community.

Although attempts to reach agreement had been made much earlier, the need to agree on common requirements for airports used by air carriers became more urgent after World War II. In compliance with Article 37 of the Convention on International Civil Aviation in Chicago in 1944, the International Civil Aviation Organisation (ICAO) adopted Annex 14-Aerodromes to the Convention on 29 May 1951. Annex 14 provides the required set of standards for aerodromes used by international civil air transport. The Annex contains information for planning, designing and operating airports. With the developments in aircraft technology described in the previous section, together with the consequent changes to airports, Annex 14 has been regularly amended and supplemented. Particular Amendments were in the majority of cases approved at sessions of the respective specialist ICAO conference on Aerodromes, Air Routes and Ground Aids (AGA). Each of the ICAO member states may propose a supplement or amendment to an Annex through its aviation authority. The proposal is usually assessed or further examined by a panel of experts. Each of the member states may nominate its experts to the panel. Within ICAO there are panels and working groups that have been dedicated to several specific issues for a long time, for example,

AWOP	All Weather Operations Panel — issues of operations under restricted meteorological conditions
VAP	Visual Aids Panel — visual aids of airports
OCP	Obstacle Clearance Panel
ACNSG	Aircraft Classification Number Study Group
ADSG	Airport Design Study Group
FASG	Frangible Aids Study Group
RFFSG	Rescue and Fire Fighting Study Group
CAEP	Committee on Aviation Environmental Protection

Other ad-hoc panels have been formed to consider a specific one-off problem, for example,

ARCP Aerodrome Reference Code Panel — method for interrelating specifications of airports

HOP Helicopter Operations Panel — operation of helicopters.

The conclusions reached by the panels are reported in the form of working papers that are sent to the states for comments. Then amendments and supplements to the Annex are usually approved at the Air Navigation Conference or at the AGA conferences.

Each of the ICAO member states is obliged to issue a national set of Standards and Recommended Practices regulating the points in question for their international airports and amplifying them as necessary. This can give rise to problems of language. The options for an ICAO member state are either to adopt one of the official ICAO languages (English, Arabic, Chinese, French, Spanish or Russian) or to translate it into its own language and notify ICAO accordingly. If there is a need, the member state may adapt some of the provisions in its national Standards and Recommended Practices if it files the differences with ICAO. The provisions in the Annex have two different levels of obligation and relevance:

Standards contain specifications for some physical characteristics, configuration, materials, performance, personnel or procedures. Their uniform acceptance is unconditional in order to ensure safety or regularity of international air navigation. In the event that a member state cannot accept the standard, it is compulsory to notify the ICAO Council of a difference between the national standard and the binding provision.

Recommendations include specifications referring to other physical characteristics, configuration, materials, performance, personnel or procedures. Their acceptance is considered as desirable in the interest of safety, regularity or economy of international air navigation. The member states should endeavour, in compliance with the Convention, to incorporate them into national regulations. The member states are not obliged to notify the differences between recommendations in the Annex and the national Standards and Recommend Practices. However it is considered helpful to do so, provided such a provision is important to the safety of air transport.

Furthermore the member states are invited to inform ICAO of any other changes that may occur. In addition, the states should publish the differences between their national regulation and the Annex by the means of the Flight Information Service.

Notes are only of an informative character and supplement or explain in more detail the Standards and Recommendations.

At present Annex 14 has two volumes; Volume I *Aerodrome Design and Operations* and Volume III *Heliports*. Besides the Annexes ICAO issues other publications. The

16 Airport Design and Operation

following manuals, which supplement Annex 14, include guidelines for aerodrome design, construction, planning and operations.

***Aerodrome Design Manual* (Doc 9157)**

Part 1 — Runways

Part 2 — Taxiways, Aprons and Holding Bays

Part 3 — Pavements

Part 4 — Visual Aids

Part 5 — Electrical Systems

Part 6 — Frangibility

***Airport Planning Manual* (Doc 9184)**

Part 1 — Master Planning

Part 2 — Land Use and Environmental Control

Part 3 — Guidelines for Consultant/Construction Services

***Airport Services Manual* (Doc 9137)**

Part 1 — Rescue and Fire Fighting

Part 2 — Pavement Surface Conditions

Part 3 — Bird Control and Reduction

Part 4 — Fog Dispersal (withdrawn)

Part 5 — Removal of Disabled Aircraft

Part 6 — Control of Obstacles

Part 7 — Airport Emergency Planning

Part 8 — Airport Operational Services

Part 9 — Airport Maintenance Practices

For other related ICAO publications see also Attachment 1.

However, airport design and operation often requires knowledge of specific parts of other ICAO Annexes and documents which are closely related to airports, that is, Annex 3 Meteorological Service for International Air Navigation; Annex 4 Aeronautical Charts; Annex 5 Units of Measurement to be Used in Air and Ground Operations; Annex 6 Operation of Aircraft; Annex 9 Facilitation; Annex 10 Aeronautical Telecommunications; Annex 11 Air Traffic Services; Annex 13

Aircraft Accident and Incident Investigation; Annex 15 Aeronautical Information Services; Annex 16 Environmental Protection; Annex 17 Security: Safeguarding International Civil Aviation Against Acts of Unlawful Interference and Annex 18 The Safe Transport of Dangerous Goods by Air. Some of these documents are discussed in this book.

Landside parts of the airports are mostly not covered in the ICAO documents, except Annex 9 — Facilitation which relates to parts of airport terminals but guidance material could be found in different IATA manuals (see Chapter 12).

This book uses these ICAO documents, which are available from the world regional distribution centres, as primary references. It is not considered necessary to repeatedly refer to them in the text.

1.2.2. National Standards and Recommended Practices

Some countries, like the United States, generate a full set of their own standards and recommendations which complement and expand on those contained in the ICAO documentation. These are published as Federal Aviation Administration (FAA) Advisory Circulars 139 and 150. Many other countries find these useful as reference material. In Europe the common standards are being developed and maintained across the European Union by the European Aviation Safety Agency (EASA) with reference to the Commission Regulation (EU) No 139/2014 (of 12 February 2014) laying down requirements and administrative procedures related to aerodromes. The rules should reflect the state of the art and the best practices in the field of aerodromes and also take into account the applicable ICAO Standards and Recommended Practices.

All signatory countries to ICAO are obliged to apply the ICAO standards to their international airports. It would be uneconomic to apply them fully to their more numerous domestic airports, though it is sensible to take note of the principles embodied in the ICAO documents. Therefore every state has the possibility of making its own national Standards and Recommended Practices dealing with specific problems of domestic airports and airfields exclusively within the territory of the particular state for aerial works in agriculture, general aviation airports as well as for limited commercial operations. Besides the various types of civilian airports, there are also military airports. Their physical characteristics, marking and equipment may be different from the characteristics recommended for civil aerodromes. In creating a national set of Standards and Recommended Practices that does not derive directly from Annex 14 or another ICAO publication, the aviation authority usually puts an expert in charge of elaborating a draft of the document. The document draft is distributed for comments from selected organisations and panels of experts. After inclusion of the comments, the new draft is once more discussed in a wider forum. The proposal is also assessed in relation to other Standards and Recommended Practices. Development of each legal document requires a considerable amount of time and effort. If the necessary amount of knowledge, together with adequate legal and technical resources, is not put into the elaboration of the standard, the consequences can be serious. The document should be supplemented, amended, re-elaborated and exceptions from it should be noted, all of which takes further time and effort.

1.3. AIRPORT DEVELOPMENT PLANNING

The rapid development of air transport in the 1980s caused the capacities of many big European airports to be fully taken up in a very short time. The increasing volumes of passengers and freight will continue to make demands for the expansion of airport facilities.

Some European airports are struggling with a lack of capacity. Conclusions of the 1992 ECAC transport ministers' conference were that each state is obliged to ensure development of ground infrastructure and to detect and eliminate bottlenecks that are limiting the capacity of the airport system. In that way it should be possible to ensure that the increased requirements for capacity of airports in the future will be met. The solutions to these capacity issues can only be approached successfully on a system basis. It is necessary to assess the capacity of each part of the airport system individually: runway, taxiway system and configuration of apron, service roads, parking lots, cargo terminal and ground access to the airport. The result of such a system study is a proposal for staging the development of airport facilities, elaborated in a master plan of the airport.

An airport master plan represents a guide as to how the airport development should be provided to meet the foreseen demand while maximising and preserving the ultimate capacity of the site. In the majority of cases it is not possible to recommend one specific dogmatic solution. It is always necessary to search for alternative solutions. The result is a compromise which, however, must never be allowed to lower safety standards.

Planning of an airport's development is usually complicated by considerable differences between types of equipment and the level of the technology of the installations that are required for ramp, passenger and freight handling, and operations on the taxiways and runways.

The Master Plan of an airport may be characterised as: '*a plan for the airport construction that considers the possibilities of maximum development of the airport in the given locality. The Master Plan of an airport may be elaborated for an existing airport as well as for an entirely new one, regardless of the size of the airport*'. It is necessary to include not only the space of the airport itself and its facilities but also other land and communities in its vicinity that are affected by the airport equipment and activities.

It must be highlighted that a master plan is only a guide for the:

1. development of facilities,
2. development and use of land in the airport vicinity,
3. determination of impacts of the airport development on the environment and
4. determination of requirements for ground access.

It is necessary to actually construct each of the planned facilities only when an increasing volume of traffic justifies it. Therefore the master plan of an airport should include the plan of the phasing of the stages of building. Table 1.2 shows what may be included in the master plan of an airport.

Table 1.2: Purpose of an airport master plan.

I. *General part*

    A. An airport master plan is a guide for:
- development of airport facilities, for aeronautical and non-aeronautical services
- development of land uses for adjacent areas
- environmental impact assessment
- establishing of access requirements for the airport

    B. Beside other uses an airport master plan is produced to:
- provide guidance for long and short term planning
- identify potential problems and opportunities
- be a tool for financial planning
- serve as basis for negotiation between the airport management and concessionaires
- for communication with local authorities and communities

II. *Types of actions during the airport master planning*

    A. Policy/co-ordinative planning:
- Setting project objectives and aims
- Preparing project work programmes, schedules and budgets
- Preparing an evaluation and decision format
- Establishing co-ordination and monitoring procedures
- Establishing data management and information system

B. Economic planning
   - Preparing market outlooks and market forecasts
   - Determining cost benefit of alternative schemes
   - Preparing assessment of catchment area impact study of alternative schemes

C. Physical planning
   - System of air traffic control and airspace organisation
   - Airfield configuration including approach zones
   - Terminal complex
   - Utility communication network and circulation
   - Supporting and service facilities
   - Ground access system
   - Overall land use patterns

D. Environmental planning
   - Preparing of an environmental impact airport assessment
   - Project development of the impact area
   - Determining neighbouring communities' attitudes and opinions

E. Financial planning
   - Determining of airport development financing
   - Preparing financial feasibility study of alternative
   - Preparing preliminary financial plans for the finally approved project alternative

III. *Steps in the planning process*

A. Preparing a master planning work programme

B. Inventory and documentation of existing conditions

C. Future air traffic demand forecast

D. Determining gross facility requirements and preliminary time-phased development of same

E. Assessing existing and potential constraints

F. Agree upon relative importance or priority of various elements:
   - Airport type
   - Constraints
   - Political and other considerations

G. Development of several conceptual or master plan alternatives for purpose of comparative analysis

H. Review and screen alternative conceptual plans. Provide all interested parties with an opportunity to test each alternative

I. Selection of the preferred alternative, development of this alternative and preparing it in final form

IV. *Plan update recommendations*

A. A Master plan and/or specific elements should be reviewed at least biennially and adjusted as appropriate to reflect conditions at the time of review

B. A Master plan should be thoroughly evaluated and modified every five years, or more often if changes in economic, operational, environmental and financial conditions indicate an earlier need for such revision

Source: Airport Planning Manual, Part 1 Master Planning, (ICAO Doc 9184-AN/902).

As it has been already emphasised, the master plan of an airport is only a guideline and not a programme of construction. Therefore it does not solve details of design. In a financial plan, which is included in a master plan, it is only possible to make approximate analyses of alternatives for development, though costs of construction over the short term do need to be estimated with some accuracy if decisions are to be made on the economic feasibility. The master plan determines the strategy of development but not a detailed plan of how to ensure financing of each of the construction stages (see also Chapter 3).

The basic objective of the elaboration of a master plan should be that the interested parties must all approve it and the public should accept it within a process of public hearing procedures. It has recently been accepted that airport planning should retain a maximum of flexibility, given the increasingly obvious inability to forecast the future with any certainty: they are almost invariably wrong, so there cannot be a single best plan to be chosen from the proposed options. Therefore, in terms of investing in built facilities, a master plan should be quite short term, the longer term futures being described as strategic options according to a range of potential developments in the air transport system and the socio/political economy. The strategic options should come from all the relevant stakeholders participating in 'bottom-up' as well as 'top-down' planning, aiming for robustness and resilience across a range of feasible futures (Burghouwt, 2007). The United Kingdom has tried to put some of this theory to work (Airports Commission, 2014). A similar structured approach to flexible strategic planning is given in Juan, Olmas, and Ashkeboussi (2012). There will remain difficulties in preserving this flexibility while satisfying the local need to use surrounding land for alternative purposes.

Attachment 1 — related ICAO publications

Heliport Manual (Doc 9261)

Stolport Manual (Doc 9150)

Manual on the ICAO Bird Strike Information System (IBIS) (Doc 9332)

Manual of Surface Movement Guidance and Control Systems (SMGCS) (Doc 9476)

Aeronautical Information Services Manual (Doc 8126)

Aircraft Type Designators (Doc 8643)

Air Traffic Services Planning Manual (Doc 9426)

Airworthiness Manual (Doc 9760)

Volume I — *Organization and Procedures*

Volume II — *Design Certification and Continuing Airworthiness*

Guidance on the Balanced Approach to Aircraft Noise Management (Doc 9829)

Human Factors Training Manual (Doc 9683)

Manual of Aircraft Ground De-icing/Anti-icing Operations (Doc 9640)

Manual on Certification of Aerodromes (Doc 9774)

Manual on Laser Emitters and Flight Safety (Doc 9815)

Manual on Simultaneous Operations on Parallel or Near-Parallel Instrument Runways (SOIR) (Doc 9643)

Procedures for Air Navigation Services — Aircraft Operations (PANS-OPS) (Doc 8168)

Volume I — *Flight Procedures*

Volume II — *Construction of Visual and Instrument Flight Procedures*

Procedures for Air Navigation Services — Air Traffic Management (PANS-ATM) (Doc 4444)

Safety Management Manual (SMM) (Doc 9859)

World Geodetic System — 1984 (WGS-84) Manual (Doc 9674)

2

PREDICTING TRAFFIC

Bob Caves

Motto: The only certainty about the future is that it will not be as forecast.

2.1. INTRODUCTION

The process of airport planning requires a large number of forecasts to be made. Most attention is given to the prediction of passenger and freight traffic. It is, however, also necessary to attempt to predict likely changes in aircraft and airport technology, productivity in passenger and freight handling, choice of access modes, number of airport workers and the number of 'meeters and greeters' per passenger, even the number of checked bags, together with airport cost and revenue per passenger. Also the airport planner is not alone within the air transport system in needing forecasts. The airlines, the manufacturers, the sub-system suppliers and the national transport planning authorities all need forecasts of air transport activity.

The technical feasibility and the strategic master planning of an airport can often be examined adequately in the imprecise light of maximum likely forecasts. However the more detailed design of facilities and financial feasibility require a set of forecasts which estimate the demand to an acceptable level of accuracy at a given point in time.

The accuracy inevitably deteriorates as the time horizon extends, yet even 25 years may not cover the expected mid-life of the project if the planning process requires 7–10 years. It should be noted that the forecasts themselves are dependent on the level of technology, average aircraft size, income per capita, population, regional planning and a number of 'external' effects; in other words, the forecasts are usually implicitly 'locked-in' to the time frame for which they have been developed.

2.2. TYPES OF FORECAST NEEDED

The most fundamental forecasts are those for the gross annual throughput of passengers and freight for the system under consideration. This allows estimates to be made of the necessary scale of the system, its impact and, in gross terms, its financial viability.

For the more detailed planning and design purposes, the most crucial parameters are the flows of passengers, cargo and aircraft in the design hour, where the latter will normally be somewhat below peak hour flows, reflecting some acceptable level of delay or congestion in the system during the worst peaks. The passenger terminal layout and scale, and the revenue generation from concessions, are dependent not just on the total flow but also on the type of traffic; that is, international/domestic, terminating/transit, charter/scheduled by class of cabin and increasingly by the percentage of traffic performed by the low-cost carriers. These flows are usually derived from similarly classified annual flows by applying calibrated ratios which allow for the change in the peakiness with airport annual throughput. Typical ratios are given in Ashford, Saleh, and Paul (2011) and in Caves and Gosling (1999). An alternative approach is to construct likely future peak aircraft schedules and to derive the flows directly from these with assumptions on aircraft size and load factor. A similar method with different parameters is usually adopted for cargo terminals, a most important parameter here being the proportion of belly hold to all-freight cargo.

The total design hour flow consists of locally originating and terminating passenger and cargo flows, together with the passengers transferring (either on one airline or interlining between carriers) and the transit traffic which usually stays on an aircraft during a short turnaround. Depending on the nature of the transfer traffic and the method of operating, it may be necessary to disaggregate it further for the design of customs and immigration facilities. Aggregation of all these traffic types allows an estimate of design hour air transport movements (atm) to be made, when combined with assumptions on the average design hour aircraft size and average design hour load factors. These aircraft movement data are essential for planning the airside capacity — runways, taxiways, aprons, gates. To complete the airside capacity planning, it is usually necessary to forecast military and General Aviation traffic, though the latter's influence on the design hour capacity requirement is being reduced at large hubs, as it is progressively priced out of the market.

2.3. METHODS OF ANALYSIS

Standard texts on demand analysis quote many possible methods for assisting in the prediction of traffic (e.g. Ashford et al., 2011; ICAO, 2006c). The choice primarily revolves around the complexity and scope of the method and the timescale of the forecast. The more temporally stable and shorter term forecasts can usually be performed adequately with quite simple trend models which may not need to be unduly concerned with causality and uncertainty. This is not the case with long-term forecasts, which must concern themselves with unpredictable events both within and outside the air transport system. The choice is all too often determined by the availability of a data base and the budget for the study. Whichever method is chosen, the data used must be auditable, the results must be understandable and presented in such a way that the users can exert their own informed judgement.

2.3.1. Informed Judgement

The simplest method of all, that of informed judgement, appeals because it needs very little data — indeed, too much data confuses the decision-maker and slows the process down. However the difficulties of making long-term judgements are illustrated by a 1960 study by the Rand Corporation which predicted that by 2015 there would be only a 10 times increase in investment in computers and that it would be possible to control weather at a regional level. The Foresight study by the UK Office of Science and Technology surveyed technical experts in their own fields. Among many quite reasonable predictions, it also concluded that multimedia teleconferencing would be preferred to business travel by 2007, that the direct operating costs of aircraft would be halved by 2008 and there would be autonomous aircraft that would not need air traffic control by 2007, these dates being the averages of the responses. Experts in the travel trade have also made some interesting predictions of events which influence transport demand and supply including a 58% probability of flying cars by 2020.

The Foresight exercise used a refined version of expert judgement called the Delphi technique, where a panel of experts has their judgements returned to them together with those of the other experts, so that they can adjust their views prior to the final collation of the results. The Delphi technique is also used by IATA in compiling airlines' views of the future.

Care should still be taken in using even the results of surveys of experts since research shows that errors of individual judgements are systematic rather than random, manifesting bias rather than confusion. Further, many errors of judgement are shared by experts and laymen alike. Studies show that, at least in the field of fund management, the experts are no better than a random number generator at choosing investments. Erroneous intuitions resemble visual illusions in an important respect: the error remains compelling even when one is fully aware of its nature (Kahneman & Tversky, 1979). Most predictions, even when using analytic methods, contain an irreducible intuitive component, for example, in the assessment of confidence intervals. Research shows that these are often assessed too optimistically, particularly before repetition provides the opportunity for feedback on predictive performance because judgements are often made on minimal samples and assumptions are too restrictive of possibilities. A classic example of this is the tendency for long-term forecasts to be unduly influenced by short-term trends.

The more data are available, the more informed the judgement can be, at the expense in time and cost of incorporating them. It is sometimes possible to take data off the shelf, but there is usually something incompatible about them. They may well fall short in terms of sample size or the extent of their disaggregation. Also, they are very likely to be out of date. Thus, when methods requiring much data are chosen, a prolonged and expensive data collection exercise will generally be necessary.

Sometimes it is possible, rather than using ready-made data, to use ready-made forecasts, such as those produced by the aircraft manufacturers and governments (Airbus Global Market Forecast, 2014; Boeing Current Market Outlook, 2014–2033; FAA, 2014b), or analogy with a similar situation. Again, the problem is usually one of compatibility in time, space or characteristics of the setting. It may well not be appropriate to use available national forecasts to predict the traffic for a small regional airport. However acceptable corrections can sometimes be made to suit the study in hand. Provided that adequate checks can be made on the validity of the base forecasts, this can often be the most cost-effective method.

2.3.2. Trend Extrapolation

It is usually necessary to generate some predictions which are unique to the situation under study. Further, the predictions have to be justified and therefore be based in some formal analysis of the historic development of the traffic. The simplest formal analytic technique is 'trend extrapolation', either in time or scale. Historic trends are derived by simple linear regression of the traffic itself or the annual growth rate. This is then projected into the future, modified by judgement to take account of changing circumstances. Short-term trends often allow for the effect of economic cycles by applying some form of Box-Jenkins technique. Surprisingly, even the relatively simple data set required for trend analysis is not always available, or, if it is, it is not possible to discern a trend. Analogy or scenario writing then become invaluable techniques.

Longer term forecasts may recognise that the growth will mature over time by fitting the data to a Gompertz or 'S' curve (ICAO, 2006c). The difficulty over the long term is that small adjustments when fitting an S-curve to historic data can result in widely differing estimates of saturation levels. Saturation can be estimated separately, to fix the size of the S-curve, by, for example, assuming that the trips per capita, or 'propensity to fly' (ptf), eventually will not exceed those made in the United States or Japan. This is a reasonable assumption since, regardless of how wealthy the travellers become, everybody has a limited time and energy budget. The relationship between air trips and wealth per capita is shown in Figure 2.1. It is, of course, necessary to compare like with like. The ptf on business for a small rural town is never likely to exceed perhaps 20% of that for a capital city, though for the same category of household, there might be nearly the same ptf for leisure. The ptf for the retired population may be very different from that for single parents. It will be much smaller in small countries and those with excellent rail services.

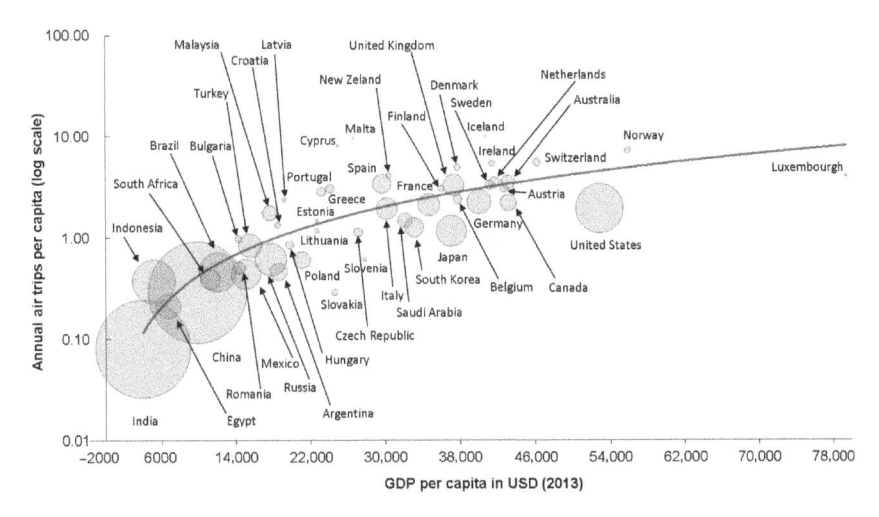

Figure 2.1: Propensity to fly. *Source*: Turiak (2015).

Another form of trend analysis is the 'step-down' procedure. This derives regional or local market shares from given national forecasts, making use of historic evidence of how the market shares have changed. The 'step-down' method is very popular in the United States, with its strong national forecasts and relatively stable traffic patterns, though it has been argued that the uncertainties created by deregulation in the distribution of traffic make it less appropriate now.

2.3.3. Econometric Models

If the above methods are deemed to be too naive, then it is necessary to bring causality into the analysis. The least complicated set of causal models is called econometric. These models try to relate the traffic to underlying economic parameters or more readily available proxies for them. Usually they are calibrated by multiple regression of historic data to derive elasticities of demand, that is, the change in demand due to a 1% change in any one of the independent variables affecting the demand.

A typical econometric model might be:

$$\ln T = -A + B \ln X + C \ln Y$$

where T is a measure of traffic, like Revenue Passenger Miles (RPM); X is a measure of income, like Gross Domestic Product (GDP); Y is a measure of price, like yield (Revenue Per Passenger km) (rpk); B is the elasticity of demand with respect to the specific measure of income; C is the elasticity of demand with respect to the specific measure of price.

28 Airport Design and Operation

Models of this type carry a number of assumptions:

- that a satisfactory explanation is possible with only a few independent variables,
- that the explanation is causal rather than co-incidental,
- that the 'independent' variables are reasonably independent rather than suffering from multicollinearity,
- that there is a constant functional relationship between the independent variables and the traffic,
- that the independent variables are easier to forecast than the traffic itself and
- that there are no significant errors in the data base.

Given that these assumptions can be accepted, the form of the model will depend on the particular circumstances. Multiplicative models are useful when the elasticities are essentially constant whereas exponential models are preferable when the elasticity varies with the independent variable. Difference models are to be preferred generally when the data set suffers from multicollinearity.

It is at least necessary to include terms to represent price and income, together with an autonomous trend term to allow for changes over time in things like quality of service. A greater number of independent variables may increase the explanation of the traffic, but the greater the number, the more that have to be forecast. Also, it becomes increasingly difficult to interpret the model results.

ICAO uses very simple econometric models to predict global scheduled traffic:

$$\log \text{PKP} = 4.63 + 1.34 \log \text{GDP} - 0.58 \log P \text{ yield} \qquad R^2 = 0.975$$
$$\log \text{FTK} = -0.37 + 1.58 \log \text{EXP} - 0.37 \log F \text{ yield} \qquad R^2 = 0.997$$

where PKP is passenger kilometres performed; GDP is world gross domestic product; P yield is revenue per passenger kilometre; FTK is freight tonne kilometres; EXP is world exports; F yield is revenue per freight tonne kilometre (ICAO, 2004). The very high R^2 show an extremely good fit to the historic data, due to the extreme aggregation of the traffic.

The use of elasticities is shown in Table 2.1. The income variable used for leisure travel was disposable income and it was predicted to grow at between 2.0% and 3.0% per annum. Its associated elasticity was calibrated to be between 1.5 and 2.5. The price (yield) was presumed to fall by 1.0% per annum and to have an associated elasticity of −1.0. There was also an autonomous trend term calibrated at between 0 and 1.5% per annum. The consequent forecast was for leisure traffic to grow at between 4% and 10% per annum.

Table 2.1: The use of an econometric model.

Determining factors	Leisure	Business
Disposable income (% p.a.)	2–3	
Income elasticity	1.5–2.5	
Trade (% p.a.)		4
Trade elasticity		0.8–1
Real airline prices (% p.a.)	−1	−1
Price elasticity	−1	0
Autonomous trend (% p.a.)	0–1.5	0
Potential growth (% p.a.)	4–10	3–4

Source: IATA Conference Montreal, Canada, 1988, adapted by the authors.

Similarly, business traffic was expected to grow at between 3% and 4% per annum, the income indicator in this case being the value of trade and the traffic having a zero elasticity with respect to price. These elasticities are quite typical of developed air transport systems. All elasticities are strongly dependent on the characteristics of the particular case. An analysis of 204 price elasticity studies showed low sensitivity (i.e. less negative values of elasticity) for long haul, for more recent studies, for short time horizons, and when the predictive equation also contains an income term. The elasticities tended to sit in two clusters: the one between −0.2 and −0.9 were mostly of business travel and the other cluster being between −1.1 and −2.0 (Brons, Pels, Nijkamp, & Rietveld, 2002). In fact, any elasticity depends on a variety of counteracting factors and borrowing a value to apply to another situation should only be done with this knowledge.

Econometric models may be applied to gross traffic, to route traffic or to generation of trips from a given geographic zone. Also they may be applied to classified subsets of the total potential travelling population in order to avoid the gross averaging process which is otherwise applied.

Any serious analysis of traffic at airports for marketing purposes, or to establish the viability of hubbing, really needs to work at the route-by-route level and to understand the competitive situation in any origin/destination market. However since the models are normally based in historic data, they deal only with demand which is revealed by the present behaviour patterns allowed by the historic characteristics of the transport offered. Also, there are many other aspects of the travellers' behaviour which contribute to their travel decisions than simply the economic characteristics of the subject modes.

2.3.4. The Travel Decisions

To begin to model behaviour in more detail, it is necessary to consider the need to travel, the mode of travel and, if the mode is air, which airport the traveller will use. The generation element of the decisions may be dealt with by market category analysis techniques as used initially by the Roskill Committee for a third London airport in 1971, or by econometric modelling. The trip generation per capita varies a great deal between industrial sectors and also by socio-economic group, with employees of small high-technology research and development companies in the United States making 150 trips per year while the average for UK citizens is one per year. In conventional transport planning, estimates of trip generation would be followed by distribution models, usually based on some analogy to the laws of gravity. This method presupposes that there is a reservoir of generated demand waiting to travel to any available destination. Simple models of this type ignore the important contribution of specific 'community-of-interest' between the origin and destination zones. A technique which distributes traffic as a function of relative attractiveness of the destination helps to overcome this drawback. Generation and distribution models can be applied to the air mode in isolation but, for shorter haul situations, it is more appropriate to model total generation and distribution and then to derive the modal split to air by use of a modal choice model.

2.3.5. Modal Shares

The availability of high-speed rail services will certainly reduce air traffic on dense short-haul routes, as demonstrated by the TGV experience. Airbus Industrie has suggested that a one hour advantage increases market share by 20% The Japanese experience, with quite similar fares for air and rail, is that air begins to take a greater share of traffic than rail at about 750 km, as shown in Table 2.2.

Certainly there is a switch of even high-yield traffic from air when door-to-door time by rail becomes less than three hours, but interlining and natural growth in traffic can still leave a significant air market: Orly to Nice, Orly to Marseille and Orly to Brussels remained as three of the densest routes in Europe (de Wit, 1995) despite the high-speed rail opportunities, and BA's shuttle from Heathrow to Manchester has nearly a million annual passengers despite trains averaging 160 km per hour over only 300 km. The resilience of air is due partly to the diffused nature of the demand within metropolitan areas (at least 50% is home-based rather than office-based) and partly to the access to down-town rail stations being often more difficult than to airports. It has been estimated that less than 10% of the European scheduled airline capacity is threatened by a future high-speed rail network. None-the-less, there are routes where rail does compete well, and in these cases it can certainly serve to reduce the strain on air capacity. The political will to encourage a switch from air to rail for environmental reasons will also have an effect on the modal split.

Table 2.2: Air and rail shares in Japan.

A: Shares of total trips				
Mode	300–500 km	500–750 km	750–1,000 km	1000 km +
Air	4.5%	13.9%	50.0%	83.6%
Rail	48.8%	64.8%	37.3%	13.2%

B: Comparison of fares and journey times		
Route	Rail — air fare (Yen)	Rail — air time (h:min)
Tokyo — Osaka	3,930	1:30
Tokyo — Fukuoka	3,780	3:23
Tokyo — Yamagata	1,990	1:32
Osaka — Sendei	4,175	3:04
Nagoya — Sendei	4,860	2:15
Fukuoka — Niigata	5,975	5:09

Source: Air Transport World (1995 July, pp. 24–32), adapted by the authors.

Telecommunications may be able to replace some physical travel, but it is generally accepted that, over short-haul distances, any substitution effects are compensated by stimulation of travel. Videoconferencing on the desktop is likely to become very common. A major Swiss bank installed 75 units worldwide, their travel department applying a 'could it be done by videoconference' test. The future impact of telecommunications is more likely to affect long-haul air travel. The differential advantages of teleconferencing must be greater at long distances, particularly with falling prices as the large capacity of fibre optic cables comes on stream. Telecommunication cost is almost constant with distance. A relatively small impact on the high-yield business market might exert a pressure for increased fares in the highly price-elastic leisure market.

2.3.6. Discrete Choice Models

Attempts to increase the richness of the behavioural content of models are often made by using a disaggregate, or discrete, approach involving the knowledge and modelling of individuals' trips. This type of modelling is usually more accurate in determining responses to changing supply or choices between alternatives because it retains the maximum richness of the available information rather than hiding some of the variety through the aggregation process. It is more important to use disaggregate information to calibrate a model: the application can often be satisfactory on aggregate data.

Most of the attempts to model the revealed choice behaviour of passengers have used individual trip data to calibrate logit models because of the models' specific properties. They express the probability of choosing an alternative in terms of the ratio of the utility to the passenger of any one of the alternative choices to the combined utility of all available alternatives, where the utility function to be maximised is a function of the attributes of the alternative and of the traveller. Almost always, the best explanation of behaviour in choosing between airports, between routes or between carriers, has been obtained by using logit models with some combination of access time, trip cost and frequency as the dependent variables. The most important choice factor, after the availability of the service, is the ease of access to the service, as shown in Table 2.3. Other factors influencing the choice decisions are the nationality mix of the airlines, the aircraft technology and, of course, fares, though it is difficult to obtain data on the fare actually paid. The effect of replacing turboprops with jets on competing routes from airports in the Midland region of the United Kingdom was estimated by comparing turboprop predictions with jet outcomes.

Table 2.3: Choice of London airport, percentage by reason.

Reason for choice of airport	Gatwick	Heathrow	City	All airports
Flights available	45.1	24.1	9.4	31.0
Nearest to home	13.4	11.4	20.0	13.3
Decision made by someone else	8.0	14.0	5.2	11.5
Connecting flights	3.3	11.7	0.1	8.3
Prefer airline	6.4	9.6	0.1	8.0
More economic/cheaper	10.4	3.5	1.3	6.0
Timing of flights	1.9	7.4	7.9	5.4
Nearest to business location	1.3	5.8	26.8	4.3
Total passengers (1000s)	17,990	36,887	171	58,357

Source: UK Civil Aviation Authority (1991), CAP 610. Adapted by the authors.

The improvement in market share due to jet service appeared to be less than 10%; indeed, it may have been less, because no attempt was made to disassociate the changes in type from the increase in aircraft size of the jets (85 seats, rather than 50–65 seats) and many in the airline industry believe that it is capacity rather than frequency which defines market share. However other evidence suggests that a balanced market with both operators using turboprops would change to 80%/20% if one operator used jets at the same frequency, and this seems to be confirmed in the experience of introducing the new 50-seat regional jets.

2.3.7. Revealed and Stated Preferences

The data for disaggregate models is usually obtained from surveys of historic travellers; that is, they are based on revealed demand. This clearly cannot account for those who have not travelled, nor for those whose travel decisions have been significantly constrained by the choices available to them; for example, charter passengers' departure airport is often constrained by flight availability. At the risk of irresponsible answers, these difficulties can be overcome by the use of 'stated preference' surveys. There are techniques for maximising the relevance of the answers obtained, the best of which are probably to impose notional budget constraints and to check against actual behaviour. The prime advantages of the technique are the cheap and rich nature of the data and the ability to consider larger scale variation in the alternatives that have actually been tried in real situations.

Stated preference surveys are used as part of market research techniques to obtain a rich picture of support for planned initiatives or to explore the best ways of tailoring a product to the needs of the potential users (Figure 2.2). Often, focus groups are employed to represent the wider public for these surveys. Care must be taken with these methods that bias is not introduced by the way in which the questions are phrased.

Figure 2.2: Market research requires various techniques of data collection. *Photo*: A. Kazda.

34   Airport Design and Operation

2.3.8. Effects of Supply Decisions

The demand can only be converted into actual travel (revealed) if an appropriate service is offered. It is becoming increasingly common for supply decisions to be uncoupled from revealed demand. This is sometimes to correct an under supplied market, but it is often a result of airline or airport strategies. An example was the KLM initiative to markedly increase frequencies and capacity at Amsterdam to encourage hubbing. The traffic may then appear at points in the system that are quite different from those of interest to the passenger or shipper. It must be anticipated that many other, and more extreme initiatives will be taken as the possibilities of liberalisation of services in Europe are realised. Also, liberalisation is continuing to spread through the rest of the world.

The most telling of recent examples of supply decisions that do not rely on existing demand patterns are those of the low-cost carriers in Europe. They often initiate routes from very small airports that have previously not had scheduled air service, the customers being prepared to travel long distances to take advantage of the low fares. Alternatively, they compete with established carriers and still stimulate equivalent levels of additional demand. The Bristol/Edinburgh route in the United Kingdom is a case in point. BA had 110,000 passengers in 2001. When EasyJet came on to the route, their traffic fell to 80,000 passengers while EasyJet carried 240,000 passengers in 2003. Again, London/Edinburgh traffic had been almost static at 2 mppa (million passengers per year). EasyJet started services from Luton in 1995, Stansted in 1998 and Gatwick in 2001, and all had generated 0.5 mppa by 2003.

2.3.9. Uncertainty

The calibration of those models which attempt to introduce causality into the understanding of passenger behaviour require a great deal of data on the mode characteristics and the demography of the potential travellers, as well as richer socio-economic information than is generally used in simpler econometric modelling. All the models' predictive ability also relies on being able to forecast the future trends of the independent variables used. Because of these difficulties, prediction based in any of the above methods suffers from likely errors which become greater as the time horizon increases. It has been said that a central case forecast is simply that outcome with a 50% likelihood of being wrong in either direction.

A study of 210 projects concluded, with very high statistical significance, that forecasters generally do a poor job of estimating demand for transport infrastructure projects, and it is not getting better with time. An average overestimate for rail projects was 106%, while 25% of road projects had errors of greater than ±40%. The worse performance of rail forecasts was primarily due to the influence of changing political inputs. The authors suggest that the solution lies in taking an outside view of the project and comparing it with a 'reference class' (Flyvbjerg, Mette, Holm, & Buhl, 2006).

There seem to be four ways in which this problem of uncertainty can be approached. The most common approach is to use sensitivity analysis on all the factors which it is felt may be in error. A rather more sophisticated approach is to assign probabilities to the forecast of the independent variables and let the output take the form of a risk analysis. In this case no preferred forecast is offered, rather the forecast's uncertainties are made transparent to the decision-maker. It should, however, be noted that there is much implicit judgement in the assignment of the probabilities, just as there is in all the quantitative methods. It is not only the purely judgemental methods that require judgement from the analyst. A third approach is the normative one of adopting policies to limit or encourage demand — in this case the models' role is to provide the understanding necessary for setting the levels of policy variables to adjust demand to achieve the planning objectives. This method was used at Amsterdam to massage the mix of traffic to achieve the maximum throughput consistent with meeting noise targets. The fourth approach is to recognise that it is not possible to predict the future but that it can be explored by writing scenarios and interpreting the consequences.

2.3.10. Scenario Writing

Scenario writing allows potential futures to be described. The forecasting techniques can then be applied to explore the traffic implications of each of those scenarios. Unfortunately, scenario writing is often used suboptimally, being limited to exploring options within the system under consideration with a view to predicting the future probabilistically. However a more productive use is to explore the range of feasible potential futures and their consequences to identify the needs to which a system might be asked to respond and the steps it would need to take in order to respond effectively. It should be emphasised that the objective is not to guess the future, even by assessing probabilities of the various potential futures and hence take a view on the most likely future. Rather, scenarios explore potential futures so that some light can be thrown on the scope and flexibility which needs to be designed into the system and on the consequences for the system's performance of not being able to meet the needs of some scenarios. Alternatively, the process can identify those possible futures which the system should not be designed to accommodate.

The essence of the method is to identify those factors which drive the business, but over which the business has little control, and condense them into several major themes. Once the consequences of the themes have been quantified using the conventional forecasting techniques, strategies may then be devised to meet each of the separate scenarios. Finally, the strategies are merged into a core strategy which is capable of responding to any one of the scenarios.

It can be taken for granted that the most likely future is the one which is presently being projected by in-house planners and system designers. It can also be taken for granted that, except in so far as the system's future use is predetermined by the closing off of future options by sunk investment, the most expected future will not coincide with any actual future state. None-the-less, this 'business as usual' case is an essential scenario because it forms the base from which the relative implications of the other optional scenarios can be explored.

There are many options for writing scenarios. The important characteristics are that the scenarios should be cohesive and should show a feasible route from the present state to the potential future state. The latter requirement is made difficult because the dynamics of change are not well understood (Keynes quoted in).

One example of the approach is to paint four potential scenarios (high and low growth, egalitarian or inegalitarian) for each of three (conservative, reformist, radical) ideologies, and to ponder the implications for international trade, political and economic stability and the need for transport. It can, for example, be inferred that a conservative high growth inegalitarian society will generate technological sophistication in the northern hemisphere and the encouragement of private transport, including aircraft. A reformist low growth more equal society would emphasise conservation and would encourage innovation in operating systems rather than in the technology of transport. The Dutch have used a consolidated version of these scenarios to inform their planning for Schiphol airport.

Once inferences have been made within each of the scenarios of the likely associated impacts on the factors driving the econometric models, they can be used to quantify the implications of the various potential scenarios, in the same way that the models would normally be used for the prediction of expected traffic. It may be that one of the scenarios would lead to a halving of the cost per pkm, in which case the consequence would be a doubling of demand if the price elasticity consistent with that scenario was judged to be 2.0. Note that the traffic itself may or may not double, depending on other consistent inferences in the same scenario with respect to available capacity, income etc.

2.4. HISTORIC TRENDS IN TRAFFIC

It is clear from the above discussion that much conventional forecasting activity relies on knowledge of past trends: with some justification, because only an unfortunate coincidence would place the forecaster at the exact moment in time where a complete discontinuity occurred.

The gross uplift has grown approximately logarithmically over the long term. This has been fuelled by increasing wealth, reducing yields and improving supply characteristics. Recently, there has been a slowing of growth in the more mature markets but strong growth in the Asia/Pacific region, a rise in importance of freight and a fall in mail. The domestic market in the developed world has matured, except for the rise in the low-cost carriers (LCC). The growth in air transport movements historically was less than that in passenger or tonne kilometres because of a trend towards longer stage length, larger aircraft and, latterly, a higher average load factor. Over the last two decades, this has been countered by an emphasis on frequency in liberalised markets, and by the introduction of over-water twin-engine aircraft and regional jets of 50 seats and less. Non-scheduled traffic, at least in Europe, has been much more sensitive to changing economic circumstances than scheduled traffic.

2.5. Factors Affecting the Trends

2.5.1. Economic Factors

In the more complex and data-hungry models, gross or discretionary income per capita tends to be used as the main causal variable. However at a local level, it is often difficult to get sufficiently local income data to be of much use for smaller airport studies. In more macro models, Gross National Product (GNP) is normally used as a surrogate for income. Modellers go to considerable lengths to interpret government views and to form their own views of future world events, but they tend to be dominated by conventional wisdom operating on common data bases. One must ask whether GNP is really any easier to forecast than traffic growth, particularly in the short term. Indeed, air traffic is often seen as an early indicator of what the economy will do. Also, one must ask whether the influence of GNP is all-pervading as indicated by simple correlation analyses at the macro level. The growth of traffic over time in Brazil suggests a more complicated relationship: the relationship between the change in the propensity to fly and the change in GNP per capita passed through levels of 5.5–1, zero and 3.5–1 as the economy has moved from the pre-industrial era through the formation of an industrial base and finally to development. Boeing Current Market Outlook (2014–2033) believes the world wide economy will grow at 3.2% for the next 20 years. The FAA takes the forecasts of a commercial organisation which also estimates that world GDP will grow on average by 3.2% through the next 20 years.

Other important economic factors include exchange rate differentials which drive all types of travel, as exemplified by the changes in North Atlantic traffic as the dollar weakened in the late 1980s. These are notoriously difficult to forecast but make a major difference in the competition between holiday destinations, as does the threat

38 Airport Design and Operation

of terrorism. There are also likely to be shifts in the global distribution of economic activity and in the distribution of wealth both globally and within nations.

2.5.2. Demographic Factors

Conventionally, air traffic increases with population, urbanisation, reduction in size of households and with younger age groups. It is generally assumed that these variables are amongst the easiest to forecast and that their relationships with traffic growth are stable. However forecasts of the US population in 2010 have varied since 1945 from 291×10^6 to 381×10^6 as the births per 1,000 women have changed from 85.9 (1945) through 122.9 (1957) to 73.4 (1972). Again, the tendency for an increased propensity to fly (ptf) may currently be explained largely by income, household make-up and job-type factors. It is quite possible that these relationships may break down. The changing structure of industry and commerce could affect air trip generation rates. Propensity to fly varies greatly with the type and size of firms. There is evidence of a reversal of domination by the top 500 Fortune companies in the United States, which could have a strong bearing on the future ptfs.

There is already a tendency to decentralisation of metropolitan areas, and this may lead to greater demand for lower density air travel.

Some feel that the information revolution or 'fourth logistic revolution' will result in fast and volatile flows of commodities, people and information, leading to a preference for the fastest and most direct mode of travel. This type of future, where a new hierarchy of cities evolves with the most powerful being those based in the four Cs of Culture, Competence, Communication and Creativity, is already visible. It should be considered together with other demographic trends, including the labour migration flows which eventually form the basis for visiting friends and relatives traffic and changes in fertility and mortality.

2.5.3. Supply Factors

2.5.3.1. Cost per passenger kilometre (pkm)

At constant levels of regularity, reliability, frequency and comfort, the primary supply factors determining demand are ticket price and generalised cost (i.e. a combination of price and value of time saved). The price is determined by the underlying cost of production and the pricing policies adopted to generate revenue. The costs of production are, in turn, determined by operating policies with respect to load factor, aircraft size, input factor costs, productivity and advances in technology. The load factor in turn depends on price, the revenue yield per seat generally falling as load factor rises. Estimates of worldwide load factors on scheduled flights reveal an increase at about a half per cent per year from 55% in 1975 to 70% by 2002. US carriers achieved load factors of 77.1% in 2005 and are predicted by the FAA to reach nearly 84% by 2033.

2.5.3.2. Aircraft size

Aircraft size had been expected to increase on the densest routes, most of the forecast demand being taken up by size rather than frequency in order to reduce costs in the competition for the low-fare passenger and to combat the effects of congestion. In this case, annual growth in seats would have been almost the same as passenger growth. In the event, even on these dense routes, frequency competition appears to be overcoming any shortage of runway capacity, resulting in more flights with the same size of aircraft. On lower density routes, and where increasing competition is unhindered by congestion, increased frequency may well cause even higher atm growth rates, led by the LCCs. Meanwhile, ever more powerful revenue management systems have allowed airlines to sell increasingly marginal seats, so increasing load factors with little increase in revenue per flight. Some airlines, notably BA, have been reducing capacity in order to reverse this trend of excessive yield dilution. It is now expected that average aircraft size will more or less stabilise at current levels. Boeing Current Market Outlook (2014–2033) expects that single aisle aircraft will make up almost 70% of the world passenger jet market by 2033, the fleet being 2,640; 29,500; 5,570; 3,680 and 790 for regional, single aisle, small wide body, medium wide body (e.g. B787-8, 787-9 Dreamliner, A 350) and very large aircraft respectively, including freighters. Airbus expects that with the emergence of new 'mega-cities', the importance of reducing costs and the continual strengthening of hub-to-hub routes will drive future demand for high-capacity transports including the A 380. Airbus forecasts growth of the number of 'mega-cities' (those handling more than 10,000 daily long-haul passengers (those flying over 2,000 nm internationally)) from 42 today, to more than double to 89 'mega-cities' by 2,032, requiring a greater emphasis on size rather than on frequency, the average aircraft size thereby increasing by 20% over the 20 years. However only some 320 A380s have been sold so far of which just over half have been delivered.

2.5.3.3. Cost of input factors

Fuel has dominated the historic changes in input factor prices. A combination of shortages and the long lead times for aircraft technology to respond to the situation has, from time to time, caused considerable problems for the industry. None of the predictions made in 2013 in any way anticipated the sudden halving of the fuel price in the second half of 2014. All the predictions are for the long-term price of oil to increase over the 2013 prices. Operational changes have helped to alleviate the situation in the past but in the longer term, the larger cost reductions to combat forecasts of fuel price increases will have to come mostly from the technology. The combined effect of improvements in labour costs, advances in technology and reductions in the first cost per seat of aircraft are unlikely to result in a total cost reduction of more than 1% per annum.

2.5.3.4. Technology

The impact of technology on future capacity is often the most neglected area of supply prediction. Heathrow's growth is a case in point, early predictions being hampered by inhibitions on the maximum feasible size of aircraft and more recent analyses by under prediction of annual runway capacity. Most studies only allow for changes in a single factor (e.g. regional jets) which is closest to the study at hand. In fact, it is clear that there will be changes in navigation, communication, guidance on approach and on the ground which will have implications for capacity, as well as changes in vehicle technology which will affect both cost and capacity. It is less clear if and when specific changes will be introduced, though they will only occur when there is a sufficient economic incentive. However it is likely that fuel efficiency will continue to improve as in the past, at the rate of 1%–2% per year for the medium term. The combination of technology, operations and alternative fuels will also reduce the environmental consequences (www.airbus.com/innovation/eco-efficiency).

2.5.3.5. Management

The technical changes will often open up new opportunities for changes in management style, as with new routes and networks for airlines, new operating procedures around airports, advanced revenue management systems and improved facilitation through the use of smart cards. However managers can also innovate without technology, as is seen by the growth of the low-cost carriers and the forming of global alliances. These and other initiatives will continue, but which of them, when and in which combinations can only be explored by scenario writing.

2.5.3.6. Capacity constraints

The above factors will only influence future traffic in the expected way if sufficient investment is made to accommodate the potential demand. Therefore any political decisions could interfere with forecasts as could any ineptitude in the planning process. In an ideal forecasting process, these effects would be internalised within the method, but political forecasting is not yet sufficiently advanced to allow this. The UK government previously encouraged the development of two more runways in the South East of England by 2030. Currently, the present government has set up a Commission to explore the best way of increasing capacity by the equivalent of a single additional runway by 2030, which would mean that a significant portion of the currently forecast unconstrained demand would not be met (Airports Commission, 2014).

2.5.3.7. Fares

All the above supply factors affect the level of fares charged. Management also can adopt policies to cross-subsidise routes or classes of traveller, so that there is often in the short term only a limited relationship between the actual price paid for a ticket and the underlying costs. Data on real air fares paid are usually only available to analysts on an aggregate basis for all operations of an airline, quoted as yield per pkm.

Forecasts of changes in fares by the pertinent carriers have to be made largely by judgement. Given the huge increases in demand levels created by the LCCs, these judgements are very important in producing reasonable estimates of traffic at an airport. Once made, they may then be used to estimate traffic by using fare elasticities. The derived historic elasticities may be modified as necessary to allow for changes in local conditions. It should be noted that none of these supply costs, fares or elasticities are necessarily appropriate for situations other than those for which they were derived; for example, a study indicated that short-haul price elasticity in the United States may be as high as 2.74, which is higher than most studies have derived even in wholly leisure markets.

Yields per passenger kilometre have in the past fallen at up to 2% per year. The FAA in the United States thinks that domestic yields will fall by only 0.5% per annum to 2034 (FAA, 2014b).

2.5.4. Economic Regulation

Since the deregulation of US domestic air transport in 1978, the liberalising ethos has spread through Europe and many other countries. It has been accompanied by a parallel trend to the privatisation of airlines and, later, of airports. Without these trends, there is less opportunity for competition, fare reductions, route development, alliances, efficiency gains and for airports to take marketing initiatives. There have been abuses of the monopoly positions that have been made possible by the new freedoms. Rather belatedly, this has led to the imposition of regulatory controls on monopoly powers. The regulatory setting should be incorporated in scenario descriptions, including free trade, open skies, foreign ownership and environmental taxes or capping as well as the regulations to control monopolies.

2.5.5. Environmental Regulation

Another aspect of regulation is the national and international stance on recovering the environmental costs induced by air travel. Specifically, it is expected that some form of carbon tax will be imposed on aviation. This was introduced in Europe in 2012 together with a carbon tax trading scheme, so that aviation can buy credits from those industries which do not use their full quotas. The charge has temporarily been withdrawn for flights to and from countries outside the EU pending an ICAO scheme to be finalised by 2016 for implementation in 2020.

2.5.6. Cargo

The analysis of the cargo market is similar in basic method and in the variables which generate traffic. Some additional factors affecting cargo growth are:

- continuing increased share of integrated carriers,
- time and cost sensitive products,
- early stage of a product's cycle,
- availability of surplus capacity on routes,
- regulation of facilitation and
- competitive advantage over other modes.

More detail on the factors driving air cargo may be found in Chapter 11.

2.6. Conclusions

The discussion has identified the requirement for forecasts of arriving and departing passengers disaggregated by purpose and type of destination, and of air transport movements by type of equipment. These forecasts are required on an annual basis but, more importantly, for the representative peak hours. Information is also needed on General Aviation and military activity, but the impact of this traffic on design hour movements is more a function of airport policy than of the unconstrained volume of traffic.

The methods available to make these forecasts have been reviewed and shown to be more or less adequate for short-term detailed planning and design, but to offer little more than a formal basis for reasoned judgement in the long-time horizons needed for airport system and master planning. Interactions between the variables which control demand are not sufficiently understood to allow more certain forecasts to be developed, so that informed judgement must be used. This judgement may be expressed explicitly as high and low forecasts, or it may be combined with formal attempts at risk analysis, or it may inform the interpretation of future scenarios.

It is clear that there should be consistency between forecasts for airports throughout the system. The obvious way to obtain this consistency would be by modelling the complete air transport net with a single methodology, a single set of assumptions and a consistent set of data. However not only is it rare to find the resources for this but there is also a real danger of constraining the output to that which follows from a dominant preference of one member of the team of analysts and advisors. There is no substitute for keeping an open mind on the future and exploring a wide range of possibilities. Consistency of methodology is undoubtedly necessary for communication and interpretation, but the inputs to the prediction process should be allowed to vary as necessary to reflect the variety in the system.

No particular model will always be the most appropriate, nor will any one model always give the most accurate prediction. It is always worth using more than one approach. Much can be learnt from a comparison of the outcomes relative to the constraints of the modelling. However the balance of cost and competence will often be best achieved by a simple econometric model, applied to a carefully selected set of markets that most affect the total traffic. If necessary, these models can be used even without a good historic data set for calibration, since, with care, appropriate elasticities can be borrowed from other studies.

Many of the important imponderables are external to the air transport system and are best dealt with by incorporation in a range of comprehensive scenarios. Inside the system, many other factors over which most airports have little control have been identified in Section 4, and these also should be included in the scenario descriptions.

Apart from economic growth, the factor which strongly influences the rate of growth of demand in the short term is the airlines' and the regulators' policies on fares. The conventional wisdom sees a continuation of a low fares policy fuelled by deregulation, leading to high load factors and larger aircraft. If this approach leads to insufficient yield to at least maintain the present airline rates of return on investment and hence leads to lack of capital to fund the needed extra capacity, the airlines may retrench and rely on the essential business traveller. If this happened, too much surplus capacity would have been provided. The low-cost carriers can make good use of some of this capacity, but experience suggests that only the most resilient LCCs survive.

The development of technology, both to combat fuel availability and rising price and also to counter the physical and environmental constraints on airport development, is important in the longer term. Further interruptions in the oil supply may increase the rate at which more fuel-efficient aircraft are developed. Advances in aircraft technology and approach guidance, designed to allow changes in operational practices, will be important in lifting the capacity constraints at large airports. Their implementation depends partly on whether the large global carriers really want capacity to be increased at their hubs.

One other air transport factor which is particularly important in determining an individual airport's traffic is the extent that the network develops as a

44   Airport Design and Operation

hub-and-spoke system rather than concentrating on direct connections. If networks predominate, it is important to forecast which airports are chosen as hubs or as spokes or are left out completely. To some extent, the airports will influence this since a hub system will not develop easily if transfers are not convenient.

Despite all these caveats, forecasts are still needed, so one current view of the future is now given in Table 2.4. A low-fare policy may result, in the short term, in emplanement growth rates higher than those forecasts as mean values. Another expected trend is for atms to grow more slowly at larger airports. The forecasts will depend on conditions at individual airports in terms of level of maturity, surface competition, marketing and the airport's share of the national traffic.

Table 2.4: Annual average predicted growth in passenger kilometres, 2014–2033.

Traffic growth to/from	China	North America	Europe	Latin America	Africa
China	6.6	6.3	6.1	7.5	6.6
North America		2.3	3.1	4.7	6.1
Europe			3.5	4.9	4.9
Latin America				6.9	6.6
Middle East					7.3
Africa					6.7

Source: Boeing Current Market Outlook (2014–2033), adapted by the authors.

Airbus predicts annual growth to 2033 as 4.7% for revenue passenger kilometres (rpk) worldwide. They expect rips per capita to rise from 0.24 to 0.69 in the Asia/Pacific region compared with 1.59–2.14 in North America. The US FAA (2014b) expects the growth rates for all US carriers from 2014 to 2034 to be revenue passenger miles (rpm) 2.8%, passengers 2.2%; the regional traffic will grow slower at 2.3% and 1.9%, respectively.

In lieu of any specific study for a particular airport, the gross regional forecasts may be used to factor up from base data, with the caveats mentioned in this section. The peak hour atms may be estimated by making assumptions about the peak hour fleet mix and load factor. These last two factors will always depend very much on the circumstances at any particular airport, not least the amount of transit traffic on the aircraft.

3

AIRPORT SITE SELECTION AND RUNWAY SYSTEM ORIENTATION

Tony Kazda and Bob Caves

3.1. SELECTION OF A SITE FOR THE AIRPORT

The construction of a new aerodrome or an enlargement of an existing one represents extensive investments and building works. It is therefore necessary to design the entire aerodrome project for a long time period. The maximum possibilities of the airport development in the proposed locality should be considered, within the limits of the airport's critical constraints. As well as ensuring that the capacity and operational requirements are met safely, the issues concerning the airport and its surroundings should be considered, particularly the impact of the airport on the nearby population and environment. The locality selected for the airport and orientation of the runway system should facilitate long-term development, and provide flexibility for future airport expansion at the lowest cost in terms of money and social impacts. The locality assessment should take a complex design approach to consider not only all parts of the airside such as the runway and taxiway system but also the location of terminals, their future expansion and the airport access system and their interfaces.

The advent of the jet era meant a substantial increase of noise impact from air transport on the neighbouring communities and significant runway extensions because of jet engine characteristics. Jet aircraft had also more demanding geometric characteristics. This resulted in greater airport land take and the need to extend airports beyond their original boundaries. Subsequent conflicts between airports and adjacent urban areas called for longer planning horizons. To assist states in planning the expansion of existing international airports and construction of new ones, the Council of ICAO on March 1967 approved a proposal for a Manual on Master Planning (ICAO Doc 9184). The prime objective of master planning is to determine the *ultimate site capacity* and then to protect it from the consequences of ill-considered development of facilities on the airport and from encroachment of incompatible land uses around the airport which might restrict either its physical expansion or result in traffic limits due to environmental impacts.

The ICAO manual defines inter alia the planning horizons: short term — three to five years in advance — provides the basis for actual development work, while medium term forecast (from 5 to 20 years, usually in five-year intervals for convenience) bridges the gap to the long term and provides interim information on probable subsequent phases of development. Similar time horizons are defined in other literature, for example IATA Airport Development Reference Manual refers to a 'projection period' for forecasting and recommends for traffic forecasts short term (≥ 1 year ≤ 5 year projection) and long term (>5 year ≤ 30 year projection) periods. As good practice many airports accepted a 20-year planning horizon as 'long' and sometimes it is still used by some states. De Neufville refers to *'ultimate vision'*, that is, a current view of the possible future a long time in the future, for example 20 years'.

The concept of ultimate limits to airport development is also included in the ICAO and FAA manuals but it is often overlooked. Referring to ICAO Doc 9184, adequate land should be acquired or protected to provide for possible *ultimate runway development*, including protection of approaches and provision for associated visual and radio navigation (non-visual) aids. According to FAA AC 150/5070-6B, 5-, 10- and 20-year time frames are typical for short-, medium- and long-term forecasts; nevertheless some studies may want to use different time frames.

At some airports it may be necessary to look beyond the 20-year time frame to protect the airport from incompatible land-use development. However, planning beyond the 20-year period should be general in nature and in much less detail than that for the short or even mid-term development. For example, if planning for a future runway, the master plan might only indicate the general location and potential length of the runway.

Airport planning practices differ significantly between countries and reflect historical experience, methods of airport financing, legal framework, political system and many other factors. For example, US airport operators prepare their plans in line with the national funding agency requirements as they can only get funds from the Federal Government for projects that are in the National Plan (the NPIAS). Furthermore, projects only get into the NPIAS if they are included in an approved master plan.

Planning airports for 20–30 years ahead could be appropriate in some states. However, in countries where the planning process requires public hearing procedures, it can seriously delay airport development, for example by environmentalist groups who can join local inhabitants to organise resistance to airport development, sometimes called NIMBY which is an acronym for the 'Not In My Back Yard', or even BANANA — 'Build Absolutely Nothing Anywhere Near Anything'. For example, it took 23 years to get the final building permission for Munich Airport. To plan for 'ultimate limits of airport development' might be proper for well-established democracies with a high standard of living with strong regulatory authorities and good control of corruption, as in those countries discussions on airport development could take many years. Some airports adopting an 'ultimate vision' development concept protect possible future airport expansion by acquiring large land areas at a time when it was possible, for example Paris-Charles de Gaulle Airport. On the other hand it would be useless to prepare a long-term airport development plan for a country with an authoritarian regime where fundamental decisions could be taken within a few hours without involvement of the public. The optimum planning horizon depends on a number of factors which are unique for any particular state or even region and include not only standard of living but also population density, price of land and regional stability.

Selection of a suitable site for more airport capacity should begin with an assessment of any existing airport and its site. It is almost always easier to modify an existing airport than to create a new one on land that has previously had a different land use. The assessment is made in the light of the prospective passenger market, its growth rate and any limitation of the growth resulting from, for example a demographic shift of population. Therefore the prognosis of the growth of a number of passengers and volume of air cargo in the catchment area of the airport is one of the key elements in planning the airport development. The methodology of forecasting traffic is considered in Chapter 2.

After the proposed airport's size and layout has been approximately determined by a preliminary study, possible sites for the development of the airport are assessed in several steps, the principal ones being:

- approximate determination of the required land area
- assessment of the factors affecting the airport location
- preliminary selection of possible localities from maps
- survey of individual sites
- assessment of impacts on the environment
- revaluation of the selection of possible sites
- production of site layout drawings
- estimate of costs, revenues and discounted cash flow
- final selection and assessment of the preferred site
- a final report and recommendations.

The same procedure is followed in principle when assessing the development of an existing airport, though some of the steps may be omitted. The number, location and orientation of the runways and of the taxiway system should meet the following criteria:

48   Airport Design and Operation

- the availability (the usability factor) of the runway system should be acceptable, considering the losses which would result from the unavailability of the airport
- obstacle restrictions defined by the obstacle clearance limits should be respected
- ideally, the final capacity of the runway system should meet the predicted demand in the typical peak hour traffic in the far future
- the site selected for the airport should be considered not only from the viewpoint of obstacle clearance limits and environmental requirements but also from the viewpoint of achieving a functional layout of the aerodrome facilities
- the airspace capacity and procedures assessment with respect to type of airport operation and to other airports in the vicinity
- the ground transport access to the airport and the urban development plans in the airport vicinity should be efficient and sustainable
- the site should be in accordance with local and regional land-use plans.

3.2.   USABILITY FACTOR

The ease with which acceptable usability of the airport can be provided depends on meteorological conditions in the selected locality and the topography of the site. The usability factor is defined in Annex 14, Volume I, Chapter 3 as: 'The percentage of time during which the use of a runway or system of runways is not restricted because of the crosswind component'. It can therefore be provided most economically by optimising the runway directions relative to the prevailing wind directions. The actual runway orientation is the result of a compromise between the usability and other factors. The sensitivity of an aircraft to crosswind depends mainly on the aircraft mass and the type of undercarriage. The greater the aircraft mass, the faster it tends to fly and the less sensitive it is to the crosswind. The crosswind limits specified in the Annex 14, for example 20 kt in the case of aircraft whose reference field length is 1500 m or over, should be considered as indicative only and are quite conservative. Modern code letter D or higher aircraft are able to operate with crosswind as high as 30 kt depending also on the runway braking action (see Section 8.4.4). Airports for large aircraft should therefore be able to meet the usability requirements without resorting to more than one direction, but the design will also be influenced by the length of runway to be provided, the associated difficulty of levelling the site, the location of areas of population, the existence of sites that encourage birds and the obstacle clearance requirements. In any case, it is usually necessary for a large airport also to cater for the needs of smaller aircraft, so it will often be necessary to install short cross runways to provide adequate usability for them.

According to Annex 14 requirements the number of runways at the airport and their orientations should provide at least a 95% usability factor for the types of aircraft which are intended to use the aerodrome, presuming very conservative crosswind capability of the aircraft. However, this should be considered as an absolute minimum for airports with scheduled transport. A basis for calculation of the usability factor are the data on percentage occurrence of the wind of the given direction and speed for a period of at least five years. The statistical data are usually published in the form of a table (see Table 3.1), or in the form of a wind rose.

Table 3.1: Wind distribution statistics by direction and speed in 0.1% meteorological station — Bratislava M.R. Štefánik airport.

Wind speed/ direction	N	NE	E	SE	S	SW	W	NW	Calm	Σ
Calm	–	–	–	–	–	–	–	–	123	123
1–2	40	82	52	31	25	23	23	65		341
3–5	45	45	34	37	26	16	19	111		333
6–10	39	3	5	15	16	4	9	90		181
11–15	4	0	–	0	1	0	1	15		21
16–20	–	–	–	–	–	–	–	1		1
>20	–	–	–	–	–	–	–	–		–
Σ	128	130	91	83	68	43	52	282	123	1000

Source: Šoltís (1982).

Determination of the usability factor may be made by several methods. It is best to choose a method which is likely to underestimate the usability rather than the opposite.

For a rapid assessment, the transparency method is frequently used (Figure 3.1). The statistical data on percentage occurrence of the wind of each combination of direction and speed are plotted in the form of a compass rose. The value of percentage occurrence is recorded in a window which corresponds to the direction and speed of the wind. The width of the overlaid transparency is equal to twice the maximum allowable crosswind speed.

The transparency is turned in such a way that it covers the maximum of the percentage wind occurrences. The same method may be repeated for each additional runway until the combined coverage meets the required usability factor of the runway system. The whole process is then repeated for other maximum crosswind criteria.

50 Airport Design and Operation

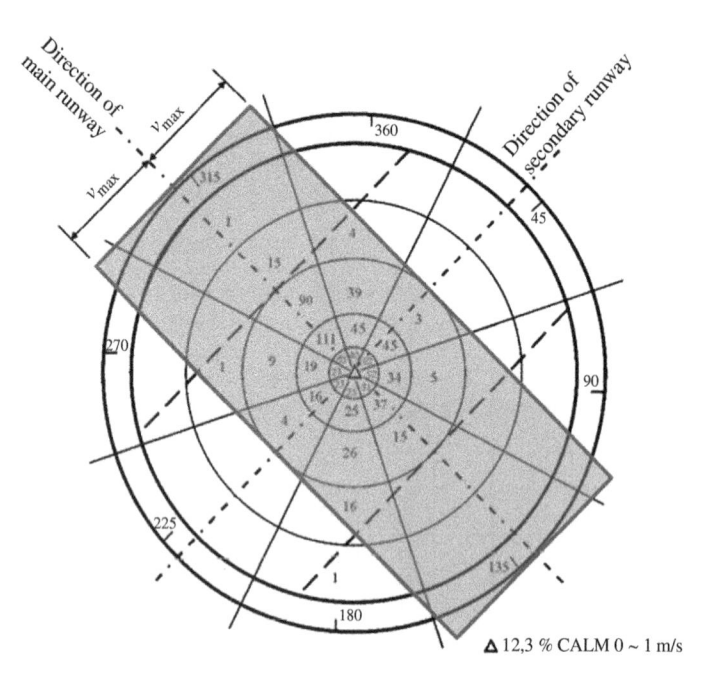

Figure 3.1: Determination of the usability factor — transparency method, Bratislava M.R. Stefánik airport.

For a more exact evaluation of the usability factor, it is convenient to use the Durst method. The basis of this method is a rectangular co-ordinate system (Figure 3.2). Directions of a compass rose are developed in the x-axis and speeds of the wind in the y-axis. In the created fields the percentage wind occurrence and speed are recorded. The following function is plotted into the grid:

$$f = v \cdot \frac{1}{\sin \alpha}$$

where v is the maximum value of the crosswind and α is the angle which is formed by the runway axis with particular directions of the compass rose.

In the direction of the runway axis, the function increases to infinity and the usability factor is limited by the maximum value of the headwind. The function is continuous within each wind data interval. The lowest point of the function is actually the value of the maximum crosswind ($\alpha = 90°$). In the direction of the runway the function grows to infinity and the usability is limited by the value of maximum head wind for particular type of aircraft and operational conditions. For example for B 737 when conducting a dual channel Cat II or Cat III landing predicated on autoland operations the max headwind is 25 knots. When the numerical values under the function are summed, the usability factor of the runway, or of the runway system, is given.

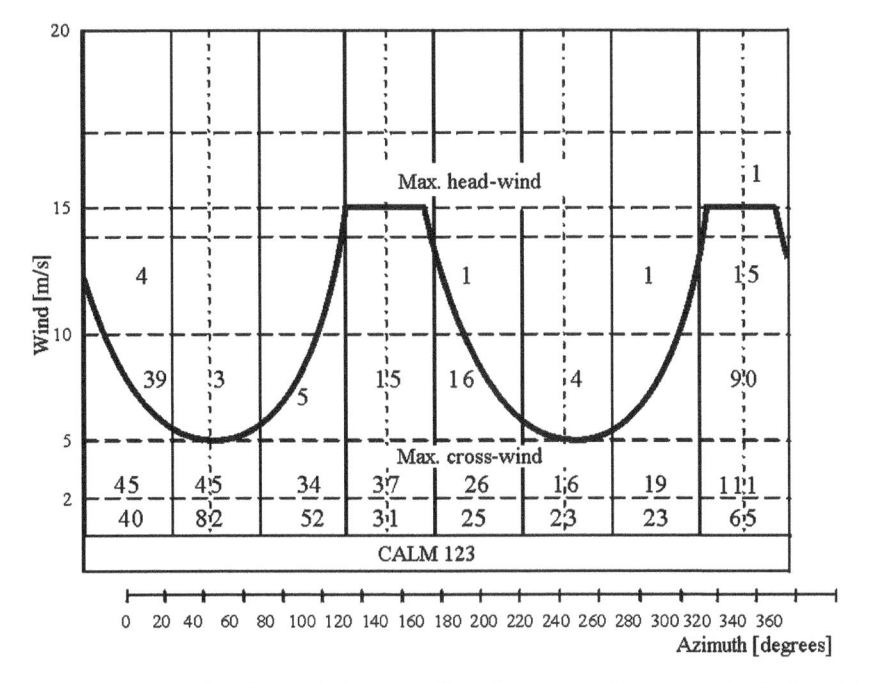

Figure 3.2: Determination of the usability factor — Durst method, Bratislava M.R. Stefánik airport.

However, sometimes it is not possible to optimise runway orientation with respect to the prevailing winds either due to the presence of obstacles, urban areas in the vicinity of airport or other environmental reasons. Airport operations could be often limited in those cases or require special procedures or qualifications of air crew (Figure 3.3).

3.3. EFFECT OF LOW-VISIBILITY OPERATIONS

Large airports are expected to continue to operate in poor weather but it is important to consider the occurrence of low visibility and low cloud base ceilings on the way the airport operates. The occurrence of Category II or III visibility (Table 3.2) represents a significant qualitative change as a consequence of considerably lowered abilities of the pilots, air traffic controllers and other persons involved to maintain the procedures and an expeditious flow of air traffic with a visual reference. A change from Category I to II and any lower category entails more than just the provision of the visual and non-visual navigation aids. Performing operations under low-visibility conditions depends also on the system of air traffic control, electrical systems back-up, meteorological equipment and services, surface movement guidance and control systems, runway incursion protection measures, security and emergency procedures and apron management. It requires also a change of the system of airport control and of the attitude of all the airport staff to fulfilling their duties. It is necessary to augment the system of airport operation quality control and Safety Management System (SMS) when operating in these lower visibility conditions.

Figure 3.3: Sometimes it is not possible to optimise runway orientation due to orography — Funchal airport. *Photo*: Z. Kazdova.

It must be emphasised that lowering the weather minima is more an economic than a technical problem today. Category II or III equipment and operations represent significant costs for airports and airlines. However, in most cases it is the only way to guarantee the reliability of operations during the whole year without significant disruptions. Diversions are expensive both for airports and airlines. For airports it is the loss of landing, parking, handling and passenger fees and a bad image of the airport. For airlines there are direct costs resulting from passenger compensations, other costs as a consequence of disruptions to the fleet planning and also a bad image for the airline. Sometimes, even if it appears sensible for an airport, some aircraft operators may not wish to upgrade their avionics or their pilot qualifications. On the other hand, many operators may only agree to operate at the airport if Category II or III is available to guarantee the regularity of flight operations.

Table 3.2: Precision approach categories.

		ICAO	FAA	EASA
CAT I	DH	200 ft ≤ DH	200 ft ≤ DH	200 ft ≤ DH
	RVR	550 m ≤ RVR 800 m ≤ VIS	1800 ft ≤ RVR 550 m ≤ RVR 1400 ft ≤ RVR with special equipment authorisation	550 m ≤ RVR
CAT II	DH	100 ft ≤ DH < 200 ft	100 ft ≤ DH < 200 ft	100 ft ≤ DH < 200 ft
	RVR	300 m ≤ RVR	1200 ft ≤ RVR < 2400 ft 350 m ≤ RVR < 800 m	300 m ≤ RVR
CAT III A	DH	No DH or DH < 100 ft	No DH or DH < 100 ft	DH < 100 ft
	RVR	175 m ≤ RVR	700 ft ≤ RVR 200 m ≤ RVR	200 m ≤ RVR
CAT III B	DH	No DH or DH < 50 ft	No DH or DH < 50 ft	No DH or DH < 100 ft
	RVR	50 m ≤ RVR < 175 m	150 ft ≤ RVR < 700 ft 50 m ≤ RVR < 200 m	75 m ≤ RVR < 200 m
CAT III C	DH	No DH	No DH	
	RVR	No RVR limitation	No RVR limitation	

Source: ICAO (2010, 2012c); FAA (2012); Commission Regulation (EU) (2012).

In some cases the selection of a site has been made without considering poor visibility conditions, thinking that the ground equipment would ensure the required regularity and allow for the aircraft to land under any conditions. However, the lower the operational minima, the bigger the capital and operational costs incurred. The ground equipment designed for operation under the conditions of the ICAO Category II or III are not only more expensive but also substantially more complicated and demanding for maintenance. The runway can be out of service during maintenance, so the total availability of the runway can be further reduced if there is 24-hour operation. Also, during poor visibility operations, the runway system capacity decreases considerably. However, the lower the minima, the less frequently do those poorer weather conditions occur (Figure 3.4).

54   Airport Design and Operation

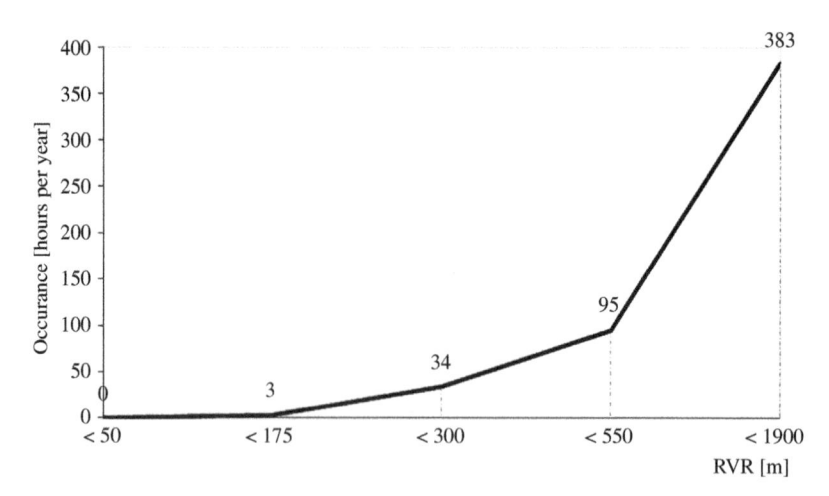

Figure 3.4: Occurrence of low-visibility conditions [hours/year] — Bratislava airport 2005. *Source*: Krollova (2005).

When an assessment of the runway equipment is made, it is necessary to determine the extent to which it is cost effective to lower the minimum. The decision is usually based on the prevailing meteorological conditions during the peak hours and the expected number of diverted flights gained or lost. For example at many busy Mediterranean airports it is quite usual that they have only non-precision approach runways as most of their traffic is concentrated to the summer season with good meteorological conditions. Therefore, when selecting a site for an airport with international or scheduled traffic, it is necessary to pay attention to an assessment of weather conditions, including occurrence of reduced visibility and low cloud base, and the associated costs and benefits.

For determination of the usability factor including the effect of low visibility, the Antonín Kazda method may be employed.[1] The method evaluates the usability factor in two steps:

1. Evaluation of the usability factor including the effect of crosswind if the visibility or the runway visual range is greater or equal to 550 m.

2. Evaluation of the usability factor considering the RVR limits without taking the wind into account if the visibility or the runway visual range drops below 550 m.

Note: If historic data of RVR measurement are not available then visibility data might be used. In this case a critical value of 400 m should be employed. This value is based

---

1. Kazda (1988).

on the general assumption that runway visual range is 1.5 times greater than the visibility.

This two step method may be employed because of a mutual dependency of the occurrence of low visibility and low velocity winds, that is 'when there is a fog, no strong wind blows', has been effectively proven with a high probability.

The evaluation of the usability factor with the wind effect may be performed by any of the known methods, for example the Durst method described above.

For evaluation of the usability factor taking account of the runway visual range limits, the method of least squares, the approximation of accumulated occurrences of visibility by the means of a polynomial, is employed. The method allows the usability factor to be evaluated with respect to the visibility limits in an arbitrary season and a time interval of a day, and may describe the time dependency (daily and yearly) of occurrences of low visibility.

The resultant usability of the aerodromes is the sum of the loss of usability due to the wind effect and the loss of usability due to the runway visual range limits.

According to Annex 14 Volume I, the ICAO precision approach runway categories are defined by decision height (DH) and runway visual range (RVR), or visibility in case of Category I (see Table 3.2). The criteria in ICAO, FAA and EASA standards are slightly different. ICAO standards Annex 6, Operation of Aircraft, Part I International Commercial Air Transport — Aeroplanes are accepted worldwide, but additional requirements or other differences could be found for instance in FAA Order 8000.94: or in the Commission Regulation (EU) No 965/2012, that is EASA. Category II weather minima permit manual landing, but it does not mean that automatic landings cannot be made. Aircraft avionics have improved significantly over the years by leveraging new and existing aircraft technologies. Some airlines practice to fly manual approaches up to Category III A using Head up Guidance System (HGS), for example Lufthansa City Line since 1994 on their CRJ fleet. During the approach procedure the auto pilot has to be switched off at 1000 feet at the latest and the captain continues manual approach with Category III A minima to landing. The Head up Display allows also for low visibility take offs since the minima with HGS is 75 metre RVR due to LOC based guidance projected into the pilots field of vision. In any case the decision height or RVR cannot be lower than specified limits. ICAO defines the decision height (or altitude) and the runway visual range as:

Decision altitude (DA) or *decision height* (DH) is the specified altitude or height in the precision approach or approach with vertical guidance at which a missed approach must be initiated if the required visual reference to continue the approach has not been established.

Runway visual range (RVR) is the range over which the pilot of an aircraft on the centre line of a runway can see the runway surface markings or the lights delineating the runway or identifying its centre line (Figure 3.5).

Figure 3.5: Part of RVR equipment in touchdown zone (ILS GP antenna behind). *Photo*: A. Kazda.

According to the European Guidance Material on Aerodrome Operations under Limited Visibility Conditions, EUR Doc 013 4th ed. (September 2012), Low-Visibility Procedures (LVP) are required for approach and landing operations in Category II or III weather conditions or for departures in RVR less than a value of 550 m (Table 3.3). Precision approaches during Category I conditions do not require Low-Visibility Procedures. However, procedures for Reduced Aerodrome Visibility Conditions (RAVC) must be introduced whenever conditions are such that all or part of the manoeuvring area, depending on the size of the aerodrome, cannot be visually monitored from the control tower (PANS-ATM Chapter 7, 7.12.1). This again means that the design, that is airport layout — tower position and its height, can significantly affect the airport operation.

For the description of different conditions of visual control of traffic from the control tower and of the pilots to avoid other traffic four different visibility conditions are specified from Visibility Condition 1 through to Visibility Condition 4. The transition from Visibility Condition 1 to Visibility Condition 2 occurs when meteorological conditions deteriorate to the point that personnel of control units are unable to maintain control over traffic on the basis of visual surveillance and in practice defines the entry to RAVC.

Table 3.3: The relationship between ICAO visibility conditions.

MET	Visibility conditions	ATC	Pilot
Aerodrome specific	Visibility Condition 1	ATC controls aerodrome ground movements visually	Pilot taxis and avoids other traffic visually
Visibility equivalent to RVR < 400 m	Visibility Condition 2 (Reduced Aerodrome Visibility Conditions)	ATC unable to control some/all of manoeuvring area visually	
	Visibility Condition 3		Pilot unable to avoid other traffic visually
RVR ≤ 75 m	Visibility Condition 4		Pilot unable to taxi visually

Source: ICAO (2012b).

For many regional airports with only moderate traffic a so-called 'full' Category I (runway equipped with inset centre line lights) may represent a good solution before introducing any lower category, as the difference in operational limits between Category I and II is only 250 m of RVR but the investment and operational costs of Category I are significantly lower. Under EASA — Commission Regulation (EU) 965, take-off on a runway where the RVR is less than 400 m is also considered as a low-visibility operation, in this case the Low-Visibility Take-Off (LVTO). In some states it is mandatory to conduct a so-called guided take off if RVR is 125 m or less (150 m for Cat D aircraft). In this type of operation a guidance system provides directional guidance information to the pilot during the take-off. However, the pilot may request to conduct a guided take off at any time. The ILS localiser guidance signal must be protected for these types of operation.

For Category I operations and non-precision approaches, RVR measurements must be provided in the touchdown zone. For Category II a second, mid-runway position must be equipped for RVR measurements and for the majority of Category III operations it is also required at the final third of the runway.

58   Airport Design and Operation

In fact, the critical value for a pilot in the final stages of approach and during flare is not RVR but the slant visual range (SVR). Slant visual range is the range over which a pilot of an aircraft can see the markings or the lights as described in the RVR definition. The pilot's ability to obtain visual reference in low-visibility conditions is further constrained by the front section of the aircraft, the cockpit cut-off angle. Minimum visual segments at DH are established for each precision approach category. Typically the values are 225 m for Category II with manual landing, 90 m for Category II with automatic landing and 60 m for Category III. In Category III conditions a pilot can see the runway lights only about 5 seconds before touchdown. The relationships between SVR, DH, cockpit cut-off angle and the visual segment are defined by a formula (Figure 3.6):

$$SRV = \sqrt{(v + (h \cot w))^2 + h^2}$$

where v is visual segment; h is pilot's eye height above the ground; and w is cockpit cut-off angle = down vision angle − pitch angle.

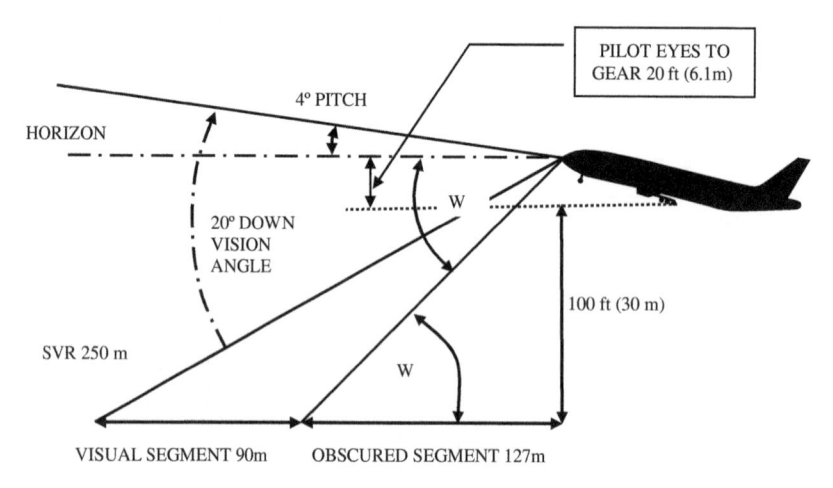

Figure 3.6:   Visual ground segments and cut-off angle — landing approach.
Source: Airbus Industry.

The main objective of Category II or III operations is to guarantee the same level of safety as during other operations, but in lower visibility and more adverse weather conditions. The desired level of safety is achieved (at the airport) by controlling:

✈ physical characteristics of runway, runway strip and the terrain before threshold
✈ obstacle clearance criteria
✈ visual aids (runway marking, lighting systems)

Airport Site Selection and Runway System Orientation **59**

↣ non-visual aids (ILS or MLS facility)
↣ RVR measurements
↣ ATC procedures
↣ airport maintenance procedures.

The details of particular airport systems or equipment with respect to the low-visibility procedures are discussed in the relevant chapters (e.g. Chapters 15 and 16).

Before the safety assessment study itself, the runway, runway strips and taxiway system layout must be examined for obstacle clearances. The most stringent area is inside the obstacle-free zone (OFZ). It is defined in ICAO Annex 14 as the airspace above the inner approach surface, inner transitional surfaces, and baulked landing surfaces and that portion of the strip bounded by these surfaces, which is not penetrated by any fixed obstacle other than a low-mass and frangibly mounted obstacle required for air transportation purposes (Figure 3.7). During Category II or III operations, taxiing or aircraft holding for take-off must also be kept out of the OFZ.

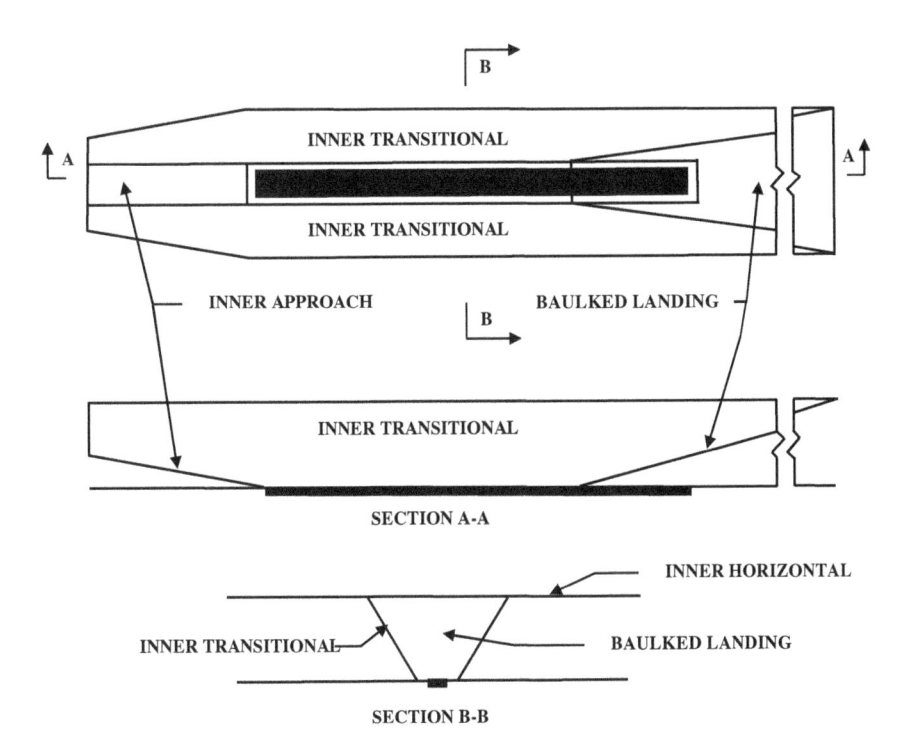

Figure 3.7: Inner approach, inner transitional and baulked landing obstacle limitation surfaces. *Source*: Annex 14, Volume I.

60   Airport Design and Operation

During low-visibility operations it is of paramount importance to ensure the movement areas' sterility against incursions in order to guarantee safety of aircraft operations and the highest quality of guidance signals for automatic landings. In most cases, Instrument Landing Systems (ILS) or, only at a few airports, Microwave Landing Systems (MLS) are used for navigation guidance (see Chapter 17). Apart from the technical parameters of the navigation installation, the quality of signals is guaranteed by protecting critical and sensitive areas.

The ILS signal can be reflected and distorted by structures at the airport and its vicinity but also by fixed or moving objects like vehicles, flying or taxiing aircraft or even pedestrians. Many potential conflicts can be eliminated by restrictions on the number and the type of movements. The ILS critical areas must be protected at all times while the ILS sensitive areas, which are larger than the critical areas, should be kept clear during Category II or III operations to protect the signal against interference caused by moving objects. The dimensions and the shape of critical and sensitive areas depend on the terrain and objects around the airport and in the approach area, as well as on the parameters, configuration and the type of ILS. At an airport licenced for Category II and III operations they are marked on taxiways as CAT II/III holding points. The MLS critical and sensitive areas are much smaller than those for ILS operations and in fact they place no restrictions on aircraft taxiing or taking-off, or on other movable objects.

During low-visibility ILS operations the runway capacity is mainly constrained by the need to protect ILS critical and sensitive areas. To ensure guidance signal integrity it is necessary to provide appropriate spacing between two successive landing or departing aircraft. Landing clearance is normally given to an aircraft on approach when the ILS localiser sensitive area is clear, usually when the aircraft on approach is not closer than two nautical miles from touchdown. The runway capacity can be further restricted during Category III operations because pilots need more time to exit the runway and critical areas, and air traffic controllers may need to ensure an additional longitudinal separation for an approaching aircraft.

MLS has smaller critical and sensitive areas around the runway. This removes some runway operational and capacity restrictions. Spacing between aircraft can be smaller and is determined by the rule that the preceding landing aircraft is clear of the so-called 'MLS Landing Clearance Trigger Line'. The MLS Landing Clearance Trigger Line is not marked on the airport movement areas but only on a controller's display, for example surface movement radar or A-SMGCS (Figure 3.8). However, aircraft spacing can never be less than the wake turbulence separation minima.

Ground operations of aircraft and ground equipment during limited visibility conditions are more demanding as visibility decreases. During low visibility operations, that is during Category II or III operations, special procedures must be activated against movement area incursion by vehicles or personnel. Safeguarding the airside sterility is assisted by security fencing around an airport and by strictly controlled access points. The airside sterility is needed not only to protect the guidance signals in critical or sensitive areas but also to ensure safety of aircraft taxiing, landing or taking-off.

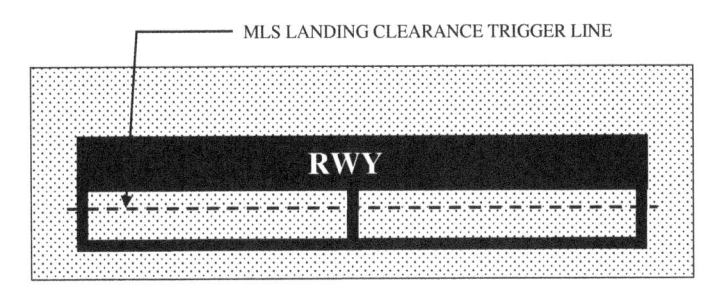

Figure 3.8: MLS landing clearance trigger line. *Source*: ICAO Doc 13.

At busy airports, surveillance of the manoeuvring area by a surface movement radar, A-SMGCS or multilateration surveillance systems is also required. At airports with light or medium traffic density, surface movements control on a system of well segregated taxiways could be based on procedural methods. The required equipment and procedures for ground movement control depend on the aerodrome operating minima and the extent of the possible conflicting traffic, the size and complexity of the manoeuvring area and the movement rate required. For Category II or III operations all personnel involved must be properly trained and regularly tested under real or simulated low-visibility conditions. This includes knowledge of the airport layout, radio-telephony and emergency procedures. All vehicles entering the manoeuvring area must obtain an authorisation from the control tower and they must maintain a permanent two-way radio communication. Special attention is given to the rescue and fire fighting personnel training for low-visibility operations (see Chapter 19).

Preparation for low-visibility procedures starts when the visibility decreases to a predetermined limit, and it is expected to fall further. The initiation criteria are defined in the airport LVP manual and is based on the prevailing meteorological and operational conditions, for example when the visibility falls below 1200 m or the cloud ceiling is below 300 ft. LVP must be running at the latest before the weather falls below Category I limits. However, if the weather deteriorates rapidly, the procedures could be started at higher limits depending on the operational experience and how long the preparation phase usually takes. The initiation and termination of low-visibility procedures are decisive from the operational point of view and they are critical for predetermining the optimum airport capacity. Most problems can occur during these phases. In particular at busy airports the capacity restrictions can have an impact long after the LVP are terminated. Close co-ordination between the airport operator, meteorological office, air traffic control and air traffic flow management is necessary to minimise delays and a smooth return to full capacity after the termination of LVP. At large airports with a considerable share of long-haul flights the commencement of low-visibility procedures can start more than 12 hours in advance.

3.4. Control of Obstacles

In the complex orographical conditions of mountainous countries, the appropriate runway orientation to minimise the crosswind is often coincident with the orientation designed with regard to obstacles. This is likely because the direction of the prevailing winds often coincides with the centre line of a valley. Even so, the obstacle limitation surfaces may well be penetrated by obstacles. Development at many airports, particularly extending the runway, is often difficult because of obstacles. The defined obstacle-free airspace is an inseparable part of each aerodrome and should permit the intended aircraft take-off, approach, landing and operation in the aerodrome circuit to be conducted safely. According to Annex 14 Volume I, obstacles are defined as all fixed (temporary or permanent) and mobile objects, or parts thereof, that are located on an area intended for the surface movement of aircraft or that extend above a defined surface intended to protect aircraft in flight.

The airspace round the airport is defined by a system of obstacle limitation surfaces. The characteristics of obstacle limitation surfaces are specified on the basis of types of airports (transport, general aviation, heliports, etc.) and are related to the intended use of the runway in terms of take-off, landing and the type of approach (non-instrument approach, non-precision or precision approach).

In cases where operations are conducted to or from both directions, then the function of certain obstacle limitation surfaces or parts of them may be nullified because of more stringent requirements of other lower obstacle limitation surfaces. Dimensions and slopes of obstacle limitation surfaces were derived from the performance characteristics of the various types of aircraft and from the statistical probability of deviations of the aircrafts' trajectories relative to the intended flight path. Characteristics of obstacle limitation surfaces correspond to the requirements of the regulations for flight operations, air traffic control, and aircraft airworthiness. Even in the most extreme foreseen position of the aircraft, the prescribed minimum horizontal and vertical distance of the aircraft from obstacles should be maintained.

Usually the significance of any object in the airport surroundings is assessed by two separate groups of basic criteria. The first one refers to the obstacle limitation surfaces for the particular runway and its intended operation. Annex 14 Aerodromes, Volume I, *Aerodrome Design and Operations* is used in many states as a legal and technical basis for this kind of assessment. The objective of the obstacle limitation surfaces is to define the airspace around aerodromes to be maintained free from obstacles so as to permit the intended aircraft operations at the aerodromes to be

Airport Site Selection and Runway System Orientation **63**

conducted safely and to prevent the aerodromes from becoming unusable by the growth of obstacles around the aerodromes.

The second set of criteria relates to the Procedures for Air Navigation Services — Aircraft Operations (PANS–OPS) Doc 8168 Volume II *Construction of Visual and Instrument Flight Procedures*. Surfaces defined by this standard guarantee obstacle-free airspace for instrument and visual flight procedures by specifying minimum safe heights/altitudes for each segment of the procedure. The limits depend on the installed navigational equipment, its type and position and also on the speed of an aircraft (see Table 6.1). Some states use also other criteria and practices in line with their national standards and recommended practices, building code or other bylaws.

The system of obstacle limitation surfaces as presented in Annex 14, Volume I, Chapter 4. 'Obstacle restriction and removal' also serves to advise a local authority of the height limits for new objects that might otherwise restrict or endanger the operations of aircraft in the future. If there is a potential conflict, the relevant Aviation Authority should be consulted.

For non-instrument runways, for instrument approach runways and for Category I precision approach runways, the following system of obstacle limitation surfaces has been specified (Figure 3.9):

✈ conical surface

✈ inner horizontal surface

✈ approach surfaces

✈ transitional surfaces

✈ take-off climb surfaces.

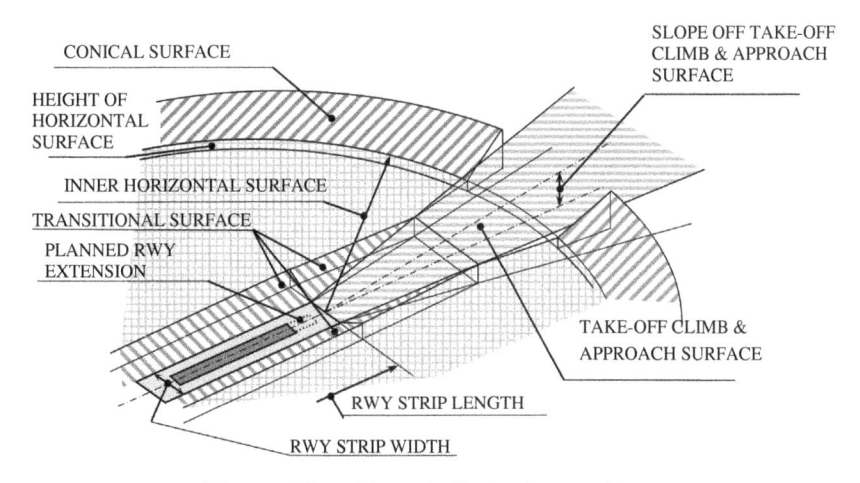

Figure 3.9: Obstacle limitation surfaces.

64   Airport Design and Operation

The most significant differences in characteristics between these classes of runway are related to the dimensions and slopes of the take-off climb and approach surfaces. The necessary data may be found in Annex 14 Volume I *Aerodromes*.

The system of obstacle limitation surfaces for precision approach runway Category II and III is supplemented by:

✈ inner approach surface

✈ inner transitional surfaces

✈ baulked landing surface.

These three surfaces define the so-called obstacle free zone for operations which are conducted in the low visibility conditions and strictly limit the existence of all fixed objects other than low-mass and frangibly mounted objects required for air navigation purposes. The dimensions and slopes of all the relevant surfaces for precision approach Category II or III are specified in Table 3.4. The determinants for assessing the consequences of a certain obstacle penetrating a surface are the type of operation on the runway in question, the position of the obstacle (which surface is obstructed by the obstacle, the distance from the runway threshold/end and how much it obstructs it) and character of the obstacle (hills, aerial transmission line poles, buildings, road profile, etc.).

As a reaction to increased regular use of helicopters in civil aviation, ICAO published Annex 14, Volume II, *Heliports*. Specifications on obstacle restriction and removal are given in Chapter 4 of these Standards and Recommended Practices. In the case of ground level heliports, the system of obstacle limits is similar to surfaces defined in Annex 14, Volume I, but it respects the specifics of helicopter operations. For non-instrument final approach and take-off area (FATO),[2] two obstacle limitation surfaces are established:

✈ take-off climb surface

✈ approach surface.

For a precision (Figure 3.10) or a non-precision approach FATO two other surfaces are defined:

✈ transitional surfaces

✈ conical surface.

---

2. Final approach and take-off area (FATO) is a defined area over which the final phase of the approach manoeuvre to hover or landing is completed and from which the take-off manoeuvre is commenced. FATO is similar to the Runway Strip of 'classic' aerodromes.

Table 3.4: Parameters of obstacle limitation surfaces for precision Category II or III RWY.

Surface	Dimensions
Conical	
✈ Slope	5%
✈ Height	100 m
Inner horizontal	
✈ Slope	45 m
✈ Height	4000 m
Inner approach	
✈ Width	120 m
✈ Distance from THR	60 m
✈ Length	900 m
✈ Slope	2%
Approach	
✈ Length of inner edge	300 m
✈ Distance from THR	60 m
✈ Divergence	15%
First section	
✈ Length	3000 m
✈ Slope	2%
Second section	
✈ Length	3600 m
✈ Slope	2.5%
Horizontal section	
✈ Length	8400 m
✈ Total length	15,000 m
Transitional	
✈ Slope	14.3%
Inner transitional	
✈ Slope	33.3%
Baulked landing surface	
✈ Length of inner edge	120 m
✈ Distance from threshold	1800 m
✈ Divergence	10%
✈ Slope	3.33%
Take-off climb	
✈ Length of inner edge	180 m
✈ Distance from RWY end	60 m
✈ Divergence	12.5%
✈ Final width	1200 (1800) m[a]
✈ Total length	15,000 m
✈ Slope	2%

[a] 1800 m when the intended track includes changes of heading greater than 15° for operation conducted in IMC, VMC by night.

66 Airport Design and Operation

The total length of approach surface is 10,000 m for a 3° standard precision approach FATO, or 8,500 m for a 6° approach. In the case of a non-instrument and non-precision FATO, the approach surface dimensions are based on a multiple of helicopter rotor diameters.

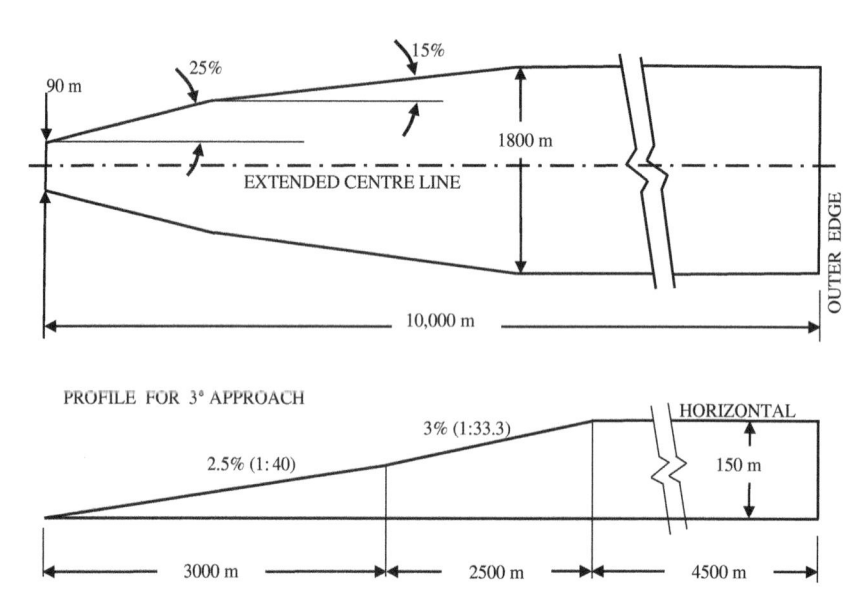

Figure 3.10: Approach surfaces for precision approach FATO — 3° approach. *Source*: Annex 14, Volume II.

The general opinion is that everything that exceeds specified limits around an airport is an obstacle and must be banned. It might be true, but not always. Another common view is that if an object does not penetrate a defined surface, the object is not considered to be an obstacle. This also might be true but also not always. The principal problem is that no two airports and no two obstacles are alike. Because of this it is not possible to give a simple rule for an objective assessment. To consider all aspects of the problem, most states require an aeronautical study as a prerequisite for obstacle assessment. The Aviation Authority may still allow operations on the basis of a study, often using the ICAO Collision Risk Model.

To evaluate the effect of obstacles on precision instrument approach operations (ILS, MLS or GNSS), the following obstacle assessment procedure is generally accepted: when ICAO Annex 14 obstacle assessment surfaces are penetrated, Obstacle Assessment Surfaces (OAS) as defined in ICAO Doc 8168 Volume II are used in lieu of ICAO Annex 14 surfaces. When even OAS are penetrated, the Collision Risk Model (CRM) is utilised. However, in the latter case, the complete obstacle situation in the vicinity of the aerodrome has to be fully transformed into the CRM format, as the risk corresponding to each individual obstacle contributes to the cumulative risk of the approach.

There is no common world-wide methodology for obstacle assessment in the vicinity of airports and states are using different practices for this purpose. ICAO issued Doc 9137-AN/898/2 Airport Services Manual, Part 6 — Control of Obstacles as guidance material. This manual contains a detailed examination of the control of obstacles at an aerodrome but it also reviews some states' practices.

Figure 3.11: If the obstacle cannot be removed (here terrain and houses in the transition surface), an exemption may be granted. However, it can result in airport operation restrictions or special requirements for crew qualification and training.
Photo: A. Kazda.

Obstacles in the take-off climb and approach surfaces and in the transitional surfaces are assessed most stringently. The construction of new obstacles or extensions of existing ones shall not be permitted if they would penetrate the Annex 14 surfaces. It is, however, not always possible to eliminate the occurrence of obstacles (Figure 3.11). Then it is necessary to determine special procedures such as offset approach or take-off trajectory, or to install special equipment for the runway, or to

68   Airport Design and Operation

limit the runway operation with higher visibility and ceiling operating limits as well as marking and lighting the obstacles.

If the obstacles only penetrate the conical surface or the inner horizontal surface, less stringent criteria are used. Even in that case, for an assessment of a specific obstacle, the location and character of the obstacle are important.

It is necessary to investigate whether the obstacle is shielded by an existing immovable object, or to prove by an aeronautical study that the object would not adversely affect the safety or the regularity of operations of aircraft.

Some objects may be considered particularly dangerous and should be removed or at least marked even if they do not obstruct any obstacle limitation surfaces. They are, in particular, isolated tall objects such as chimneys, poles and posts, or aerial high and extra-high tension transmission lines in the approach and take-off climb surfaces. The Aviation Authority may order removal (e.g. trees) or marking (e.g. aerial high tension transmission line — Figure 3.12) of any object that might according to an aeronautical study endanger aircraft on the movement area or in the air. The object should generally be removed by its owner, not by the aerodrome operator. Obstacle marking of objects which extend to a height of 150 m or more above ground elevation is compulsory.

Figure 3.12:   Marking of high voltage lines. *Photo*: T. Kazda.

3.5. OTHER FACTORS

The orientation of take-off and landing runways is always linked with the airspace requirements. In fact, there are usually few options for runway and taxiway layout solutions. Therefore, in planning an airport, the take-off and landing runways and taxiways are usually designed first. The capacity of the runway system is a limiting factor for further development of the largest airports. In many cases, further expansion of the airport is impossible due to public opposition. Even if approval is eventually given, it may take decades to build an airport or even an additional runway.

The capacity of the terminal airspace and of the runway system of the airport may be assessed with computer simulations, analytical models or handbook methods. There are several different simulation models and each of them has some special features. The most known are the following:

The Airport and Airspace Simulation Model (SIMMOD™), was validated by the Federal Aviation Administration (FAA) as a standard analysis tool for airport planners, airport operators, airline companies, airspace designers and air traffic control authorities for airport and airspace operations.

Total Airspace and Airport Modeller (TAAM) was developed by Preston Aviation Solutions (a Boeing Company subsidiary) in cooperation with the Australian Civil Aviation Authority (CAA). TAAM is a large scale fast-time simulation package for modelling entire air traffic systems. It can be used for ATM analysis and feasibility studies or for planning purposes. The advantage of this tool is that it can simulate complete gate-to-gate operations and in more detail than comparable models, but at considerable expense.

The Airport Machine is a simulation tool from the Airport Simulation International (ASI) company. It can be used for detailed simulations of movement areas operations. It is based on a node-link structure similar to that of SIMMOD, and it covers all aircraft activities from a few minutes before landing until a few minutes after take-off.

The Total AirportSim was developed by IATA. It is designed for simulations of the whole airport operations including passenger terminal, gate, apron, runway and airspace. It provides realistic results. The model is designed to offer a comprehensive solution that covers a wide variety of demand/capacity and level of service applications for the simulation of aircraft and passenger flows.

Rams Plus solution is a product from ISA Software. This company is a part of the ISA Research Group, which offers ATM analysis to customers like FAA and EUROCONTROL. The RAMS Plus gate-to-gate system has two main parts, the Airside and the Groundside, which function together in the same environment to provide a global macro/micro gate-to-gate view of the air traffic system.

70   Airport Design and Operation

It is also possible to carry out simulations of parts of the manoeuvring area using for example Monte Carlo techniques. The crucial factor, whether the models are simple or comprehensive, is the quality of the data used for calibrating the model, for example runway occupancy time by type of aircraft and meteorological condition. The budget is another factor. Some of the tools can take weeks to load all the airport data, even after the prolonged process of capturing the operational data.

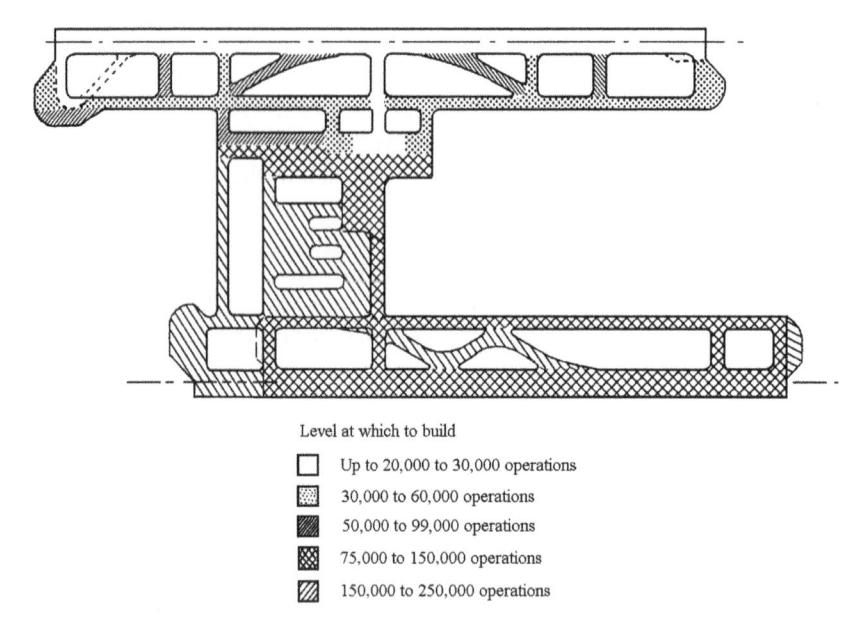

Level at which to build
- ☐ Up to 20,000 to 30,000 operations
- 30,000 to 60,000 operations
- 50,000 to 99,000 operations
- 75,000 to 150,000 operations
- 150,000 to 250,000 operations

Figure 3.13:   Typical phased development of airport. *Source*: ICAO Airport Planning Manual.

FAA Airfield Capacity Model is a handbook method for assessing the runway and the manoeuvring areas. According to the FAA, it is possible to define the runway capacity as the maximum number of movements within an hour once the tolerable delay has been defined. It tends to give better estimates of capacity for operations similar to those in the United States.

The magnitude of an acceptable delay differs at individual aerodromes. From the maximum number of movements in an hour, it is possible to derive the daily and yearly capacity of the runway system. Figure 3.13 illustrates how ICAO suggests that capacity should be enhanced with the growth of the airport in terms of the number of movements in a year. It should be emphasised that this was developed for US airports. With different economic environments, operational practices and mixes of traffic the runway capacity might well expand more slowly.

Number	Runway use configuration	Hourly capacity ops/h VFR	Hourly capacity ops/h IFR	Annual service volume
1	 (single runway)	51 - 98	50 - 59	195 000 - 240 000
2	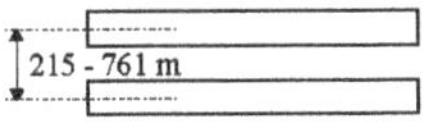 215 - 761 m	94 - 197	56 - 60	260 000 - 355 000
3	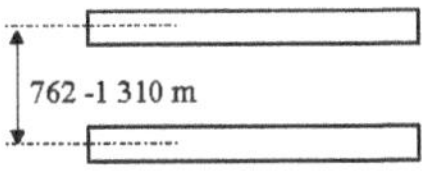 762 - 1 310 m	103 - 197	62 - 75	275 000 - 365 000
4	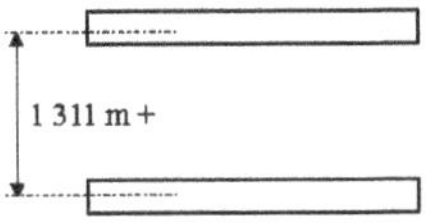 1 311 m +	103 - 197	99 - 119	305 000 - 370 000
5	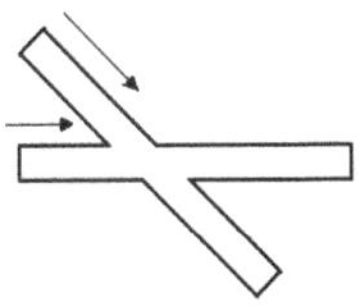 (intersecting runways)	72 - 98	56 - 60	200 000 - 265 000
6	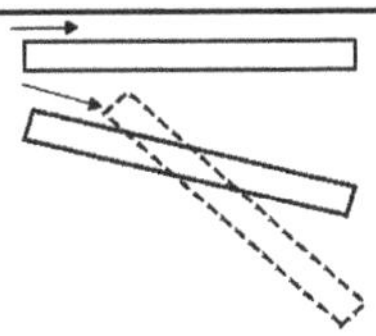 (open-V, same direction)	73 - 150	56 - 60	220 000 - 270 000
7	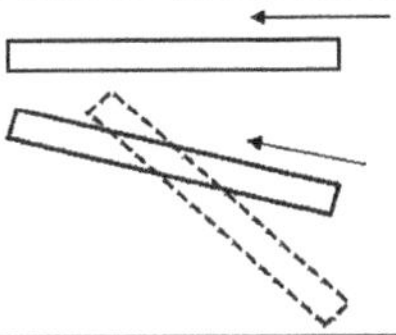 (open-V, opposite direction)	73 - 132	56 - 60	215 000 - 265 000

Figure 3.14: Hourly capacity and annual service volume. *Source*: ICAO Airport Planning Manual.

Figure 3.14 illustrates how the capacity varies with the runway configuration. On one runway with an optimum layout of taxiways and with a sufficient capacity of the apron, it is possible to reach up to 250,000 air transport movements in a year, as at London Gatwick in 2013, even using Instrument Flight Rules (IFR).

The decision to construct another runway depends not only on the annual runway capacity but also on the magnitude of the peak hour traffic and the type of operation ('hub and spoke' or 'point to point'), as well as on an assessment of other factors which, if they occurred, would bring about a complete shut down of an airport with one runway. These might be an accident, an unlawful act of intervention in an airport or an aircraft, winter maintenance of the runway, etc.

The design of systems of taxiways is dealt with in Chapter 6.

At many airports there are some operational limitations for example night ban or other restrictions. These night curfews cause further intensification of peak hour traffic. The pressure is quite intense for some major hubs and, in some cases, at secondary airports which are located in very densely populated areas. In many cases it was inadequate land-use management which allowed urban expansion around airports, resulting in an increase of the number of people significantly exposed to aircraft noise. Operational airport restrictions result largely in response to the pressures and concerns highlighted above. As of mid-2012, approximately 250 domestic and international airports worldwide imposed some form of night time operational restrictions. Of the 161 international airports, 66% are located in Europe, 16% in North America, 8% in Asia/Pacific, 5% in Latin America and the Caribbean, 3% in Africa and 2% in the Middle East.[3] There are also limitations of the number of movements to limit the level of air pollution as in Zürich and Stockholm's-Arlanda, and other administrative limitations. Chapter 20 of this book deals with the problems of environment protection and ecological limitations.

---

3. ICAO (2013d).

4

RUNWAYS

Tony Kazda and Bob Caves

4.1. AERODROME REFERENCE CODE

As with other Annexes, the main objective of Annex 14 is to determine criteria for providing world-wide safety, regularity and economy of international air transport.

Many provisions which are included in Annex 14 are of a universal character and are applicable to the majority of personnel in international air transport, its procedures and equipment. However, some of the provisions are applicable only to specific types of operation, or depend on critical dimensions or performance characteristics of the given aircraft.

The main task is to achieve an effective and efficient relationship between particular facilities and aircraft. Since there are many types and variants of aircraft in operation, it would be practically impossible to develop specific facilities for each of the types of aircraft in common use, with all the variations in size and in numbers and location of engines. In Annex 14, there are therefore many facility sizing criteria related to the most demanding (critical) type of aircraft or group of aircraft that may be used in the airport. The established system of aerodrome reference code (Table 4.1) results from that principle in Annex 14.

Establishment of the Aerodrome Reference Code in Annex 14 pursues three purposes with regards to:

1. Design	**to provide aerodrome designers with guidelines** on how to plan and design an aerodrome by **relating** rational design criteria with current and future aircraft requirements
2. Standard	**the aerodrome reference code** is a simple method for **interrelating the numerous specifications** concerning the characteristics of aerodromes
3. Operation	the aerodrome reference code serves as an indication of the characteristics of the ground facilities, so the prospective **aircraft operator may contemplate whether aerodrome facilities are suitable** for that aircraft to operate there, provided that Annex 14 was complied with during its construction (this is only valid for movement areas)

Table 4.1: Aerodrome reference code.

Code element 1			Code element 2		
Code number	Aircraft reference field length	Code letter	Wingspan	Outer main gear wheel span[a]
1	Less than 800 m	A	Up to but not including 15 m	Up to but not including 4.5 m
2	800 m up to but not including 1200 m	B	15 m up to but not including 24 m	4.5 m up to but not including 6 m
3	1200 m up to but not including 1800 m	C	24 m up to but not including 36 m	6 m up to but not including 9 m
4	1800 m and over	D	36 m up to but not including 52 m	9 m up to but not including 14 m
		E	52 m up to but not including 65 m	9 m up to but not including 14 m
		F	65 m up to but not including 80 m	14 m up to but not including 16 m

Source: ICAO (2013b).
Note: Some states (e.g. the Czech Republic) already introduced in their national SARP code letter G with the wingspan over 80 m and outer main gear span over 16 m.
[a] Distance between the outside edges of the main gear wheels.

When the aerodrome reference code was designed, it was decided that it had to consider the aircraft operational (performance) characteristics and geometrical dimensions.

The following parameters may be given as examples of the operational characteristics:

✈ take-off and landing distances
✈ approach speed
✈ nose wheel lifting speed
✈ glide slope angle
✈ climb rate
✈ taxiing speed

✈ turning radius

✈ engine exhaust jet blast velocity.

Examples of geometrical dimensions of aircraft:

✈ wingspan

✈ distance between the engines and the main undercarriage

✈ outer main gear wheel span

✈ distance between the nose wheel and the main undercarriage

✈ height of pilot's eyes above the ground

✈ aircraft total length

✈ aircraft height.

The aerodrome reference code is composed of two elements, and ICAO chose to use take-off distance, wingspan and gear span as the aircraft indicators, as follows:

The first code element (indicated by numbers 1–4) is based on the aircraft reference field length.[1] It refers to the critical aircraft's normalised take-off requirements, and relates in particular to the runway geometry and obstacle clearance limit aspects of the design. However, it does not control the actual required length of the runway.

The second element is determined by wingspan and/or outer main gear wheel span, whichever is the more critical. It relates in particular with the provisions referring to taxiway and apron sizes and spacing. Characteristics of some aircraft and their classification by code number and code letter are shown in Table 4.2. Some provisions specifying the characteristics of movement areas are determined by a combination of both elements of the aerodrome reference code.

The FAA airport coding system uses the approach speed instead of the reference field length to represent this dynamic part of the aircraft characteristics. Grouping of aircraft is based on the indicated airspeed at threshold.

---

1. Aircraft reference field length is the minimum length required for the take-off at maximum certificated take-off mass at sea level, standard atmospheric conditions, still air and zero runway slope, as shown in the appropriate aircraft flight manual prescribed by the certificating authority or equivalent data from the aircraft manufacturer. Field length means balanced field length (the term balanced field length is explained hereinafter) if applicable, or a take-off distance in other cases. The ICAO documentation provides a listing of Reference Field Lengths.

Table 4.2: Aircraft characteristics and classification by code number and code letter.

Aircraft make	Model	Code	Aircraft reference filed length (m)	Wingspan (m)	Outer main gear wheel span (m)
Cessna	182 S	1A	462	11.0	2.9
Pilatus	PC-12	1B	452	16.2	4.5
Lear Jet	24F	2A	1005	10.9	2.5
LET	L410 UVP-E	2B	920	20.0	4.0
Dassault Aviation	Falcon 10	3A	1615	13.1	3.0
Bombardier Aero	CRJ 200	3B	1440	21.2	4.0
Boeing	B737-600	3C	1690	34.3	7.0
SAAB	340A	3C	1220	21.4	7.3
DeHavilland Canada	Dash 8 Q400	3D	1300	28.4	9.55
Airbus	A300 B2	3D	1676	44.8	10.9
Bombardier Aero	CRJ 200LR	4B	1850	21.2	4.0
Embraer	EMB-145 LR	4B	2269	20.0	4.1
Embraer	EMB-190 LR	4C	1890	28.7	11.4
Airbus	A320-200	4C	2480	33.9	8.7
Boeing	B737-500	4C	2470	28.9	6.4
Airbus	A310-200	4D	1845	44.8	10.9
Airbus	A330-600	4D	2332	44.8	10.9
Tupolev	TU154	4D	2160	37.6	12.4
Airbus	A350-900	4E	2438	64.7	24.8
Boeing	B787-8	4E	3139	60.1	23.6
Boeing	B747-300	4E	3292	59.6	12.4
Boeing	B777-300	4E	3140	60.9	12.9
Airbus	A380	4F	3350	79.8	14.3

Source: ICAO (2006a, Appendix 1 — Airplane classification by code number and letter); Airbus, Boeing, Embraer and DeHavilland airplane manuals.
Note: The second code element (indicated by letters A to F) is based on the aircraft geometrical characteristics.

The groups are the same as in PANS-OPS Doc 8168 Volume I, ICAO.[2] The second parameter representing geometric characteristics in the FAA Airplane Design Group is based on wingspans tail and heights where always the more demanding parameter determines the relevant group (see AC 150/5300-13, FAA). The wingspan intervals correspond to those defined in the Annex 14 code letter groups.

4.2. RUNWAY LENGTH

Determination of the runway length to be provided is one of the most important decisions both for designing an aerodrome, but also to allow economic operation of a particular type of aircraft to specific destinations. The runway length usually determines the types of aircraft that may use the aerodrome, their allowable take-off mass and hence the distance they may fly.

The number of new aerodromes for which runways have to be designed is limited. In the majority of cases the task is to extend existing runways or to provide supplementary runways but also and even more frequently to analyse existing runways with respect to new destinations or new aircraft types. The basic requirements for the runway parameters may be specified from market research into the types of aircraft, the networks of the operating airlines and prognoses of further market development at the airport in question.

A runway extension often makes it possible to open the aerodrome to larger aircraft, and to longer flight sectors, so extending the market of the aerodrome. There is usually considerable flexibility in how an aircraft can reach its maximum operating mass, between the extremes of maximising fuel load with limited payload and maximising payload at the expense of fuel and hence range. Additionally, it may be that the sector to be flown does not require a full load of fuel, or that the passenger load factor is expected to be low, or there is likely to be little cargo. In these cases, the aircraft will not be at maximum structural take-off mass (MTOM), and hence may gain no economic advantage from a runway extension.

A payload range diagram is shown in Figure 4.1. At the point A an aircraft has maximum payload, but it has no range as it has zero fuel. The range of the aircraft increases from A to B by loading fuel while the payload remains the same. The greater the fuel added to the aircraft, payload being constant, the greater is the aircraft mass. At the point B the aircraft has the maximum take-off mass. From this point the range could be increased only by replacing each kilogram of payload by

---

2. See Chapter 6, Table 6.1. Categories of aircraft according to their speed overhead the threshold.

78   Airport Design and Operation

the same amount of fuel. At the point C the aircraft has maximum capacity of fuel and the range could be further increased only by reducing the payload, but this is not used in practice. The aircraft mass decreases between C and D.

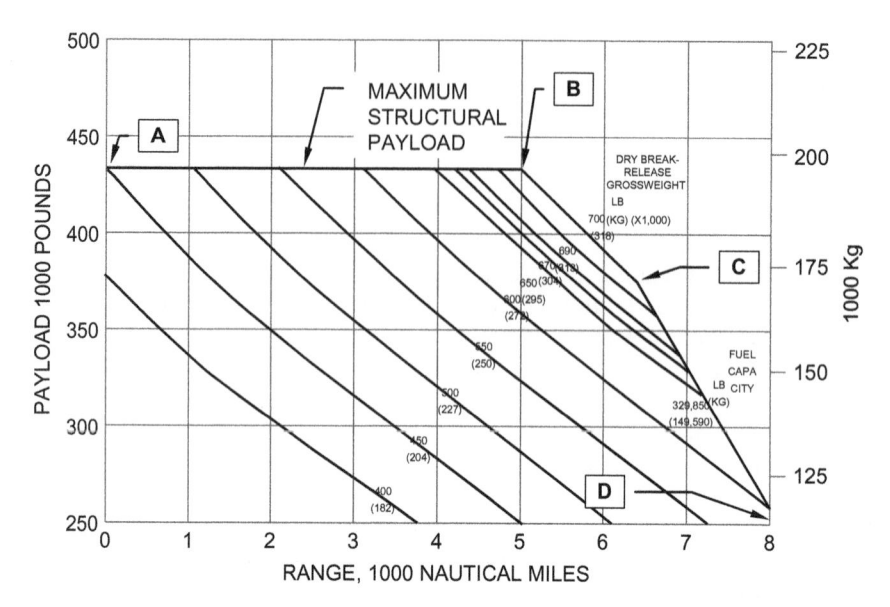

Figure 4.1: Payload/range diagram.

The first step in specifying the necessary runway length is to create a list of the aircraft that may wish to use the aerodrome after the runway extension and their likely destinations from the aerodrome. It is advisable to divide the aircraft into groups which are characterised by *take-off mass* and *payload*, each of the aircraft groups requiring approximately the same runway length. In the majority of cases, only a small group of aircraft requires the longest class of runway, or even only one aircraft: the 'critical' aircraft.

The number of movements of a critical aircraft is sometimes so small that there is no economic justification for extending the runway to meet its needs. In that case the operation of a critical type is usually still possible from the existing runway, though with a lowered payload, or reduced flying range, or both at some times of the day and year. However, the airlines will not use such an aerodrome if it would frequently require considerable reduction of the payload or limitation of the flying range.

A runway extension may also be economically unjustifiable because it is too technically demanding to construct it, because there are practically immovable obstacles in the take-off and approach areas or because the safety area would encroach on populated areas. In that case, a combination of runway and stopway, or runway and starter strip may allow the same take-off length to be provided. It is clear from the above that the determination of the required runway length is exacting, and any case of runway lengthening should be considered specifically for each case, taking all factors into account. It is necessary to consider not only the future development of the air transport market, but also changes in the prices of land, building materials, labour and economic development as a whole. It is often sensible to buy land and make at least a territorial provision for a prospective extension of the runway. Another consideration is that airlines appreciate excess runway length so that they can operate reduced thrust take-offs and so reduce engine wear and brake wear at airports where reverse thrust is banned. Also, the longer the runway the more it is possible to displace landing thresholds to keep aircraft higher over communities under the approach path to reduce noise and improve safety. In any case, the overriding criterion is to provide the defined degree of safety of the aerodrome operation which is derived from the EASA Certification Specifications and Acceptable Means of Compliance for Large Aircraft CS-25; Amendment 14; 19 December 2013 (see Table 4.3).

Table 4.3: Failure condition classifications and probability terms.

Description of failure conditions	Estimate of frequency	Average probability per flight hour
Probable	Anticipated to occur one or more times during the entire operational life of each aircraft.	More than 10^{-5}
Remote	Unlikely to occur to each aircraft during its total life, but which may occur several times when considering the total operational life of a number of aircraft of the type.	Between 10^{-5} and 10^{-7}
Extremely remote	Unlikely to occur to each aircraft during its total life but which may occur a few times when considering the total operational life of all aircraft of the type.	Between 10^{-7} and 10^{-9}
Extremely improbable	So unlikely that they are not anticipated to occur during the entire operational life of all aircraft of one type.	Less than 10^{-9}

Source: EASA Certification Specifications and Acceptable Means of Compliance for Large Aeroplanes CS-25 (2013).

80   Airport Design and Operation

The runway performance characteristics of the critical type of aircraft may be obtained in several ways:

- using the parameters given in the flight manual of the aircraft type
- using the parameters given in the aircraft type's airport planning manual (Aircraft Characteristics for Airport Planning) issued by all manufacturers of commercial aircraft
- take-off simulation.

Aircraft performance characteristics are usually described as **measured**, **gross** and **net**.

Measured performance is obtained from tests of pre-production aircraft. However, these data are obtained from new aircraft flown by test pilots and must be further adjusted to match everyday airline operation.

Gross performance is the average performance of an aircraft fleet in an airline company provided that the fleet is maintained, operated and flown in line with the flight manual and standards and recommended practices. The gross performance could be used as a reference but with no safety margins. Sometimes the gross performance is referred to as demonstrated (e.g. demonstrated landing distance).

Table 4.4: EU No 965/2012[a] Aircraft performance classification.

Parameter	Engine specification		
	Multi-engined jet	Propeller driven	
		Multi-engined turboprop	Piston
MTOM: Greater than 5700 kg Passenger: More than 9	A	A	C
MTOM: 5700 kg or less Passenger: 9 or less	A	B	B

[a]Commission Regulation (EU) (2012); hereinafter EU 965/2012.

Net performance is the gross performance reduced by a safety margin specified for aircraft public transport operation because of for example variations in piloting technique, temporary below average performance, etc. The safety margin required for public transport is based on an incident probability of one in one million flights (10^{-6}) making an incident a 'remote' possibility. According to EU No 965/2012 different criteria for performance assessment are used for different *aircraft classes* (Table 4.4):

Performance class A aircraft means multi-engined aircraft powered by turbo-propeller engines with an maximum operational passenger seating configuration

(MOPSC) of more than nine or a maximum take-off mass exceeding 5700 kg, and all multi-engined turbo-jet powered aircraft.

Performance class B aircraft means aircraft powered by propeller engines with an MOPSC of nine or less and a maximum take-off mass of 5700 kg or less.

Performance class C aircraft means aircraft powered by reciprocating engines with an MOPSC of more than nine or a maximum take-off mass exceeding 5700 kg. According to the EASA AMC CS 25, Performance Class A aircraft have the strictest airworthiness but the least stringent operating regulations. These aircraft are allowed to operate in all weather conditions. Performance Class C aircraft are large piston powered aircraft which are not used frequently nowadays and mostly belong among vintage aircraft.

From the performance characteristics drawn in the charts (or tabulated) given in the flight manual of the critical aircraft type, the landing distance, take-off distance and accelerate-stop distance at the various feasible configurations of flap settings and airframe system availability can all be derived. The data are chosen to represent the given parameters of the aerodrome and the proposed operation. The parameters include the aerodrome altitude, outside air temperature, runway longitudinal slope, as well as the maximum likely aircraft mass.

The gross take-off distance of a single engine class B aircraft can be read from graph in Figure 4.2. The use of the graph is illustrated by arrowed broken lines.

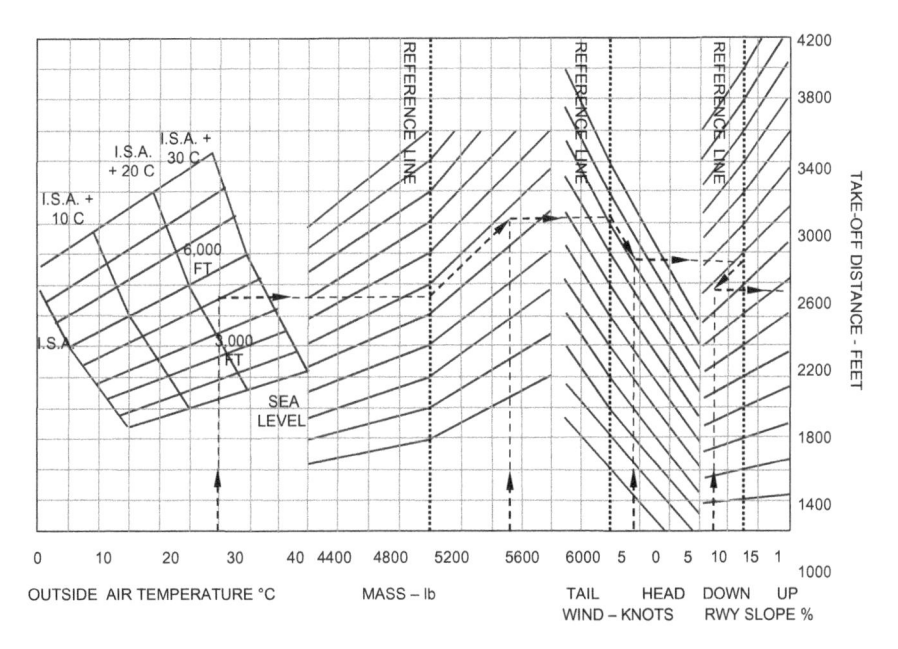

Figure 4.2: Single engine piston aircraft take-off distance.

82 Airport Design and Operation

The performance characteristics of the critical aircraft at a particular aerodrome have to allow the aircraft, with regards to the runway characteristics and actual meteorological conditions, either to complete the take-off safely, or to stop after aborted take-off. For all multi-engine aircraft for each take-off, a speed V_1 is defined, which is called the decision speed (sometimes called also critical speed), see Figure 4.3.

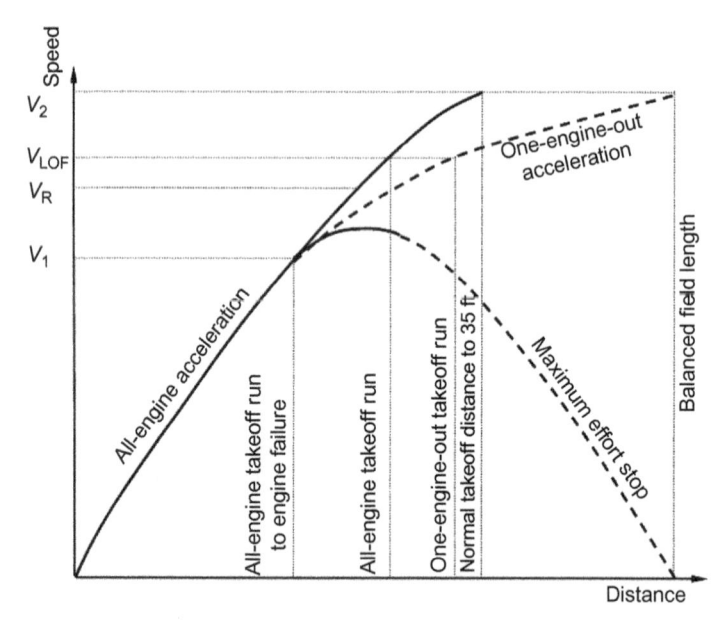

Figure 4.3: Take-off field length demonstration requirements for multiengine aircraft.

If an engine failure occurs at a speed lower than V_1, the pilot shall abort the take-off. If the aircraft has a greater speed than V_1 when the emergency is recognised the pilot shall continue the take-off with one engine inoperative. Trying to continue the take-off with an engine failed before V_1 would require a longer take-off distance due to the reduced power available after an engine failure.

On the other hand, there must not be problem in stopping the aircraft in the remaining portion of the runway. With an engine becoming inoperative only after exceeding the speed V_1, the aircraft has a relatively high speed and the remaining power is sufficient for completing the take-off on the remaining runway. If the decision to abort the take-off is taken after exceeding the V_1 speed, the aircraft speed would be so high that the aircraft would need an excessive distance to stop safely in the remaining runway.

Decision speed is the speed determined by the captain according to the respective limitations of the aircraft, the airline operator rules and procedures, runway characteristics and actual meteorological conditions. The accelerate-stop distance d_2 (Figure 4.4) is the sum of the take-off distance to reach the speed V_1 with all engines operating, the distance which the aircraft travels during the pilot's reaction time (assumed reaction time for the purposes of performance calculations is 1 second) and the braking distance until the aircraft comes to a full stop.

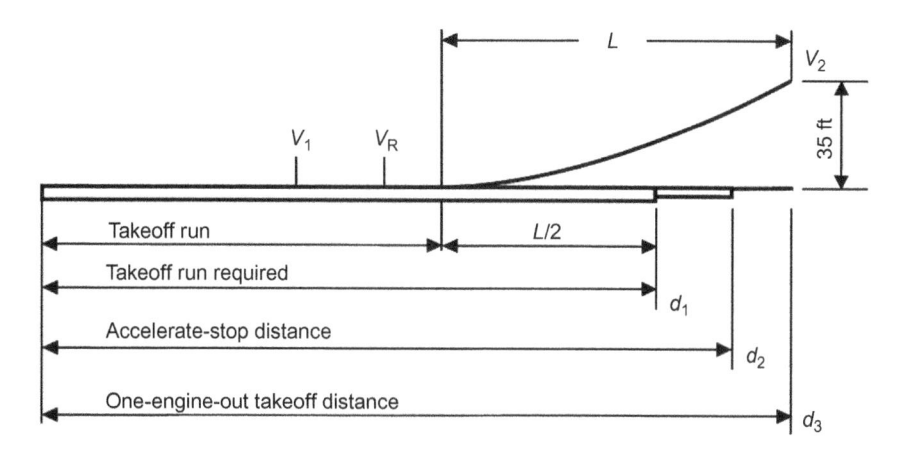

Figure 4.4: One engine-out take-off/aborted take-off.

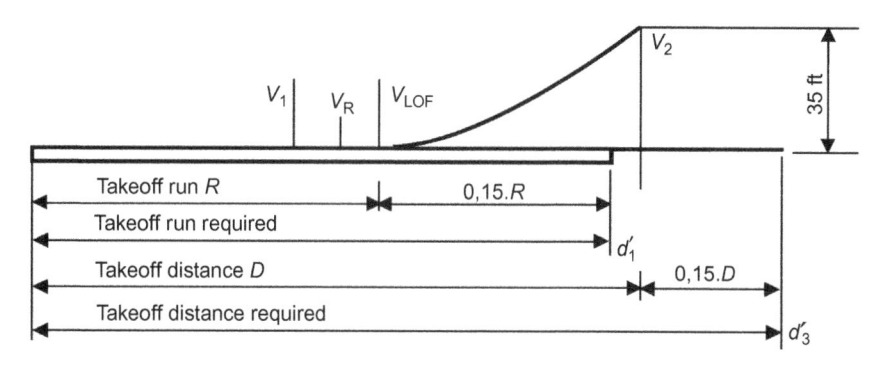

Figure 4.5: Take-off — all engines.

The take-off length with one engine inoperative, d_3, is the sum of the distance with all engines operating until reaching the speed V_1, the length of a further distance to lift off with one engine inoperative and the climb to the height 10.7 m (35 ft) (Figure 4.4) where the particular speeds are defined as (Figure 4.6):

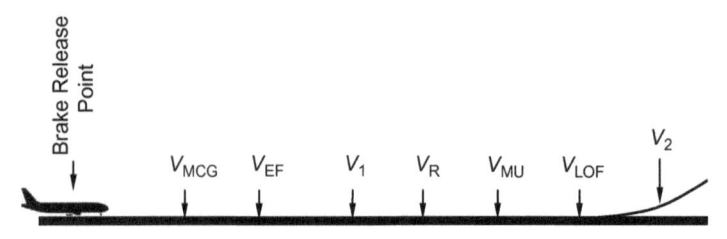

Figure 4.6: The take-off speeds.

V_{EF} The speed at which the critical engine is assumed to fail to be afterwards recognised as 'an engine failure' at V_1. This means that V_1 is always higher than V_{EF}. It is presumed that for a trained pilot it takes one second to recognise the engine failure and initiate the first action.

V_1 Decision speed is defined as the maximum speed at which the pilot must take the first action in order to stop the aircraft within the remaining accelerate-stop distance. V_1 should better be called action speed because the decision must be taken before the action. It is also the minimum speed after engine failure that the pilot is able to complete take-off within the remaining take-off distance. It is assumed that the aircraft accelerates for two more seconds after the take-off abortion action due to the momentum of the aircraft and change of its configuration (application of brakes, speed brakes and thrust reverser).

V_R Rotation speed or speed at which the pilot initiates the action to lift the nose wheel. The aircraft 'rotates' about its lateral axis to raise the nose gear off the ground.

V_{MU} Minimum unstick speed is minimum speed which allows safe continuation of take-off. The lowest speed at which an aircraft can safely lift-off and continue the take-off climb. Because the V_{MU} is close to the stall speed in actual operational conditions the aircraft does not lift off at V_{MU} but slightly higher speed.

V_{LOF} Lift-off speed, main gear lift-off speed. V_{LOF} should be greater than the minimum unstick speed V_{MU}.

$V_{MC(A)}$ Air minimum control speed/minimum control speed in the air, minimum speed at which critical engine failure may occur and still a straight flight is possible at this speed and is fully controlled. During the critical engine failure and with the remaining operating engine/engines at take-off setting causes asymmetric thrust which causes the yaw. An angle of bank must not exceed 5°.

V_2 Take-off climb (or safety) speed 1.2 V_s, (where V_s is a stalling speed for the take-off configuration), that must be reached at or prior to the screen height of at least 35 ft above the RWY surface in case of engine failure. In fact the V_2 is the slowest speed which will enable the aircraft to have sufficient excess thrust to climb more than the minimum required climb gradients.

Other important limits and speeds which can be found in flight manuals and are related to aircraft performance during take-off and climb are as follows:

V_{MBE} Maximum brake energy speed — the maximum speed from which a stop can be accomplished within the energy capabilities of the brakes.

V_{STOP} The highest speed from which a take-off can be safely rejected for a given ASDA (for ASDA see Section 4.3).

V_{GO} The slowest speed from which a take-off can be safely continued after an engine failure for a given TODA (for TODA see Section 4.3).

The required safe distance necessary for a take-off is considered to be whichever is the greater of:

✈ take-off distance with one engine inoperative — d_3 (Figure 4.4)

✈ 115% of the demonstrated take-off distance with all engines operating d'_3 (Figure 4.5).

If a lower V_1 is chosen, the take-off distance with one engine inoperative — d_3 (Figure 4.7; Point 1) determines the necessary runway length; while with a higher V_1 chosen, the accelerate-stop distance d_2 (Point 2) is taken. If the 115% of the take-off distance with all engines operating is so small that it lies under the Point 3 (Line a), the distance d'_3 has no effect upon the necessary take-off distance. However, if the 115% of the take-off distance with all engines operating lies over the Points 1 and 2 (Line c), it is the only determinant for the take-off distance (Figure 4.7).

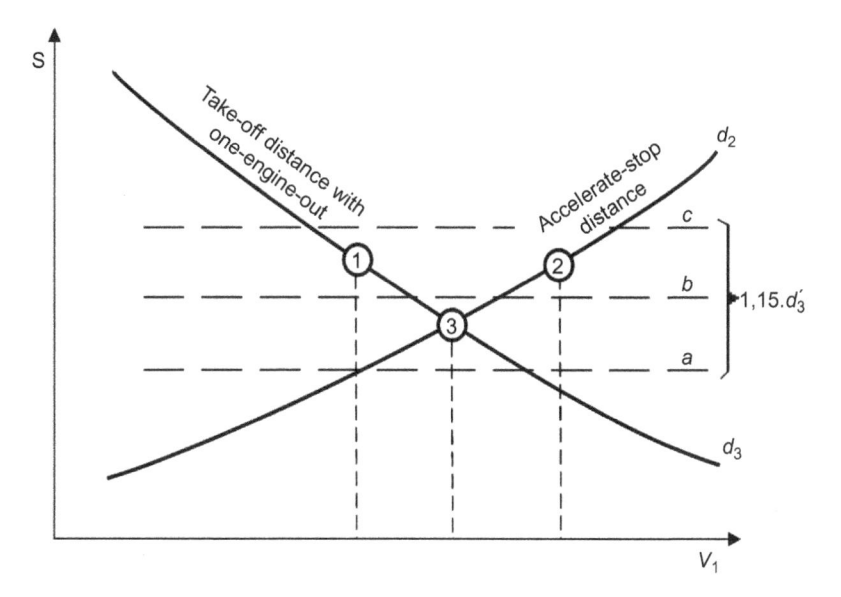

Figure 4.7: Effect of decision speed V_1 on take-off distance, accelerate-stop distance and take-off run.

86   Airport Design and Operation

If it lies between the Points 3 and 1, or 2 (Line b), all the above named events may occur. The aircraft Airport Planning Manuals normally report only the most dominant of these criteria and only consider the balanced engine-out field length where V_1 is chosen so the distance for take-off and rejected take-off are equal (Figure 4.7).

For a take-off with one engine inoperative, the hard runway part of the total available distance should be longer than the actual required take-off run. That distance shall be determined by the sum of the distance with all engines operating up to the speed V_1, a further distance run with one engine inoperative up to the lift-off speed V_{LOF} and a half of the horizontal distance to climb up to 10.7 m (35 ft) (Figure 4.4).

The distance necessary for a take-off run is considered to be whichever is the greater of:

✈ take-off run distance with one engine inoperative plus half of the horizontal distance to climb to 10.7 m (35 ft) — d_1 (Figure 4.4)
✈ 115% of the take-off run with all engines operating — d'_1 (Figure 4.5).

When considering an aircraft operation on a particular runway there are also other factors and limitations. Generally, six areas that have take-off performance requirements can be identified. They are the following:

✈ field length
✈ climb
✈ obstacle clearance
✈ tyre speed
✈ brake energy
✈ inoperative systems according to Minimum Equipment List (MEL) or Configuration Deviation List (e.g. nonfunctional Antiskid).

The last three factors, tyre speed, brake energy limitations and inoperative systems are wholly operational and are not further discussed.

For a quick determination of an aircraft take-off mass (performance class A aircraft) a chart could be used. In this graph no clearway or stopway is taken into account. This means that in this special case the field length is the take-off run (TORA) which equals the accelerated-stop distance (ASDA) and also equals the take-off distance (TODA). This gives more strict conditions and restricts more the aircraft take-off mass but substantially simplifies the determination of the take-off mass (Figure 4.8).

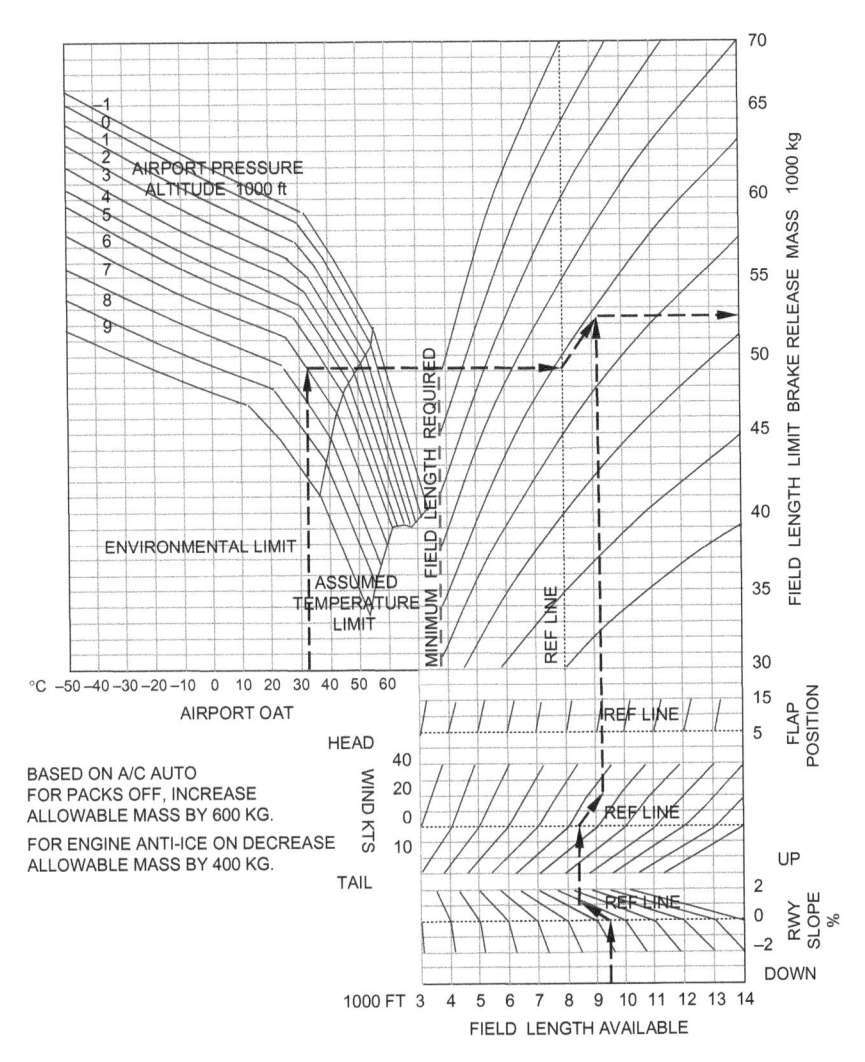

Figure 4.8: Take-off performance field limit, class A aircraft.

For take-off calculations the runway length must be reduced by runway alignment distance. The amount of reduction depends on the runway turn pad geometry (see Section 4.4.3) and aircraft type. The distances are referred to in aircraft flight manuals for 90° turn-on and for 180° turn-on. For example for B 737-400 the 90° turn-on alignment distance is 33.0 m by which the take-off distance must be reduced.

When calculating take-off length the runway conditions must be also considered. In most aircraft flight manuals information can be found on contaminated runway effect usually for 3 mm of standing water; 15 mm of wet snow or slush; 60 mm of dry snow or 80 mm of very dry snow.

88 Airport Design and Operation

```
                    ①         ②         ③      ④     ⑤

ELEVATION  2356 FT                          RUNWAY 27       LZTT

*** FLAPS 01 ***     AIR COND AUT   ANTI-ICE OFF    TATRY
NORMAL CONFIG                                       POPRAD
737-700       CFM56-7B22                            DATED 16-AUG-2005
*A* INDICATES OAT OUTSIDE ENVIRONMENTAL ENVELOPE
OAT   CLIMB              WIND COMPONENT IN KNOTS  (MINUS DENOTES TAILWIND)
 C    100KG        -10                 0               5             10

 40    617    549*/26-27-31     567*/30-30-34    570*/30-30-34    573*/31-31-34
              563**/33-35-39    589**/40-42-45   593**/41-43-46   597**/43-44-47
 39    624    555*/27-28-32     573*/30-30-34    576*/31-31-35    579*/31-31-35
              568**/33-35-39    594**/40-42-46   598**/42-43-47   602**/43-44-48
 38    630    560*/27-28-33     579*/31-31-35    582*/31-31-35    585*/32-32-36
              573**/33-35-39    599**/41-42-46   603**/42-43-47   607**/43-44-48
 37    637    566*/28-29-33     584*/32-32-36    587*/32-32-36    590*/33-33-37
              578**/34-35-40    604**/41-42-46   608**/42-44-47   613**/43-45-48
 36    643    571*/29-30-34     590*/32-32-37    593*/33-33-37    596*/33-33-37
              583**/34-36-40    609**/41-43-47   614**/42-44-48   618**/44-45-49
 35    650    577*/29-30-35     596*/33-33-37    599*/33-33-38    602*/34-34-38
              587**/34-36-40    614**/41-43-47   619**/43-44-48   623**/44-45-49
 34    656    582*/30-31-36     601*/33-34-38    605*/34-34-38    608*/34-34-39
              592**/34-36-41    619**/42-43-47   624**/43-44-48   626**/43-44-48
 33    663    588*/30-32-36     607*/34-34-39    610*/34-35-39    614*/35-35-39
              597**/34-36-41    624**/42-43-48   626**/42-43-47   626**/40-41-45
 32    669    593*/31-32-37     613*/34-35-39    616*/35-35-40    619*/36-36-40
              602**/35-37-41    626**/40-42-46   626**/39-40-44   626**/38-39-43
 31    676    599*/31-33-38     618*/35-36-40    622*/36-36-40    625*/36-36-41
              607**/35-37-42    626**/38-39-44   626**/37-38-42   626**/37-37-41
 30    682    604*/32-33-38     624*/35-36-41    627*/36-36-41    630*/36-36-41
              611**/35-37-42    626**/36-37-42
 25    713    629*/34-36-41     648*/35-36-41    651*/35-36-41    655*/35-36-41
 20    714    630*/34-36-41     650*/35-36-41    653*/35-36-41    656*/35-36-41
 15    715    632*/34-36-41     651*/35-36-41    654*/35-36-41    658*/35-36-41
 10    716    633*/34-36-41     652*/35-36-41    656*/35-36-41    659*/35-36-41
  5    717    634*/34-36-41     654*/35-36-41    657*/35-36-41    660*/35-36-41
  0    718    635*/34-36-41     655*/35-36-41    658*/35-36-41    662*/35-36-41
 -5    719    636*/34-36-41     656*/35-36-41    659*/35-36-41    663*/35-36-41
-10    720    637*/34-36-41     657*/35-36-41    660*/35-36-41    664*/35-36-41

MAX BRAKE RELEASE WT MUST NOT EXCEED MAX CERT TAKEOFF WT OF    62600 KG
MINIMUM FLAP RETRACTION HEIGHT IS  1000 FT
LIMIT CODE IS F=FIELD, T=TIRE SPEED, B=BRAKE ENERGY, V=VMCG,       ⑥
           *=OBSTACLE/LEVEL-OFF, **=IMPROVED CLIMB
TORA IS  2600 M , TODA IS    2700 M , ASDA IS    2600 M
RUNWAY SLOPES ARE   0.89 PERCENT FOR TODA  AND    0.89 PERCENT FOR ASDA
LINE-UP DISTANCES:      23 M  FOR TODA,     23 M  FOR ASDA    OBS FROM LO-FT/M
RUNWAY          HT   DIST  OFFSET     HT   DIST  OFFSET      HT   DIST  OFFSET
27              26   410      0      194   2505      0      210   2815      0
                486  5670     0      663   9220      0
ENG-OUT PROCEDURE:
Climb on 271°. At 3400 turn left to PPD HP. D112.1 PPD HP: Inbound 271°,
left turn.
```

Figure 4.9: Take-off performance limitations — 737-700 at Poprad — Tatry airport. *Courtesy*: Sky-Europe Airlines. *Notes: (1) Maximum theoretical take-off mass with respect to outside air temperature (OAT) × 100 kg, e.g. 617 = 61,700 kg; (2) obstacle in climb sector limited MTOW (calm); (3) V_1 — 130 kts; (4) V_R — 130 kts; (5) V_2 — 134 kts and (6) other limitations codes.*

Most of the airline companies have computerised programs which allow the calculation of the exact parameters by the dispatcher for particular airport conditions, and also for the pilot to recalculate them on laptop or tablet when in full possession of the weather situation, runway and aircraft status shortly before take-off. Figure 4.9 gives an example of printed output for Poprad — Tatry airport from the STAS programme.

In the majority of cases, the aircraft landing distance is smaller than the take-off distance. The final phase of landing begins at an altitude of 50 ft and ends with the full stop of the aircraft. In practice, the aircraft often vacates the runway, for example through a high-speed exit taxiway, rather than coming to a full stop, but the regulation distances are based on the former case.

The necessary runway length for landing should be greater than the actual minimum landing distance demonstrated by the manufacturer, so that day to day variations in pilot behaviour and effects of meteorological conditions may be taken into account. Regulations vary considerably with respect to the margins required over the determined landing distance, depending on the country and the state of the runway. According to the EU Regulation No 965/2012/FAR requirements for a dry runway the mandatory distance is 167% of the demonstrated stopping distance.

For a wet runway there are additional safety factors; the distance required is 115% of the dry runway distance from a height 50 ft above the runway, no reverse thrust, maximum manual braking. The safety factor for Class A turboprop and all Class B aircraft is smaller: in their case, the demonstrated landing distance required is 143% for a dry runway (Figure 4.10).

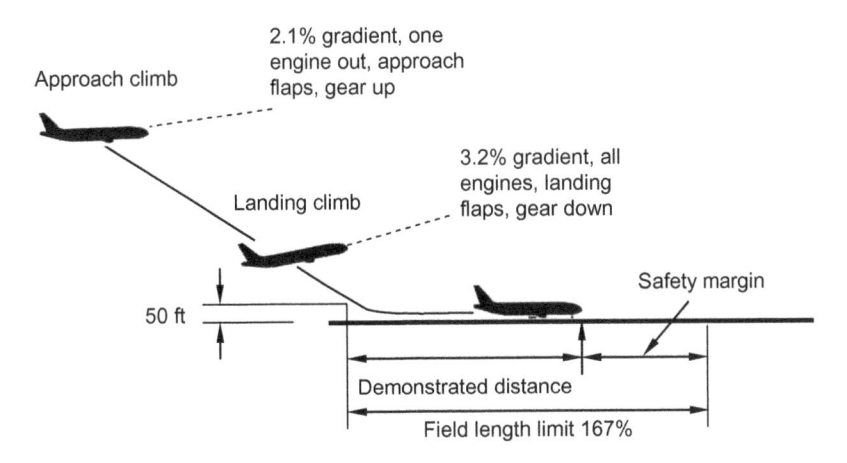

Figure 4.10: Landing performance requirements.

Most regulations for the determination of landing performance now require the demonstration to be made on a standard wet runway with an appropriate flying technique for the conditions. An additional safety factor is often applied for low visibility landings, which can result in the landing distance being greater than the take-off distance required, particularly for modern twin-engined aircraft.

The decision as to whether the runway is to be extended or only a stopway and a clearway are to be provided beyond the original runway ends depends, among other things, on the characteristics of the terrain beyond the runway end and on the occurrence of possible obstacles.

In line with Annex 6, Part I, Operation of Aircraft, Attachment C, 'Aircraft performance operating limitations', and also with EU No 965/2012, there are specified take-off obstacle clearance limitations (compare also with Annex 14, Volume I, Chapter 4 — 'Obstacle restriction and removal'). The net take-off flight path must clear all obstacles vertically by 35 ft or laterally by at least 90 m plus $0.125\,D$, where D is the horizontal distance the aircraft has travelled from the end of the take-off distance available on both sides of extended RWY centreline. The sector widens with distance until becoming parallel after a specific distance.

If the flight path does not include track changes of more than 15° the maximum sector width is 300 m on each side from the centreline under VMC by day or when using navigation aids such that the pilot can maintain the aircraft on the track, and 600 m for flights under all other conditions.

Where the flight path includes track changes of more than 15° the maximum sector width is 600 m on each side from the centreline under VMC by day or when using navigation aids such that pilot can maintain the aircraft on the track, and 900 m for flights under all other conditions (Figure 4.11).

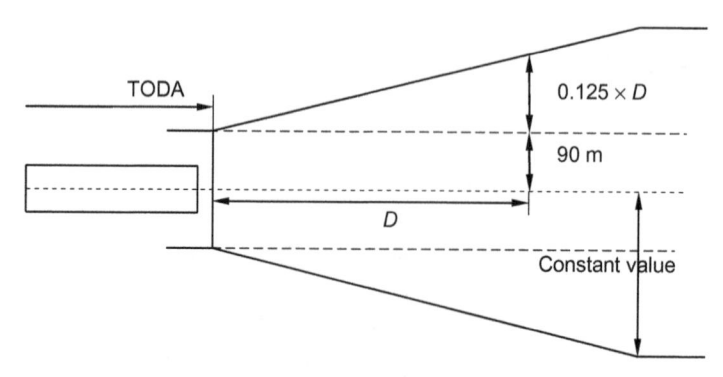

Figure 4.11: Take-off climb obstacle clearance.

For aircraft with a wingspan less than 60 m (i.e. for most of commercial transport aircraft) the initial width of the sector is limited to 60 m + half of the wingspan.

Besides the field and structural limits a take-off also involves climb limits. If an engine fails after V_1, the reduction of thrust may cause some obstacle clearance in the take-off sector to become critical.

The net take-off climb path begins at the screen height of 35 ft and ends at a minimum of 1500 ft above the aerodrome level. However, if the take-off is performed from wet or contaminated (low friction) runway the take-off climb begins at 15 ft. The point on the ground immediately below the 35 ft screen height is called 'reference zero'. For every segment the gross gradient must be calculated first, assuming that the critical engine failed at V_{EF}. Then the net gradient is the gross gradient reduced by:

- 0.8% for 2-engined aircraft
- 0.9% for 3-engined aircraft
- 1.0% for 4-engined aircraft.

The net take-off path is used for calculation of the obstacle limited mass of the aircraft. Minimum gradient requirements are specified for every climb segment. Each of these will determine an aircraft mass for the airport altitude and outside air temperature. The lowest mass will determine the MTOM. All obstacles in the take-off climb sector must be cleared by at least 35 ft (Figure 4.12). Any part of the net take-off flight path in which the aircraft is banked by more than 15° must clear all obstacles by a vertical distance of at least 50 ft. The take-off climb is usually divided into four segments. Each segment is defined by the aircraft configuration, speed, thrust and climb gradients.

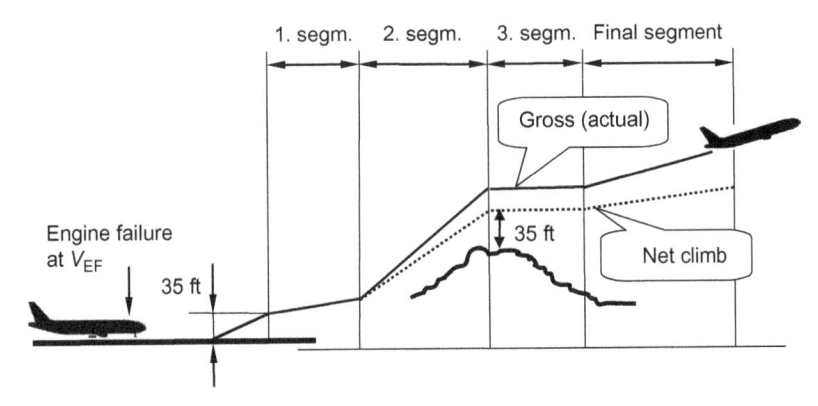

Figure 4.12: Take-off performance requirements.

First Segment	*From 35 ft to gear-up*: take-off (TO) slats and flaps, TOGA (take-off and go-around thrust), V_2 speed from 35 ft to gear-up. The aircraft is with one engine inoperative and should climb as fast as possible. This is difficult due to drag created by the gear. Because of that the aim is to retract the gear as soon as possible. The aircraft must maintain positive rate of climb during the first segment.
Second Segment	*From gear-up to level-off flap retraction*: TO slats and flaps, TOGA thrust, V_2 speed to minimum 400 ft or minimum acceleration height given by the aircraft operator. Flap retraction is not permitted below 400 ft. Therefore the action of the pilot in the second segment is to climb until 400 ft or acceleration altitude is reached. The minimum climb gradient is 2.4% which is the most severe regulatory requirement for a twin engine jet or turboprop aircraft.
Third Segment	*Level flight, flap retraction*: Slats/flaps retracted TOGA. TOGA is certified for use for a maximum of 10 minutes, in case of an engine failure at take-off, and for a maximum of a 5 minutes (FAA) or 10 minutes (EASA) with all engines operating. As a result, the *en route* configuration (at the end of the third segment) must be achieved within a maximum of 10 minutes after take-off, thus enabling the determination of a maximum acceleration height. Acceleration from V_2 to final segment speed, or V_{ZF} (V_{ZF} = minimum speed with flaps up, or 'Green dot' speed[3] in Airbus). The aircraft must accelerate before flap retraction from V_2 to V_{ZF} and then to the speed which gives the best rate of climb with one engine inoperative. Then the TOGA could be reduced to Maximum Continuous Thrust (MCT).
Final Segment	*Final climb segment*: Clear configuration, MCT, final segment speed (V_{ZF} or 'Green dot' speed). The pilot climbs to 1500 ft where the take-off climb ends. The captain than takes the decision to return to land at the airport of departure or to continue to the alternate airport. The rate of climb during the last segment must be at least 1.2%.

If there are obstacles to be considered in the take-off path, the flap setting that results in the shortest take-off distance does not always result in the maximum possible take-off mass. The higher flap setting will result in short take-off distance, but because of increased drag the climb gradient after take-off may be not sufficient to clear the obstacles (Figure 14.13).

---

3. Green dot is an optimised speed. It is an approximation of the best lift to drag ratio. Flying at green dot speed provides the best climb performance.

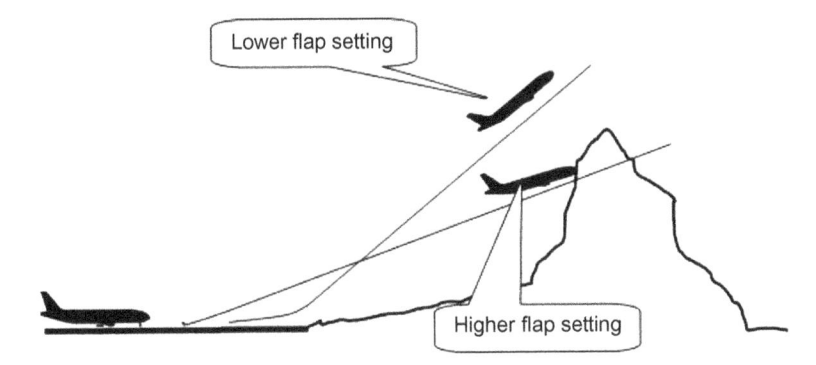

Figure 4.13: Flaps setting for obstacle clearance.

4.3. DECLARED DISTANCES

In order that the pilots and flight operators may be informed about the lengths of runway, clearway, stopway and runway strip, the airport operator has to publish the respective information in the Air Information Publication (AIP).

The information is published in a standardised form, and the declared distances determined for each runway direction include:

- TORA (take-off run available)
- TODA (take-off distance available)
- ASDA (accelerate-stop distance available)
- LDA (landing distance available).

The particular lengths shall not be determined only by the length of runway or other movement surfaces, but also by their actual condition and contingent obstacles in the take-off or approach areas.

Figures 4.14a–c give several examples of declared distances. In general, it may be stated that the take-off run and landing may be performed only on a full strength runway, while the remaining part of the take-off may be completed over a runway strip or a clearway.

In many cases the runway will be full strength for its whole length with thresholds located at the runway ends, runway strips with specified dimensions shall extend beyond the end as well as along each side of the runway. In that case the declared distances will be the same in both directions, see Figure 4.14.

94   *Airport Design and Operation*

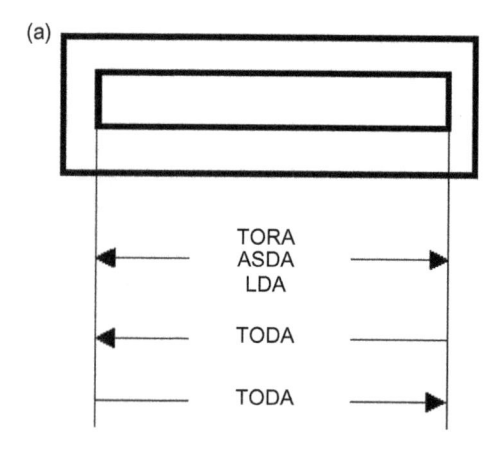

Figure 4.14:   (a) Declared distances.

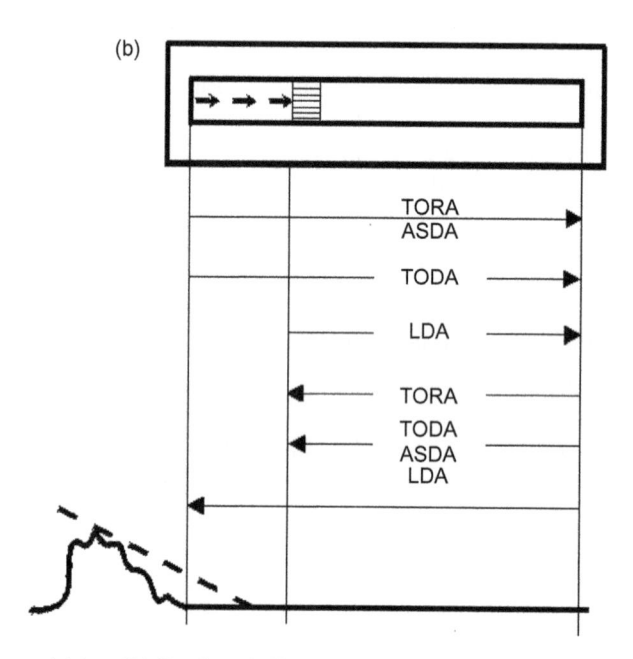

Figure 4.14:   (b) Declared distances with a displaced threshold.

Alternatively, the runway may have the landing threshold shifted, usually due to obstacles in the approach area. However, the portion of the full strength runway before the displaced threshold may be used for take-off, see Figure 4.14b.

Figure 4.14c illustrates a runway where the threshold was displaced due to a runway section in maintenance (left). Beyond the end of the runway a stopway was established (right). It is not possible to extend the runway in that direction because an obstacle occurs in the take-off area. The stopway is full strength and is surrounded by runway strips.

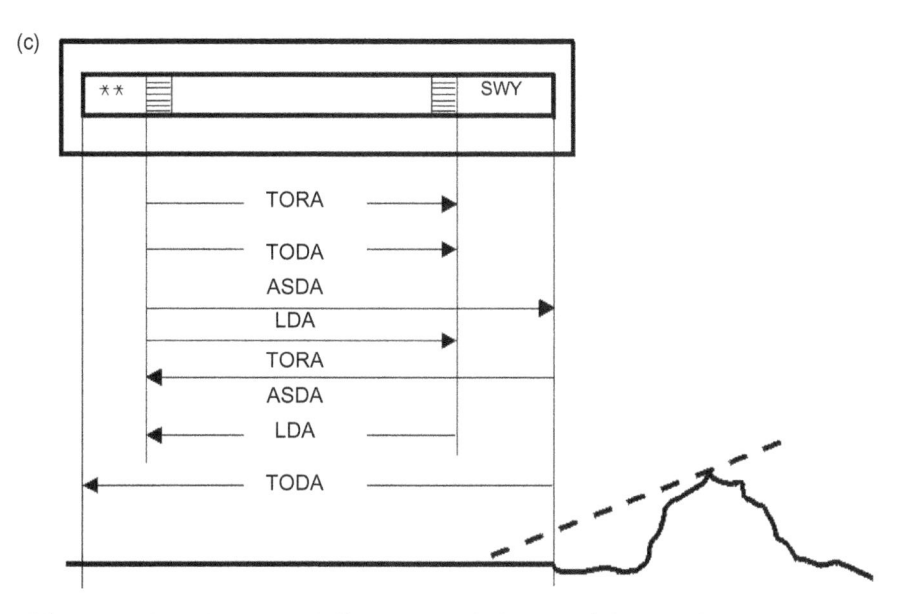

Figure 4.14: (c) Declared distances with displaced threshold and stopway.

4.4. RUNWAY WIDTH

4.4.1. Runway Width Requirements

As already stated in the above Section 4.1 on the Aerodrome Reference Code, the width of a runway is one of the elements that is affected by several geometrical characteristics of aircraft:

✈ the distance between the outside edges of the main gear wheels
✈ the distance between wing mounted engines and the longitudinal axis of an aircraft
✈ the wingspan.

However, the required runway width is also affected by the operational elements:

✈ the approach speed of the aircraft
✈ the prevailing meteorological conditions.

96 Airport Design and Operation

Lack of sufficient width will cause constraints on the operations. The minimum runway width is therefore specified in Annex 14 by interrelating both of the code elements, see Table 4.5.

Table 4.5: Minimum runway width.

Code number	Code letter					
	A	B	C	D	E	F
1[a]	18 m	18 m	23 m	–	–	–
2[a]	23 m	23 m	30 m	–	–	–
3	30 m	30 m	30 m	45 m	–	–
4	–	–	45 m	45 m	45 m	60 m

Source: ICAO (2013b).
[a]The width of a precision approach runway should be not less than 30 m where the code number is 1 or 2.

Under normal conditions, the width of a runway should ensure that an aircraft does not run off from the side of the runway during the take-off or landing, even after a critical engine failure causing the aircraft to yaw towards the failed engine.

Deviation from the runway axis during take-off and landing usually occurs due to a lack of ability in tracking the centreline. With impaired visibility, adverse cross — wind conditions, or poor braking action, the probability of an off-centre touchdown or a slide sideways from the runway axis during take-off increases. Therefore a wider runway is specified for precision approach runways where the code number is 1 and 2. Under some regulations, it is possible for the first part of the take-off roll to be on a reduced width runway (referred to as a starter strip).

The width of runway is critical during an aborted take-off. In case of engine failure the aircraft tends to veer-off the runway due to asymmetric thrust and fast reaction of the pilot in command is necessary to be able to keep the aircraft on the runway. This is the main reason why some airline operators and civil aviation authorities do not allow B 737 and A 320 operations from 30 m wide runways or require specific crew experience or training even though the main gear width of those aircraft types allows the operation under normal conditions.

4.4.2. Runway Shoulders

According to Annex 14, Volume I, Chapter 3, a width of 45 m is sufficient for runways where the code letter is E, that width of pavement conforming with the undercarriage gauge of the largest aircraft. In the event that an aircraft touches down off the centreline of a runway, even though the undercarriage is on a full strength pavement, the wing mounted engines overlap its edge. Thus they could take in loose material, such as small stones, with a possibility of consequent Foreign Object Damage (FOD) to the engine. Therefore runway shoulders should be provided for a runway where the code letter is D and E so that the over-all width of the runway and its shoulders is not less than 60 m. If the code letter is F, the width of the runway itself must be 60 m wide.

The shoulders may be turf over stabilised earth, but are usually designed as light asphalt pavements with a load bearing capacity that will support the loads of the ground equipment and reduce the probability of damage to an aircraft veering off the runway. Runway edge lights are located in the runway shoulders. The winter maintenance of the edges of a runway is then considerably simpler than where the lights are installed in the grass strip.

A light asphalt shoulder is not designed for the very rare event of an aircraft running off the runway, so when it happens it damages the pavement. The repair of an asphalt pavement is not demanding. Nevertheless, its load bearing capacity should prevent the wheels from being bogged down, which may cause serious damage to the aircraft. At the same time, a shoulder provides a gradual change of the load bearing capacity between a runway and its associated strip. If a runway is made of cement concrete and a shoulder is made of asphalt, the passage between the runway edge and the shoulder is visually distinctive and clear enough. If both the runway and the shoulder have asphalt surfaces, the edge of the runway should be indicated by a runway side stripe marking. The shoulder should be vertically flush with the runway edge. An example of a runway design with a shoulder is given in Figure 4.15.

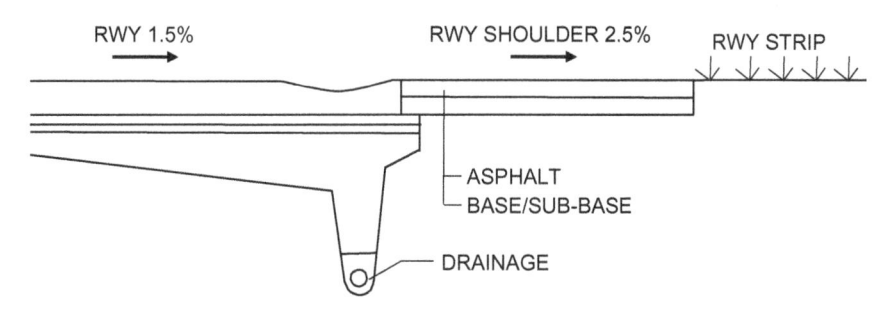

Figure 4.15: Runway with runway shoulder. *Source*: Čihař (1973).

4.4.3. Runway Turn Pads

At the end of a runway with a code letter of D, E or F, a turn pad must be provided if the runway end is not served by a taxiway or a taxiway turnaround to enable a 180° turn of aircraft (Figure 4.16). It is useful to have such turn pads also

98   Airport Design and Operation

along the runway to reduce taxiing time for aircraft which do not require full runway length. The shape, location and placement of the turn pad are not obligatory but it is an advantage to locate the turn pad on the left side of the runway, since the standard pilot-in-command position is on the left seat. There are also no binding requirements with respect to the turn pads' dimensions except that the intersection angle of the pad with the runway should not exceed 30° and the aircraft nose wheel steering angle to be used in the design of the runway turn pad should not exceed 45°.

Figure 4.16:   Turn pad at the runway end, Kos airport. *Photo*: A. Kazda.

Depending on the code letter of the runway and the aircraft wheel base dimensions, minimum margins of safety between 1.5 m and 4.5 m should be provided between any wheel of the aircraft landing gear and the edge of the turn pad when the cockpit of the aircraft is over the turn pad marking. The requirements on the turn pad slopes, bearing strength of the pavement and the pavement surface are similar to those of the adjacent runway.

4.5.   RUNWAY SLOPES

4.5.1.   Transverse Slopes

Annex 14 recommends that runways be constructed with a roof-like cambered transverse slope, the slope providing quick rainwater drainage particularly in heavy rain and side winds. In spite of that, there are cases when a designer is forced or prefers to design a one-sided transverse slope of a runway to reduce costs of earthwork (Figure 4.17), or to avoid too thick a runway construction when a roof-like profile is selected, or to reduce cost by providing a drainage channel on only one side of the runway. In order that drainage may be ensured, the transverse slope of a runway should be at least 1% but should not exceed 2%, depending on the code letter. The transverse slopes should normally facilitate drainage and minimise the layer of water accumulated on the runway. In some cases the drainage is improved by a transverse grooving of the pavement which, however, in no case may replace the prescribed slopes.

Figure 4.17: One-sided transverse runway slope and changes of longitudinal slopes are clearly visible, Poprad-Tatry airport. *Photo*: A. Kazda.

4.5.2. Longitudinal Slopes

Few runways are level throughout their length in the longitudinal direction. Often there are several slope changes in its length. A completely horizontal runway could be constructed, but this would generally require considerable earthwork except of a few airports constructed on artificial islands or atolls. On the other hand, there is a limitation on the allowable longitudinal slope due to the performance characteristics of particular types of aircraft and the requirements of the operation. The Annex 14 standard therefore limits the so-called average slope of a runway, which is calculated by dividing the difference between the highest and the lowest points of a runway by its all-over length; and also limits the maximum slope of any arbitrary portion of a runway, the profiles usually being controlled in segments of 50–100 m.

For example, the longitudinal slope of the precision approach runways where the code number is 3 and 4 is limited to 0.8% in the first and last quarter. This strict limitation of the slope is determined either in order to facilitate the final phase of flare in impaired meteorological conditions or by technical parameters of the equipment for a precision instrument approach, particularly the radio-altimeter. The longitudinal slope of non-precision runways of civil airports is limited to 1–2%, depending on the code number.

In special cases, a longitudinal slope up to 8% is permitted for one-way runways in airfields for agricultural activities, for special airfields for mountain rescue services, or, indeed, for commercial service in mountainous areas. However, there are some exceptional airports with respect to longitudinal slope as for example Courchevel (France) where the slope reaches 18.5% Tenzing-Hillary (Lukla) Airport in eastern Nepal with a 12% gradient.

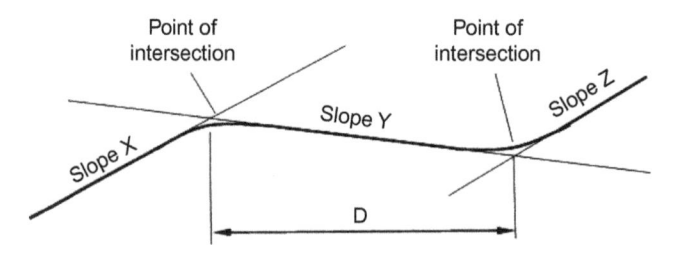

Figure 4.18: Profile of centreline of runway. Source: ICAO Annex 14, Aerodromes, Volume I, Aerodrome Design and Operation; 6th edition, July 2013 (ICAO, 2013b).

Longitudinal slope changes are also limited. The objective of limiting local undulation is to confine dynamic load of the undercarriage system of an aircraft when it moves at a high speed on the runway, and to provide permanent and safe contact of the tyres with the surface of the runway. Longer slope change limitations are to ensure the pilot has an adequate view of the runway. A change from one slope to another should be achieved by a minimum radius curvature of 7500–30,000 m, depending on the code number of the runway. Similarly, the distance between the apexes of two neighbouring curvatures shall be limited. The distance should not be less than the sum of the absolute numerical values of the difference of longitudinal slopes multiplied by a factor, depending on the magnitude of the curvature, in turn depending on the code number of the runway, or 45 m, whichever is the greater (see Figure 4.18).

$$D = k\bigl[|x-y| + |y-z|\bigr]$$

where D is minimum distance between point of intersection of slope changes in metres; k is minimum radius of curvature (between 7,500 and 30,000 with respect to the RWY code number) and x, y, z are longitudinal slopes per mile.

The longitudinal slope changes are furthermore limited by the requirement of mutual visibility between two points at a height of 1.5–3 m (according to the code number of a runway) and at a distance which is equal to at least one half of the runway length.

5

RUNWAY STRIPS AND OTHER AREAS

Tony Kazda and Bob Caves

5.1. RUNWAY STRIPS

Each runway should be surrounded by a runway strip. The runway strip is intended to ensure the safety of an aircraft and its occupants in the event of an aircraft:

→ undershooting, overrunning or veering-off the runway during landing or take-off and

→ deviating from the runway centreline during a missed approach.

The physical characteristics and other requirements for runway strips are derived from the need to allow for these events.

The required width of a runway strip depends on the degree of approach guidance to the runway and also on the runway reference code number. The width of a non-instrument runway strip should extend for 30–75 m on each side of the centreline of the runway. A strip of an instrument and a precision approach runway should extend for 75–150 m. The runway strip should extend past the runway end by 30–60 m. No equipment or constructions which may create an obstacle or endanger aircraft should be situated on a runway strip. The only exception is radio-navigation, visual and other equipments required for aircraft safety which should be installed on the runway strip to support the operation of aircraft; for example, the Instrument Landing System (ILS) glide path antenna, Precision Approach Radar (PAR), Runway Visual Range (RVR) equipment and the like. Such equipments shall be mounted on frangible fittings and marked with an obstacle marking, and the antenna masts must be of frangible constructions. Also no mobile equipment is permitted within a specified part of the runway strip of an active runway. The specified distance from the runway centreline is related to RWY code number, code letter and precision approach category. The distance is also linked to the mode/category of operation. During VFR and Cat I. operations mobile equipment could operate up to within 75 m of the centreline with a height limit defined by the transition obstacle limits 1:7. During the Low Visibility Procedures (i.e. Cat II/III operations) a strip up to 150 m from the centreline is closed to all obstructions.

104 Airport Design and Operation

In the majority of cases the surface of a runway strip should be grass, soil or gravel. It is intended that, in the event of an aircraft running off the runway, the undercarriage of the aircraft should gradually sink into the runway strip, thereby providing effective deceleration of the aircraft. The aircraft's structure should not suffer serious damage which could lead to casualties (Figure 5.1).

Figure 5.1: Cessna Citation after runway overrun stopped safely at the end of the RWY strip. *Photo*: J. Stehlík.

It is probable that in the event of an aircraft veering-off the runway, the aircraft will not need the whole width of the runway strip for protection. Therefore the surface of a runway strip need not have the same quality over all its area. Depending on the category of approach guidance of a runway and its reference code number, the surface of a runway strip should be able to cater for the event of an aircraft veering-off to a distance of 30–75 m from the centreline of the runway with little damage to the aircraft. This part should have a bearing strength of approximately CBR 20, which requires the top 20 cm of soil be compacted as well as complying with the specified slopes and quality of the surface to minimise hazards arising from differences in load-bearing capacity to aircraft in the event of running off the runway. Also, over the whole portion of a runway strip to the specified distance from the runway, there must not be any solid vertical constructions even under the surface of the runway strip to a depth of at least 30 cm which a sinking wheel of the aircraft might impact. Such obstacles may, for example, be foundations of lighting installations and airside signs, drainage gutter inlets or the edges of taxiways within the runway strips. At least the upper 30 cm of such a construction should be chamfered. Other equipment which need not be installed at the surface level should be sunk to a depth of at least 30 cm.

The strength of a runway strip will depend on geological conditions and the height of the water table. It may be necessary to put in drains to reclaim the land. In exceptional cases, a runway strip should be strengthened in the same way as a runway with a grass surface at general aviation airports. In addition, it is necessary to guarantee a gradual rate of change in the reduction of strength from the edge of a runway or runway shoulders to facilitate the smooth deceleration. If the edge of a runway directly abuts a grass strip, it is possible to design wedges of compacted gravel or fine gravel upon which a layer of topsoil is spread for a distance of approximately 5 m from the edge of the runway (see Figure 5.2). There are also new solutions for increasing the runway strip strength based on a mixture of natural constituents and inert polymer fibres which guarantee three-dimensional strength and stability of the root zone. The polymer fibres reinforce and protect the natural turf. A gradual change of the strength is provided also by runway shoulders if they are constructed, as described in Chapter 4.

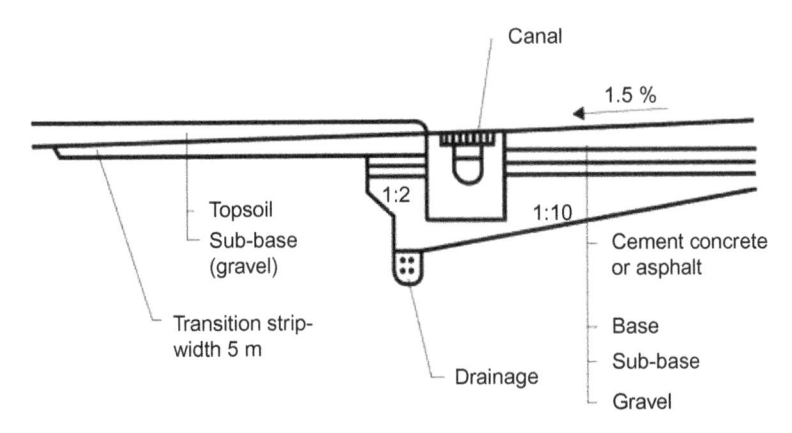

Figure 5.2: Transition strip — gravel and topsoil. *Source*: Čihař (1973).

There are firm recommendations for the control of temporary obstacles in the strip. The extent to which hazards may be created by work in progress should be controlled by the relevant authority. It should take account of the type of aircraft, the runway width, the meteorological conditions and the possibility of using other runways. If work is allowed, the resulting hazards must be promulgated by NOTAM. In Zone 1, within 21 m of the runway edge, work may only take place on one side of the runway. The area under work must be of limited size, and the clearance of engines and propellers must not be compromised. No plant or machinery should operate in this area when aircraft are using the runway. If an aircraft is disabled in this zone, the runway must be closed. Less severe restrictions apply in Zone 2, which extends to the edge of the graded portion of the strip.

If an airport is regularly used by aircraft with jet engines, the portion of a runway strip within a distance of at least 30 m before a threshold should be sealed over the width of the runway to avoid blast erosion. This prevents erosion by the action of exhaust gases and exposure of an edge of the runway pavement. The construction of a light pavement for such a purpose may be similar to that of runway shoulders.

As with other airside dimensions, those specified for a runway strip should be understood as the minimum requirements, and in some cases it may be necessary to widen a runway strip. On the other hand, there may be cases where a smaller width of a runway strip is sufficient; for example, for the portion of a runway intended only for take-off, the so called 'starter-strip'.

5.2. CLEARWAYS

A clearway may be provided to extend the take-off distance available (TODA) beyond the end of the hard surface which defines the declarable length of the available take-off run (TORA). It implies that the ground in the clearway should not project above a plane having an upward slope of 1.25% between the end of the hard runway and the end of the clearway. A clearway is feasible because the last part of a take-off takes place in flight rather than on the ground. It is usual to also provide a stopway, often rather shorter than the clearway but not too short, since aircraft cannot normally use a very unbalanced take-off. To the extent that the aircraft's performance characteristics permit it, the clearway allows a pilot to choose a relatively low decision speed, after which it would be advisable to fly rather than stop if an emergency occurred. Then a relatively shorter accelerate-stop distance is needed and hence a lower cost of prepared surface. A clearway should extend to a width of at least 75 m on each side of the centreline of the runway and its length should not exceed half the length of the runway length available for take-off. It is subject to the same possible limitation as a more normal provision of TODA, that the imaginary take-off surface which starts at the end of the clearway must not be broken by obstructions.

5.3. RUNWAY END SAFETY AREAS

Operational experience and statistical data of accidents where the aircraft has landed short or overshot a runway have shown that the 30–60 m portion of the runway strip which is located off the ends of runways does not provide sufficient protection for aircraft in these circumstances. Some of the most serious overrun accidents occur in the event of an aircraft running off the runway end when the decision to abort a take-off has been taken after the speed has already exceeded the V_1 speed, after which take-off should normally be continued. Such cases occur mostly when a pilot judges that, even at that speed, the take-off cannot be continued because of the nature of an engine failure, or serious vibrations, or when a failure occurs after the normal lift-off speed (V_{LOF}) which makes it impossible to fly, such as an incorrectly adjusted stabiliser trim setting. In that case the aircraft has considerably greater speed than, for example, in the event of an aircraft running off beyond the runway end after landing under poor braking conditions or when a pilot touches down well beyond the touchdown zone, which is normally some 300 m from the threshold.

For those reasons, a runway end safety area (RESA)[1] of at least 90 m length shall be provided at each end of a runway strip of a runway where the reference code number is 3 or 4 and the runway is an instrument one. RESA is defined as an area symmetrical about the extended runway centreline and adjacent to the end of the runway strip primarily intended to reduce the risk of damage to an aircraft undershooting or overrunning the runway.

The latest 6th edition of Annex 14 (July 2013) recommends that, where possible, 240 m of RESA should be provided for code 3 and 4 runways, making together with the runway strip a total of 300 m available to contain an overrun and 120 m for instrument runways where the code number is 1 or 2 or 30 m for a non-instrument runway. The width of RESA must be at least twice that of the respective runway. The length of RESA could be reduced when an arresting system (i.e. EMAS) is installed at the runway end.

Dimensions of runway end safety areas are determined on the basis of statistical data on an occurrence of accidents beyond the runway end. The ICAO accident database shows that about 80% of all overrun and undershoot accidents would be contained within a runway end safety area extending 90 m beyond the end of the runway and double the width of the runway, surrounded by a normal runway strip. Ninety per cent of all such accidents would be contained within a runway end safety area 150 m wide and 300 m long. However the data only give the actual distance that aircraft have overshot the runway, regardless of the available length, the expected distance required for the take-off or landing, and the characteristics of the runway and overrun area. Figure 5.3a shows the difference between the ICAO data and the real extent of the excess distance used up compared with the nominal flight manual performance in a small sample of landing accidents.

A similar set of results is obtained for rejected take-off overruns, as in Figure 5.3b. These differences should be taken into account in the assessment of risk at airports,

---

1. In FAA documents are referred to as Runway Safety Areas.

108   Airport Design and Operation

as it is clear that those airports where a large proportion of landings using the full runway length will be more likely to experience an overrun, other things being equal.

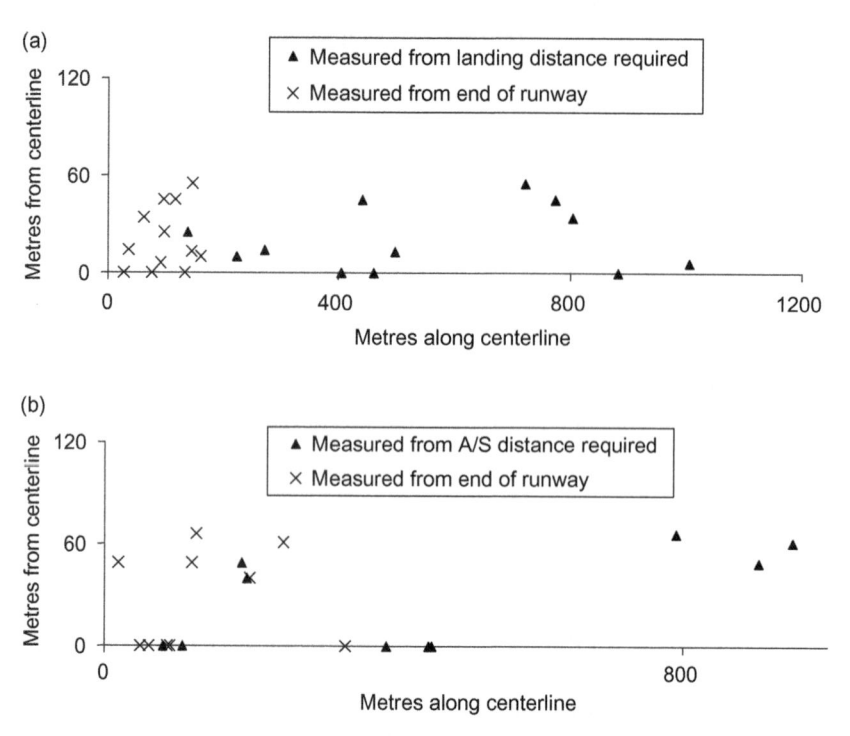

Figure 5.3: (a) Wreckage location relative to runway end (landing). (b) Wreckage location relative to landing distance required. *Source*: Kirkland and Caves (1998).

Further enlargement of a runway end safety area is practically impossible due to the fact that at the majority of airports, the localiser antenna of the ILS is located not further than 300 m beyond the end of the runway. In the event of an aircraft running off into the runway end safety area, it should not sustain such damage as it would have a serious consequence for the occupants; for example, breach of fuel tanks, breakage of fuselage or fracture of undercarriages. Therefore in the runway end safety areas, there must not be any embankments or ditches for roads, railways or water courses. There are also limits to the positive and negative slopes which are allowed.

In some cases there is not enough space to provide a runway end safety area without reducing the declared runway length. Neither is it possible to use for civil purposes the type of equipment commonly used in the Air Forces for stopping aircraft that run off beyond the runway end. The reasons are both the different procedures, particularly the legal position of the pilot-in-command, and the dimensions, mass and the resulting moment of inertia of the commercial aircraft, for which it is technically difficult to provide equipment that would be able to stop them safely.

Many airports built before the current RESA standards were adopted in 1999. Sometimes it is not possible to build up RESA because of limited space (Figure 5.6). There may also be some obstacles such as water courses, roads, railroads and urban areas or steep terrain drop-off. In these cases RESA construction would not be possible, or practical and could require extensive earthworks. Sometimes there are also environmental reasons limiting construction of long enough RESA. In those cases, arrestor beds using an Engineered Material Arresting System (EMAS) could be constructed and be a cheaper solution. EMAS is a 'soft ground arresting system' consisting of a crushable cellular cement material installed on the runway overrun to decelerate an aircraft in an emergency and provides capability to stop an aircraft before reaching an existing hazard (road, railroad, waterway, steep embankment etc.).

When an aircraft is unable to stop on the runway or the runway strip behind it, the aircraft rolls into the EMAS arrestor bed, and is decelerated by the loads applied to the aircraft landing gear as the aircraft wheels travel through the EMAS. As tyres crush the material, it provides a decelerative load, the drag load rather than friction slowing the aircraft down (Figure 5.4).

Figure 5.4: The Dassault Falcon 900 business jet successfully arrested by the ZODIAC-ESCO EMAS system at Greenville Downtown airport on 17 July 2006. *Courtesy*: ESCO ZODIAC; *Photo*: David J. Heald, C.M.

110   Airport Design and Operation

The EMAS characteristics are designed specifically for each particular airport and with respect to the critical aircraft. One of the biggest advantages of EMAS over RESA is its predictable static properties as opposed to RESA which could be influenced significantly by adverse weather conditions. The performance of an overrunning aircraft on a grassy RESA with length of 240 m is unpredictable, as the grass could be wet, muddy or frozen. As a result, even a recommended RESA in poor weather conditions might fail to accommodate an overrunning aircraft because the properties of the ground's surface have been compromised such that it will not support the weight of the aircraft. The depth of the EMAS gradually increases as the aircraft travels into the arrestor bed, providing increasing deceleration when required by heavier or faster aircraft. Aircraft overrun distance is determined by the aircraft size, mass, speed and bed configuration. The cellular cement blocks of the arrestor bed are produced in factory as pre-cast blocks and are then transported to the airport for installation.

There are no problems for aircraft landing short onto the EMAS, because most of the weight is being taken by aerodynamic forces and the tyres will not break through the surface. It would, however, be a problem if the aircraft touched down prior to the EMAS because it would then encounter the thick section first and probably

Figure 5.5:   EMAS installation at Boston Logan Airport. *Courtesy*: ESCO ZODIAC Company; *Photo*: David J. Heald, C.M.

shear off the landing gear. Therefore the EMAS should be as far off the end of the runway as possible and be marked as non-load-bearing pre-threshold surfaces by yellow chevrons (Figure 5.5). The EMAS bed matches up with the width of the corresponding runway, plus 5–9 m for the stepped sides, to facilitate emergency vehicle access and passenger egress, depending on the maximum depth of bed material. EMAS includes a paved rigid ramp, usually 22.5 m long, in front of the arrestor bed.

FAA identifies EMAS as an equivalent to a 1000-foot long runway end safety area and provides guidelines on comparing various runway safety area (RSA) improvement alternatives to the EMAS option. A standard EMAS installation extends 600 feet from the end of the runway. EMAS performance is varied by increasing the depth and the length of the bed and variations of the bed material strength. The FAA specifies the design goal as 70 knots runway exit speed, including the effects of deceleration on the paved portion of the EMAS in front of the arrestor bed.

Snow removal from the EMAS bed is necessary if snow obscures approach lights or affects the ILS localiser (LLZ) performance. If a large, asymmetrical snow accumulation occurs on one side of the bed, it may affect the LZZ signal and the snow must be removed. For snow clearance, a special snow blower tracked vehicle with very low surface pressure must be used to avoid EMAS damage.

According to the FAA[2] by 20 August 2014 EMAS was installed at 79 runway ends at 49 airports in the United States, with plans to install 13 EMAS systems at nine additional US airports, but there are also installations at four airports outside the United States at (two systems at each location): Jiuzhai Huanglong Airport (Sichuan Province China), Barajas-Madrid (Spain), Taipei City (Taiwan) and Kristiansand (Norway) and more are planned in Europe. By September 2014 there have been nine incidents where EMAS has safely stopped overrunning aircraft with a total of 243 crew and passengers aboard those flights, without any casualties. Other types of crushable arrestor beds, similar to truck arrestor beds, have been installed in other countries, including at Manchester airport, that is, arrestor bed made up of crushable pellets.

If safety is to be ensured, it is necessary to consider not only the runway strip and the runway end safety areas. Attention must be given also to the near surroundings of the airport, particularly under the approach or take-of trajectories.

Here there should not be any elevated cement-concrete constructions (even though not penetrating obstacle limitations), sharp embankments for the containment of sewage plants, canals, or transport links. For example, when a Boeing 737 made an emergency approach to the East Midlands Airport in Great Britain on 8 January

---

2. Fact Sheet — Engineered Material Arresting System (EMAS).

112 Airport Design and Operation

1989, the aircraft impacted the embankment before the runway where the M-1 motorway runs through a cutting. The severe impact loads broke the fuselage and the aircraft wreckage blocked the motorway. Forty-seven people died in the subsequent fire. If the motorway had been covered, the accident probably would not have claimed victims. There is a whole range of airports with similar dangerous obstacles which fall outside the direct responsibility of the airport and its designers. Yet this is the location of most remaining fatal accidents, largely because the on-airport protection and rescue facilities are so good.

Figure 5.6: At some airports it is sometimes not possible to build up RESA — Heraklion Airport. *Photo*: A. Kazda.

It is also the area where third parties are likely to be killed or injured. Some countries therefore require that a public safety zone (PSZ) be established off the end of busy commercial runways with the purpose of limiting the allowable land uses. The United Kingdom has recently adopted a new form of PSZ based on research which shows the area of constant risk to be described by contours of similar shape to the noise contours. They have been approximated to thin triangles in the UK regulations, where new dwellings will not be permitted if the individual risk of death is greater than one in 100,000 per year.

6

TAXIWAYS

Tony Kazda and Bob Caves

6.1. FUNCTIONAL CRITERIA AND TAXIWAY SYSTEM DESIGN

A taxiway is a part of movement areas used for aircraft taxiing and is intended to provide a link between one part of the aerodrome and another.

The design of a taxiway width is such that, when the cockpit of the aircraft for which the taxiway is intended remains over the taxiway centreline markings, the clearance distance between the outer main wheel of the aircraft and the edge of the taxiway should be not less than a specified clearance. The clearance depends on the particular code letter and is graduated. It is greater for larger aircraft since the ability of a pilot to follow the centreline is decreased in larger aircraft because of the cockpit height.

It is often difficult to design an optimum system of taxiways. The taxiway system may have a decisive influence on the capacity of the runway system, and thereby also the overall capacity of the aerodrome. At airports with just a few movements an hour a basic taxiway system connecting apron and runway could satisfy the operational needs. However, the taxiway system capacity must be expanded so as not to be a limiting factor of the airport throughput. This is particularly important in the case of extreme runway capacity saturation. The greater the complexity of the runway and taxiway system, the greater the possibility for reducing operating costs through a comparison of alternative taxiway systems. The surface area of taxiways may be greater than the area of runways. For example, in the case of the Osaka's Kansai International Airport, opened in 1994, the taxiways have an overall length of 11.3 km and the runway has a length of 3500 m. Considering that the load bearing strength of a taxiway should be equal to or greater than the load bearing strength of a runway, the construction of taxiways represents an important item in the total investment costs. Therefore, it is necessary to optimise the taxiway system layout to provide efficient taxiing without undue expense.

The taxiways should permit safe, fluent and expeditious movement of aircraft. They should provide the shortest and most expeditious connection of the runway with the apron and other areas in the airport. This minimises time and also the fuel consumption of the aircraft, which has a positive effect upon the environment. The safety of aircraft is enhanced if the taxiways are designed to allow one-way operation, and if crossing other taxiways, and particularly runways, is minimised.

114   Airport Design and Operation

At larger airports the safety issues are also linked with the taxiing distances. If an aircraft is taxied at maximum take-off mass over a distance from 4 to 7 km, depending on aircraft type and outside air temperature, the tyre carcass temperature during take-off can exceed a critical value of 120°C. This can affect the tyre cord strength and increases the risk of tyre failure in particular during an aborted take-off. Because of this every airport master plan, irrespective of the size of the airport development, should recognise the need to minimise taxi distances, in particular for departing aircraft, for both economy and safety. The FAA has issued guidance for the provision of end-around taxiways in order to minimise the need to cross runways. These taxiways are to be provided with a screen to hide aircraft on the taxiway from pilots lined up for take-off so that they concentrate only on observing aircraft which may be actually on the runway (Airports International, April 2006).

In those aerodromes where the number of aircraft movements during the peak hour traffic is relatively small, it is usually sufficient to provide only a short taxiway at right angles to the runway to connect it to the apron. To cope with larger aircraft, it is then usually necessary to provide additional pavement at the ends of the runway to allow the aircraft to turn round. The runway occupancy time is then considerable (Figure 6.1).

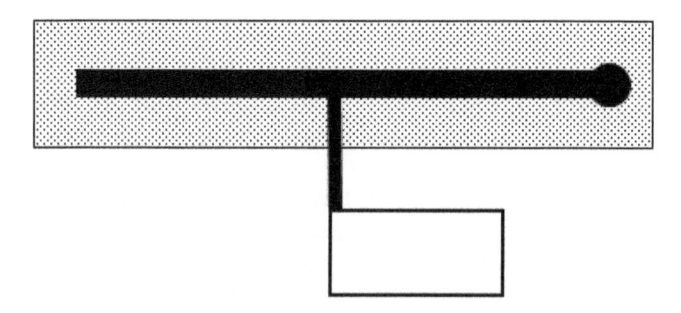

Figure 6.1:   Runway and apron connected with short right-angle taxiway.

If the number of movements during the peak hour traffic exceeds about 12, consideration may have to be given to construction of a taxiway parallel to the runway and right angle connecting taxiways at the ends of the runway. In addition, in the event of a longer runway, several right angle connecting taxiways may be constructed, usually at one third or quarter of the runway length.

The system of a parallel taxiway with right-angle connections may be sufficient for up to 25 movements during the peak hour (Figure 6.2). Take-offs from taxiway/runway junctions and the use of rapid exit taxiways after landing not only reduce taxi distances and runway occupancy time but also increase runway capacity.

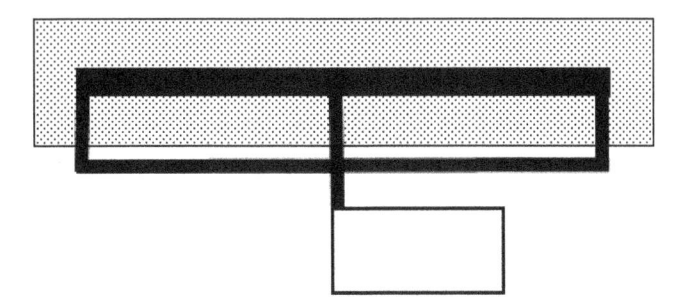

Figure 6.2: System of a parallel taxiway with right-angle connections.

6.2. Rapid Exit Taxiways

To improve the capacity further, it is necessary to construct one or more rapid exit (high-speed exit) taxiways, usually from the preferred direction of the main runway, whose parameters and location need to correspond to the type of operation on the given runway (Figure 6.3).

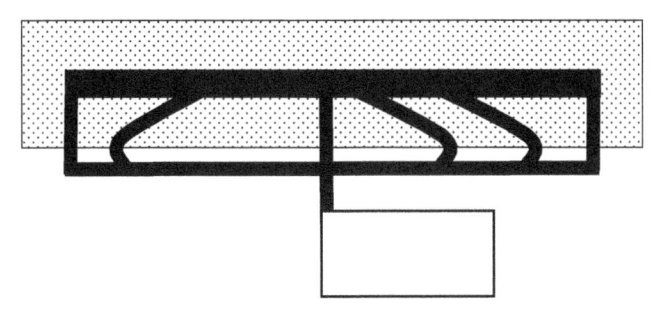

Figure 6.3: System of a parallel taxiway with right-angle connections and rapid exit taxiways.

This type of taxiway is connected to a runway at an acute angle and designed to allow landing aircraft to turn-off at higher speeds than are achieved on other exit (right angle) taxiways, thereby minimising runway occupancy times.

The parameters of rapid exit taxiways, their dimensions and geometrical characteristics are standardised in Annex 14 and in the *Aerodrome Design Manual*, Part 2. The purpose of the standardisation is to allow the pilot to anticipate the conditions at a less familiar aerodrome in order to avoid prolonging the time of occupation of the runway occupancy time by failing to make the best use of the turn-off.

To increase situational awareness, in particular, in low visibility conditions the start of turn could be designated by the Rapid Exit Taxiway Indicator Lights (RETILs).

116   Airport Design and Operation

They usually consist of six yellow lights adjacent to the runway centreline, configured as a three-two-one sequence spaced 100 m apart with the single light positioned at 100 m from the start of the turn for the rapid exit taxiway. The parameters of a rapid exit taxiway, particularly the radius of the turn-off curve, should permit turning off the runway at a speed of up to 93 km/h if the runway code number is 3 or 4 and 65 km/h if the runway code number is 1 or 2, even if the surface of the runway is wet. The correct location of the beginning of the rapid exit taxiway from the landing threshold of the runway is important in obtaining optimum use of the facility. It depends on:

✈ the speed of the aircraft crossing the threshold of the runway,

✈ the deceleration of the aircraft after its touch-down and

✈ initial speed of turn-off.

In line with Doc 8168/I, Part III, Chapter 1, the aircraft have been divided into four categories based on 1.3 times the stall speed in the landing configuration at maximum certificated landing mass overhead the threshold of the runway, as shown in Table 6.1.

Table 6.1: Categories of aircraft according to their speed overhead the threshold.

Category	Aircraft speed overhead the threshold V_{at}[a]
A	Less than 169 km/h (91 KT) IAS
B	169 km/h (91 KT) or more but less than 224 km/h (121 KT) IAS
C	224 km/h (121 KT) or more but less than 261 km/h (141 KT) IAS
D	261 km/h (141 KT) or more but less than 307 km/h (166 KT) IAS

Source: ICAO: Aircraft Operations; Procedures for Air Navigation Services, Volume I, Flight procedures, Doc 8168, OPS /611; 5th ed 2006.
Note: After Concorde ceased operation the current Category E aircrafts are not normally a civil transport aircraft and their dimensions are not necessarily related to V_{at} at maximum landing mass. For this reason, they should be treated separately on an individual basis.
[a] V_{at} — speed at threshold based on 1.3 times stall speed in the landing configuration at maximum certificated landing mass.

For the purpose of rapid exit taxiway design, the aircraft is assumed to cross the threshold at a speed that is equal to 1.3 V_S, where V_S is the stalling speed of the aircraft in the landing configuration with an average gross landing mass (85% of the maximum landing mass). Table 6.2 gives examples of ranking of the aircraft into groups on the basis of their threshold speed.

Table 6.2: Examples of ranking aircraft into categories on the basis of their threshold speed.

Category V_{at}	Aircraft types	Typical aircraft in this category
A	Small single engine	Cessna 170; Dornier Do 228
B	Small multi engine	Beechcraft King Air 100; Fokker F 27; ATR — 72, 42; Saab 340; Dash 8-300
C	Airline jet	Embraer 190; Bombardier CRJ900; B-737; B767-200ER/300ER; B777-200ER; B787; A-320; A-330; A-340-200/300; A-380[a]
D	Large jet	A-340-500/600; A-350-900; A-380[a]; B767-400; B777-300ER; B-747

[a]Depends on the aircraft mass.

The final design of rapid exit taxiways, their number and location, will depend also on other factors, such as aircraft mix, runway slope, aerodrome elevation, meteorological conditions — in particular wind direction, number of movements in peak hour and reference temperature.

It must be emphasised that the number of rapid exit taxiways depends on the types and number of aircraft intended to utilise the runway during the peak period. The location of the start of a rapid exit taxiway may be derived by assuming a constant retardation 'a'. The following shall apply for such a movement:

$$\frac{dv}{dt} = \frac{d^2 s}{dt^2} = a; \quad v(t) = \int a \cdot dt = a \cdot t + C_1$$

$$s(t) = \int (a \cdot t + C_1) \cdot dt = \frac{1}{2} \cdot a \cdot t^2 + C_1 \cdot t + C_2$$

With the beginning of path $s=0$ to the point with speed $v=0$ in the moment of time $t=0$ (see Figure 6.5) then $C_1 = C_2 = 0$ and the following expression is generally valid:

$$v(t) = a \cdot t; \quad s(t) = \frac{1}{2} \cdot a \cdot t^2$$

if the direction to the right of the full-stop point (i.e. $s(0)=0$) is considered to be positive and at the same time the 'acceleration' 'a' negative. This is illustrated in Figure 6.4, the meaning of which is expressed by the following:

118 Airport Design and Operation

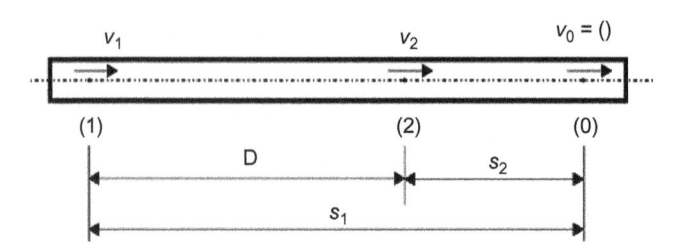

Figure 6.4: Rapid exit taxiway distance from a threshold.

After substitution in s_1 we will get

$$s_1 = \frac{1}{2} \cdot a \cdot t_1^2; \quad v_1 = a \cdot t_1; \quad t_1 = \frac{v_1}{a}$$

$$s_1 = \frac{1}{2} \cdot (a \cdot t_1) \cdot t_1 = \frac{1}{2} \cdot v_1 \cdot t_1 = \frac{1}{2} \cdot \frac{v_1^2}{a}$$

As an analogy we can derive

$$s_2 = \frac{1}{2} \cdot \frac{v_2^2}{a}$$

The final expression is

$$D = s_1 - s_2 = \frac{v_1^2 - v_2^2}{2 \cdot a}$$

where:

A	Mean deceleration of the aircraft after landing (m/s^{-2})
S_1	Total braking distance from the touch-down (1) to the full-stop point (0) (m)
$S(t)$	Distance at time (t) when the beginning of the path is in (0) (m)
$v(t)$	Speed at time (t) (m/s^{-1})
v_1	Speed over a threshold of the runway (m/s^{-1})
v_2	Design speed of turn-off at location (2) (m/s^{-1})
T	Time (s)
t_1	Time required to stop in distance s_1 from the approach speed v_1
t_2	Time required to stop in distance s_2 from the speed v_2
D	Rapid exit taxiway distance from a threshold

The deceleration should not exceed 1.5 m/s^{-2} in order to avoid passenger discomfort. The intersection angle of a rapid exit taxiway with the runway should not be greater than 45° or less than 25° and preferably should be 30°. It is a common practice for the entry to take the form of a spiral, so that the lateral deceleration requirement is not too severe while the aircraft is moving at high speed.

This makes it more likely that the exit will be used when the runway friction is reduced. A rapid exit taxiway should include a sufficient straight portion after the turn-off curve so that the pilot can stop the aircraft before the next taxiway intersection. Figure 6.5 illustrates an example of a rapid exit taxiway.

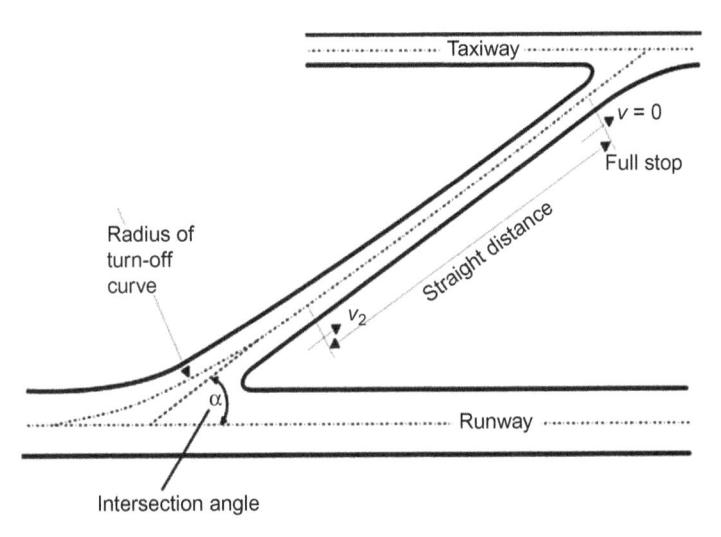

Figure 6.5: Characteristics of rapid exit taxiway.

To specify the beginning of the rapid exit taxiway turn-off point from the threshold, methodology, known as the Three Segment Method, was developed. It is based on the landing performance analysis and empirical assumptions. The first segment covers the flare part from the threshold to touch-down. The second — transition is the distance from the main gear touchdown to stabilised braking configuration. The third is the distance needed for deceleration in normal braking conditions to nominal turn-off speed.

The taxiway system design and positions of rapid taxiway exits should be evaluated in the complex airport layout in particular, if other runways are positioned at an angle to the main runway; but also apron and terminal position could lead to compromises in the design of taxiways.

It is often necessary, where the traffic volume is high, to establish two parallel taxiways to facilitate one-way flow, together with holding bays located near the ends of

120   Airport Design and Operation

the runway. Holding bays allow the flight controller to bypass aircraft and thus optimise the sequence of take-offs depending on the aircrafts' speeds, weights and departure routings.

6.3. Taxiway Separations

The minimum safe separation distance between the centreline of a taxiway and the centreline of a runway is defined as a standard in Annex 14. The actual distance depends on the code number of the runway and the category of its approach aids, and it is such that the wingtip of a taxiing aircraft will not encroach into the runway strip.

With the introduction of a Code F in Annex 14, the distance might have to be as much as 190 m. Similarly, the minimum safety separation distances are specified between parallel taxiways; these should be 97.5 m to allow unimpeded use by aircraft of up to 80 m wingspan (see Table 6.3).

The effect of these separation standards for Code F is shown by the total investment of £450 million that Heathrow made in order to accept the A380. Many of the piers had to be shortened and many of the stands closest to the runways were lost because the inner taxiway clearance line had to be moved 40 m closer to the buildings. The runways were widened and new shoulders constructed for the runways and taxiways. A new pier was built with four A380 stands. This may be contrasted with the A380 acceptance at Hong Kong where, because it is a third-generation jet airport rather than a pre-jet one, only minor changes have been necessary to the airside, like moving objects from the Code F taxiway strips and changing stop-bars. The separation requirement between parallel taxiways is based on the wingtip clearance requirement when an aircraft has deviated from the taxiway centreline (Figure 6.6).

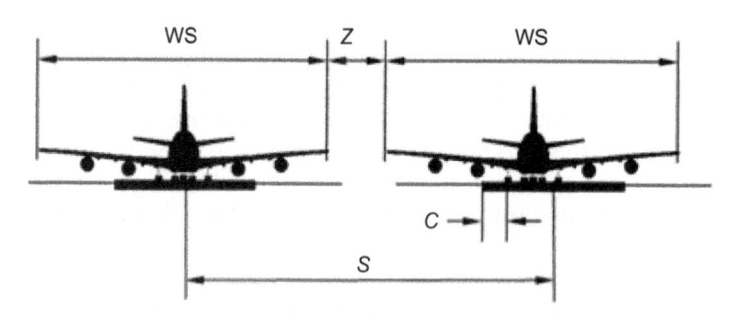

Figure 6.6:   Parallel taxiways separation. *Source*: ICAO Doc 9157 Part 2.

The formula for the separation distance in this case is:

$$S = WS + C + Z$$

Table 6.3: Taxiway minimum separation distances.

Code letter	Distance between TWY centreline and RWY centreline [m]				TWY centreline to TWY centreline [m]	TWY, other than aircraft stand taxilane, centreline to object [m]	Aircraft stand taxilane centreline to object [m]
	Instrument runways code number							
	1	2	3	4			
A	82.5	82.5	–	–	23.75	16.25	12
B	87	87	–	–	33.5	21.5	16.5
C	–	–	168	–	44	26	24.5
D	–	–	176	176	66.5	40.5	36
E	–	–	–	182.5	80	47.5	42.5
F	–	–	–	190	97.5	57.5	50.5

Source: ICAO Annex 14, Aerodromes, Volume I, *Aerodrome Design and Operation* 6th edition.
Note: Data for non-instrument RWYs are not shown.

122 Airport Design and Operation

where WS is wingspan; C is the clearance between the outer main gear wheel and the taxiway edge (maximum allowable lateral deviation) and Z is wingtip clearance. Separation distances in FAA taxiway design standards are different from the ICAO ones. They are based on allocation of aircraft into 'airplane design groups' and FAA uses different formulas from ICAO for calculations.

In line with the ICAO and FAA standards, taxiways on aprons and stands are split into two groups: *apron taxiways* which are located directly on aprons and *taxilanes* providing access to aircraft stands (Figure 6.7). Standards for both of them are generally the same as for the 'classic' taxiways except the separation criteria. Separation can be less stringent in the case of apron taxiways and taxilanes because the aircraft taxi more slowly in the apron areas.

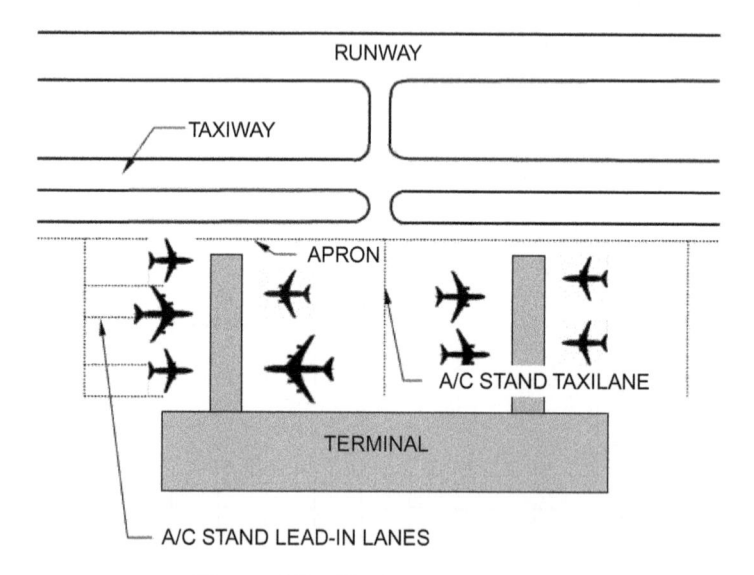

Figure 6.7: Taxiways on apron.

The category of approach aid affects the protection which must be given to those electronic aids. ILS signals may suffer interference from a stopped or taxiing aircraft in ILS critical and sensitive areas particularly during Cat II or Cat III system operations (see also Chapter 3, part Low Visibility Procedures). The information on the ILS and MLS critical and sensitive areas are given in Annex 10, Volume I, Attachments C and G.

6.4. TAXIWAY GEOMETRY

Depending on the code letter of the runway, minimum margins of safety between 1.5 and 4.5 m should be provided between the outer main gear wheel edge and the taxiway edge when the cockpit is over the taxiway centreline.

The resulting taxiway width of the taxiway is less than the width of corresponding runway because aircraft taxi speed is considerably slower than the speed on take-off or landing and the aircraft is always in contact with the ground. A straight portion of a taxiway should have a width of 7.5 m (code letter A) to 25 m (code letter F). On future aircraft the outer main gear wheel span is expected to increase up to 20 m. With wheel-to-edge clearance of 4.5 m the future taxiway width for planning purposes reaches 29 m.

The taxiway width, W_T is based on a formula:

$$W_T = T_M + 2C$$

where W_T is the taxiway width on the straight parts of the taxiway; T_M is the outer main gear span; C is the clearance between the outer main gear wheel and the taxiway edge.

The clearance value depends on the taxiway code letter (Figure 6.8).

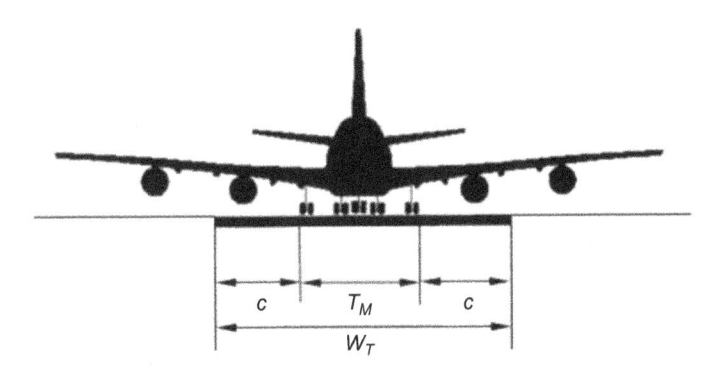

Figure 6.8: Taxiway width requirements. *Source*: ICAO Doc 9157 Part 2.

Taxiways need to be widened with fillets where they have sharp curves so that the necessary safe separation distance between the outer main gear wheel edge and the runway edge may be maintained when the nosewheel is tracking the centreline (see Figure 6.9).

There are commercial programmes available that simulate the tracks of all the wheels for all the main types of aircraft. Occasional movements of an aircraft with greater gear track than that for which the taxiway was designed can be accepted by steering the nosewheel outwards from the centreline (Figure 6.10).

As with runways, taxiways may also be provided with shoulders. Since the speed of aircraft moving on a taxiway is quite slow, the main function of the shoulders is to prevent ingestion of stones, debris and other foreign objects by an aircraft's engines that overhang the edge of a taxiway. A taxiway shoulder is a light pavement providing a transition between the full strength pavement and taxiway strip. It also allows easier snow removal if taxiway edge lights are installed, and also enables the occasional passage of emergency vehicles. The width of a taxiway shoulder is determined by the need to meet these requirements and the aircraft's characteristics.

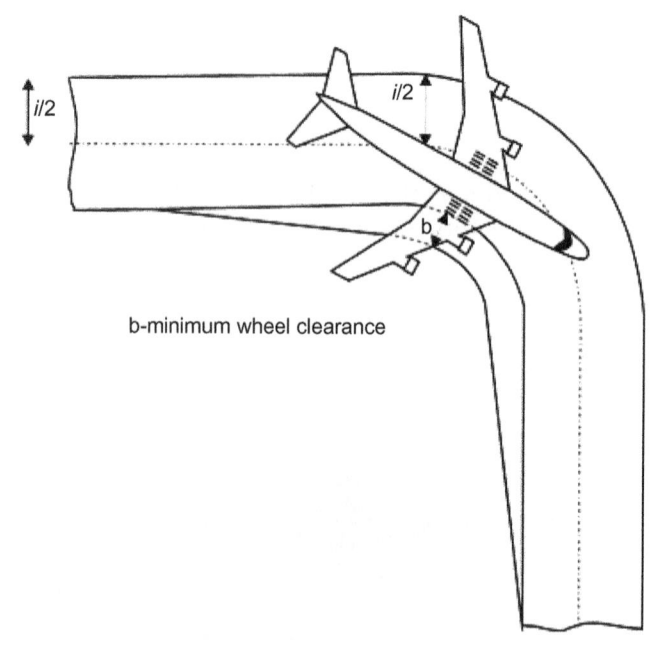

Figure 6.9: Taxiway widening to achieve minimum wheel clearance on curve.

Taxiway strips extend from the edge of a taxiway. Their function is similar to that of a runway strip. It should provide an area clear of objects to protect an aircraft operating on the taxiway and reduce the risk of damage to an aircraft accidentally running off the taxiway. For a specified distance from the centreline of a taxiway, the grass surface of the taxiway strip should be well-maintained, without obstacles and the prescribed slopes of the strip should be maintained.

Other physical characteristics of a taxiway, including longitudinal slopes, transverse slopes and a change of the transverse slope, also depend on the code letter of the associated runway. A further requirement is for clear visibility from any point on the taxiway surface to a point on the taxiway 150 or 300 m away, measured at a height of 1.5 or 3 m, the distances and heights depending on the code letter of the respective runway.

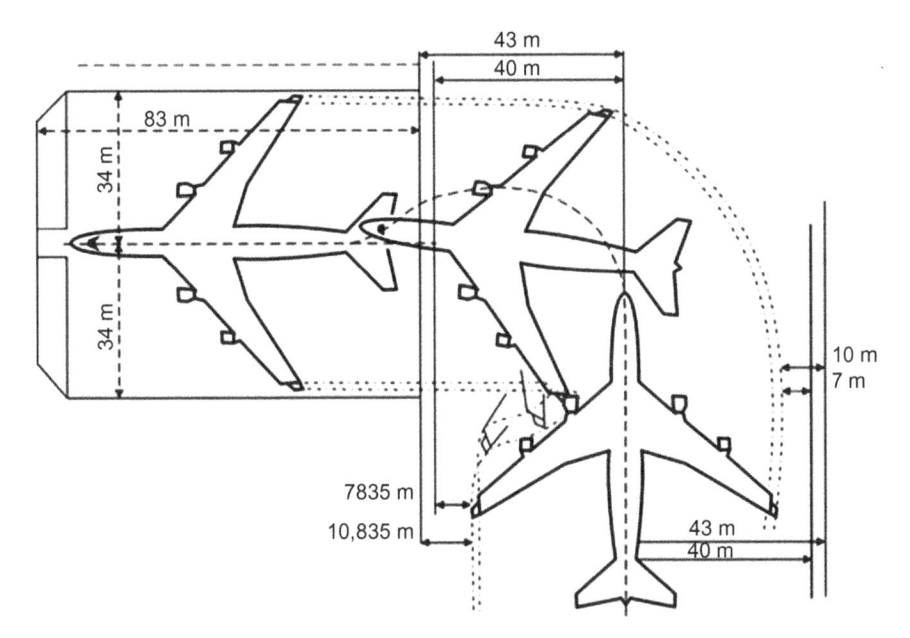

Figure 6.10: Realigning the front undercarriage guide line outwards from the centreline for B-747-400.

At many large airports there are limited possibilities for runway or taxiway system expansion. This sometimes leads to concepts where taxiways must bridge other transport infrastructure such as motorways, roads, railway lines etc. Besides the basic requirements on the bridge structural design, the taxiway layout must guarantee that taxiing on bridges does not impose any difficulties on pilots during day or night operations, low visibility or other adverse conditions.

A taxiway on a bridge must fulfil all requirements for standard taxiways. Besides this, taxiways on bridges should also:

+ Be always located on a straight part of the taxiway. The length of the straight sections should be at least twice the wheel base of the largest aircraft in addition to the length of the bridge. This facilitates a pilot in aligning the aircraft on the taxiway centreline before crossing the bridge.

+ All surface modes that must be bridged should be concentrated in one place to minimise the number of taxiway bridges.

+ Protection against jet engine blast should be provided on places where a taxiway bridge crosses other transport mode structures. For this purpose, light cover structures such as grid or bar constructions which reduce initial jet blast are usually used.

If a runway requires a bridge, rapid exit taxiways should not be located on the bridge.

Sometimes it is not clear which aircraft types will use the airport in the future. To prevent a costly reconstruction once a larger aircraft begins to operate on the airport the taxiway bridge should be designed for a higher code letter.

The taxiway strip on the bridge structure has the same characteristics and bearing strength as the taxiway pavement. This facilitates not only snow clearance during the winter season but also emergency vehicles' access to the bridge.

7

APRONS

Tony Kazda and Bob Caves

7.1. APRON REQUIREMENTS

Aprons are designed for parking aircraft and turning them around between flights. They should permit the on and off loading of passengers, baggage and cargo, and the technical servicing of aircraft including refuelling.

The requirements for the construction of aprons are similar to those of the other reinforced surfaces. It is mentioned in Chapter 8 that the aprons are the most heavily loaded of all the movement area pavements. The aircraft has its maximum mass on the apron just prior to departure. The surface is subject to concentrated point loads from the wheels of standing or slowly moving aircraft. In addition, it is dynamically stressed by vibrations after starting up the engines. Therefore, when calculating the apron pavement thickness, a safety factor of 1.1 is used.

In the past, concrete was used almost exclusively for apron pavements. The original concrete aprons are often renovated by covering them with layers of asphalt. Asphalt is also commonly used for aprons on aerodromes intended for use by small or medium-sized aircraft. However, asphalt pavements are frequently damaged by spilled kerosene or gasoline if it remains on the surface for even a short period of time. This can be partially solved by treating the top asphalt layer with special sealants. It is mentioned in Chapter 8 that there is increased use of block paving for aprons.

In order that aircraft can park and taxi out under their own power, the slopes should be minimised but still be sufficient to allow the surface water to drain adequately. In the aircraft parking area, the maximum slope of the apron should not exceed 1%. To enhance apron fire safety, in particular with respect to aircraft fuelling, apron pavements should slope away from a terminal to prevent the spread of the fuel fires in the direction of the terminal building.

The apron is a bridging point between the runway system and the terminal building. The location of the apron and its aircraft stands should allow convenient access. The following are some of the basic requirements that should be objectives when designing the apron:

- Location of the apron to minimise the length and complexity of taxiing between the runway and the stands

128   Airport Design and Operation

- The apron should permit mutually independent movements of the aircraft on to and off stands with minimum delay
- On the apron it must be possible to locate a sufficient number of stands to cope with the maximum number of aircraft expected during the peak hour
- The apron should be adequate to allow quick loading and unloading of passengers and cargo
- The apron should be designed so that there is sufficient space for the turnround activities to be performed independently of activities on an adjacent stand
- The apron area should be adequate to provide sufficient space for parking and manoeuvring of the handling equipment and also for the technical personnel
- There should be a safe and effective system of airside roads for technical equipment to access the stands, preferably avoiding the need for aircraft to cross them to access the taxiway system
- They should be clearly marked with the width of each lane able to accommodate the widest piece of ground equipment
- The negative impacts on the workers' environment, particularly safety, noise and exhaust gases should be minimised, the emphasis being the health and safety of the staff and passengers accessing aircraft across the apron
- The possibility of further extension of the runway system, aprons and buildings should be considered.

Each of these requirements has an effect upon the final design of the apron.

In addition to the passenger terminal aprons, there are also other special purpose aprons such as general aviation and cargo terminal aprons satisfying particular requirements. In addition to them, some airports may also need a certain number of stands located at a remote parking area where aircraft can be placed for extended periods. These aprons can be used for periodic servicing or maintenance of temporarily grounded aircraft. Another requirement is for a isolated aircraft parking position where aircraft suffering a terrorist threat may be isolated (see Section 7.7). Regardless of the apron function, apron physical characteristics relating to safety and geometry are universal for all apron types.

7.2.   APRON SIZING

The above objectives can only be achieved if the apron size and the positioning of the stands are adequate to permit expeditious handling of the aerodrome operation during predicted peak hour traffic levels.

The appropriate apron size depends on the types of aircraft which are intended to use the apron, but the shape and apron layout also depends upon a number of other geometric considerations. In addition to the aircraft stands, the over-all apron area also includes space needed for apron taxiways and taxilanes, safety clearances, space for blast fences, service roads and areas designated for ground equipment and vehicle placement.

The size and shape of land available also influence the stand types and apron layout concept. The diagram (Figure 7.1) explains how the Annex 14, Volume I separation standards are derived. The minimum separation distance between the aircraft wing tip and an object when the aircraft deviates from the apron taxiway centreline is defined by a formula:

$$S = \frac{WS}{2} + D + Z$$

where WS is the wing span; D is the lateral deviation of the aircraft's longitudinal axis from the taxilane centreline; Z is the wing tip clearance; C is the clearance between the outer main gear wheel and the taxiway edge at the maximum allowable lateral deviation from the taxiway centreline.

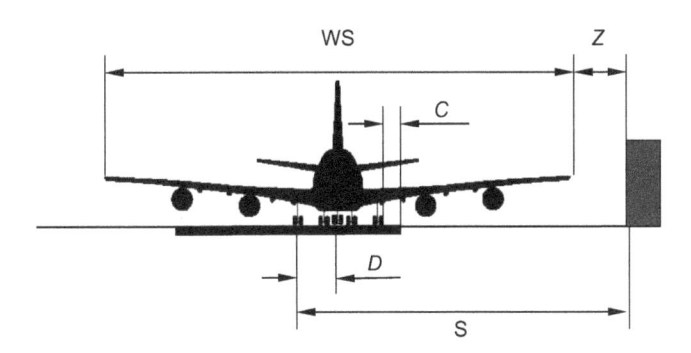

Figure 7.1: Apron taxiway to object geometry.

Areas required for contact stands vary from 2,200 m² for Code B regional jets to 15,000 m² for Code F A380. The stand depths required vary from 30 to 85 m for Codes B and F, respectively.

Each aircraft type needs sufficient stands in the correct position, and possibly a specific manner of stand operation, either nose-in or inclined for self-manoeuvring. There may also be specific requirements for technical servicing (e.g. refuelling), for technical equipment, and for guidance to the parking position when nose-in operation is used. The accuracy of the aircraft guidance on the apron may have an effect upon the size of the apron.

130   Airport Design and Operation

The total apron area depends also on the type of the operation prevailing on the aerodrome and the occupancy time of stands. For some airline operators, the aerodrome may serve as a base; for others, as a departure/arrival aerodrome or only as a transit stop. The based aircraft normally have a greater occupancy time between arrival and departure, so much so that it may be better to consider towing them off to remote stands.

In order that all the factors that influence the apron size may be considered, particularly with large aerodromes, it is advisable to simulate the operations on the apron. Similarly, the operational control of the apron, for example stand allocation, is computerised at large airports.

7.3. APRON LOCATION

Theoretically, the most efficient location for the apron is at a distance of 1/3 length of the runway from the main runway threshold (Figure 7.2), based on a normal split of use of the two runway thresholds.

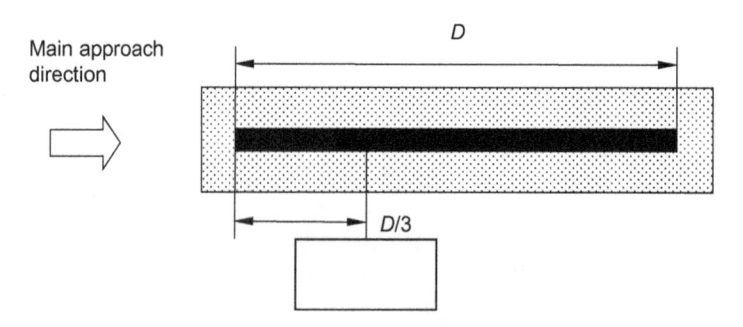

Figure 7.2:   Apron location.

When determining the apron location and orientation, the following objectives should be taken into account:

✈ Minimum length of taxiing of the aircraft

✈ The shortest distance possible from the walkway in front of the terminal building to and from the aircraft

✈ Minimum impact of the engine exhaust on the terminal building

✈ Possible expansion of the apron as well as the terminal building.

In the majority of cases, the apron is constructed directly in front of the terminal building, allowing the stands to be served by airbridges if they are deemed to be necessary. If bridges are not provided, some other safe way of getting the passengers to and from the aircraft will be needed, using well-policed walkways or buses. Preferably, the airside service road runs under the airbridges; otherwise it has to run between the back of the stands and the taxiways.

In some other cases, the whole apron, or a section of it, is constructed at a considerable distance from the terminal building. Then it is necessary to provide some form of transport of passengers between the terminal building and aircraft. This increases the operating costs. There are many reasons why an apron might be located remotely.

It may be to try to create a comfortable ambience in the terminal building and to reduce walking distances as with busses at Milan Linate or the mobile lounges at Montreal's Mirabel airport, a need to earmark a section of the apron for longer-time parking of aircraft as at Washington Dulles International Airport, to generate the space to handle large aircraft as at Los Angeles, or to provide a cost-effective way of handling the additional peak hour traffic as at Munich or Rio Galleao. In the majority of cases, it has been the practice for the airport operator to allocate the contact stands (those directly adjacent to the terminal building) for high-capacity aircraft and scheduled airlines, particularly if they are part of a hubbing operation.

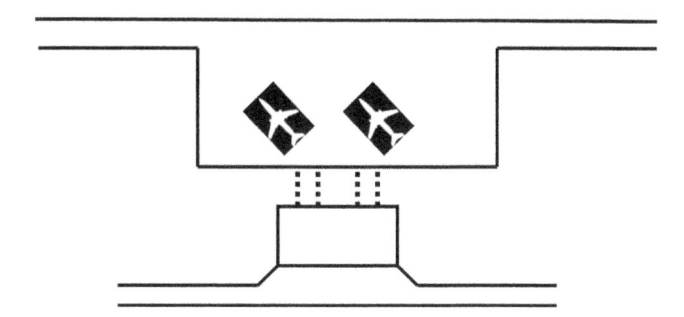

Figure 7.3: Simple concept. *Source: ICAO: Aerodrome Design Manual, Part 2, Taxiways, Aprons and Holding Bays, Doc 9157-AN/901, 4th ed. 2005.*

7.4. APRON CONCEPTS

The geometric and manoeuvring characteristics of aircraft make it practically impossible in most cases to locate all the stands required for peak traffic directly adjacent to the central processing part of the terminal building. It is therefore necessary to generate other solutions. Several basic concepts that have developed over time may be identified, depending on the total size of the airport. Each concept has its advantages and disadvantages, so the solution is often a compromise and a combination of the basic concepts discussed below. Apron design must be consistent

132   Airport Design and Operation

with the adjacent terminal. Apron and terminal design is an iterative process where the optimum combination of apron and terminal concepts are analysed at the same time.

7.4.1.   Simple Concept

This concept is used normally at very small airports with a few movements of commercial aircraft a day. Stands are always the 'self-manoeuvring' type. Aircraft parking positions (angled nose-in or nose-out) usually depend on the terminal frontage and the slope of the apron to minimise engine jet blast or propwash on the terminal building (Figure 7.3).

7.4.2.   Linear Concept

At many airports the simple concept develops gradually to the linear concept. Individual stands are located along the terminal building (Figure 7.4) as at Munich and Roissy-Charles de Gaulle-Aérogare 2 and 3. A modification of that concept may be found in large airports where the stands are placed along several parallel passenger loading piers (or satellites) that are connected with one another and also with the central terminal by a transport system, as at Atlanta Hartsfield, and Heathrow Terminal 5 or Brussels airport. An advantage of the linear apron concept is the simple access from the terminal building to the aircraft, a simple installation of the passenger loading bridges and sufficient space for technical handling equipment and staff at the level of the apron.

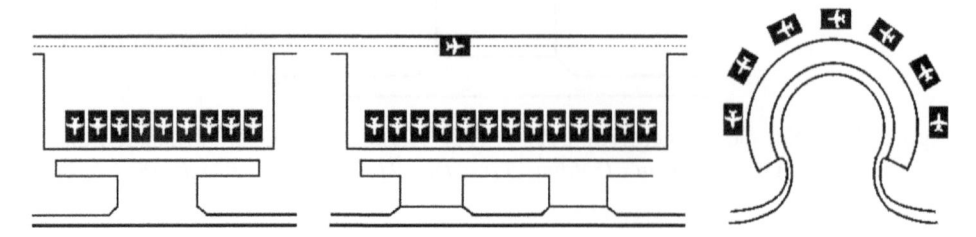

Figure 7.4:   Linear concept. *Source: ICAO: Aerodrome Design Manual, Part 2, Taxiways, Aprons and Holding Bays, Doc 9157-AN/901, 4th ed. 2005.*

A service road is usually placed between the terminal frontage and the apron, for example at Brussels Airport to facilitate ground vehicle and handling equipment circulation. A disadvantage in larger airports may be the large distance between the extreme stands and the central processing point in the terminal building, and sometimes an even larger distance to another stand for transfers between airlines.

The latter problem may be solved by people movers. There is now pressure on airports to rearrange the stand allocations to ensure short transfer distances between flights within an airline grouping, so giving the carrier's hub a competitive advantage. Generally the linear concept provides a very flexible solution where the apron could be easily expanded without disrupting the flow of traffic. Modifications of the linear concept are also successfully implemented at large airports as London Heathrow Terminal 5 with its piers/satellites T5B and T5C, Washington Dulles International Airport with the mid-field terminals or Hartsfield−Jackson Atlanta International Airport but effective ground transportation allowing passenger transfers between terminals is inevitable for those airports.

7.4.3. Open Concept

In this concept, the stands are located on one or more rows in front of the building (Figure 7.5). One of the rows may be close-in, but most will be a long way from the terminal. The transport of passengers to the distant stands is provided by buses or mobile lounges, with only a short walk for passengers. Milan Linate is an example where almost all the aircraft stands are on open aprons. The concept allows many aircraft to be served from a very short terminal frontage. Apron location can be optimised with respect to aircraft operations, that is close to the runway to minimise taxi distances and fuel burned. Expansion is easy. Servicing is usually done from islands set out in the middle of the aprons.

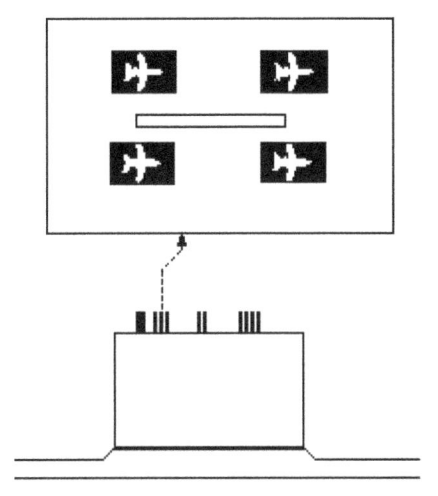

Figure 7.5: Open concept. *Source: ICAO: Aerodrome Design Manual, Part 2, Taxiways, Aprons and Holding Bays, Doc 9157-AN/901, 4th ed. 2005.*

134 Airport Design and Operation

A main disadvantage is the need to provide transport to distant stands for all passengers, requiring a large workforce and fleet of buses or mobile lounges. The length and lack of reliability of these bus trips makes the concept unsuitable for hub operations with transfer passengers. Flights are subject to a very early close-out. Another disadvantage is the large number of additional movements on the apron, increasing the possibility of accidents with aircraft and other ground vehicles.

7.4.4. Pier Concept

In many large airports, the introduction or extension of piers was the most convenient way of providing a greater number of contact stands and to increase the capacity of the airport while providing weather protection for the passengers.

The shape of the passenger loading piers varies and depends on the space available at the airport (Figure 7.6). Amsterdam Schiphol and Frankfurt airport, are classic examples of such a solution. Piers have the advantage of keeping all the gates under one roof, allowing direct contact with the central processing area and a relatively simple navigating task for transferring passengers. The footprint of the terminal and apron complex can be kept quite compact, particularly if the piers are double-sided. For this apron concept, only nose-in stand types are used. Aircraft can be parked at the stand either angled to reduce stand depth, or more normally perpendicular to the pier.

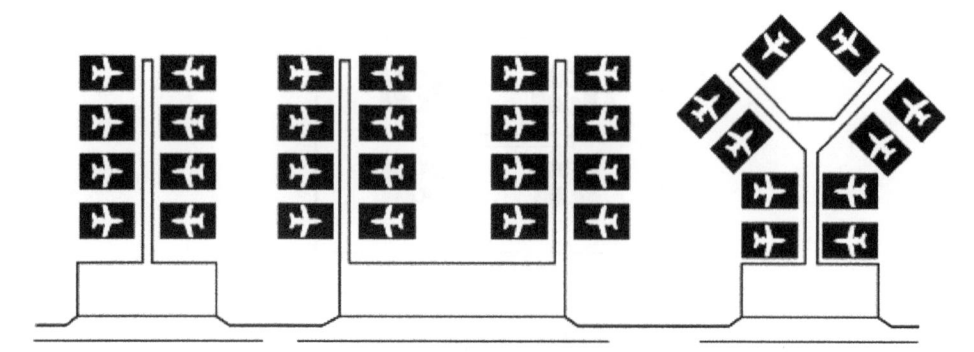

Figure 7.6: Pier concept. *Source*: *ICAO*: *Aerodrome Design Manual, Part 2, Taxiways, Aprons and Holding Bays, Doc 9157-AN/901, 4th ed. 2005.*

Piers generally involve the aircraft having to taxi into cul-de-sacs to get to the stands (Figure 7.7). The limited space leaves little extra room for aircraft handling.

There is usually only a small parking bay for equipment and service personnel under the piers. In the confined space, the effects of noise and exhaust fumes create poor working conditions for the staff. Any increase in aircraft size tends to result in congestion on the cul-de-sac taxilanes, particularly if they have only a single lane. It is imperative to have dual lanes if there are more than 10 or 12 stands in the cul-de-sac. Future larger aircraft operation must be considered during the planning process to avoid costly reconstructions. The concept is probably able to cope up to 45 mppa, but distances become very long.

Figure 7.7: Pier system at London Heathrow Airport with cul-de-sacs — status in 2008. *Photo*: A. Kazda.

7.4.5. Satellite Concept

In this concept, each of the remote passenger loading satellites is connected with the terminal building by underground tunnels or by overhead corridors, as in Figure 7.8. The satellites may be any shape from linear as at Atlanta Hartsfield to circular as with Charles de Gaulle Terminal One (CDG T1) or Genève Aéroport. Typically, the number of stands at circular satellites varies between four and eight aircraft but the linear ones may well have 20 stands per side. The larger ones are usually sited in mid-field, their length being limited by the separation between the runways. Satellites, as opposed to unit terminals, imply that the processing takes place centrally, but in some cases there is a degree of decentralised processing, particularly with security screening. Given a suitable gate assignment system, this concept is good for hubbing, except for international/domestic transfers that require clearing through the central terminal.

There is ample space for aircraft servicing, particularly around a circular satellite. Also push-back operations are simple and safe. The satellite concept avoids

136   Airport Design and Operation

cul-de-sacs and their disadvantages. However, it requires a larger total apron space. Since the distance from the terminal building to the aircraft is considerably extended, it is necessary to transport passengers between the terminal building and the satellites, usually with a high frequency automated people-mover, though sloping moving walkways are used at CDG T1. Connections are often below the surface to allow the whole perimeter to be used for parking, but sometimes the connection is over-ground, allowing some contact stands to be located along the connectors. This can be an efficient use of space. Depending on the size and the shape of the satellite, aircraft can be parked radially or parallel to the sides of the satellite.

Figure 7.8: Satellite concept. *Source: ICAO: Aerodrome Design Manual, Part 2, Taxiways, Aprons and Holding Bays, Doc 9157-AN/901, 4th ed. 2005.*

7.4.6. Hybrid Concept

At many airports combination of two or more above mentioned concepts is usual. During the summer peak season it is quite common to park some, especially charter aircraft, on the remote apron and transport passengers by busses or transporters to the aircraft stands. A combination of satellite and linear concept could be found at Genève Aéroport or pier concept and linear concept at Frankfurt Airport Terminal 1 (pier) and Terminal 2 (linear).

7.5. STAND TYPES

There are several types of aircraft stands. The two most common are 'taxi out' and 'nose-in' stands. The 'taxi out' stand is also sometimes referred to as 'self-manoeuvring'. The aircraft taxis in and out under its own power. The alternative is a 'nose-in' or 'tractor assisted' stand. The aircraft taxis in under its own power, but to leave the gate it must be pushed back into a specific position usually through a 90° turn onto the adjacent taxilane. In most cases, the engines may not be started until the push-back is complete. The separation clearance between the aircraft parked on the stands and the adjacent taxilanes is defined by formula (Figure 7.9).

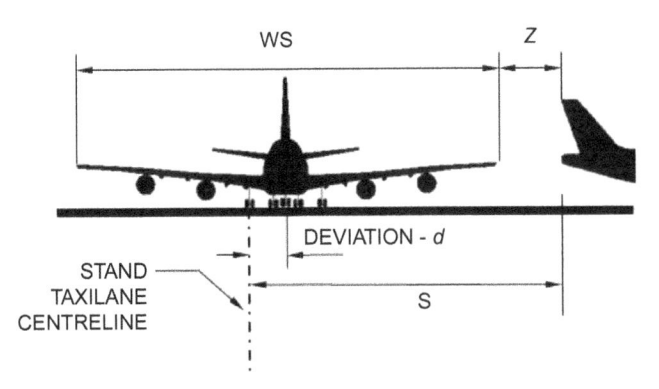

Figure 7.9: Aircraft stand taxilane to object geometry. *Source*: ICAO: *Aerodrome Design Manual, Part 2, Taxiways, Aprons and Holding Bays*, Doc 9157-AN/901, 4th ed. 2005.

Taxi out (or *self-manoeuvring*) aircraft stands are mostly used at small airports with a small number of movements. Advantages of the taxi out aircraft stands are their low operating costs, in the majority of instances daily inspection of the stand surface and marking of the apron is sufficient. They have good flexibility and only require a marshaller, whereas the nose-in arrangement is usually equipped with a docking guidance system and needs the push-back equipment and staff for the push-back operation. This type of operation makes it almost impossible to use a passenger loading bridge, approximately 20% more space is required, and jet blast during the aircraft turn can be a problem.

$$S = \frac{WS}{2} + d + Z$$

where WS is the wing span; d is the lateral deviation from the taxilane centreline; Z is the wing tip clearance.

Approximately 10 stands of the nose-in type may be positioned in the same space as eight taxi out aircraft stands, thereby increasing the apron utilisation. The self-manoeuvring stands may overlap one another but, if they do, simultaneous movements of the aircraft on adjacent stands are not possible, as shown in Figure 7.10 where d is the distance between the stands turning centres; R is the aircraft turning radius; b is the stands overlap; y is the alignment distance.

138   Airport Design and Operation

The size of the taxi out aircraft stand depends on the turning radius, wing span and the fuselage length of the aircraft. When the aircraft turns, a so called safety distance between the wing tip and an obstacle should be ensured, as shown in Figure 7.11. When the aircraft is turning, particularly if it has a multi-wheel main undercarriage, there is a considerable shear stress on the pavement. Pavements with an asphalt surface are often damaged in this way in particular during high summer temperatures. Another disadvantage of the power-out stands is the effect on the terminal building, on other objects, or on the traffic on the apron produced by the engine's exhaust plume at break-away thrust and during turning. In order to start the aircraft rolling, the break-away thrust requires up to 50%–60% of the maximum continuous thrust, while turning requires approximately 25%–30%.

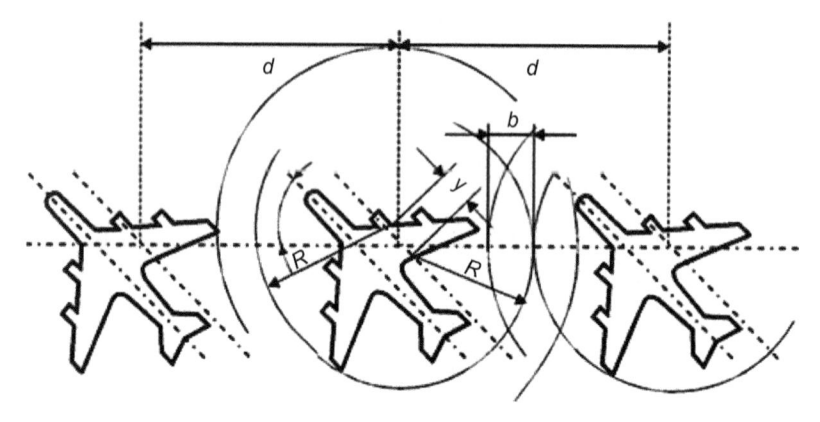

Figure 7.10:   Individual stands overlap.

In contrast, in normal taxiing of aircraft with high bypass ratio engines, there is often too much power even at ground idle settings. Figure 7.12 gives velocities of the exhaust gases of the aircraft B-767-200 at break-away on the stand.

Nose-in aircraft stands are normal at airports with a high density of operations. Besides being space-efficient, they also facilitate the use of passenger loading bridges, which may be apron drive bridges with 3° of freedom or nose loaders which do not have articulation in azimuth, these latter requiring a more accurate positioning of the aircraft. They can all mate with the second door on the left side of the aircraft if there is sufficient vertical flexibility to reach down to the smaller jets.

At some airports more airbridges are used to speed up the movement of passengers, even in some cases reaching over the wing to the rear door, though IATA does not recommend this. For example the third bridge to the A380 could save about 19 minutes on boarding and 12 minutes on disembarking depending on the seat configuration. IATA does suggest that a world class airport will serve 90% of passengers by airbridges. Using push-back tractors allows closer spacing of this type of stand, thus reducing the apron size.

Normally the push-back takes in average 2–4 minutes depending on the size of aircraft, type of push-back tractor (with tow-bar or towbarless) and on the position where the engines are started-up (at the stand or after push-back). They cost more to equip and operate than the self-manoeuvring type. As well as the air bridge, if it is provided, there are the tractors and parking guidance systems to be provided. There may well also be hydrant fuel systems and fixed power installations, which are more appropriate for push-back stands than for the more flexible open-apron power-out operations.

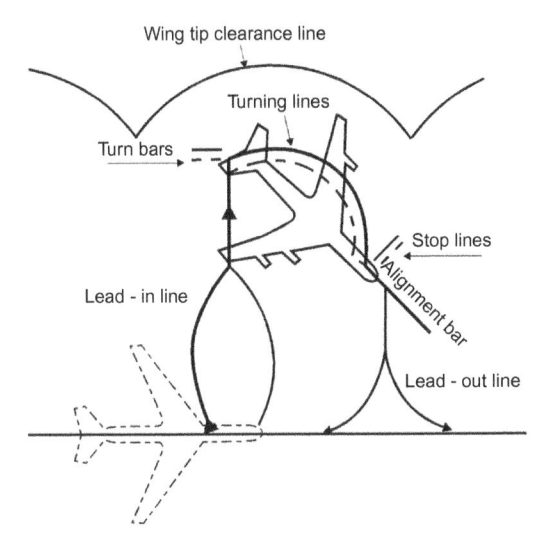

Figure 7.11: Aircraft stand markings.

The details related to technical servicing of aircraft on both stand types are given in Chapter 9.

On the stands, the aircraft may be positioned nose inwards, at an angle nose inwards, nose outwards, at an angle nose outwards, and parallel to the terminal building, as shown in Figure 7.13. Each solution has its advantages and disadvantages.

On the nose-in push-back aircraft stands (Figure 7.14), the aircraft are located either directly nose inwards, or at an angle between 30° and 60°.

The angle depends on the size, shape and layout of the apron. Fundamental requirements are for an aircraft on stand not to overlap the required separation from the adjacent stand, not to block the apron taxiway behind it, for the tail not to conflict with the transitional protection surface associated with a runway and for the tow tractor not to block the service road in front of the aircraft when initiating push-back.

140   *Airport Design and Operation*

On self-manoeuvring stands, the aircraft may also be theoretically parked in any of the above positions. However, the 'nose inwards' or the 'nose outwards' placement, that is perpendicular to the terminal building edge are normally not used because the stands cannot overlap one another (see Figure 7.10), and thus the aircraft would occupy the maximum area.

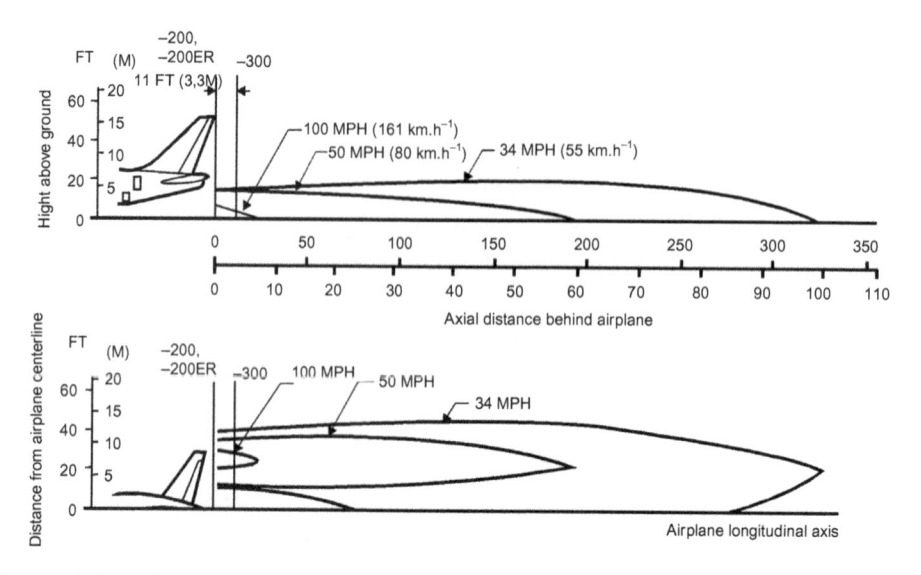

Figure 7.12:   Jet engine exhaust velocity contours — break-away thrust B-767-200 (CF6-80A engine). Conditions: sea level; still air; stationary aircraft; thrust 38.6 kN (8500 lb), each of the engines; both engines in operation.

Placing the aircraft parallel to the terminal building is convenient particularly due to the fact that the passenger loading bridges may be installed to both the front and the rear door of the aircraft. A disadvantage is that after starting up the engines, it is practically impossible for the aircraft on the stand behind to be serviced.

Therefore this parking arrangement is also used only rarely, and only at airports with a sufficiently large apron, a low frequency of flights or with the circular satellite stand concept for example at Genève Aéroport. On self-manoeuvring stands the aircraft are most often parked at an angle of between 30° and 45° inwards or outwards from the terminal building frontage where the maximum overlap is possible, depending on the aircraft type. In spite of the fact that individual stands overlap one another, the prescribed safety distance should be provided between the wing tip of the moving aircraft and the aircraft on an adjacent stand. However, the aircraft on the adjacent stand must be at a standstill. The orientation of the aircraft inwards or outwards depends also on the slope of the apron. Although the slope of the apron is strictly limited, its effect on the break-away thrust required for an aircraft with the maximum mass is significant. Therefore, at the beginning of the power-out manoeuvre, the aircraft should face in the direction of the declining slope.

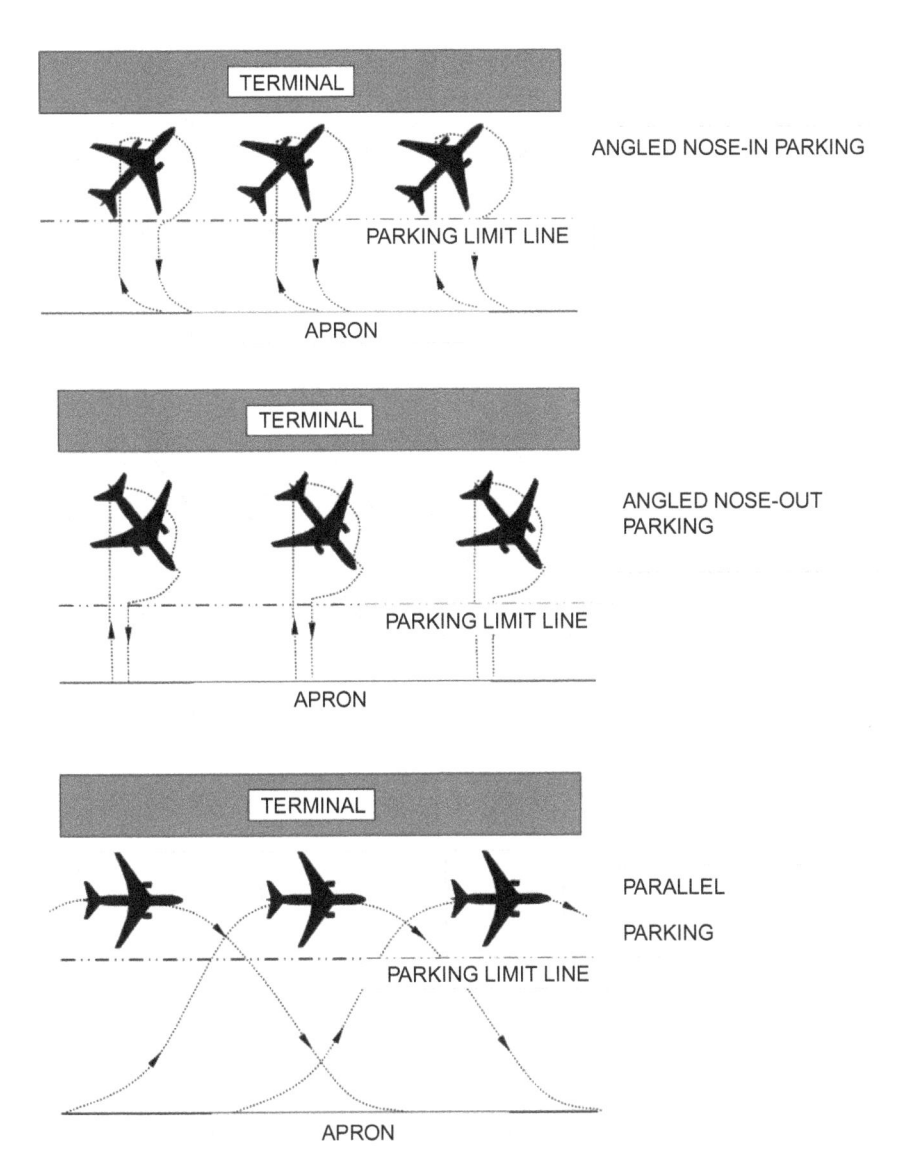

Figure 7.13: Positioning of aircraft on the stands. *Source*: *ICAO*: *Aerodrome Design Manual, Part 2, Taxiways, Aprons and Holding Bays, Doc 9157-AN/901, 4th ed. 2005.*

142 Airport Design and Operation

To relieve the dynamic effect of jet blast on neighbouring taxi out stands and the terminal building during the aircraft turning manoeuvring, blast fences may be placed on the edges of the stand to reduce the blast speed to less than 56 km/h. The fences usually consist of sets of steel vanes which turn the jet exhaust upwards (Figure 7.15). The height of the walls on the apron is limited and in most cases does not exceed 3 m. Because of this they are effective for under wing mounted engines only. Much larger blast fences are often installed to protect public areas around aprons, as at Heathrow Terminal Four. They may be in the form of a latticework rather than vanes, as at East Midlands airport in the United Kingdom.

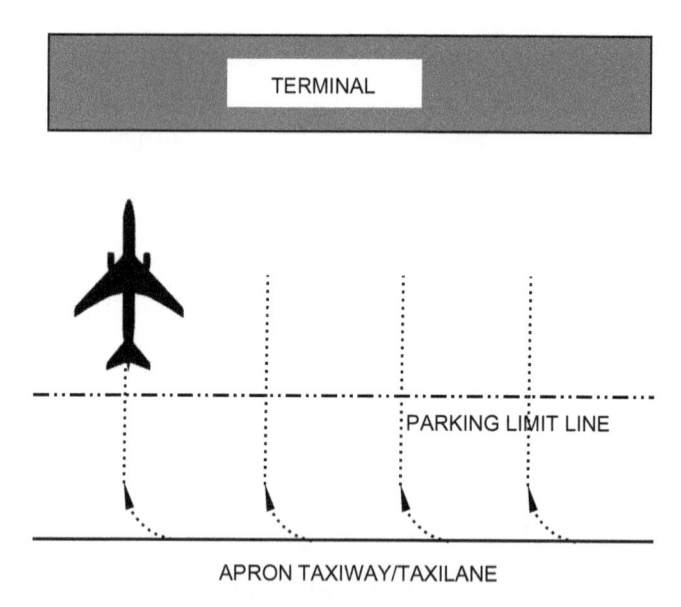

Figure 7.14: Nose-in stands for pushback operations. *Source: ICAO: Aerodrome Design Manual, Part 2, Taxiways, Aprons and Holding Bays, Doc 9157-AN/901, 4th ed. 2005.*

7.6. APRON CAPACITY

At a properly dimensioned aerodrome, the capacity of all parts of its system is in approximate equilibrium. There are then no critical bottlenecks in the system. Some portions of the aerodrome can bear a short-term overload, others cannot. The apron should be dimensioned in such a way that peak hour delay is an acceptable minimum, say no more than 2% of flights.

To cope with these small number of movements, and to ensure that delays do not back up onto the taxiways or into the air, the capacity of the apron may be increased by establishing parking lots away from the terminal building. Such stands are usually not equipped with any technology. In the majority of cases, they are used for charter flights and for aircraft waiting for on-line repairs, but are extremely useful to accommodate aircraft that would otherwise block taxiways while waiting for specific stands to become available. Typically, the number of remote stands might amount to 10% of the number of stands on the apron, depending on the types of operation at the airport (Figure 7.16). Cargo aprons can often be brought into play for this purpose, since the daily peaking characteristics are different.

Figure 7.15: Blast fence. *Photo*: A. Kazda.

The number of stands on the apron depends not only on the number of movements of the aircraft, but also on their distribution during the day, the length of stay on stand, and on the aircrafts' dimensions and seat capacity. Theoretically, the required number of stands may be expressed by a formula:

$$N = k \cdot \frac{t \cdot n}{120}$$

where N is the number of stands; k is the coefficient of variability of use of the stand. It depends on the total number of stands and their types, on the structure of flight timetables, and allows for the time taken to clear a stand after the doors close

144   Airport Design and Operation

and the time the stand becomes available for the next aircraft. It ranges between 1.3 and 2.0; t is the aircraft turn around time in minutes; n is the number of movements in the peak hour.

A determination of the required number of stands by simulation is more appropriate and precise, but requires the construction of an accurate representative schedule.

Figure 7.16: Remote parking stands operations with mobile lounges allows balancing apron peak capacity. *Photo*: A. Kazda.

7.7. Isolated Aircraft Parking Position

An isolated aircraft parking position should be established on each aerodrome for parking any aircraft which is known or believed to be the subject of unlawful interference. The isolated parking position should be located at the minimum distance of 100 m from other parking positions, buildings or public areas. The position should not be located over underground utilities (fuel, gas, electrical, light-current cables).

8

PAVEMENTS

Tony Kazda and Bob Caves

8.1. BACKGROUND

The choice of the type of pavement depends on the characteristics of the aircraft which are intended to use the aerodrome or the respective runway, operational requirements (particularly with respect to reconstruction of the runway) and geological conditions. Requirements for pavement bearing strength, longitudinal and transverse slopes of runways and other movement areas, pavement texture and braking action are all specified by Annex-14, Aerodromes, Volume I and amplified in the *Aerodrome Design Manual*, Part 3, Pavements. Operational regulations of individual airport administrations complement these requirements and offer guidance on the occasional excessive loading of pavements.

It is not an objective of this chapter to give instructions on how to design the pavement structure, which is the construction engineer's responsibility. But nevertheless, the staff from the operational department of an airport should have at least a basic knowledge about designing aerodrome pavements so as to be able, based on their understanding, to monitor the pavement condition and manage maintenance and reconstruction. Some characteristics of existing aerodrome pavements differ from those of road pavements. To be able to understand the transmission mechanism of high loading forces from the heaviest aircraft, full-scale pavement fatigue tests were carried out by Airbus and Boeing. They included both flexible and rigid type of pavements and resulted in specialised computer programs for airport pavement calculations — French ALIZE and FAA FAARFIELD. The whole aerodrome pavement should comply with four basic requirements:

✈ its bearing strength must be appropriate to the operation of the aircraft which are intended to use the aerodrome

✈ it should provide good ride capability of an aircraft during its movement on the runway by preserving a smooth pavement

✈ it should provide good braking action even on a wet surface

✈ it should offer good drainage capability.

146   Airport Design and Operation

The first requirement refers to the pavement construction, the second to geometrical characteristics of the surface, the third one to the texture of the pavement surface and the fourth one to both the geometry and the texture.

All the four criteria are fundamental and complement one another. It is the only way for a pavement to fulfil the operational requirements. From the operational viewpoint, the most important are the third and fourth requirements because they have a direct impact upon the safety and regularity of the aerodrome operation. Thus the requirement for a long-term provision of good longitudinal friction coefficient (f_P) may affect the choice of a pavement construction and surface.

8.2. PAVEMENT TYPES

The choice of movement area surfaces as a unpaved or paved construction type is influenced by many factors.

In particular, the following factors are important:

✈ types of aircraft from the viewpoint of the maximum point load for which the aerodrome is intended, and other operational requirements

✈ availability and price of suppliers, materials and works

✈ geological conditions

✈ prevailing climatic conditions

✈ number of movements per year, movement forecast and traffic split.

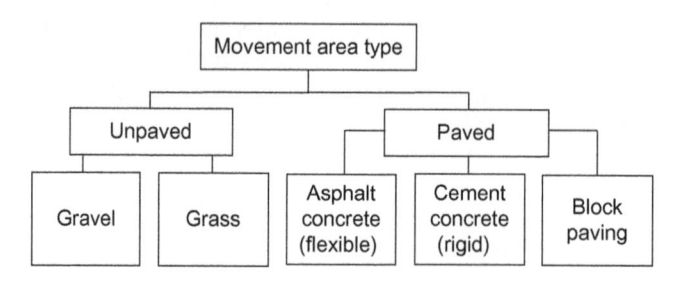

Figure 8.1:   Movement area types.

The basic division of pavement types is given in Figure 8.1. However, various combinations of pavement types and stabilised layers in complex pavements could be classified between flexible and rigid constructions. In hot climates like Africa, Asia, South America or in the United States but also in high-altitude airports, compacted earth strips are also used at general aviation airports. Those airfields are many times also used by mid-size turboprops for passenger transportation. The Brazilian ones are formed from what they call pre-laterite soil.

8.2.1. Unpaved Movement Areas

Grass strips which may be employed in most climatic conditions year-round are suitable only for the lightest types of general aviation aircraft. Their suitability depends on appropriate composition and good drainage characteristics of the subsurface. In an ideal case, such an aerodrome would be located on flat land, on a natural layer of gravel covered with approximately a 20 cm layer of topsoil. The surface should be covered with a grass carpet. The type of grass should be chosen so that the roots are sufficiently dense to create a thick and strong carpet in order to reinforce the soil layer. The majority of grasses used are slow growing and therefore do not require frequent mowing. It is unusual to find such an ideal interplay of all factors. The problem is an appropriate water regime. In spring, after a period of melting snow, and in autumn, after longer-lasting rains, the surface may be sodden and its bearing strength reduced. Even if artificial drainage is provided, it is hard to justify the use of grass aerodromes for a business activity because it is difficult to provide the required regularity of operation. In summer, or in a period of dry weather and with an intensive operation, the grass carpet could be damaged, sometimes even destroying the roots which would not be able to bind the upper soil layer properly.

If the type of airport operations allows using grass runway, it is always the cheapest solutions both from investment and operational costs. In some cases, it is worth increasing the bearing strength of a grass runway. During World War II, the Allies used steel grills to build temporary field aerodromes, interconnecting several fields with them. The same method of surface strengthening is still in use in the army today, particularly for heavy equipment parking areas.

A modification of that principle consists in laying down geonets/geogrids made from synthetic fibres over the upper stratum of topsoil when the grass carpet is put in place. The netting helps to anchor the roots of individual grass turfs better. It carries part of the load and, at the same time, takes the friction load from braking of the aircraft's tyres. Alternatively, the same system used for improving the runway strip bearing strength (see Chapter 5, Section 5.1) could also be used to increase the strength of the grass runway. The technology called 'fibre-turf' is based on a mixture of natural constituents and inert polymer fibres which guarantee a three-dimensional strength and stability of the grass rootzone. The polymer fibres reinforce and protect the natural turf.

In some cases it is necessary to increase the bearing strength of a runway but at the same time to maintain its grass surface. Such construction is often designed only for temporary use and later must be either returned to its original condition, or used in a different way for agricultural purposes, for example, aerodromes for agricultural operations or as a temporary general aviation airport. A runway may be temporarily strengthened with lime in its bearing layer. Attempts have been made to apply lime even in the upper layer of a pavement. The advantage of that method is its simplicity as well as the possibility of returning the runway to agricultural use by deep ploughing. Calcium is added to the earth either in the form of quicklime — CaO or calcium hydrate — Ca(OH)$_2$. The cementing bonds among the particles of earth are reversible, that is to say, the stabilisation may be broken and made compact again when the bonds reconstitute in the earth. If it is necessary to drain the pavement, it is appropriate to use powdered lime with a high percentage of active CaO with excellent hydration ability.

Occasionally it might be required to increase the bearing strength of a runway with a grass surface by constructing the bottom layers in a similar way to those under pavements with an asphalt surface. Then a 15–20 cm layer of soil is spread on the sub-layer and a grass carpet is put on top.

8.2.2. Pavements

8.2.2.1. *Use of hard surface pavements*

In an aerodrome intended for a year-round regular operation of aircraft with mass greater than approximately 2,000 kg in European climatic conditions, the use of a hard surface is essential. The most common solution for manoeuvring areas (i.e. runway or taxiway) is the use of asphalt concrete or cement-concrete pavements.

Concrete is defined as any composite material composed of mineral aggregate adhered with a binder. If the mixture consists of mineral aggregate bound together with asphalt we refer to the term 'asphalt (or asphaltic) concrete' commonly called asphalt or blacktop.[1] Later in the text we will use the term 'asphalt' for asphalt concrete pavement. If the composite material is a mixture of broken stone or gravel, sand, cement and water we refer to the term 'cement concrete'. Later in the text we will use term 'concrete' for cement-concrete pavement.

A choice of the pavement construction is influenced by many factors and the final design is often a compromise. It is possible to cater for even the heaviest aircraft with asphalt runways.

It is appropriate to use concrete — rigid pavements for some movement areas while, in other cases, asphalt flexible pavements are suitable. In general, the types of construction do not differ from constructions used in road building except in the required thickness, which is considerably greater in the aerodrome pavements designed for transport aircraft than in the roads as a consequence of larger point loads.

---

1. Asphalt concrete pavements at the airport are sometimes called tarmac for historical reasons, although they do not contain tar and are not constructed using the macadam process.

The construction of a pavement consists of subgrade, sub-base, bearing course and wearing course.

8.2.2.1.1. Subgrade. The area chosen for an aerodrome should have soil with suitable mechanical properties. The earth in the subgrade should have a sufficient bearing strength but at the same time it must be impermeable and its volume should not be affected by frost or changes in humidity. A soil analysis determines which kinds of soil need to be removed from the aerodrome pavement subgrade, which may be improved by adding different soil, and which soil and minerals in the aerodrome vicinity may be used in its construction.

The bearing strength of the subgrade has a considerable effect on the choice of type of pavement construction. In simple terms, the greater the bearing strength of the subgrade, the relatively thinner and cheaper may be the entire construction of the pavement. The type of pavement is affected also by the availability of suitable building material in the aerodrome vicinity. The choice of type of pavement is decided by an appraisal of life cycle costs, taking into account not only the costs of the pavement construction but also maintenance, and reconstruction of the runway system in a monitored time horizon, including an assessment of the losses due to closing the aerodrome during its repairs and reconstruction.

To declare the bearing strength of aerodrome pavements by means of the ACN-PCN method, it is necessary to know the subgrade bearing strength. For rigid pavements dimensioned according to Westergaard, the bearing strength is expressed by a modulus of reaction of the subgrade 'k'. The modulus of reaction of the subgrade is the contact pressure that is required for pressing a standard loading plate into the subgrade. Either of two methods may be used:

➤ the loading plate is weighed down with a pressure $p = 0.07$ MPa and then depth 'z' is determined

➤ the plate is pressed into the depth of 1.27 mm and the contact stress is determined.

The modulus of reaction can then be determined by the relation:

$$k = \frac{p}{z} (\text{MN m}^{-3})$$

where k, modulus of reaction; p, contact stress (N m^{-2}) and z, loading of the plate (m). For flexible pavements, the subgrade bearing strength is expressed under the ACN-PCN method by the California Bearing Ratio (CBR). The CBR is the ratio of the tested material's bearing strength to the bearing strength of a standard sample of crushed limestone, expressed as a percentage. The standard material for this test is crushed California limestone which has a value of 100. It is determined in a special apparatus by loading tests in which a steel pin with diameter of 5 cm is pressed into the soil.

The dynamic modulus of elasticity (E) is determined by the method of damped impact to simulate running a wheel over the pavement. The effects are produced by a 100 kg weight falling onto a rubber pad of prescribed hardness. The deflection of the plate is measured.

The finished and compacted subgrade may be covered with a geotextile which prevents contingent penetration of the subsoil stratum and any infiltration of water into the pavement construction. The bearing strength of the finished subgrade must be uniform.

8.2.2.1.2. Sub-base. The finished subgrade, which may be protected by a geotextile, is covered by a layer of granular material: stabilised granular material such as gravel, crushed stone/gravel or humidified untreated graded aggregate. That layer is supposed to fulfil a draining and filtration function. It drains the condensation created by temperature variations from the pavement construction and also catches any capillary water. The water is led away by means of catch drains to collectors.

8.2.2.1.3. Bearing course/base course. The role of a bearing course is to receive and distribute the pressures from the aircraft undercarriage to an appropriately large area of sub-base and subgrade. The bearing course usually has several layers. The thickness and composition of individual stratums depend on the subgrade bearing strength and on the construction of the wearing course. In order that the pavement may be designed economically, the upper layer should always have greater bearing strength than the layer under it. The bearing course consists of a variety of different materials, which generally fall into two main classes, untreated and treated. An untreated bearing layer normally consists of compacted crushed or uncrushed aggregates. A treated base normally consists of crushed or uncrushed aggregates (crushed stone/gravel) mixed with a stabiliser such as cement or bitumen. It is also possible to use clay or argilliferous stabilisation.

8.2.2.2. Flexible (asphalt) pavements

In the late 1970s the use of asphalt pavements became popular. They usually consist of hot mix asphalt crushed material mixtures. They are now used for all types of movement areas and loads for which in the past only concrete pavements were used. At comparable costs, the asphalt pavements have several advantages. The construction of asphalt pavements is less demanding.

It is simpler and less expensive to carry out repairs and renovation on asphalt pavements. The reconstruction of asphalt pavements may be carried out even without interrupting airport operations if the works are performed by night. Asphalt pavements also withstand winter maintenance better when chemical de-icing materials are used. The asphalt pavements have the further advantage that their surface is even and without joints.

On the other hand, the asphalt pavements are less used on military aerodromes due to their reduced resistance against the impact of hot exhaust gases from military jet engines. Their resistance to spilt fuel is also lower.

The relatively lower bearing strength of asphalt pavements is due to the different manner of transmitting the load. With asphalt pavements, the load is transmitted by an interaction of individual material particles under the effect of physical bonds of asphalt. The bearing strength is limited to the load that causes a permanent deformation of the flexible asphalt layer.

The upper part of the asphalt pavement is usually composed of two asphalt layers which have different functions. A supporting asphalt layer containing coarse gravel fractions is put in place on the bearing course. Its role is to transmit the load on the bearing layers. Its thickness depends on the required resultant bearing strength, or when reconstruction is carried out, also on the condition and profile of the pavement underneath (Figure 8.2). Depending on the total thickness, which is usually within the range from 10 to 40 cm, the asphalt layer may be spread several times by a asphalt paver — finisher in order that the required compactness can be obtained.

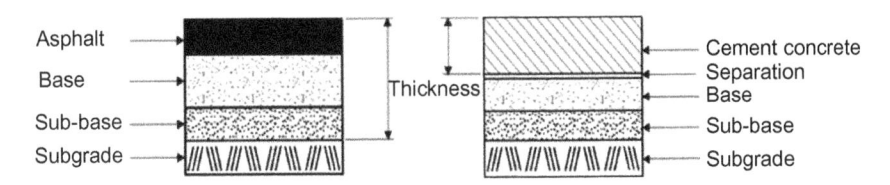

Figure 8.2: Typical construction of asphalt and cement-concrete pavement. *Source*: *Aerodrome Design Manual*, Part 3, Pavements, ICAO Doc 9157-AN/901.

The upper wearing layer contains finer fractions of high quality — hard aggregate. The function of the wearing layer is to resist friction forces that are created by braking on landing, or during a rejected take-off and turning of the aircraft. The pavement surface roughness is designed to ensure appropriate braking action. In order that the wearing layer can resist these forces, it should be at least 4 cm thick. Its second function is to create an impermeable surface. It must perfectly seal the entire construction of the pavement. If water penetrates into the subgrade, it will gradually erode, lose bearing strength, and subsequently lead to breakdown of the bearing courses. The wearing course also transmits the load to the layers underneath.

Besides other things, it is important to monitor the temperature of the asphalt hot mixture composition during the construction. Maintenance of the prescribed temperature, which should not drop under roughly 150 °C, is a precondition of achieving the required compactness of the strata and the bonding of the strips. It is often at badly treated joints that ruptures appear.

When considering the bearing strength of the asphalt pavement, one of the decisive factors is the overall thickness 'h' of the pavement including the bearing course.

8.2.2.3. Rigid (concrete) pavements

The main advantage of rigid pavements is their higher bearing strength which is derived, among other things, from the different transmission of the load. The upper layer, a concrete plate, rests on a semi-flexible subgrade. The rigidity of the concrete plate depends on the quality of the mixture.

The plate transmits the load to a considerably greater area. Inasmuch as the rigid wearing courses are not flexible enough to be able to follow even the slightest deformations of the subgrade, they only may be designed on a quality subgrade with an even bearing strength.

Another advantage is a longer design life of the concrete plate. The design life of a rigid pavement on an appropriate subgrade and with proper maintenance may be 20–30 years.

A concrete plate is usually made from plain concrete 20–30 cm thick. The thickness of the concrete slab is designed basically with respect to the maximum bending tensile stress resulting out of maximum wheel load stress and critical environmental stress should be less than the flexural strength of the concrete. The pavement should withstand the predicted number of loads during its design life. The thickness of the plate is limited by the possibilities of regular compacting of the concrete mixture and by the fact that, with an increasing thickness of the plate, the magnitude of the internal stress increases due to temperature variations causing differential expansion of the upper and the lower parts of the plate. The temperature gradient is about 0.5 °C per 1 cm. As a matter of fact, it is considerably greater in the upper part of the plate up to a depth of about 3 cm.

The different expansion of the upper and the lower parts of the plate as a consequence of temperature variations manifests itself in a tendency to deform the concrete plate into a convex or concave shape. The internal forces reduce the bearing strength of the plate in comparison with its normal condition and are taken into account during design. A sliding asphalt intermediate layer, or different separation layer, can be put in place on the subgrade, which allows the dilation of the concrete plate.

The concrete plate is laid by means of a concrete slipform paver. On aerodromes, slipform pavers are used that are capable of putting a 15 m wide or smaller plate in place. The treated and compacted concrete mixture behind the slipform paver has a smooth surface. Therefore the flaccid concrete mixture should be roughened generally with plastic brushes or burlap brushes which are pulled perpendicularly to the longitudinal centre line of the runway.

In order to prevent the concrete plate from developing irregular cracking under the effects of the internal stress at differential expansion of the upper and the lower parts of the plate, the cement strip should be divided by contraction and expansion joints into individual plates.

The contraction joints (Figure 8.3) with a width of approximately 5 mm are cut in the hard-set concrete mixture to approximately 1/4 of the plate thickness. In the subsequent setting, as a consequence of the internal stresses, it will break under the incised joint. The contact surfaces are uneven with irregularities which anchor the neighbouring plates and provide them with a support. Thus the concrete strip is divided by the contraction joints into plates which are varying between 4 and 7.5 m, depending on the pavement width, its thickness and quality of the concrete mixture.

Figure 8.3: Contraction joint. *Photo*: A. Kazda.

The trend has been to move from short undowelled slabs which were used in the late sixties to the larger slabs used today with dowelling according to the specific movement area. The advantage of short slabs is that they are less sensitive to thermal unsteadiness, but need more maintenance of the joints.

The pavement design is a complex problem and this leads to many configurations considering slab dimensions, the subgrade and type of joints. Joints are categorised according to their function. There are contraction, construction and isolation joints.

154   Airport Design and Operation

Contraction joints provide controlled cracking of the pavement when the pavement contracts due to a decrease in moisture content or a temperature drop. Contraction joints also decrease stresses caused by slab warping. Construction joints are between concrete strips which were laid down at different times, so the earlier one is partially hardened. Isolation joints isolate intersecting pavements or pavement from other structures (e.g. channel/drainage systems).

When the dowels are designed, their alignment and elevation is extremely important to achieve an adequate joint. The dowels improve the joint's behaviour and load transfer and interaction across concrete joints between slabs. They are placed across transverse joints of concrete pavement to allow movements to take place. Across longitudinal joints tie bars are usually provided. In contrast to dowel bars, tie bars are not load transfer devices, but serve as a means to tie two slabs. Hence tie bars must be deformed or hooked and must be firmly anchored into the concrete to function properly. They are smaller than dowel bars and placed at large intervals.

Where movement is purposely designed for longitudinal joints, dowel bars can be adopted as well. The dowels are smooth, round and straight, coated with an anti-adhesive film to prevent their fixing in concrete (completely for dowels placed on wire cage or basket which holds dowels in position (Figure 8.4); unsealed part for dowels when installed by drilling).

Figure 8.4:   Dowel bar on basket. *Photo*: A. Kazda.

A plastic cap is placed on the end of the expansion joint dowels on the coated side. An alternate procedure for placing dowels in the transverse joint is to use a slipform paver equipped with an automated dowel bar inserter (Figure 8.5). The joints should be perfectly sealed (Figure 8.6). If water penetrates and erodes the bearing courses and the subgrade, the concrete plate will lose its support and can crack under load. The subsequent repair would be expensive. The joints are sealed with different types of rubber and then by a joint sealant. The sealant should perfectly adhere to the joint walls and must be permanently flexible. In winter during freezing it must not crack, and in summer with high temperatures it must not flow. The base of sealants consists of asphalt with additives of rubber and plasticising agents.

Figure 8.5: Slipform paver equipped with an automated dowel bar inserter. *Photo*: A. Kazda.

The quality of the concrete mixture must be controlled during the construction, samples being tested. The concrete will reach the required strength after 28 days (Figure 8.7).

Aprons are critical points of the movement areas. There the aircraft has its maximum mass. Similarly, the taxiways are highly stressed by a slowly moving aircraft

156 Airport Design and Operation

with the engines running. It is often thought that the runway is stressed maximally in the touch-down area.

Figure 8.6: Damaged sealing compound on a joint. *Photo*: A. Kazda.

In fact, at touchdown, the aircraft has still a lift from the wings so that the load seldom exceeds 40% of the aircraft maximum mass. Taking into account the probability of transverse distribution of the aircraft movements on the runway, it is possible to dimension the maximum load only in the central part of the pavement (Figures 8.8 and 8.9).

As with other types of construction, the aerodrome pavement has a fatigue life. The pavement is designed, for example, for 10,000 movements of a given limit load. The strength declines as the pavement is used. As is apparent from Figure 8.10, if it is required to cope with a greater number of critical aircraft movements, the built-in safety factor would have to be even higher. On the other hand the strength of the concrete increases with time as the concrete ages.

The rigid pavement design in the FAARFIELD programme is based on the Cumulative Damage Factor (CDF) concept, in which the contribution of each aircraft in a given traffic mix to total damage is separately analysed. Therefore not only one — critical aircraft but the entire traffic mix of aircraft using a particular movement area should be analysed.

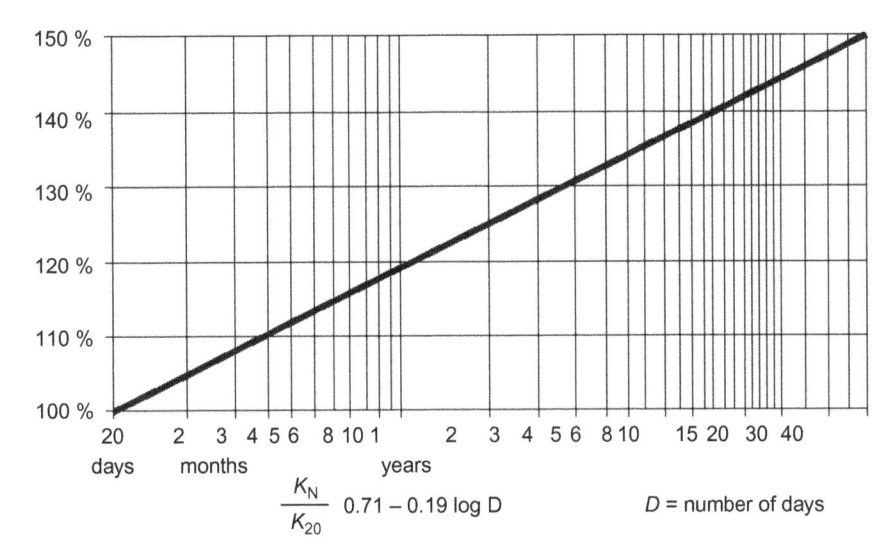

Figure 8.7: Increase of concrete strength with time.

The resultant characteristics of the concrete plate, when evaluating the construction of the pavement in relation to the load characteristics of the aircraft undercarriage, is expressed by the radius of its relative stiffness 'l' by the relation:

$$ 1 = \sqrt[4]{\frac{D}{k}} = \sqrt[4]{\frac{E \cdot h^3}{12(1-\mu)k}} \quad [m] $$

where D is bending rigidity of the plate $D = E \cdot h^3/12(1-\mu^2)$; E is Young's modulus of concrete elasticity in tension and compression [N m^{-2}]; h is thickness of the cement-concrete plate [m]; μ is Poisson constant, usually $\mu = 0.15$ (non-dimensional); k is modulus of cubic compressibility or 'modulus of subgrade reaction' (8.2.3.1.1) [N m^{-3}].

The radius of relative stiffness is graphically illustrated in Figure 8.11. The pavement bearing strength is influenced significantly by:

✈ quality of the concrete expressed in the modulus of concrete elasticity, E
✈ thickness of the cement-concrete plate, h
✈ quality of the subgrade characterised by 'modulus of subgrade reaction', k.

In practice the real bearing strength differs from theoretical values and is influenced by many factors. Therefore the real bearing strength in practice is determined by loading tests currently used in the building industry.

158   Airport Design and Operation

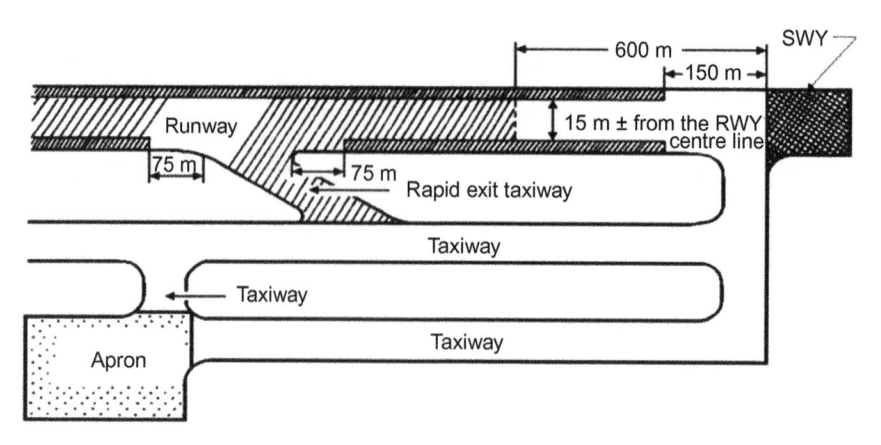

Figure 8.8: Optimising pavement design — RWY equipped with a parallel TWY. *Source: Aerodrome Design Manual*, Part 3, Pavements, ICAO Doc 9157-AN/901.

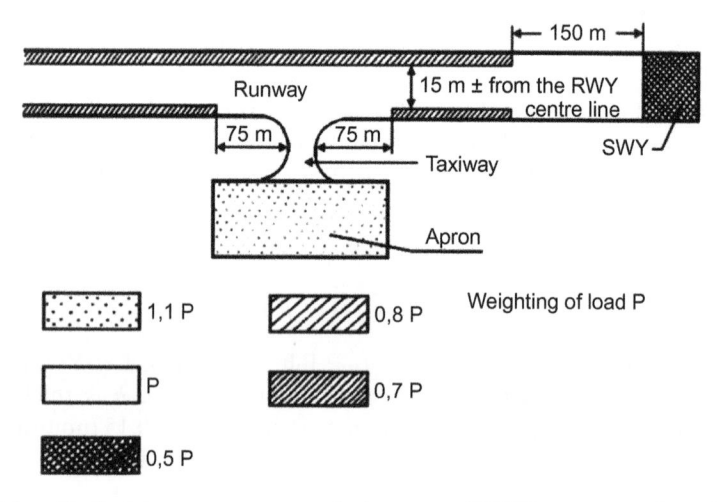

Figure 8.9: Optimising pavement design — RWY not equipped with a parallel TWY. *Source: Aerodrome Design Manual*, Part 3, Pavements, ICAO Doc 9157-AN/901.

8.2.2.4. Combined pavements

There are often cases in practice where reconstruction of a rigid pavement is performed by laying a new cover of several layers of asphalt. This type of reconstruction has both advantages and disadvantages which must always be assessed from case to case.

8.2.2.5. Block paving

For aprons, and parking areas in particular, the use of special concrete block paving has become more common. The first use of concrete block pavements on an airport was at Luton International Airport in the early 1980s (Figure 8.12). Rectangular concrete block pavers pressed from high-strength concrete of dimensions 100 mm × 200 mm by 80 mm thick are most common, but other shapes are used as well. The blocks are usually laid in a 45 degree herringbone pattern into coarse-grained and sharp sand.

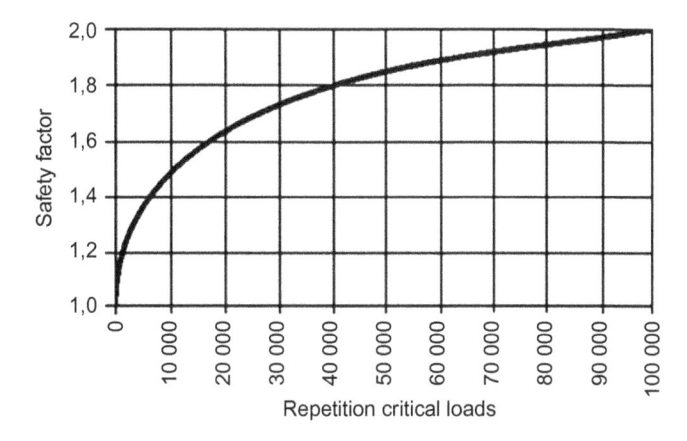

Figure 8.10: Effect of critical loads repetition on pavement tensile strength.

Finer sand is vibrated into the joints by which the blocks are fixed one to another. By mutual 'interlocking' of the blocks, the pavement transmits vertical loads on a considerably greater area. In order that the fine sand does not blow out from the joints or that the bedding sand layer is not damaged if an oil spillage occurs, the block paving should be sealed. A liquid sealant is used which, after being spread into the joints, polymerises and creates a permanently flexible sealing of the block paving.

The main advantages of the block paving are as follows:

✈ they may be laid even in winter in sub-zero temperatures
✈ they may be used by aircraft immediately on completion of laying
✈ maintenance of pavements is speedy and simple.

The main disadvantage is that block paving is labor-intensive.

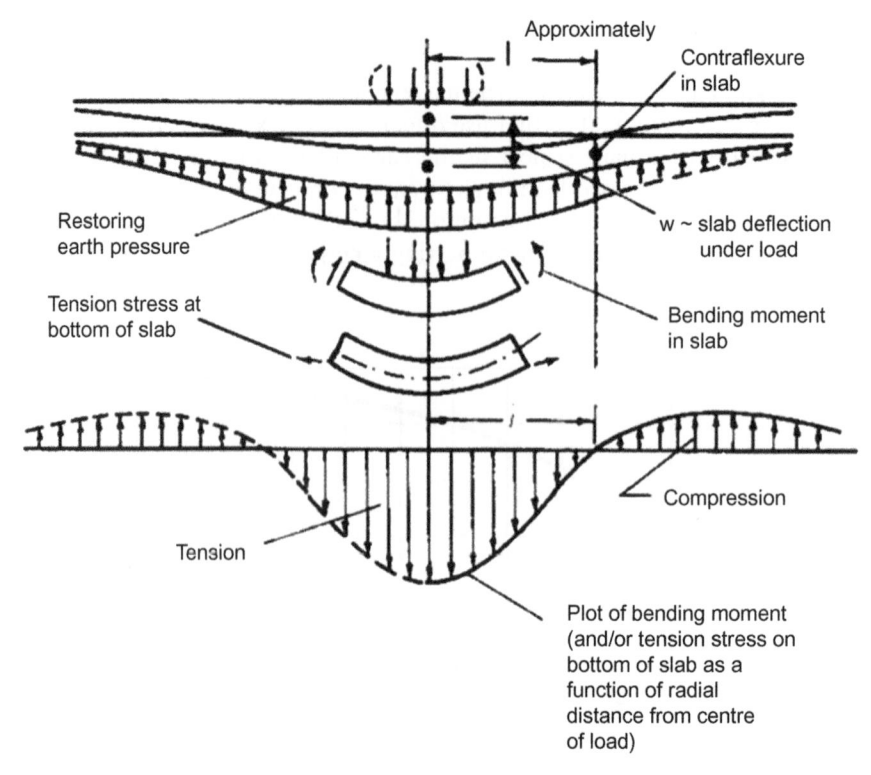

Figure 8.11: Physical meaning of Westergaards 'radius of relative stiffness', 'l'. Source: *Aerodrome Design Manual*, Part 3 Pavements, ICAO Doc 9157-AN/901.

8.3. PAVEMENT STRENGTH

8.3.1. Pavements-Aircraft Loads

There is a general rule that the bearing strength of runways, taxiways and aprons should be capable of withstanding the maximum load derived from the aircraft which the aerodrome may expect to serve. Such an aircraft is called a critical aircraft and the derived load is the critical load.

It is difficult to determine the bearing strength of grass areas. The bearing strength is dependent on the quality of the grass surface and the bearing strength of the soil. The bearing strength of the soil may vary very quickly, depending on the humidity. In practice, the capability of a grass area may be assessed visually, or even better, by repeated rides through the area in a car. In this way it is possible to discover places with locally lower bearing strength. A greater measure of objectivity may be provided by means of a metal stick, with a scaled taper on the end. According to the depth the taper penetrates into the soil when the stick is dropped from a specified height, an approximate bearing strength of the grass surface may be determined.

Figure 8.12: Apron block paving Luton airport. *Photo*: A. Kazda.

The load an aircraft applies to a pavement depends not only on the total mass of the aircraft but also on other factors, in particular on:

✈ type of undercarriage

✈ number of wheels on the main undercarriage leg

✈ geometric configuration of the wheels in the undercarriage

✈ tyre pressure.

The critical loading is generated by the main undercarriage leg. In the simplest case, a main leg has only one wheel. Then it is possible to determine the load magnitude from the size of the contact area of the tyre with the pavement surface, the proportion of the aircraft mass which is applied to one undercarriage leg and on tyre pressure:

$$A = \frac{Q}{P_0}$$

where A is contact area of tyre [m^{-2}]; Q is load transmitted by the main undercarriage leg [N]; P_0 is tyre pressure p [Pa] multiplied by a contraction coefficient of the tyre m which expresses the resistance generated by its rigidity (values within the range from 1.03 to 1.1) $P_0 = p \cdot m$.

The contact area for aircraft tyres has approximately the shape of an ellipse.

A more complicated event occurs if the main undercarriage leg has two, four or more wheels. Then the loading generated by individual wheels must be partially summed, depending on the distance of one wheel from another, as well as on the pavement properties. The manner of summing the loads is schematically illustrated in Figure 8.13. At the same time it is apparent from Figure 8.13 that the resultant load is dependent on the pavement construction thickness. With a more detailed analysis it is possible to determine also the dependence on the quality of individual courses of the pavement.

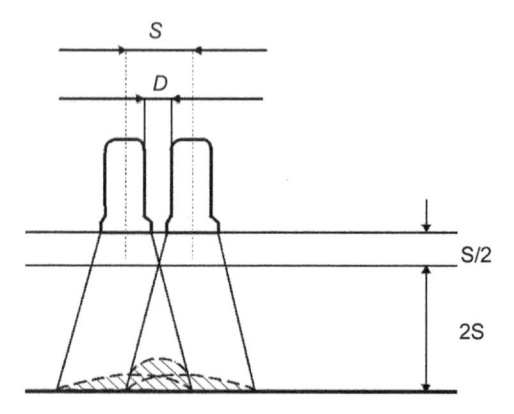

Figure 8.13: Scheme of load distribution from dual wheel undercarriage in the pavement layers.

The effect of a multi-wheel undercarriage is expressed by the Equivalent Single Isolated Wheel Load. This is the theoretical load generated by a single wheel which generates the same effects in the pavement as a real multi-wheel undercarriage. The tyre pressure is the same in the two cases.

8.3.2. Pavement Strength Reporting

The information on the bearing strength of an aerodrome pavement is necessary:

✈ to ensure the integrity of the pavement to its optimum design life

✈ to determine the types of aircraft that may use the pavement and their maximum operational masses

✈ to design the aircraft undercarriage so that it may be used in the majority of the current aerodromes.

Annex 14, Volume I, Chapter 2 prescribes the consistent manner in which the bearing strength of movement areas is to be reported, thus encouraging its world-wide standardisation. This method includes two procedures.

The first procedure is related to the pavements intended for aircraft up to maximum mass of 5,700 kg. The bearing strength is determined by maximum allowable mass and maximum tire pressure, see Figure 8.14.

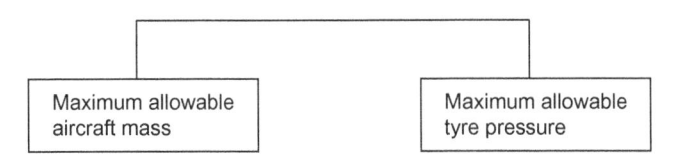

Figure 8.14: Reporting of the bearing strength of a pavement intended for aircraft of apron mass equal or less than 5700 kg.

The second procedure is called ACN-PCN (Aircraft Classification Number — Pavement Classification Number). ACN is the number indicating the relative effect of an aircraft on the pavement resting on any of four types of subgrade of standard quality. PCN is the number indicating the bearing strength of a pavement with an unlimited number of aircraft movements.

The ACN-PCN method expresses the bearing strength of a pavement in whole numbers between zero and an unspecified maximum. The same scale is used for expressing the load characteristics of aircraft. A single wheel with a weight of 500 kg and tire pressure of 1.25 MPa represents a load of 1 ACN. The bearing strength of a pavement with a bearing strength that just corresponds to that load registers 1 PCN. In other words, a PCN for an unlimited number of movements should be equal or higher that the ACN.

For example, the ACN of Airbus A320-200 Dual with MTOM 73,500 kg resting on flexible pavement with high category of subgrade (CBR 15%) is ACN 38/F/A/X (see Table 8.1). It can use pavements with characteristics PCN 48/F/A/X/T without any limitations as the PCN is greater than the ACN.

The ACN-PCN method is a very simple method of reporting pavement strength from an operational viewpoint. It does, however, require the ACN of individual aircraft to be published. The ACN-PCN is not intended for designing pavements, neither does it determine a method by which the bearing strength of a pavement should be specified. On the contrary, an airport operator may use an arbitrary method for PCN determination.

The ACN of an aircraft is determined by means of two mathematical models. One of them is applicable for rigid, the other for flexible pavements.

164   Airport Design and Operation

Table 8.1: ACNs for several aircraft types.

Aircraft type	All-up mass/ empty [kg]	Load on one main gear leg [%]	Tyre pressure [MPa]	ACN for rigid pavement subgrades [MN/m³]					ACN for flexible pavement subgrades [CBR]			
				High 150	Medium 80	Low 40	Ultra low 20	High 15	Medium 10	Low 6	Very low 3
(1)	(2)	(3)	(4)	(5)	(6)	(7)	(8)	(9)	(10)	(11)	(12)
DC-8-55	148,778	47.0	1.30	45	53	62	69	46	53	63	78
	62,716			15	16	19	22	15	16	18	24
DC-8-61/71	160,121	46.5	1.29	47	56	65	73	49	56	67	83
	65,025			15	16	19	22	16	16	18	24
DC-8-62/72	162,386	47.6	1.34	50	60	69	78	52	59	71	87
	72,002			17	19	23	26	18	19	22	29
A 350-900	268,900	46.84	1.66	63	70	82	95	63	70	79	109
	115,700			26	27	28	31	24	25	27	31
DC-9-21	45,813	47.15	0.98	27	29	30	32	24	26	29	32
	23,879			12	13	14	15	11	12	13	15
DC-9-2	49,442	46.2	1.05	29	31	33	34	26	28	31	34
	25,789			14	15	15	16	12	13	14	16
DC-10-40	253,105	37.7	1.17	44	53	64	75	53	59	70	97
	120,742			20	21	24	28	22	23	25	32
A 320-200 Dual	73,500	47.0	1.45	44	46	48	50	38	40	44	50
	39,748			20	22	23	25	19	19	20	24
Concorde	185,066	48.00	1.26	61	71	82	91	65	72	81	98
	78,698			21	22	25	29	21	22	26	32
DHC 7 DASH 7	19,867	46.75	0.74	11	12	13	13	10	11	12	14
	11,793			6	6	7	7	5	6	6	8
B 757-200	109,316	45.2	1.17	27	32	38	44	29	32	39	52
	60,260			12	14	17	19	14	14	17	22

Source: Aerodrome Design Manual, Part 3 Pavements, ICAO Doc 9157-AN/901; Aircraft manuals.

A computer programme was elaborated by the Portland Cement Association of the United States and was provided to ICAO to determine the ACN of the majority of aircraft in service. The ACN values are given in an *Aerodrome Design Manual, Part 3, Pavements.* Similarly, in the case of new types of aircraft, ICAO may determine ACN for the given type of aircraft based on the data provided.

For calculation of the actual ACN for arbitrary mass between the maximum and empty mass of an aircraft, a linear function is applied (Figure 8.15). An overview of the ACN-PCN method of data coding is given in Table 8.2.

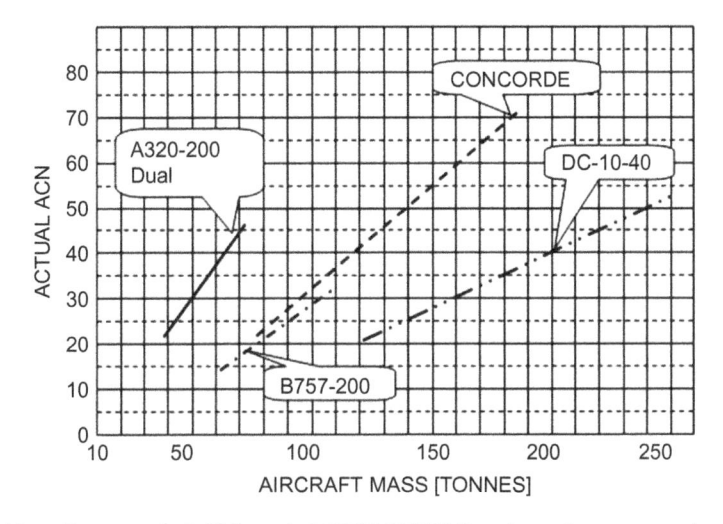

Figure 8.15: Range of ACN and MTOM/OEM values for some aircraft: rigid pavement; $k = 80$ MN/m^3 (relevant data are shaded in Table 8.1). *Source*: A/c manuals.

To find for example the actual ACN of an Airbus A 320-200 Dual with a mass of 60,500 kg on a flexible pavement resting on a medium strength subgrade (CBR — 10%) and with a tyre pressure 1.45 MPa the following formula could be used:

$$ACN = ACN_{MAX} - \frac{(MTOM - AOM)}{(MTOM - OEM)} \times (ACN_{MAX} - ACN_{EMPTY})$$

where ACN_{MAX}, ACN for the maximum take-off mass; MTOM, maximum take-off mass; AOM, actual operating mass; OEM, operating empty mass and ACN_{EMPTY}, ACN for the operating empty mass.

$$ACN = 40 - \frac{(73,500 - 60,500)}{(73,500 - 39,748)} \times (40 - 19)$$

$$ACN = 32$$

Note: ACN must be always a whole number.

166 *Airport Design and Operation*

Table 8.2: Reporting of the bearing strength of a pavement intended for aircraft of apron mass greater than 5700 kg.

PNC	Pavement classification number	Code	Pavement type	Code	Subgrade strength category	Code	Tyre pressure	Code	Evaluation method
	Bearing strength without limitation of a/c movements	R	Rigid	A	High $K = 150$ MN m^{-3} CBR = 15%	W	High (no pressure limit)	T	Technical evaluation
		F	Flexible	B	Medium $K = 80$ MN m^{-3} CBR = 10%	X	Medium up to 1.50 MPa	U	Using aircraft experience
				C	Low $K = 40$ MN m^{-3} CBR = 6%	Y	Low up to 1.00 MPa		
				D	Ultra low $K = 20$ MN m^{-3} CBR = 3%	Z	Very low up to 0.50 MPa		

The ACN-PCN method publishes the required data in the form of a code. The data are given for two different types of pavements, rigid R, and flexible F, for eight standard categories of subgrade (four values of 'k' for rigid pavements and four values of CBR for flexible ones). Then the system uses four categories of tyre pressure. Since some modern aircraft have tire pressures exceeding 1.5 MPa (i.e. current 'X' tyre pressure limit) Airbus in cooperation with other partners conducted a full-scale test campaign High Tire Pressure Test to assess the effect of new aircraft on pavements. The tests focused on pressures from 1.5 MPa to 1.75 MPa. The primary objective was to explore whether the new proposed tyre pressure limit for code letter X — 1.75 MPa was a reasonable upper limit for typical pavements. This objective was successfully achieved and the experiment allowed additional lessons which could be of interest for further investigation on this topic.

The code contains also information on the method of determining PCN. When PCN is determined on the basis of operational experience, according to the ACN of the critical aircraft, a U (Using Aircraft Experience) code is used. If PCN is determined by any of the more sophisticated methods used in civil engineering, a T code (Technical Evaluation) is used. Table 8.1 gives data of several types of aircraft as they are published by ICAO and aircraft manufactures.

8.3.3. Overload Operations

Under certain conditions, the pavement can bear even greater load than that calculated. With the exception of massive overloading, pavements in their structural behaviour are not subject to a specific limiting load above which they will break up suddenly. Pavement can sustain a designed load for an expected number of repetitions during its service life. Loads which are larger than the defined loads shorten the design life of the pavement, whilst smaller loads extend it. Overloading of pavements can result from loads being too large, or from a substantially increased number of movements. With the exception of massive overloading, pavements are not subject to a particular limiting load when under static loading. Their behaviour is such that a pavement can sustain a definable load for an expected number of repetitions during its design life. Occasional minor overloading only results in limited loss in pavement life expectancy. Therefore, for such cases, the following criteria are suggested:

- For flexible pavements, occasional movements by aircraft with ACN not exceeding 10% above the reported PCN should not adversely affect the pavement.

- For rigid or composite pavements, in which a cement-concrete plate provides the crucial transmission of the load, occasional movements by aircraft with ACN not exceeding 5% above the reported PCN should not adversely affect the pavement.

168 Airport Design and Operation

If the pavement structure is unknown, the 5% limitation should apply. The annual number of overload movements should not exceed 5% of the total annual aircraft movements. Such overload should not be permitted on pavements that are damaged or exhibit other signs of distress, or after strong freezing, or in cases when the bearing strength of the pavement or its subgrade could be weakened by water. Overload movements are subject to the approval of the airport operating authority. However, the airport authority should be aware of overload operation consequences in terms of the accelerated structural deterioration and the reduction in pavement service life which may occur. If overloads are allowed, the pavement should be inspected regularly to ensure that unacceptable structural damage is not taking place.

8.4. RUNWAY SURFACE

8.4.1. Runway Surface Quality Requirements

The runway surface quality requirements have gradually increased with implementation of ever heavier jet aircraft and the increasing speed of aircraft movements on runways. The runway surface must not show unevenness sufficient to cause a loss of braking action, it must provide good braking action even when the surface is wet and should be cleaned of any foreign objects which might damage the engine. Annex 14, Volume I, Chapter 3 describes in detail the standards which specify the particular physical characteristics of the surface.

The smoothness of the runway surface is defined by several standards, which complement one another. The basic requirement for cement-concrete pavements is a test with a 4 m long board under which no unevenness larger than 5 mm may appear with the exception of drainage channels. In order that an adequate smoothness of the surface can be ensured, it is necessary to choose the appropriate types of inset lights, back-inlet gulleys and the like.

The manual detection of unevenness by means of a 4 m long board is lengthy, time consuming and rather inaccurate. An 'automatic' board, or Planograph, equipment with automatic registration of unevenness (Figure 8.16) is often used instead. However, it can only register unevenness up to a length of approximately 2 m.

8.4.2. Methods of Runway Surface Unevenness Assessment by the Dual Mass Method

At higher speeds of aircraft movement on the runway, unevenness with small wave lengths loses its effect, and unevenness with wave lengths of the order of tens of meters acquires importance. That unevenness may be detected by means of a precise survey which is carried out after construction or reconstruction of the runway. Thus the supplier generally proves that the construction of the pavement complies with the respective standards within the permitted tolerances. However, once the aerodrome is in operation, it is impractical to make exact surveys of the runway.

Figure 8.16: Planograph for surface unevenness measurement. *Photo*: A. Kazda.

From the operational viewpoint, it is essential to know whether some unevenness complies or not with a certain constructional standard. For example, transversal grooving of the runway represents unevenness that does not conform to the standard 4 m board. However, not only does the grooving not impede the aircraft's operation but it generally improves pavement drainage characteristics (Figure 8.17).

Therefore the conclusive criterion should be the effect of the pavement on the aircraft that is moving on the runway at a certain speed. The condition of the pavement will have been gradually deteriorating since commissioning. Differential settlement of the subgrade, the effect of the operation of aircraft, breakdowns of the pavement and their repairs all progressively degrade the pavement unevenness. Small deviations from the specified tolerances have no substantial effect on the operation, but above some critical level, unevenness of the runway surface will endanger the safety of the aircraft during take-off or landing.

170   Airport Design and Operation

Unevenness up to approximately 3 cm over a distance of 50 m is considered acceptable. However it must always be judged in relation to the type of operation on the runway, and in particular, to the speed of the aircraft on the runway. The frequency of oscillation of the undercarriage wheel depends on the speed of movement of the aircraft on the runway at the given length of unevenness. It can be expressed by:

$$f = \frac{v}{\lambda} \ [s^{-1}]$$

where f is frequency of oscillation of the undercarriage wheel; v is speed of movement of the aircraft on the runway and λ is length of the wave in the pavement.

Figure 8.17:   Runway asphalt pavement transversal grooving, open macro-texture.
Photo: A. Kazda.

An oscillation with a frequency of more than 20 Hz is not significant because it is damped by dampers. Similarly, an oscillation less than 1 Hz is not dangerous because it develops only small vertical acceleration. The oscillation frequencies between 2 and 10 Hz are particularly dangerous. The measurement and evaluation of unevenness of the runway surface can be performed by means of a truck pulled behind a car (first mass) and an accelerometer (second mass) located on the truck which reacts to the surface unevenness. Permanent contact of the system with the surface must be ensured. The DM Method is based on the use of the theory of accidental functions. The unevenness of a length of wave up of 100 m may be measured.

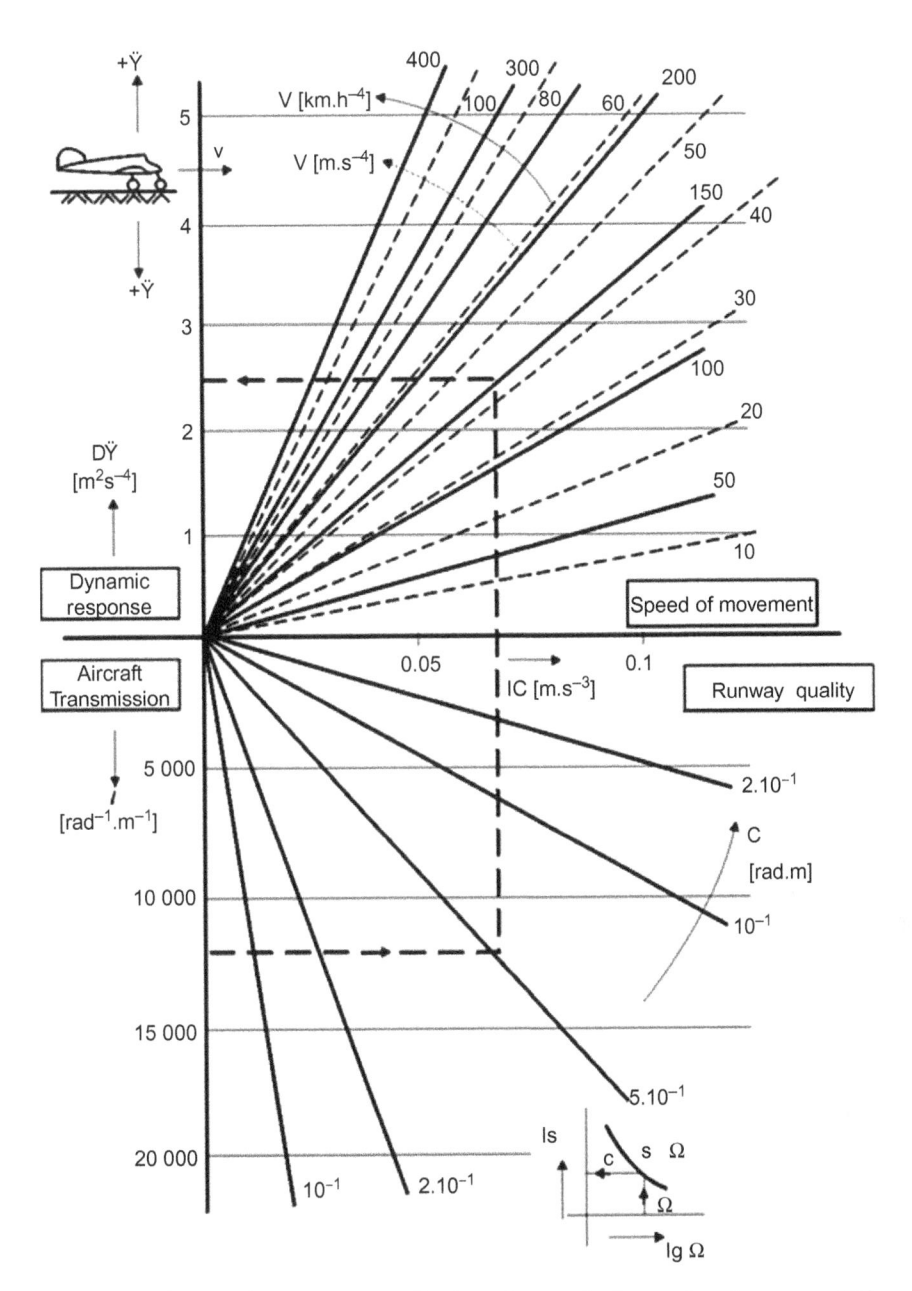

Figure 8.18: Dynamic response of aircraft vertical acceleration on different pavement quality. *Where $D\ddot{Y}$ is square os acceleration* [$m^2\ s^{-4}$]; *IC is variation of* [$m\ s^{-3}$]. *Source*: Čihař (1973).

The chosen speed of movement of the measuring system may be from 5 to 40 m/s (i.e. from 18 to 144 km/h). The equipment measures the vertical acceleration and provides information on the spectrum of the wave lengths (from 2.7 to 100 m) Figure 8.18.

The measurement should be carried out on a portion 500 m long as a minimum which is homogeneous from the viewpoint of the kind of covering, treatment and surface condition. The basic statistical characteristic of the vertical unevenness is its scatter. Information on the intensity with which it occurs within the entire scatter of unevenness of particular wave lengths is given by the power spectral density of unevenness.

When an assessment is made of the dynamic response to the aircraft moving on the runway, the following three factors should be assessed:

- impact of unevenness on the crew
- impact of unevenness on the passengers
- impact of unevenness on the aircraft construction from the viewpoint of the strength and fatigue stress and from the viewpoint of the aircraft's controllability and the safety of its movement on the runway.

Other assessment criteria of the DM Method are the parameter C and zone classification included in the standard. The parameters determined by the DM Method are not mutually compatible with the results obtained by other methods.

8.4.3. Pavement Texture

The pavement surface should have specific antiskid properties, the properties being given by its micro-texture and macro-texture.

If the pavement is clean and dry, it generally has a good coefficient of longitudinal friction whatever the kind of surface (asphalt or concrete-cement pavement) or its drainage characteristics (pavement slope). In addition, the coefficient of friction of a dry pavement only changes very slightly with an increase of the aircraft speed.

Whenever water in any quantity is present on a runway, the situation changes considerably. The coefficient of friction then depends substantially on the quality of pavement surface in terms of the surface texture and drainage characteristics of the pavement, particularly any tendency to pooling of water. At high aircraft speeds the coefficient of friction may drop considerably as a consequence of aquaplaning. Even so, the pavement must have a good coefficient of friction even when wet in order that the safety of aircraft operations is not endangered. The coefficient of longitudinal friction of the pavement depends on the following:

Pavements **173**

→ macro-texture created by the binding of individual gravel fractions in the mixture, this controlling contact between the tyre and the pavement surface

→ micro-texture characterised by the surface roughness of the grains of aggregates or mortar. Micro-texture is important for the so called 'dry contact' of a tyre on a wet pavement.

The macro-texture of a pavement is supposed to ensure good water drainage from the area in contact with the tyre. Therefore it is particularly important at high aircraft speeds. Macro-texture of a common road pavement is smoother, while that of an aerodrome pavement is more open. This is due to different drainage characteristics of car and aircraft tyres, and the higher speeds that are reached by aircraft on a runway. A correct macro-texture may be obtained in a concrete surface as well as in an asphalt pavement surface. In cement-concrete pavements it is obtained by transversal brushing of the setting concrete. In asphalt pavements it is achieved by using correct fractions of the aggregates in the wearing layer and the correct procedure in laying the pavement.

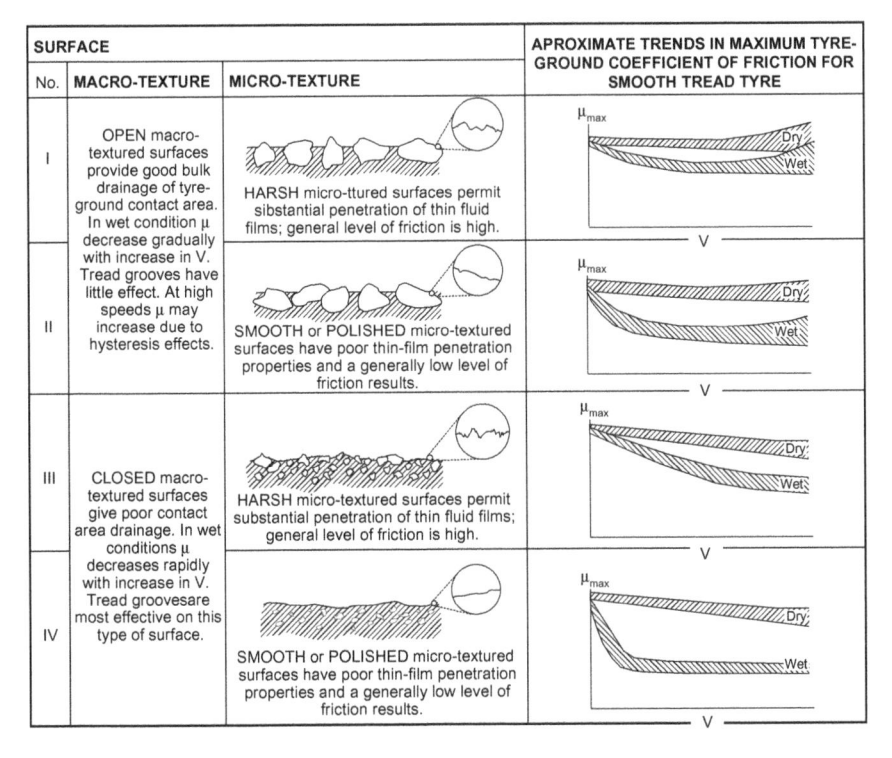

SURFACE			APROXIMATE TRENDS IN MAXIMUM TYRE-GROUND COEFFICIENT OF FRICTION FOR SMOOTH TREAD TYRE
No.	MACRO-TEXTURE	MICRO-TEXTURE	
I	OPEN macro-textured surfaces provide good bulk drainage of tyre-ground contact area. In wet condition μ decrease gradually with increase in V. Tread grooves have little effect. At high speeds μ may increase due to hysteresis effects.	HARSH micro-ttured surfaces permit sibstantial penetration of thin fluid films; general level of friction is high.	μ_{max} Dry/Wet vs V
II		SMOOTH or POLISHED micro-textured surfaces have poor thin-film penetration properties and a generally low level of friction results.	μ_{max} Dry/Wet vs V
III	CLOSED macro-textured surfaces give poor contact area drainage. In wet conditions μ decreases rapidly with increase in V. Tread groovesare most effective on this type of surface.	HARSH micro-textured surfaces permit substantial penetration of thin fluid films; general level of friction is high.	μ_{max} Dry/Wet vs V
IV		SMOOTH or POLISHED micro-textured surfaces have poor thin-film penetration properties and a generally low level of friction results.	μ_{max} Dry/Wet vs V

Figure 8.19: Effect of surface texture on tyre-surface coefficient of friction. *Source*: *Aerodrome Design Manual*, Part 3, Pavements, ICAO Doc 9157-AN/901.

174   Airport Design and Operation

The micro-texture of a pavement is formed by fine, but sharp irregularities on the grain surface which are capable of penetrating through the thin viscous film of water between the tyre and the pavement. It provides dry contact with the tyre even at higher speeds. The micro-texture is created by the 'micro-roughness' of the individual surface particles. It is difficult to discern with the naked eye, but is discernible by touch. Therefore only quality, hard and rough aggregates should be used in the pavement wearing layer. The effect of the macro-texture and micro-texture on the coefficient of longitudinal friction depends on the aircraft speed as illustrated in Figure 8.19.

The greatest problem of the micro-texture is the fact that it may change in a relatively short time. A typical example is rubberising of the runway in the touch-down zone. The rubber deposits may cover the micro-texture without any change of macro-texture. As a result there is a substantial reduction of the coefficient of longitudinal friction when the pavement is wet.

The pavement texture depth represents the rate of the pavement surface deviation from a true planar surface within the wavelength ranges defined for macro- and micro-texture. The deviations are measured within the surface area, which corresponds to a tyre-pavement interface (Figure 8.20). While the macro-texture can be characterised either by its mean texture depth (MTD), obtained by the volumetric patch method or mean profile depth (MPD), obtained by the use of laser-measured surface profiles, no objective criterion has been specified to determine the micro-texture.

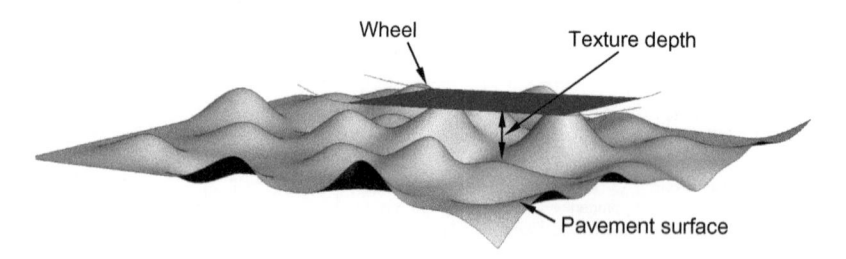

Figure 8.20:   Pavement texture depth. *Author*: M. Kováč.

Based on experience it may be stated that the individual particles must be rough and have sharp edges. It is indirectly assessed by the coefficient of longitudinal friction, the pendulum test or other methods.

At the time of construction the roughness should be optimum. The antiskid properties of the pavement surface can be determined by several standard methods:

✈ measurement of pavement surface macrotexture depth using a volumetric patch technique (sand method)

↛ method of measurement of skid resistance of a surface — the pendulum test

↛ determining the coefficient of longitudinal friction by means of a dynamometer.

The 'sand method' is generally used during commissioning, while the dynamometer method is used for assessing the operational capability of the surface. In the sand method, an average depth of the potholes is calculated from the volume of sand spread out on a pavement surface. The sand is spread to form a circular patch in which all the unevenness is filled, the diameter of which is measured. By dividing the volume of sand with the area covered, a value is obtained which represents the average depth of the sand layer, that is an average 'texture depth'. For tests in line with the ASTM E 965-87 Standard[2] the sand is replaced with solid glass spheres.

The mean texture depth is calculated from the cylinder volume (we know the volume and measure the patch diameter). The MTD for 90% of all measurements should comply with the criterion MTD ≥ 0.8 mm.

The pendulum test is an indirect form of measurement of pavement micro-texture. It is a routine method of checking the resistance of wet and dry surfaces to slipping and skidding, both in the lab and on site by measuring the frictional resistance between a rubber slider mounted on the end of a pendulum arm and the test surface. The pendulum is released from a horizontal position so that it strikes the sample surface with a constant velocity. The distance travelled by the head after striking the sample is determined by the friction of the sample surface.

In the dynamometer method the coefficient of longitudinal friction is calculated from the force required for pulling the braked measuring wheel on a wet pavement surface. The friction ought to be determined:

↛ on a periodical basis on a runway where the code RWY number is 3 or 4 for calibration

↛ on a runway which is slippery when wet

↛ on a runway which has bad drainage characteristics.

The calibration measurement should be performed in good meteorological conditions on a dry and clean runway. In order to ensure homogenous conditions, the measuring device must be provided with self-wetting equipment. Thus the measurement is effectively carried out on a wet surface. The runway surface is wetted before measuring, and the quantity of water should be the same at all speeds. The measuring is performed with a braked wheel (100% slip) at the speeds of 20, 40, 60, 80 and 100 km/h, sometimes up to 135 km/h.

---

2. ASTM E 965-87, Standard test method of measuring surface macrotexture depth using a volumetric technique. American Society for Testing and Materials.

176   Airport Design and Operation

The intervals of measurements should be determined as a function of the type of aircraft, the number of movements, the type of pavement and its maintenance, climatic conditions, etc.

To obtain detailed and accurate information on the pavement surface characteristics laser scanning is increasingly used. The laser scanner is able to obtain data with accuracy up to 0.3 mm. Usually also distance data, crack data, transverse profile data and longitudinal profile data are recorded.

If there is any suspicion that the braking characteristics may be impaired because the runway has a transverse slope or other inconvenient drainage characteristics, an additional measurement should be carried out when it is actually raining.

The measurements are performed with special vehicles, for example Tatra Friction Tester, Saab Friction Tester or by dynamometric trailers pulled behind a car, for example Mu-meter. Each of these use different principles of measurement of the coefficient of longitudinal friction.

8.4.4. Runway Braking Action

The braking action is connected with the antiskid properties of the runway surface. Whilst the macro and micro-textures are the characteristics of the pavement surface monitored during commissioning, the braking action is an operational characteristic of the pavement surface.

The braking action is influenced by:

- speed of the aircraft on the runway
- design of the aircraft
- design of the brakes
- type of automated antilocking device
- dimensions and number of tyres
- tyre tread pattern and wear
- aircraft mass
- meteorological conditions
- as well as the coefficient of longitudinal friction on the runway surface.

Whilst the longitudinal friction coefficient of a given pavement decreases slowly over several years, the braking action may change from minute to minute.

Determination of the braking action is a part of the runway daily checks to assess the operational capability. The checks should be carried out by the airport operator. Besides the frequent periodical inspections, additional checks should be performed, for example each time the climatic condition changes. The results should be reported to Air Traffic Control. Research suggests that 5-minute rainstorms create the most critical drainage challenge. Since the braking action is affected by many factors, it is important to understand what is happening in the contact surface between the tyre and the pavement surface.

In the contact between the tyre and the runway surface, a footprint is created. Its size depends on the radial load and the tyre dimensions. When the tyre is loaded only by a radial force, the elementary forces in the contact area are in static equilibrium. If a force other than the radial one begins to exert its influence on the wheel, that is if the wheel begins to transmit peripheral or lateral force due to braking or crosswind (Figure 8.21), the distribution of the elementary forces in the tyre footprint will change. The resultant force in the contact area of the tyre is called a force of adhesion. The force bond between the tyre and the runway surface is determined by the adhesion. Therefore it is important from the viewpoint of the safety of operations to create good adhesion conditions.

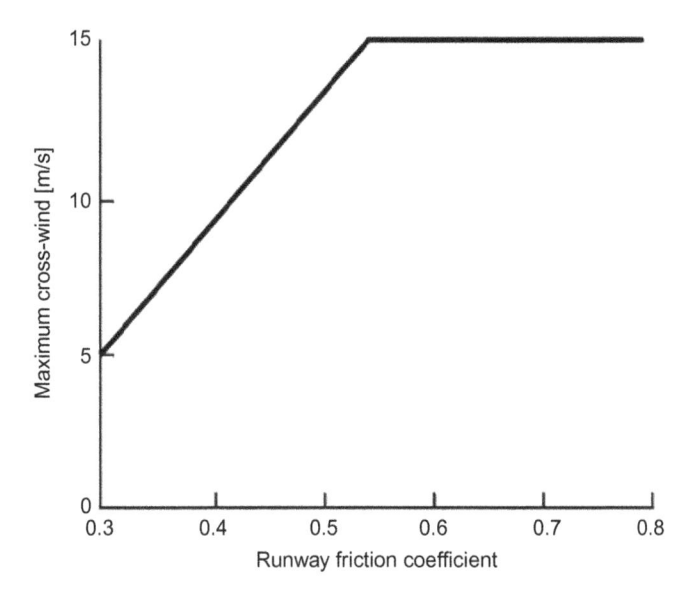

Figure 8.21: Relationship of runway friction coefficient and the allowable crosswind.

Another important characteristic is the wheel slip. An absolutely rigid wheel would roll on the pavement on a radius R (Figure 8.22). If a real wheel with a tyre is loaded, it will deform. The radius from the centre of the wheel to the footprint will diminish to a static radius R_S. When the wheel is rolled on a pavement, a fictitious radius

178   Airport Design and Operation

called a rolling radius R_R may be considered. It is the radius of a perfectly rigid wheel whose peripheral velocity and rolling velocity are the same as the peripheral velocity and rolling velocity of a real tyre. Here $R > R_R > R_S$ applies. From the comparison R and R_R, the so called relative slippage of the wheel may be determined.

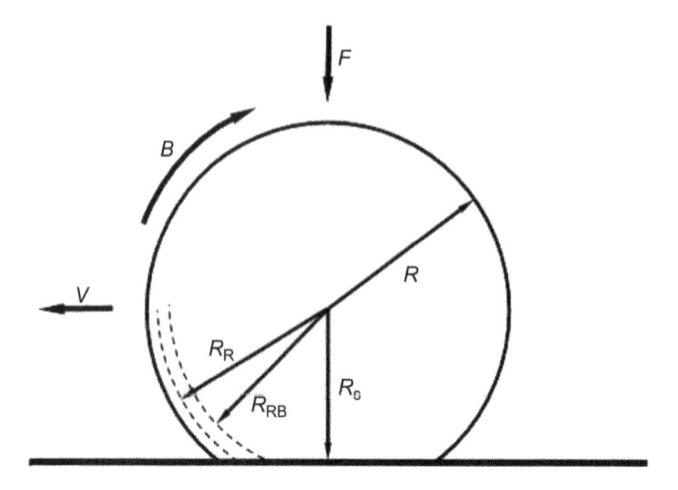

Figure 8.22:   Relative slippage of the wheel — parameters.

If we load the wheel with an additional tangential braking force, it will deform further, and the wheel will roll on the radius R_{RB}, generating another tangential slippage. The following applies for the assessment of a relative slippage:

$$S = \frac{R_R - R_{RB}}{R_R} \cdot 100 \ (\%)$$

where S is relative slippage [m]; R_R is rolling radius of the tyre in free rolling (m) and R_{RB} is rolling radius of the tyre when loaded with braking torque (m).

It must be emphasised that the relation applies for a relative slippage, that is to say while the wheel is turning rather than being locked up.

In free rolling the relative slippage is zero. The relationship between the force of adhesion and the slip determines the most efficient manner of braking. The relationship is of the general form shown in Figure 8.23.

The maximum adhesion occurs with modern aircraft between 15% and 30% of slippage, and the brake antilocking systems are generally adjusted to that value. The shape of the slip characteristic depends on the runway surface condition. If the runway is covered with water, snow or ice, the relationship is rather more abrupt, and the maximum adhesion occurs at higher slip values.

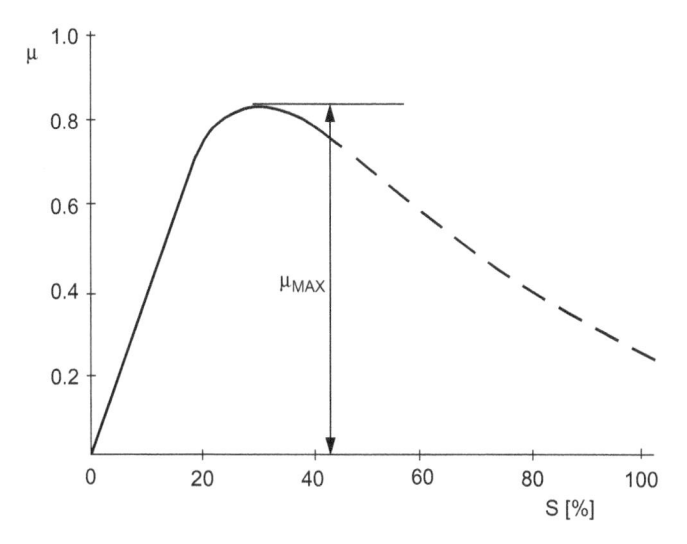

Figure 8.23: Relationship between percentage of slip and friction coefficient, μ.

The braking action may be determined in several ways:

↣ by estimation

↣ by measuring the distance or time required for the vehicle to come to a full stop

↣ by braking the vehicle equipped with an accelerometer

↣ by special devices for continual measurement (Skiddometer, Surface Friction Tester, Mu-meter, Runway Friction Tester, GRIPTESTER).

The data obtained from the measurements carried out by particular devices cannot be used directly to quote a braking action. In order that the braking action can be reported, operational experience is also needed to interpret them.

When the braking action is determined on the basis of measurement of the distance or time required for complete stoppage of the braked vehicle, the testing vehicle without ABS is accelerated to a specified speed. The vehicle is usually occupied by two persons. At a chosen moment, the driver locks the brakes, giving 100% slip. Therefore the data obtained by that method are worse than the real conditions and it creates the necessary safety margin.

The deceleration meter may only be used under certain conditions, for example on a runway covered with compacted snow, ice or a very thin layer of dry snow. The deceleration meter or a braked vehicle should not be used when the runway surface covering is not continuous. Reliable reporting of the braking action may be obtained from those devices that measure friction at high speed. They each function on different principles.

180   Airport Design and Operation

The Mu-meter (skid meter) evaluates a braking coefficient from the value of lateral force of the tested wheel mounted at an angle of 7° towards the movement of the vehicle moving at a speed of 66 km/h (±4.8 km/h) (Figure 8.24).

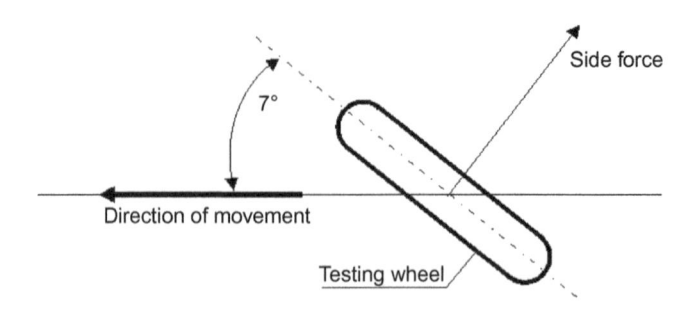

Figure 8.24:   The principle of measuring friction coefficient by Mu-meter.

Saab Friction Tester (slip meter) represents an application of the anti-locking equipment used in the aircraft. The braking coefficient is determined on the basis of the value of peripheral force (braking force) measured at a tested wheel with the slippage adjusted to a constant 15%. The results from the measurements at a speed of 130 km/h approximate to real conditions of an aircraft moving on the runway.

The results from these methods are clearly not directly comparable. Therefore it is always necessary to state by which method, device and at which speed the braking action was determined.

Table 8.3 includes the data on the measured or calculated braking coefficients and their coding. The data were determined on the basis of experience and measurements with one type of deceleration meter and one type of continual measuring device at constant speeds. The braking action must be measured for every third part of the runway separately and coded results are reported in sequence for the touchdown zone, mid-runway position and the final third of the runway.

The runway surface may not always be uniform. Part of the runway could be covered by snow, ice or water and parts could be free of any deposits. A runway is considered to be contaminated when more than 25% of the runway surface area within the required length and width being used, is covered by surface water more than 3 mm deep or by slush or loose snow, equivalent to more than 3 mm of water.

8.5.   PAVEMENT MANAGEMENT SYSTEM

In the past, most airport operators have based their decisions about runway maintenance more on their previous experience and/or an urgent need, rather than on some sophisticated methods. This type of pavement maintenance was neither optimal nor effective. Later some enlightened engineers decided to use note cards to better prioritise the maintenance activities. However, these practices usually led to the selection of a few preferred maintenance strategies and an ineffective use of resources.

Table 8.3: Friction data and descriptive terms of braking action.

Measured coefficient	Estimated braking action	Code
0.40 and above	Good	5
0.39–0.36	Medium to good	4
0.35–0.30	Medium	3
0.29–0.26	Medium to poor	2
0.25 and below	Poor	1

The Pavement Management Systems (PMS) have been used since the 1980s, initially in motorway and road maintenance. Although airport pavement design, operational conditions, pavement loading and maintenance needs differ considerably from the road industry, most states are still using pavement management systems originally designed for roads. In the United States the Airport Pavement Management Program is used. It was developed by the FAA (Airport Pavement Management Program AC No: 150/5380-7A) and represents a special solution for the airport industry.

The basic aim of the pavement management system is to determine the type of pavement maintenance and rehabilitation needed. The pavement management system is a set of tools and methods that support decision making to find a cost effective strategy for pavement maintenance. Each pavement management system should consist of two basic components. The first is a database, which contains current and historical information on pavement condition, pavement structure and traffic. The second part is a set of methods and tools that allow the determination of the existing and future pavement conditions, the assessment of financial needs and pavement maintenance technologies.

Each pavement usually maintains its characteristics and performs well for the majority of its life but after reaching a critical point it starts to deteriorate rapidly. By a complex evaluation of the pavement conditions, defining traffic loading and climatic effects, it is possible to determine an optimum time for interim improvements and major rehabilitation just before a pavement's rate of deterioration begins to increase (Figure 8.25).

182   Airport Design and Operation

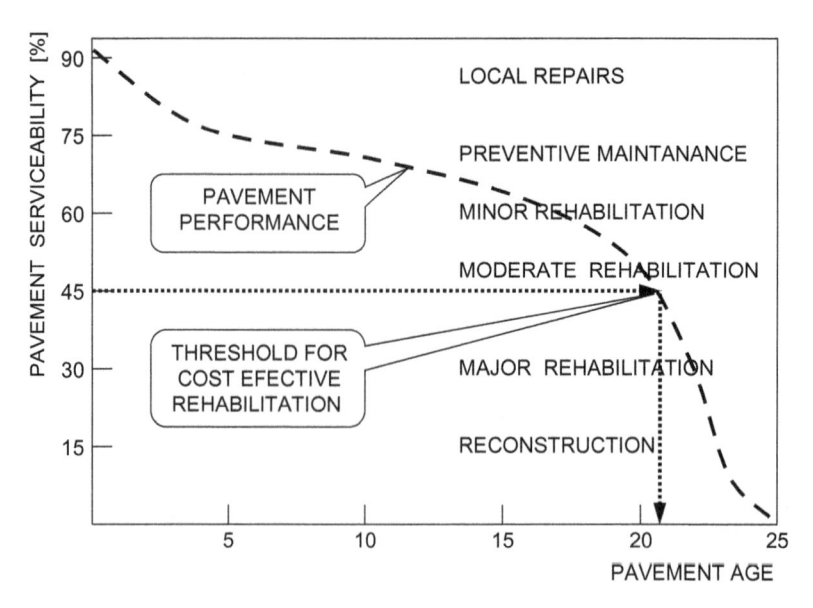

Figure 8.25:   Pavement serviceability and the pavement age.

A pavement management system generally includes:

✈ the information database
✈ structural pavement evaluation
✈ decision making systems
✈ financial requirements
✈ optimisation tools.

The information database is usually divided into subsections containing different types of information. It must contain data on the pavement structure history, pavement construction, materials used, composition of structural layers and thickness. The part of the database with the maintenance history forms a basis for evaluating the maintenance practices and for the cost-benefit analysis. An operational database contains information on the number of movements and type of aircraft using the pavement. Pavement condition data are based on routine and systematic inspections and pavement testing. The database usually contains results of pavement visual inspections, pavement friction characteristics, data from longitudinal and transversal pavement roughness measurements, pavement bearing capacities from deflection tests, etc. For the structural pavement evaluation a number of standard methods are used, but there are also new methods, like ground-penetrating radar, which complement or replace standard tests.

Although pavement management systems have been used since the 1980s, many changes have evolved in the field of pavement management, leading to the continued development and refinement of computerised capabilities and analysis tools.

9

AIRCRAFT GROUND HANDLING

Tony Kazda and Bob Caves

9.1. AIRCRAFT HANDLING METHODS AND SAFETY

Airline companies are the most important customers of any airport. Airlines can minimise their extra investment for growth by increasing aircraft daily utilisation. Many low-cost carriers define the standards for time-effective handling quite strictly. Most of them require the turnaround time not to exceed 25 minutes for a standard Boeing 737 or Airbus 320 en-route operation with the aim of further reduction. A well-known phrase 'the aircraft earns only when flying' holds true. Therefore, the basic requirements all airlines place on the ground handling are:

✈ to ensure safety of the aircraft, avoiding damage to it,

✈ to reduce ground time,

✈ to ensure high reliability of handling activities, avoiding delays.

The airport administration authorities are also interested in reducing the ground time. In order to cope with an expected growth of air transport, most large European international airports will have to invest in further development of their infrastructure. Apron capacity is the limiting factor for some of them. It would, therefore, be more advantageous for them to reduce the ground time and thus increase apron and terminal productivity without significant investments. A possible solution could be an introduction of modern technologies for aircraft ground handling and new concepts of how to cut down delays at the airports. But there are many regional airports with one or only a few aircraft movements a day. When the requirements for technical handling at such airports are compared with those having busy air traffic, a considerable difference can be seen.

The process of ground handling consists of a series of highly specialised activities. In order to carry them out, both highly skilled staff and sophisticated technical equipment are needed.

Four basic approaches to ground handling can be identified:

✈ The aircraft's own technical equipment is used as much as possible

✈ Mobile technical equipment of the airport is used — a 'classical approach'

186   Airport Design and Operation

- Airport fixed installations such as air bridges, fuel hydrant systems, air-conditioning, engine start air supply and electric power connections with a minimum of mobile facilities
- 'vehicle free apron' with most systems buried under apron or installed in underground tunnels and the rest installed on the air bridges.

Each of the above approaches has its advantages and disadvantages. Individual approaches can also be combined. The choice of a particular approach to ground handling is basically influenced by the following:

- airport type, that is regional, hub, low cost, holiday
- airport size and its throughput
- type of flight, short and long-distance
- availability of capacity at the airport
- the intensity of utilisation of a particular stand
- aircraft size
- extent of ground handling required, depending on whether it is a line or end station
- environmental concern of the population.

For short-haul flights which are served with modern jet or turboprop aircraft it may well be required that the total time allotted to the ground handling during a turnaround without refuelling sometimes do not exceed 10–15 minutes. In such cases the aircraft's on-board equipment such as airstairs and the Auxiliary Power Unit (APU) is used for the ground handling as much as possible. The number of ground handling activities is reduced to a minimum. Most turboprop aircraft serving short distance flights cannot use stands equipped with common passenger bridges owing to the distance of the left power plant from the bridge construction and a low door sill height.

At most medium-sized airports aircraft are serviced either by means of mobile equipment or through a combination of mobile systems and airport fixed ground systems. Figure 9.1 shows ground support systems and mobile equipment of the B-777 for a typical turnaround. When compared with a complex installation of the fixed ground systems, the former solution requires less investment. This way of servicing an aircraft is preferred also at those airports where competition among several ground handling companies exists. In such a case only a few companies are ready to invest in ground support systems which will have to be shared with competitors, for example fuel hydrant systems. The largest vehicle on the stand is usually the fuel tanker. The location of the tanker should be clearly marked on the apron, and an exit route should always be available in a forward direction away from the aircraft. The APU exhaust efflux or other dangerous areas should be avoided.

Aircraft Ground Handling **187**

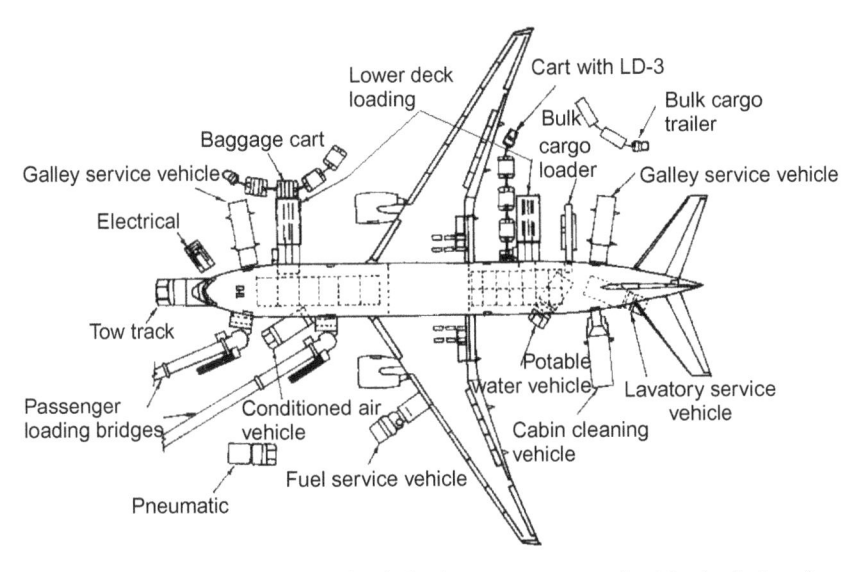

Figure 9.1: The B-777 being serviced during a turnaround with the help of ground support systems and mobile equipment. *Source*: Boeing 777 airplane characteristics for airport planning.

The apron at a large modern airport is not only the concrete or tarmac surface itself. In order to reduce surface traffic across the apron, to improve the apron utilisation and to reduce the ground time, the systems needed for servicing an aircraft are either buried in the surface or located under passenger bridges. The fixed ground systems considerably improve the apron safety by reducing the number of service vehicles on the apron and especially in the vicinity of the aircraft. But even when fixed ground systems are used, around 21 different pieces of ground support equipment might be employed for a standard Airbus 380 turnaround, or for other wide-body aircraft.

Apron accidents and incidents are sometimes not defined as aviation accidents and are investigated according to different standards. This practice contributes neither to apron safety enhancement nor to the creation of a positive safety culture. The work environment on the apron is often not ideal for safe and speedy ground operations. The workers are exposed to congestion, stress, noise, jet-blast, extremes of weather and sometimes to low visibility conditions. The open apron space and differences in aircraft size may lead to a vehicle driver's misperceptions of the distance or the speed. Sometimes the workers are penalised for turnaround delays or schedule disruptions. Handling companies frequently employ low-paid and poorly motivated workers with a high rate of staff turnover. All these factors create potential hazards for apron accidents and incidents.

Apron accidents sometimes involve minor damage only, but when unnoticed or unreported, they may result in an aircraft emergency. Minor damage can also be

188   Airport Design and Operation

expensive for airline companies. Even if the aircraft repair may not be costly, the indirect costs resulting from the schedule changes and consequent disruptions are much higher.

The rate of apron incidents is shown in Figure 9.2. The survey was conducted by ACI in 2004 with 193 participating airports. It was the sixth year for which data has been collected. The main goal of the survey was to help to prevent future apron incidents and accidents. The survey reported 3,233 apron accidents and incidents during handling of 15,119,020 aircraft movements, giving a rate of 0.214 incidents/accidents per 1,000 movements or one incident per 4,676 aircraft movements. The accident/incident rate differs considerably according to which continent is involved. According to the ACI Survey (2005)[1] the highest rate is observed in Africa (0.574) and the lowest in Latin America and Caribbean (0.062 incidents/accidents per 1,000 movements). The pie chart (Figure 9.2) illustrates the percentage distribution of types of accidents and incidents in the year 2004. It shows the types of accidents/incidents which occur most frequently and need special attention. Apron incidents involving aircraft are most serious. These are often caused by passenger handling or aircraft servicing equipment. Damage either to aircraft by moving equipment or to moving equipment by other equipment can be considerably reduced by installing fixed services.

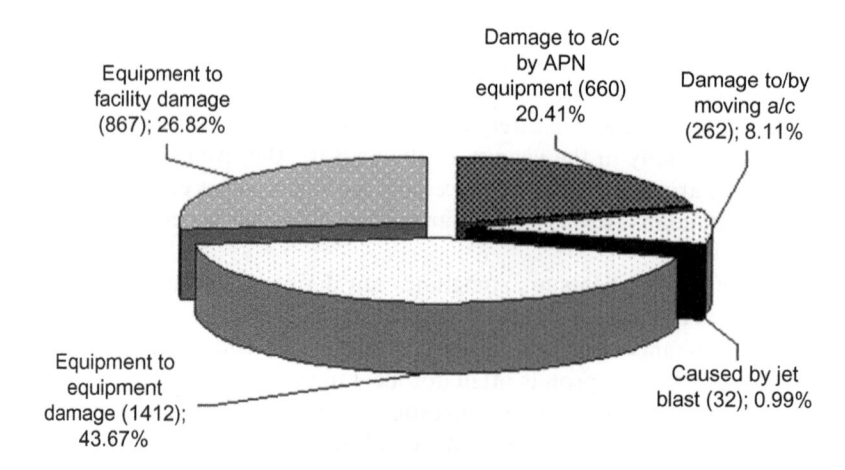

Figure 9.2:   Causes of apron incidents/accidents year 2004. *Source*: ACI survey on apron incidents/accidents, December 2005.

---

1. ACI stopped Apron Safety Survey in 2005. They are going to start a different collection in 2015, based on coordinated definitions and with a wider scope — airfield operational safety in general.

Apron incidents involving modern aircraft, with a high content of Carbon Fibre Reinforced Polymer (CFRP) structures, such as Boeing 787 or Airbus 350, are even more serious. Solid laminate aircraft composites can be internally damaged by impacts in a way that is not visible on the outer surface. Wide-area, or blunt, impact damage from collisions with ground vehicles remains a major issue for those composite structures. Of particular interest are impact sources causing considerable internal damage with minimal visual detectability.

The UK Health and Safety Executive, together with the industry, have been collecting and analysing data under an initiative starting in 2002 to reduce accidents during turn-round. They found that, of those accidents requiring more than 3 day off work, 56% were caused by manual handling, 19% were slips or trips, 9% were hit by a moving object, 6% hit by a stationary object, 3% were falls from height. They have made it a target to limit manual handling of bags to a maximum of 23–25 kg (International Airport Review, 2006).

An apron equipped with fixed ground systems has a lower impact on the environment due to lower noise and exhaust emissions. Equally important is reduced traffic in the manoeuvring area. It also leads to saving on personnel which is significant especially in countries where the cost of labour is high. However, the latest initiative, the so called 'free ramp concept' is also installed at China — Beijing airport. The equipment and networks buried in the apron require higher initial investment costs but, in general, reduce operational costs. It is, therefore, more suitable especially for airports with high daily utilisation of stands. Before the decision concerning a particular type of ground handling is taken, it is necessary to carry out a detailed analysis.

With fixed distribution networks buried either in the apron or fixed on passenger bridges, it is possible to provide the following:

- aircraft refuelling
- electrical power to aircraft systems
- telephone and data connection
- air-conditioning
- compressed air for the engine start
- potable water
- lavatory service
- push back of aircraft
- transport of baggage to/from aircraft
- waste removal.

190   Airport Design and Operation

The ducts for air-conditioning are light and of large diameter, so can be caught by the wind and blown about. This makes it difficult to control when trying to make a connection with the aircraft.

In fixed systems, fuel is always supplied to hydrants through pipes buried in the apron, while the systems on the passenger bridges provide air-conditioning, power and compressed air supplies. It is important that servicing lines between the aircraft and the installation should be as short as possible and they should not cross one another. In the future, especially at airports which accommodate wide-bodied aircraft, a further expansion of ground support systems is to be expected. An example of a 'vehicle free airport' where most of the systems are fixed and buried under apron are Stockholm Arlanda, Beijing and Dubai Airports (see Figure 9.3 showing air-conditioning connections at Dubai Airport).

Figure 9.3:   Fixed ground handling systems 'vehicle free apron' Dubai airport.
Photo: Prabakaran Rao.

Figure 9.4 shows individual technical servicing activities of the B 747 during a turn-around within a time limit of 60 minutes. All manufacturers supply Airport Planning Manuals that indicate the notional time needed for turnrounds without any operational difficulties, and also the location of all the servicing points and the way in which simultaneous activities can be carried out without mutual interference.

Aircraft Ground Handling 191

Next, some other activities arising from aircraft servicing are described, particularly those which implement technologies different from those used at small or medium-sized airports. Aircraft refuelling will be described in Chapter 10.

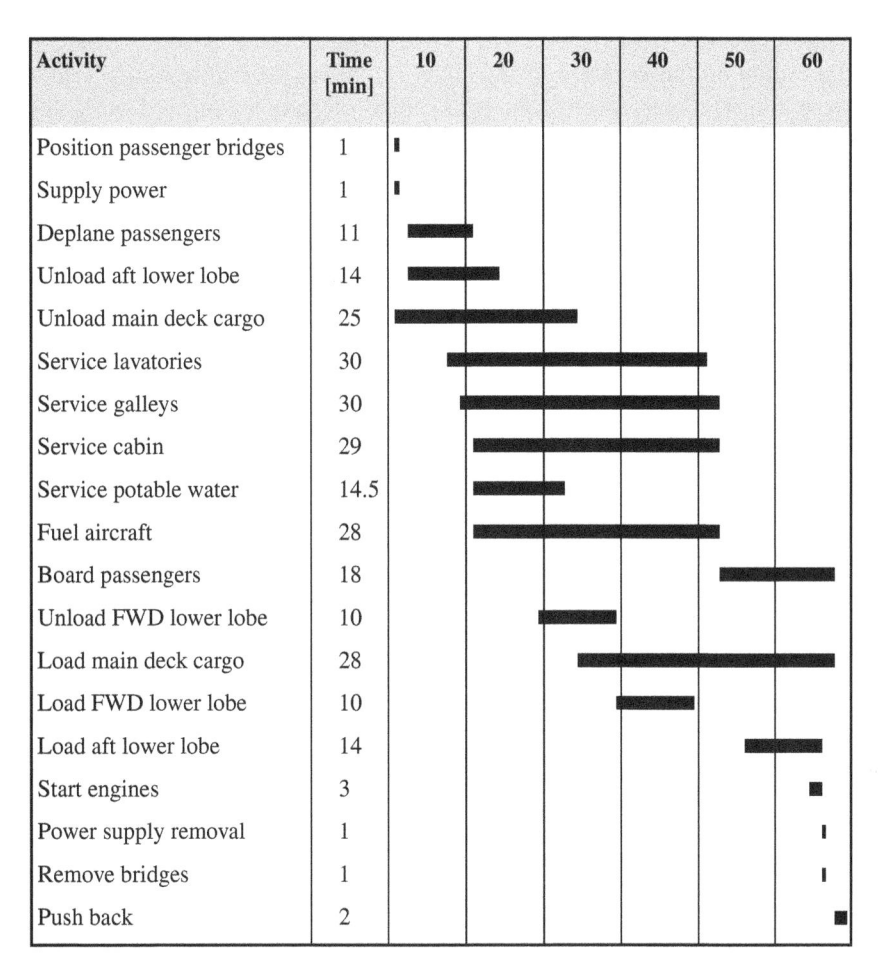

Activity	Time [min]	10	20	30	40	50	60
Position passenger bridges	1						
Supply power	1						
Deplane passengers	11						
Unload aft lower lobe	14						
Unload main deck cargo	25						
Service lavatories	30						
Service galleys	30						
Service cabin	29						
Service potable water	14.5						
Fuel aircraft	28						
Board passengers	18						
Unload FWD lower lobe	10						
Load main deck cargo	28						
Load FWD lower lobe	10						
Load aft lower lobe	14						
Start engines	3						
Power supply removal	1						
Remove bridges	1						
Push back	2						

Figure 9.4: An example of the B-747 servicing at an end station turnaround.

9.2. AIRCRAFT GROUND HANDLING ACTIVITIES

9.2.1. Deplaning and Boarding

Deplaning and boarding can be provided by means of:

✈ stairs carried by the aircraft
✈ mobile stairs

192   Airport Design and Operation

✈ mobile lounges

✈ passenger bridges.

A combination is also possible, for example a bridge attached to the aircraft front door and mobile stairs for the rear exit. The aircraft door sill height varies considerably depending on the aircraft type (Table 9.1). Therefore, it is sometimes necessary to have a range of passenger stairs or a range of stands equipped with different bridges at the ramp.

Table 9.1: Door sill heights of some aircraft types.

Aircraft type	Door sill height (mm)
DC — 92	2130
B — 737	3100
Tu — 154	3200
DC — 8	3200
B — 767 — 300	4160
IL — 62	4200
A — 300	4500
B — 747	4700
DC — 10	4800
B — 777	5280
A — 380 — main doors	5100–5400[a]
A — 380 — upper deck doors	7870–8150[a]

[a]Depending on aircraft actual mass, door position and centre of gravity position.

Although passenger bridges are more costly than mobile stairs, the former are increasingly used also at medium-sized airports. One of their advantages is that passengers can change between aircraft more quickly. While the aircraft is being boarded or deplaned, other servicing activities, including the aircraft refuelling, can be carried out simultaneously. The movement of the servicing vehicles across the apron is not obstructed. The safety of passengers is also ensured as the contact of passengers with servicing vehicles is avoided.

The flow path of the passengers is straight-forward, and they reach their aircraft safely without getting lost. The quality of the whole process is improved as passengers are protected against bad weather. No significant change in level is experienced at the majority of airports and for most types of aircraft, as there are maximum allowable slopes on the bridge ramps.

At a few airports (e.g. Amsterdam Schiphol) the passenger bridges are also used at the aircraft rear door, by extending over the aircraft wing. This speeds up boarding and deplaning and helps to shorten the turn-round time, but is not recommended by IATA. Some airports where the Airbus 380 is handled guarantee simultaneous passenger boarding to the main and upper deck. However, for the 90 minutes turn-round time two bridges on the main deck are sufficient. The upper deck boarding bridge must be able to reach the 8 m (26.2 ft) door sill height (Figure 9.5). In this configuration it is planned to achieve a turnaround time of 90 minutes which is also the case of B 747-400 with fewer passengers on board. The average boarding time is five, and for disembarking eight, passengers per minute per tunnel. However, passenger bridges are more costly than mobile stairs and airline companies are usually charged more for using them. Because of this, low-cost carriers usually prefer mobile stairs or even the aircraft's own stairs.

Figure 9.5: The upper deck boarding bridge for Airbus 380 must be able to reach the 8 m (26.2 ft) door sill height. *Photo*: Courtesy Airbus company.

If the passenger bridge installed has a limited manoeuvrability, the stand has less flexibility. On the other hand, bridges of this type are less expensive than bridges

194 Airport Design and Operation

with apron drive which have the ability to move in azimuth and can be fully adjusted for all aircraft types.

Passenger boarding bridges with two telescopically connected parts consist of the following main parts:

- Rotunda
- Telescopic tunnel
- Pedestal with undercarriage
- Crew cab
- Aircraft servicing equipment.

The bridge can usually be extended by about 10 m and the height can be adjusted within a range of almost 4 m. The maximum incline of a bridge does not exceed 10% (Figure 9.6).

Figure 9.6: A part of the passenger boarding bridge crosses the service road. *Photo*: A. Kazda.

9.2.2. Supplies of Power, Air-Conditioning and Compressed Air

Most aircraft can meet their energy requirements on a ramp with an Auxiliary Power Unit (APU). Advantages of such a solution are independence of the ground source and a small saving on time required to connect/disconnect the ground source. However, the total costs of APU operations, including maintenance costs, are often higher than the charges levied by the airport for using external ground sources. Other disadvantages are low efficiency of the APU (only about 30%), and high noise and environmental pollution due to exhaust gases. At some airports, APUs have to be turned off within a short time after the plane docks. After that, the aircraft has to be supplied with external electrical power of 400 Hz, 115/200 V. The distribution is provided either from the central power source or from a static (non-rotating) converter installed on the loading bridge. Banning APU usage at airports can reduce airport pollution by almost 10%.

If there is no ground support system supplying air-conditioning, the pilot cannot avoid using APU in order to provide ventilating, heating or cooling in the cabin. In this case a ground source of electricity is not usually requested. Before the start of a long-haul rotation, the aircraft needs power so that the maintenance engineers can check the systems. Most modern aircraft have in-flight entertainment (IFE) systems and personal televisions installed in the seatbacks which also require power supply for maintenance during the turn-round.

Airports with Airbus 380 operations must ensure increased requirements for handling equipment. The A380 needs at least 3×90 kVA supplies, and possibly 4×90 kVA, compared with 2×90 kVA for the B 747-400. The ground power connections are 2.6 m above ground, requiring access steps. It also needs four air connections rather than the B 747's two connections. Catering vehicles need to be able to reach the upper deck, which is possible with a double-scissors lift. All other servicing requirements are high but within the capability of existing equipment, though dedicated de-icing equipment is necessary to reach the top of the fin and four underwing pressure connectors are needed to fuel sometimes more than 350,000 l of Jet A 1 fuel in 90 minutes.

Air-conditioning has to be provided even in mild latitudes. Due to the high concentration of people aboard, cooling of the cabin has to be provided when the temperature rises above 10 °C. When the temperature is between 4.4 °C and 10 °C air has to be circulated, and when it drops below 4.4 °C the cabin has to be heated. Central distribution systems can supply air of −6.5 °C in summer and +66 °C in winter. In order to start the engines compressed air of 230 °C is supplied.

9.2.3. Cargo and Baggage Loading

Cargo and baggage loading is a time-critical activity especially when handling wide-bodied aircraft serving long-haul routes and on charter flights. When servicing

196   Airport Design and Operation

the B 747, it is necessary to load and unload about 6500 kg of baggage and in the case of the A 380 this will rise towards 10,000 kg. An advantage of modern wide-bodied aircraft is that baggage is transported in containers. This reduces the number of bags going astray as the baggage for any one destination is stored in the same container. The containers are towed to the bag hall by tugs, in trains with a maximum of five dollies per train, or three in low friction conditions due to the greater braking distances.

Figure 9.7: Power stow enables baggage and cargo to be transported directly between the worker at the stack in the hold and the worker at the baggage cart.
Photo: Courtesy Power Stow A/S.

Handling loose cargo and baggage is the most serious problem because such handling is time-consuming. Some airports use transporters to handle them, or they build systems of conveyers to transport the baggage from the departure hall directly to the aircraft. One possible improvement is the Power Stow technology; a belt loader extension that is built into a mobile belt conveyor in order to facilitate the loading and offloading of passenger baggage and cargo in the aircraft cargo hold. It allows faster and efficient loading and unloading of bulk baggage and cargo, while improving working conditions for handlers (Figure 9.7).

It is becoming desirable to power the tugs electrically, otherwise the bag hall has to be ventilated and the total of airport emissions increases. More detailed information on cargo handling can be found in the Chapter 11.

9.2.4. Push Back Operations

In order to improve apron utilisation, most of the large international airports are equipped with air bridges and nose-in aircraft stands. When ready for taxiing, the aircraft has to be pushed back into a position where it can use its own engines. Aircraft tugs of different designs are being used to push back the aircraft.

The original push back unit was a traditional combination of a tractor and a towbar. An advantage of such a unit is a relatively lower initial cost when compared with other modern equipment. A disadvantage, especially at major airports, is the need for keeping different towbars for each type of aircraft. The towbar costs increase the overall investment needed for handling the nose-in stands. One towbar tractor requires usually two people to operate it (Figure 9.8).

Figure 9.8: A towbar tractor being connected to an aircraft. *Photo*: A. Kazda.

When compared with some other methods, the towbar method has a higher labour requirement. Other disadvantages, from an operational point of view, are low speeds of operational towing, for example between apron and a hangar, and problems arising from pushing back heavier aircraft under poor braking conditions on the apron. In order to provide the necessary adhesion, the tractors designed to handle the heaviest aircraft must also be very heavy. This results in their higher fuel consumption. At some airports electrically driven tractors are also used where adhesion is guaranteed by the weight of batteries.

198 Airport Design and Operation

There are several types of towbarless tractors designed to push back or tow aircraft. PowerPush from the German company Schopf has developed a push back vehicle on a fundamentally different basis (Figure 9.9). The PowerPush is a towbarless push back tractor that can handle aircraft up to 100 metric tons. Unlike other vehicles for moving aircraft, this unit connects to the main landing gear, rather than to the nosewheel of the aircraft. Profiled friction rollers are placed hydraulically in contact with the tyres of the main gear wheels.

Figure 9.9: A towbarless PowerPush tractor, Schopf company. *Photo*: A. Kazda.

The rollers are driven by a hydrostatic engine and make the wheels of the main gear rotate. The pilot controls the aircraft according to directives received from the PowerPush operator. The aircraft's brakes can be used without damaging either the aircraft or the push back vehicle. When compared with conventional vehicles, this unit is very light, powered by only a 58 kW (78 HP) engine, it is simple, and both its operational and initial costs are lower.

It can be remotely controlled. In such a case the operator of the vehicle has a direct visual contact with the pilot of the aircraft. Its disadvantage is that it can be used only for push back operations on an apron over a distance of from 50 to 150 m, and it is designed only for aircraft with one row of wheels on the main landing gear. The airport therefore needs to have other vehicles designed for operational towing.

Many companies currently offer towbarless tractors designed for push back or operational towing at high speeds. In most cases the tractor picks up and lifts the aircraft's front wheel off the ground hydraulically (Figure 9.10). The wheel is clamped firmly. When compared with a conventional type, this tractor can be lighter as the required adhesion force is induced by the aircraft itself. The shocks which are transmitted to the nose undercarriage leg when a towbar is used for towing or pushing back, are, in this design, absorbed partially by the aircraft tyre. The main benefit of this system would come when towing an aircraft from the apron to the end of the runway before take-off. The economic benefits are fuel savings on taxiing, for example a B 747 at Frankfurt needs on average 607 kg of fuel for taxiing to the end of the runway. Consequently, there is a cleaner environment at the airport and its surroundings. For countries with high labour costs it is also of interest that the tractors of this design are operated by a single person. The disadvantage is a higher initial cost of the unit. In fact, due to persistent operational and legislative problems, it has not yet been generally possible to introduce operational towing.

Figure 9.10: Heavy Goldhofer towbarless tractor. *Photo*: A. Kazda.

The design of the aircraft's nose gear legs have not been designed to transfer loads when towing an aircraft at higher speeds for long distances, but only for short distances on push back. It would be necessary to build new special tow tractor roads for the return journeys to the apron from the holding bays. There is also resistance from the airlines, who fear that any problem with starting the engines could mean a long return trip to the stand. Despite all the questions mentioned so far, the price of the vehicles was already reduced and they are commonly used in many airports.

A new emerging technology in the phase of testing is a WheelTug. The WheelTug or e-taxi system is a high torque electric motor mounted into the hub of the nosewheel and powered by the aircraft's APU. It allows backwards movement without the use of push back tugs and also forward movements without using the aircraft's engines. However, there are operational limitations due to the engine start-up procedures and cooling of engines after landing. This might limit usage of WheelTug to only airports with long taxi times.

9.3. Collaborative Decison-Making (CDM)

The biggest advantage which air transport has compared with other transport modes is speed. However, even though the planes get faster, some airlines have increased the scheduled flight times on their routes to accommodate growing delays. Since 2000 the en-route delays have dropped but the delays at the airports has grown considerably. These disruptions in the airport operations have a domino effect and hence a widespread impact. At an airport, for example, some of the handling partners services might be delayed or not delivered, there could be a failure found on the vital airport systems after the engine start up etc. Small disruptions of 2–5 minutes could result in hours at the end of a day.

Sharing the latest information on the operational issues, data exchange and the enhancement of the operational decision making capabilities make air traffic more predictable and result in minimising disruptions. However, some partners may be reluctant to release certain information when they consider them as 'commercially sensitive'. Also the information systems of the main CDM partners were usually built without co-ordination and are incompatible.

To increase the airspace and control capacity while maintaining safety, as well as to meet future air transport needs in the ECAC states, a new Gate-to-Gate and ATM Strategy 2000+ was adopted. Principal points of this new concept among others are the following:

✈ Flight management from gate-to-gate
✈ Enhanced flexibility and efficiency

✈ Collaborative decision making (CDM)

✈ Collaborative airspace management.

The CDM concept was successfully tested at San Francisco International Airport at the beginning of 1998 in order to cope with large reductions in capacity and then it was extended to all US airports by the end of 1998.

According to Airport CDM (A-CDM) Operational Concept Document (Eurocontrol CDM, May 2006) Airport Collaborative Decision Making is a concept which aims at improving Air Traffic Flow and Capacity Management (ATFCM) at airports by reducing delays, improving the predictability of events and optimising the utilisation of resources. In this operational framework all airport CDM partners (at least airport and aircraft operators, handling companies, ATM and CFMU) can optimise their decisions in collaboration with other A-CDM partners to improve the overall efficiency of operations at the airport, with a particular focus on the aircraft turn-round and pre-departure sequence. This is possible due to the fact that everyone knows the preferences and constraints of the other shareholders as well as the actual and predicted situation. The decision making is achieved by the sharing of information, by common procedures, mechanisms and tools. The A-CDM concept is divided into the following elements:

✈ Airport CDM Information Sharing

✈ CDM Turn-round Process — Milestones Approach

✈ Variable Taxi Time Calculation

✈ Collaborative Management of Flight Updates

✈ Collaborative Pre-departure Sequence

✈ CDM in Adverse Conditions

✈ Advanced CDM.

CDM Information Sharing is a basic step for implementing of any other element and for this reason it must be put into practice first. It is also a pre-requisite for creating the common situational awareness (CSA) of each stakeholder and for improving the co-ordination, traffic predictability and operational planning of the A-CDM partners. It is also the core A-CDM Element and the foundation for the other Airport CDM Elements. The *CDM Turn-round Process (Milestone Approach)* is aimed at particular turn-round processes and the prediction of their termination by describing progress of a flight from the initial planning to the take-off from a A-CDM by defining Milestones to enable close monitoring of significant events. The goal is to ensure a common situational awareness and to predict the imminent events for each flight. The interrelations of the processes and their linking to the flight planning and CFMU must be ensured. The milestones contribute substantially to the pre-flight and post-flight predictability. The CDM Turn-round Process and the A-CDM Information Sharing Element are the foundations for the other A-CDM elements.

By implementing the *Variable Taxi Time Calculations* the accuracy of taxi-in and taxi-out times to improve the estimates of in-block and take off times, so improving the traffic predictability.

Collaborative Management of Flight Updates provides the required operational flexibility of ATFM to facilitate modifications of departure times with respect to traffic changes and operators' preferences to provide estimates for arriving flights to CDM Airports and improve the ATFM slot management process for departing flights. Variable Taxi Time Calculation and the CDM Turn-round Process elements must be operational before implementing this level of CDM.

Collaborative Predeparture Sequence is a pre-flight 'count down' sequence of aircraft which are planned to depart from their stands (push off blocks). It contributes to flexibility, predictability and to the optimisation of the airport partner's facilities and equipment and for expressing their preferences.

Airport capacity during low visibility operations (see also Section 3.3) could be enhanced by introducing *CDM in Adverse Conditions*. This element facilitates the dissemination of capacity changes and recovery after operational disruptions. CDM in Adverse Conditions also ensures flexibility and optimum use of airport CDM partner resources. The *Advanced Concept Elements* are not defined yet. These Elements will further improve A-CDM partners' situational awareness and increase the collaboration by introducing airport advanced control and surveillance technologies and tools like A-SMGCS, AMAN/DMAN.

The most important CDM goal is focussed on the improvement of flight operations by participation of airline companies, airport operators and handling companies in the process of Air Traffic Management (ATM). This is achieved by information management systems and procedures and implementing functions which take into account internal priorities of particular aircraft operators and/or other stakeholders to make full use of the shared data. CDM partners can in this way synchronise their operation but also support other shareholders by feedback.

A-CDM not only improves time efficiency but has also significant impact on the environment. Paris Charles de Gaulle reports not only a reduction of about 7% in average taxi times but also savings of 14.5 tonnes of fuel per day. Brussels Airport reports reduced taxi-out time by a quarter, and cut taxi-in time 15% since it introduced A-CDM. Fuel savings for the airlines amount to 7200 tons a year, while reducing carbon emissions by 22,700 tons and nitrogen oxide by 30 tons.

The CDM concept generates very high added value with relatively low costs. However, the full benefits will only be enjoyed after implementation of the project scope on a wide European basis. The whole CDM project is planned in four basic stages:

Level 1   Airport CDM Information Sharing and CDM Turn-round Process
Level 2   Collaborative Management of Flight Updates and Variable Taxi Time Calculation
Level 3   Collaborative Pre-departure Sequence and CDM in Adverse Conditions
Level 4   Other CDM applications.

The number of fully implemented A-CDM airports in Europe in December 2014 stands at 14; Munich, Brussels, Paris CDG, Frankfurt, London Heathrow, Helsinki, Düsseldorf, Zurich, Oslo, Rome, Berlin Schönefeld, Madrid, Stuttgart and Milan Malpensa.

Airport CDM control and also airport operation control at smaller airports are performed by the airport dispatching the flight. At very small airports up to one million passengers a year it could be just a site manager with mobile phone and walkie-talkie who can spend most of his time at the apron to control the operation or, in some cases, a simple control room with necessary planning software and one working position. For airports with throughput of more than two million passengers/year it might be necessary to interconnect airside and terminal control to allow advanced planning in case of disruptions. At large airports with more than 10 million passengers/year the airport dispatching the flight has usually more working positions for different sections of operation, for example apron, security, baggage handling systems, terminal and airport access (Figure 9.11).

Figure 9.11:   Airport dispatching at a large airport has usually more working positions. *Photo*: B. Kandera.

9.4. VISUAL GUIDANCE SYSTEMS

Semi-fixed passenger loading systems with a small range of adjustment are less expensive than the fully mobile bridges. They do, however, have less flexibility and need precise positioning of the aircraft in order to avoid damage and also to attach the equipment needed for the aircraft servicing.

It is, therefore, particularly necessary to install visual guidance systems to facilitate the positioning, though guidance systems are usually installed also at stands equipped with fully mobile bridges. Exact positioning is possible also when the aircraft is guided manually by a marshaller, but when an airport accommodates wide-bodied aircraft, the pilot often cannot see the marshaller easily from the cockpit, for example if the aircraft stand is close to the terminal building. When the aircraft is damaged, it is often difficult to decide whether it is the marshaller or the pilot who is to be blamed for the damage. At large airports where passenger boarding bridges are usually installed it is therefore vital to implement guidance systems.

One of the first guidance systems was installed at the airport in Atlanta, Georgia in 1956. The simple and cheap equipment consisted of a stick attached on the passenger loading bridge to a tennis ball with a string. It was installed at stands designed for the Douglas DC-6. The aircraft was parked correctly when the tennis ball touched the windscreen. A disadvantage was that the guidance system could be used for one aircraft type only or it would be necessary to move the tennis ball.

A modern but more advanced modification of the above mentioned stick, rope and tennis ball is a system with a hydraulically controlled arm. Data about types of aircraft using the stand are stored in a computer memory. The computer adjusts the position of a robotic arm to the position required for the pre-selected aircraft type. At the end of the arm there is a light tube containing sensors. The aircraft has to touch the tube with its windscreen. If the aircraft does not stop, the arm keeps moving with the aircraft.

The Swedish company Inogon International AG makes use of two indicators working on a principle of interference of light to guide the aircraft along the centre-line and to indicate distance. In a box-like indicator there is a fluorescent light below which two grids are located and where interference of light takes place. The resultant patterns are called moiré (Figure 9.12). Their shape depends on the line of vision. The centreline beacon provides directional guidance for the aircraft. Another indicator located obliquely from the pilot's line of vision gives the distance of the aircraft from the stop position at the stand. Black lines shaped like arrows direct a pilot to the correct position on the stand. In the stop position the pilot can see black perpendicular lines in the centre of both indicators. The system has no moving parts, and its production and operation are simple. Another modern system installed at Stockholm Arlanda, is a system which uses data from radar. The main advantage of this system called APIS (Aircraft Parking and Information System) is that it requires no sensors located on the tarmac. The radar measures the distance of the aircraft to its final position where it is to stop. The pilot maintains direction by following the indicator which makes use of interference of light on a grid, as already mentioned.

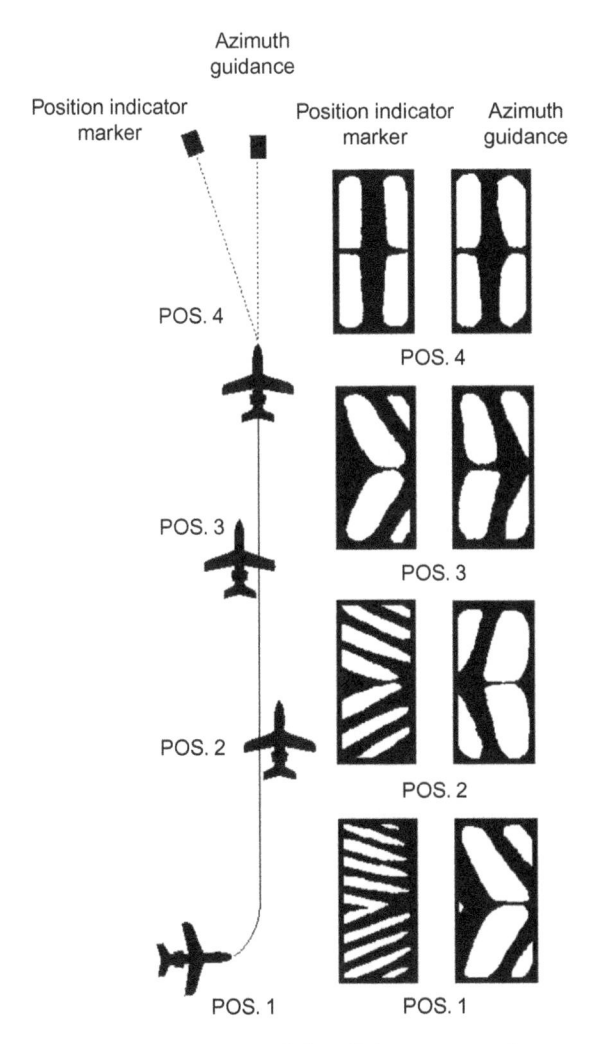

Figure 9.12: Visual guidance system of Swedish company Inogon working on the principle of interference of light.

206   Airport Design and Operation

Exact parking of the aircraft is enabled also by another parking system produced by Safegate SATT Electronics AB (Figures 9.13 and 9.14). Its advantage is that information about direction and distance is provided at a single source. Centreline guidance is provided by a vertical illuminated green bar which is, from the pilot's line of vision, displayed in a yellow symbol of an aircraft painted on the surface of the guidance panel. The information about forward motion of the aircraft is presented on a vertical panel through sequential activation of green lights.

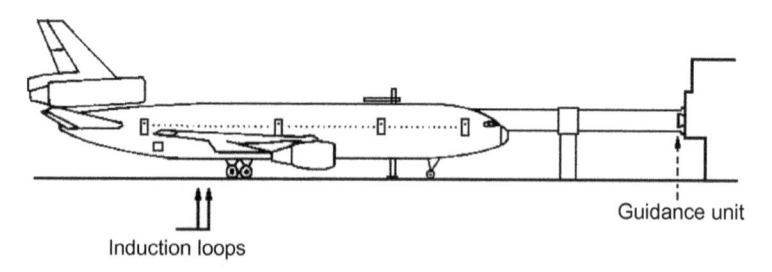

Figure 9.13:   Principle of Safegate visual guidance system.

Exact, final positioning of the aircraft is indicated by flashing red lights; however, depending on the light intensity, the red light will appear in any "Fifty Shades of Grey" on black and white photo, similarly as perceived by color blind people (Figure 9.14). The position of the aircraft nosewheel is sensed through induction loops placed under the apron surface or through a different system of nosewheel distance measuring. The data concerning the aircraft position are processed by means of a microprocessor. Before the manoeuvre starts, the data about the aircraft type have to be recalled from the memory of the computer. They are, for confirmation, displayed in the top part of the guidance panel.

Different technology is used in Honeywell's Visual Docking Guidance System (VDGS) (Figures 9.15 and 9.16). This fail-safe aircraft guidance system utilises a high dynamic range video sensor unit and a powerful image processing system based on 3D aircraft models. The computer-assisted VDGS calculates precise information on an aircraft's location and transforms it into precise guidance information for both pilot and co-pilot.

Its video sensor and 3D model-based processing system are able to recognise the outline of an approaching aircraft at distances of up to 100 m. The Sensor Unit is a high dynamic range, passive video camera which could be also used for the aircraft stand permanent monitoring and thus it can enhance the safety and security of the stand operations. The system provides continuous data registration and could be used as an input for automatic reporting of ON-BLOCK/OFFBLOCK times in the CDM concept (see Section 9.3).

Aircraft Ground Handling 207

Figure 9.14: Typical installation of the Safegate guidance system. *Photo*: A. Kazda.

Some airports today prefer relatively low-cost and simple systems. The airport authorities are less keen to use equipment requiring sensors and cables which need to be installed on the tarmac. When new types of aircraft are later introduced and, consequently, the stand configuration is to be changed, relocation of sensors is a costly procedure.

Among these low-cost, simple and reliable systems is one called AGNIS (Azimuth Guidance for Nose-in Stands) which is widely used at many airports for directional guidance of aircraft to stands, with a Side Marker Board (SMB) or Parallax Aircraft Parking Aid (PAPA) to indicate the distance of the aircraft from its final position. Any skilled carpenter or electrician is able to construct SMB and PAPA.

The SMB indicator consists of a white indicator placed on the pilot's left-hand side, where the passenger boarding bridge is. Perpendicular boards indicating a particular aircraft type are installed on the indicator vertically in distances which correspond to each aircraft type. The front part of the board, seen by the pilot, is painted green, the edge is black and the reverse side is red.

208   Airport Design and Operation

The pilot has to stop so that he can see only the black edge of the board which corresponds to the particular aircraft type. A disadvantage of the system is that the accuracy of parking depends on the position of the pilot's head. When the pilot has long legs and his seat is well back, he parks 'long' and the pilot with short legs parks 'short'. Some cases were reported with a B 737 that, when the pilot was too short, a stall speed indicator placed near the aircraft door was damaged.

The AGNIS indicator is placed on the stand centreline. Two vertical illuminated light tubes emit green and/or red light through openings, thereby indicating a position of the aircraft on or away from the parking centreline. A disadvantage of this system is that the pilot has to refer to two sources for information.

Figure 9.15: Honeywell's Visual Docking Guidance System (VDGS). *Photo:* Courtesy Honeywell.

The Parallax Aircraft Parking Apparatus (PAPA) (Figure 9.17) is located to the side of the stand centreline. The indicator consists of a black board with a longitudinal opening. Aircraft positions are indicated through white perpendicular lines showing a particular aircraft type. At a certain distance behind the board there is a fluorescent light tube placed vertically. When taxiing to the stand the fluorescent light in the opening seems to move. When the aircraft is parked accurately, the pilot can see the fluorescent light tube behind the white line which corresponds to the particular aircraft type.

The principle on which the Burroughs Optical Lens Docking System (BOLDS) system operates, used mostly in the United States, is similar to that of the PAPA system. A difference is that the distance is indicated by means of a tube placed horizontally, and the direction through a vertical fluorescent light. Optics consisting of a Fresnel lens are placed in front of the light source. An advantage of the BOLDS, shared with the Safegate Docking System and UCRAFT, is that it provides information on both direction and distance from one source. When compared with the latter systems, the former is simpler and cheaper.

Figure 9.16: Honeywell's VDGS installation at Brussels airport. *Photo*: A. Kazda.

Also simple and low cost is the equipment installed at Gatwick in London for exact positioning of aircraft. Positions of the nosewheels of different aircraft types are painted on the apron surface. In front of the stand, on the left from the pilot's line of vision, there is a mirror in which the pilot can see the nosewheel of the aircraft when parking to a desired position. In winter the mirror is heated.

210   Airport Design and Operation

Figure 9.17:   Parallax Aircraft Parking Apparatus (PAPA). *Photo*: A. Kazda.

This survey of visual guidance systems is not complete. The choice of a suitable system depends on many factors, mainly traffic density, number of aircraft types using the particular stand and also on prevailing meteorological conditions. Future development of guidance systems will follow two paths which can be called, respectively, 'hi-tech' and 'beauty is in simplicity'. Meanwhile, the proliferation of types of guidance system can cause problems for pilots due to lack of consistency.

10

AIRCRAFT REFUELLING

Tony Kazda and Bob Caves

Motto: Fuel is the blood of aviation

10.1. BACKGROUND

In the early days of aviation, storing of fuel and aircraft refuelling was an undemanding activity. Piston engines had low consumption and the aircraft was not able to carry a large quantity of fuel. The aircraft was refuelled from barrels or tankers. It was not necessary to build fuel stores at the airport. The same fuel as for cars was used for aircraft.

Initially, the requirements for quantity and quality of fuel gradually increased. Introduction of the jet aircraft in the late 1950s changed overnight the requirements, both qualitatively and quantitatively, for storing and supplying the aircraft with fuel. New aircraft required greater quantities of fuel, yet had to be refuelled within the normal turnround time of less than 1 hour. Another great change occurred in 1970, when the wide body Boeing 747 was introduced into operation. About 150,000 l of fuel had to be transported through four refuelling openings with a diameter of 64 mm into the tanks of that aircraft at a height of 4.5 m within approximately 45 minutes. If there was no hydrant fuel system at the airport, refuelling was provided by tankers with a capacity of up to 80,000 l of fuel. At present the aircraft kerosene Jet A1 is the most common kind of fuel. For comparison, Table 10.1 gives volumes of the tanks of the selected types of aircraft.

According to Index Mundi (2011) the fuel consumption of jet fuel in barrels in the air transport sector increased from 670 million in 1984 to 1900 million in 2010. This represents 4.38% growth per annum. However, according to the World Bank data (2013) the passenger average growth for the same period was 5.25% per annum. Even more interesting is the trend for the period 2000−2010 with fuel consumption growth 1.67% per annum and average passenger growth 4.95%, for the same period. Thus consumption of fuel has risen less rapidly than the air transport traffic.[1]

---

1. Hromádka (2014).

Table 10.1: Fuel tanks capacity of selected types of aircraft.

Aircraft	Maximum fuel tank capacity (standard) [l]	Maximum fuel tank capacity (long-range version) [l]
Turboprop		
ATR 42	5,625	
ATR 72	6,173	
Dash 8-Q400	6,616	
Jet		
BAe 146	11,728	12,901
Embraer 170/190	11,525	16,013
B 737-500	20,105	23,170
A 320-200	24,210	30,190
B 737NG 600-900ER	26,020	
A 300-600 R	68,150	73,000
B 767-200/-200ER	63,000	91,400
DC 10-30	138,294	153,434
A 340-200	155,040	
B 787-8/-9	126,210	138,700
A 350-800/1000	129,000	156,000
A 330-200	139,090	
B 747-100B	183,380	
B 777-200LR	202,270	
A 340-500HGW		222,000
B 747-400/400 ER	216,840	241,140
B 747-8I	243,120	
A 380-800	323,546	352,000

Source: Aircraft manufacturers data.

The latest piston engine airliners of the late 1950s such as the Lockheed L-1649 Starliner and Douglas DC-7 were as efficient as current jet and thus 70% more efficient than the first jet aircraft, which replaced them in the airline fleets. It must be noted that at the time piston engines were completing their life cycle, while emerging jets were at the beginning of their cycle.

Nonetheless, jets have about twice the cruise speed and therefore higher drag. Aircraft fuel consumption per passenger kilometres declined by 70% between 1960 and 2000.

This improvement in fuel productivity was achieved by jet engines efficiency (69%) and aircraft aerodynamics (27%) but also:

↳ a gradual growth of the percentage of aircraft seats filled (load factor), by the European airlines in particular

↳ a gradual substitution of new modern types of aircraft with lower unit consumption than the old ones

↳ improvement of flight procedures (vertical flight profile and cruising speed optimisation; clean configuration and continuous descent approaches; aircraft take-off mass optimisation, reduced thrust reversal, etc.) and wide use of flight management systems.

Fuel efficiency of the latest generation of jets goes even further. Fuel efficiency of most of them is better than most modern small passenger cars in terms of fuel economy per passenger kilometre (Table 10.2). However, it is difficult to compare aircraft fuel efficiency. For calculations different underlying assumptions, especially the average flight length and the number of passengers on board are used, but it is influenced also by the actual take-off mass and load factor. It is therefore possible to find different values of fuel consumption per 100 passenger kilometres for the same type of aircraft.

Table 10.2: Aircraft fuel efficiency.

Aircraft	First flight	No. of seats	Fuel efficiency [l/100 PAXkm]
Medium haul — 5600 km sector (except 767-400ER: 6,047 km)			
Boeing 767-400ER	1999	304	2.43
Boeing 747-8	2011	467	2.59
Boeing 777-300	1997	368	2.61
Boeing 777-200ER	1996	301	2.89
Airbus A330-300	1992	262	2.98
Airbus A330-200	1997	241	3.11
Airbus A340-300	1992	262	3.25
Long haul — 11,000 km sector (except A380: average)			
Boeing 747-8	2011	467	2.75
Boeing 777-300ER	2003	365	2.84
Airbus A380	2005	555	2.9
Boeing 777-200ER	1996	301	3.08
Boeing 747-400	1988	416	3.26
Airbus A330-200	1997	241	3.32
Airbus A340-300	1992	262	3.49

Source: Aircraft manufacturers data and Wikipedia (Retrieved from http://en.wikipedia.org/wiki/Fuel_economy_in_aircraft. Accessed on 30 August 2014).

10.2. FUEL — REQUIREMENTS

The basic requirements for the provision of fuel at an airport are to ensure:

✈ high purity of fuel
✈ sufficient supply of fuel
✈ economic and rapid supply of fuel to aircraft
✈ environmental protection
✈ high safety standards of the operation.

10.2.1. Requirements for Fuel Quality

In order that the stringent safety requirements can be observed, the international organisations in co-operation with the major oil companies have prescribed criteria that the aircraft kerosene Jet A1 must comply with.

During the flight of an aircraft at high altitudes, the temperature in the aircraft tanks drops under −40 °C. Therefore the fuel must not freeze even during long-time exposure to very low temperatures and must have the required viscosity. A resistance against freezing depends in particular on the proportion of paraffin in the fuel, which crystallises at low temperatures. With Jet A1, crystals of paraffin may not occur in the fuel above a temperature of −47 °C.

Fuel is used in the aircraft for cooling the on board equipment, and therefore it may be exposed to high temperatures. Even at high temperatures the fuel must be thermally stable and must not create deposits. Thermal instability of fuel may be caused by the presence of trace elements. Their precipitation is obtained by the use of additives to the fuel.

Fuel must not be corrosive so that it cannot damage parts of the fuel system and the aircraft engines. Corrosion is caused mostly by the presence of sulphur in the fuel. Sulphur is removed from fuel during its production. Some compounds of sulphur, for example hydrogen (mono)sulphide, may be created by the activity of micro-organisms.

Fuel must not contain active surfactants. Surfactants, or surface active agents, are chemical compounds present in the fuel which are attracted to surfaces and interfaces. These can be present in the fuel from refinery carry over or from cross-contamination in multi-product pipelines. The active surfactants act on the water as emulsifiers and prevent its removal from the fuel. The active surfactants are removed from the fuel by means of clay filters which are located before the micro filters.

The fuel must not contain more than 0.003% of free water. With higher contents of free water, the ice crystals might block the filters after a drop in temperature. In addition, the presence of water supports the growth of micro-organisms in fuel. The free and dissolved water is removed from the fuel during its production in refineries. It leaves the refinery within its specification. However, during its transport and storage, water gets into it by the so-called 'breathing' of the storage tanks. 'Big breathing' indicates a penetration of fresh air into the storage tank while it is being emptied. The atmospheric humidity condenses when there is a variation of temperature on the walls of the storage tank, and water gets into the fuel. 'Small breathing' occurs in the storage tanks during a change in the volume of fuel and air, as a consequence of variations of temperature between day and night. In happens in railway containers in particular. Kerosene is hygroscopic, and in a damp atmosphere it readily absorbs water. In addition to free water jet fuel contains some quantity of dissolved water. This amount expressed in millionths of its volume is approximately equal to the fuel temperature in °F. For example in jet fuel at 70 °F (21.1 °C) there is approximately 70 ppm of dissolved water. When the fuel is cooled to, for example 20 °F (−6.7 °C), it contains 20 ppm of dissolved and 50 ppm of free water. Thus, if the aircraft is exposed to low temperatures, for example at high altitudes, fuel cooling causes condensation of water dissolved in the fuel. If the temperature drops below freezing point (32 °F or 0 °C), condensed water droplets begin to form ice crystals which can block fuel filters. For example, wide body jet refuelled at Dubai airport at an outside temperature of 40 °C and a relative humidity of 100% with 200,000 l of Jet A1 could release 20 l of free water in the aircraft tanks during flight, in average. Therefore, fuel additives in the form of so-called icing inhibitors are added into fuel.[2]

Therefore water should be removed from the fuel every time it is transferred from one container to another. Similarly, solid impurities get into fuel during its transport and storage. They are mostly rust and dust from the walls of the containers. As it enters the aircraft's tanks, the fuel must not contain more than 3 g of solid impurities per 1,000 kg of fuel with the size of impurities not greater than 5 µm.

Chemical contamination of kerosene usually occurs accidentally when mixing jet fuel with other petroleum products. This can significantly affect the chemical and physical properties of the fuel and can be detected by laboratory tests only. Contamination can be avoided by isolating fuel from other products; primarily in the process of delivery to the airport as the airport itself operates a dedicated system planned exclusively for the specific type of fuel. Contamination is most often caused by transportation in boats, wagons and tankers insufficiently cleaned from the previous product, that is during transport in so-called multi-product delivery.

---

2. NAVAIR (1992).

216   Airport Design and Operation

It can be difficult to detect contamination without laboratory tests. However, unusual odour, colour or appearance may be some indicator for an experienced worker.

Large volumes of aircraft kerosene must be pumped into the tanks of an aircraft within a relatively short time while ensuring its high, almost pharmaceutical quality. This is possible only with correct construction and maintenance of the equipment and by following the prescribed procedures.

The construction of storage tanks must eliminate any possibility of water or impurities getting into the storage tanks. The bottom of the storage tank must be given a gradient, and at the lowest point there must be a drain to drain off the water and sediment. Sedimentation, sludge and water discharge are possible due to the conical bottom of the tank. This should, according to the Airport Development Reference Manual, have a slope of 1:50 (for horizontal tanks) or 1:30 (vertical tanks). Although a number of new concepts of shaped tanks were proposed, tanks with conical bottoms are still the most used in the airline industry as they provide the shortest way for water to flow to the low-point of the tank. This is particularly important in cases when minimum settling time is applied.

Similarly, the pipework must also have a minimum gradient and at its lowest point be equipped with sediment discharge valves. Pipes are usually designed with slope minimum 0.5% thus facilitating the fuel flow and allowing installation of low points. Accumulated water and other debris are then vacuum-sucked by special equipment. The low points are located along the pipeline in easily accessible special pits usually protected by aluminium hatches. The high points are designed for pipeline venting and are usually placed where pipes enter into the ground. Low points are subject to monthly maintenance, and high points on a quarterly basis. The pipeline is cleaned by a special foam cartridge. The charge is pushed from one maintenance point to another and pushes dirt and water ahead of it.

Treatment of the fuel during its transport and storage must ensure that, when used, the quality of fuel approaches the same quality it had when it was leaving the refinery. The delivery of fuel must be accompanied by a certificate guaranteeing its quality. The certificate should be presented whenever the fuel is handled. Techniques to maintain the quality include sedimentation, discharge of the water and sediment, suction from the surface, filtration and separation.

After delivering the fuel into storage tanks in the airport, the fuel must rest quiescent for a prescribed period of time in order that the impurities and free water can settle to the bottom of the storage tank. The time for sedimentation of the impurities depends on the height of the fuel column in the storage tank, and is about 3 hours per metre. Therefore, in the airport there should be at least three storage tanks so that, at any time, one of them is being filled, another resting quiescent, and aircraft refuelling is done from the third.

Suction of the fuel from the surface ensures that the relatively cleanest fuel is drawn out. A suction pipe is divided into a fixed part and a movable part which are connected with a joint. In the end of the movable part there is a float which maintains the intake of the suction pipe just below the surface. However, this technology is not used in new, modern fuel stores where the fuel quality is maintained by other hi-tech solutions.

Filtration is a process by means of which solid impurities are removed from fuel. The filters operate with a gauge of 3–10 µm. If the filter has a filtration gauge of, for example, 5 µm, it does not mean that larger particles do not pass through the filter. Elongated impurities may pass through the filtering pore even if their length is more than 5 µm. In the majority of cases centrifugal separators are used for water separation, or a double-stage filter. Separators remove free water and dirt. Separators compound small droplets into large, which are subsequently separated. Separators also eliminate the air bubbles so they are considered a decisive part of the entire fuel infrastructure in terms of ensuring fuel quality. Sometimes, mostly in subtropical and tropical zones, micro-organisms may be observed in the fuel system filters or tanks. In the majority of cases, they are micro-organisms which use the aircraft kerosene as a power source for their metabolism. In some fuel stores *Cladosporium resinae* mould was discovered. That species of micro-organism is very small and easily passes even through fine filters. The protection consists in an exhaustive cleaning and disinfecting of the affected sections of the fuel system including replacement of filters.

10.2.2. Fuel Deliveries and Storage

Fuel is transported to fuel terminals or airport stores from refinery in the following ways:

- by a pipeline directly from the production plant or the oil company's storage facility
- by ships or barges
- by railway containers
- by lorry-mounted containers.

Some airports use more than one way of fuel transportation. Diversification is useful in particular for large airports in case of delivery failure from one of the suppliers or, in case of system maintenance (pipeline, rail).

Transport by pipeline is suitable especially for large airports, or if the airport is situated near a refinery. In such cases, and if a reliable supplementary supply of fuel is available, the requirement for constructing large stores of fuel directly on the airport is reduced. A pipeline supplies the fuel stores at Heathrow, Gatwick, Stansted and Vienna airports. In some cases it is advantageous to transport the fuel by pipeline even for long distances. For example, a 100 km long pipeline supplies the Jet A1 from the refinery at São José dos Campos in Brazil to the São Paulo Guarulhos airport. It is also necessary to provide a temporary storage for the supplied fuel where individual deliveries of fuel are stored. They are subjected to a test which, unless they satisfy, the fuel shall be sent back to the supplier and not mixed with already stored fuel that has passed quality tests.

For large airports fuel transportation by ship or barge tankers is also practical. For example Frankfurt airport is supplied by barges. Shannon and Hong Kong airports are supplied by sea tankers. In the case of Hong Kong airport fuel is transported to Sha Chau island depot, from where it is delivered to airport storage by pipeline.

In the case of rail deliveries an airport does not have to be connected to public rail track. In many cases an airport has just a railway siding for example Prague and Bratislava airport. In the case of Prague Airport there are five different companies providing rail deliveries. According to Prague Airport a whole train of fuel that is 14–28 railway containers can deliver 0.9–1.5 million litres of fuel. In summer peak period the airport receives daily one full train of Jet A1.

Supplying fuel stores by truck is more suitable for small airports. However, there are a few exceptions. Some airports with throughput of approximately 5 mppa (million passengers per annum) supplied their fuel stores exclusively by truck, with about 100 containers per day. A problem with ensuring the fuel stays within specification is that the demand varies considerably during the year. The consumption in the peak is approximately 60% higher than in the off-peak months. In some airports there are also significant weekly variations in the demand. The storage capacities of fuel in the airport should be determined depending upon the reliability of fuel supplies and of variations in the demand. When determining the number of individual storage tanks and their capacities, consideration should be given to the peak daily consumption, the already mentioned reliability of fuel supplies, the state regulations and the required reserves of fuel. For example according to Council Directive 2006/67/EC of 24 July 2006 EU Member States have an obligation to maintain minimum stocks of crude oil and/or petroleum products to at least 90 days' average daily internal consumption in the preceding calendar year. It is expected that this amount will increase in the future.

The required reserve is the quantity of fuel that is in the store, which may not be used at that time to refuel the aircraft. Fuel must be treated (see the preceding section) and depending upon the manner of transport of the fuel to the airport, it is sometimes necessary to analyse it. In the event that the fuel samples do not correspond to the specification, the fuel may be returned to the supplier. Therefore the required reserve must cover a sufficient quantity so as to ensure a continuity of supply for airlines.

Fuel storage capacity usually corresponds to three to seven days fuel demand. Insufficient capacity may lead to fuel shortage, resulting in serious consequences for airlines. Conversely, too large capacity results in higher costs, which are reflected in fuel costs. Guidance on fuel storage capacity planning could be found in the IATA Guidance on Airport Fuel Storage Capacity manual. When calculating the total amount of storage required, it is recommended that airports take account of the fact that it is unlikely that the worst combination will be encountered on every parameter at the same time. For example the sum of the extreme cases for all parameters would be two days for Customs clearance plus three days for supply plus two days for fuel quality control, making a total of seven days, but users are advised to think about how likely it is that maximum stock would need to be held for all three parameters at the same time. As a 'rule of thumb', if there are more parameters which have to be taken in account, the total difference is unlikely to need to be more than 85% of the sum of each of the individual differences. In our case, the airport should be satisfied that, for instance, six days of fuel reserve would in practice be sufficient. However, there are still other regulations that must be taken into account.

The quantity of fuel reserves at the airport must take into account also emergency situations and possible alternatives of fuel deliveries. Heathrow Airport diversifies the deliveries considerably and in normal operations gets 35% from the Buncefield Fuel Depot, about a quarter from a depot at Fawley near Southampton and a further quarter from a facility at Walton-on-Thames, Surrey. Another 15% of fuel comes by rail. When the Buncefield Fuel Depot caught fire on the 11 December 2005 Heathrow airport lost 35% of its fuel supply. The airport company has increased the amount of fuel from other suppliers according to contingency plans. The amounts of fuel available for uplift by some carriers were reduced by up to 60% compared with normal operations. At the beginning, the preliminary measures differentiated between base and visiting carriers with base carriers being given higher fuel allocations. London Heathrow Airport storage capacity is 56 million litres of Jet A1 fuel in total and plans to build two additional tanks with a capacity of 13 million litres.

According to the IATA Airport Development Reference Manual the fuel stores should be located airside and at least 100 m from public roads, if practicable. If the distance is less than 100 m solid barriers of brick or reinforced concrete up to 1.8 m shall be installed. Fuel store location must also take into account the effects on radar performance (reflections). Placements in remote parts, far from terminals and public areas, are preferable.

Figure 10.1 shows a scheme of the fuel system at Charles de Gaulle airport in Paris. Before filling the fuel into the main storage tank, water is separated from the fuel in a separator. After sedimentation and the discharging of water and sediments from the storage tank, the fuel passes through a clay filter with a gauge of 20 µm. Then the fuel passes through the fuel system. Static electricity accumulates in it due to the friction from the piping walls, so the system has been designed in such a manner that the velocity of fuel does not exceed 3 m/s, and an antistatic additive is added.

220 *Airport Design and Operation*

Static electricity is discharged from the fuel during its temporary storage in a delivery storage tank. Before aircraft refuelling, the fuel is once more cleaned in a filter with a gauge of 2 μm and for the last time in a filter directly in the delivery dispenser.

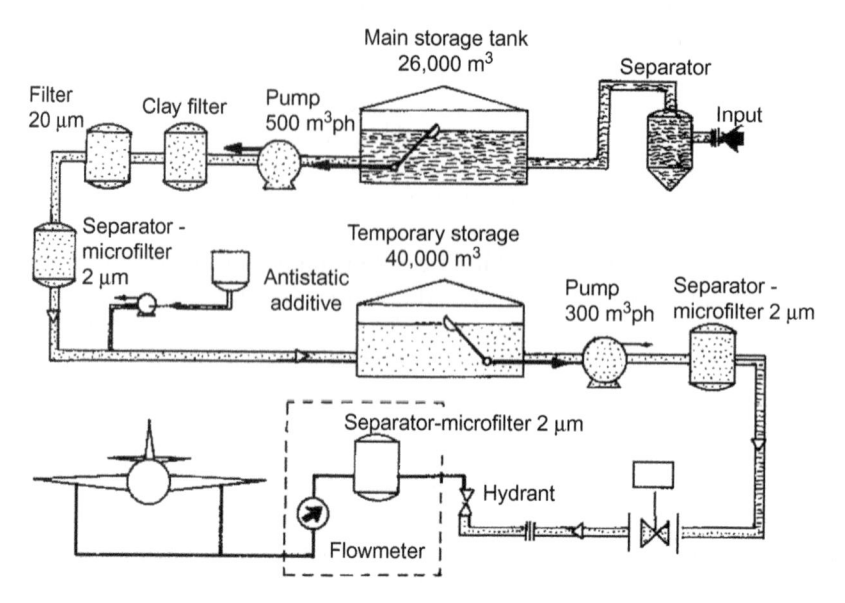

Figure 10.1: The fuel system at Charles de Gaulle airport.

Fuel storage creates a potential danger in that leaks may affect the quality of underground and surface waters, and the soil environment. The danger may be limited by obeying the fuel technology and construction measures.

Storage tanks may be underground or overground, the decision being made on the basis of a special study for each case. However, with respect to environmental regulations and the resulting costs overground tanks are the most common solution even for small airports today. Underground tanks are built in confined spaces where required fire breaks cannot be guaranteed. The underground storage tanks are still installed in some smaller airports with volumes of individual storage tanks up to 120 m^3, rarely up to 300 m^3. The underground storage tanks are, with respect to national standards, usually designed with double jackets, or with a single jacket located in an impermeable caisson. When double-jacket storage tanks are used, the cost of the technology is higher, while when single-jacket ones are used the price of the construction works is higher. In some countries, a cathode protection of the storage tank against corrosion is considered to be sufficient to secure the storage tanks against fuel leakage. The price of the overground storage tanks themselves is usually lower than that of the underground ones. However, the fire regulations require the overground storage tanks to have greater safety distances from other objects and prescribe the availability of considerable fire fighting equipment. The overground storage tanks are situated in pits which are able to retain the fuel in the event of an accident. The storage tanks should be cleaned on a regular basis, usually once in three years, and are subject to prescribed tests.

The fuel storage facilities at Prague Ruzyně airport have six storage tanks and a total volume of 4.8 million litres and the airport plans the fuel storage expansion to cover needs of future long-haul transport. This is a relatively very small amount compared with fuel storage capacity at big airports which can reach a 100 million litres of fuel with daily throughput of more than 20 million litres.

10.3. Fuel Distribution

Distribution of fuel from the fuel storage on the airports into aircraft may be executed basically in two ways. One is a distribution to aircraft by tanker, the other is a distribution by fixed systems directly into aircraft. Both of the systems have their advantages and disadvantages, and both of them, naturally to a different degree, are used at small as well as large airports.

In addition to the two most common ways of refuelling in small airports with a large number of general aviation aircraft, it is often more appropriate to install a fixed fuel dispensing point delivery system, to which the aircraft will taxi. Similarly, it is advantageous to install such a fixed dispensing point for example at a heliport with more stands. In both cases, the principal advantage is a reduction of operating costs and an increase of safety and reduced danger of contamination during aircraft refuelling.

At smaller airports the larger aircraft is usually refuelled by tanker, since at airports with low volumes of fuel deliveries, it is not economical or efficient to install a fixed system to each individual stand. Within a certain time, fuel in the systems degrades, its quality is not ensured, and part of the fuel should be sucked out before aircraft refuelling. At large airports with the hydrant system the peak flow deliveries can reach 35,000–90,000 l/min depending on aircraft types and the number of movements.

At the present time, the administrations of most large international airports prefer hydrant fuel distribution. The decision in favour of fixed installations is not only made on cost grounds but also for operational reasons. The hydrant system should be designed as a closed loop under the apron or around a terminal. From this main pipeline side lines are led to each hydrant. This arrangement provides sufficient flexibility with respect of possible future development, fuel system maintenance, cleaning and repair of any faults. At points where the side lines join the main pipeline, service pits with isolation valves are usually installed.

At an airport with wide body Boeing 747 turnrounds, the fuel supply to a single aircraft usually amounts to 100 m^3 or 80 tonnes of fuel. In the case of Airbus A380 turnrounds and European flights from 6.5 to 7.5 hour duration the aircraft takes from 80 to 100 tonnes and for ultra-long flights 140 tonnes for 11 hour flight or 200 tonnes of fuel for 14.5 hour flight could be expected.

With a hydrant distribution system, in the majority of cases a single worker with one dispenser, which is only a light vehicle, can perform the aircraft refuelling (Figure 10.2). The dispenser has equipment for reducing the operating pressure of fuel in the system, which reduces from 2 MPa, to a value of 0.35 MPa, the pressure of aircraft refuelling. Besides that, there are filters and fuel quantity measuring instruments, and a height adjustable manipulation platform. A safety valve, most often of the Dead Man's Handle type, is installed between the hydrant (Figure 10.3) and the dispenser, which in the event of a failure closes the fuel intake. In comparison, two or three tankers with operators are required for refuelling a high-capacity aircraft (Figure 10.4). The tankers obstruct a large space around the aircraft and impede other activities. Another advantage of the hydrant system is an increase in fire and environmental safety. There are no tankers on the apron with a large quantity of highly inflammable fuel, but only a dispenser with fuel in hoses from the hydrant to the aircraft and in the filters.

Figure 10.2: Aircraft refuelling by fuel dispenser blocks off a minimum of space during turnround operation. *Photo*: A. Kazda.

Tankers and dispensers must be parked somewhere. The operator prefers to park and fill tankers in the closest proximity possible to the apron in order to increase their utilisation and reduce costs. But the areas near the apron are commercially the most valued, and with the large size of vehicles, the price of the lease will be high. If the parking lot is located further away, the tankers add to the traffic on the service roads and in the event of an accident, they increase a possibility of spillage and fire. In contrast, the main parking area for dispensers may be situated even outside the central part of the airport. Only those dispensers that are immediately needed will be on the apron. The other equipment of the hydraulic system, storage tanks, pumps, filters, etc., is generally located off the apron.

Figure 10.3: Detail of the fuel hydrant. *Photo*: A. Kazda.

Unless they are fully drivable, passenger loading bridges limit to some extent the flexibility in where the aircraft can park in relation to the hydrants, but the hydrants also limit the location of the aircraft due to the limitation on the length of the hose — maximum 10 m, from the hydrant to the aircraft refuelling point. However, at the same time, this type of airbridge ensures that the location of the stand for a particular aircraft type will not be changed frequently. If there is a need, the hydrant may be repositioned by several metres overnight, if the work is well organised.

224 Airport Design and Operation

Another possibility is the so-called 'free ramp concept', where a hydrant system is installed without any elements of it being under the apron. Refuelling is provided directly from delivery points built into the airport apron, which are mounted flush with the level of the apron when not being used. Examples of such a solution could be found at Stockholm-Arlanda Airport (see also Chapter 9), Dubai International Airport and Beijing Capital International Airport.

Figure 10.4: B 747 refuelling requires sometimes more than one tanker. *Photo*: A. Kazda.

At airports with a density of operations up to 10 million passengers per annum, the decision to be made between tankers or a hydrant system is often not simple. The investment costs of constructing the hydrant system may differ even between different locations on the same airport, depending on the extent of the system and the possibility of construction in stages. Consideration should be given to the following:

- length of the distribution piping
- building works, including embedding of the distribution system under the apron, taxiways or runways
- number of stands and hydrant pits
- pay-back period
- operational costs.

The investment costs of constructing a hydrant system are several times higher than those of a purchase of several tankers. Nevertheless, operating costs of the system are lower and its service life longer. When projecting and constructing a hydrant system, it is necessary to ensure a high quality of works and materials because the costs of repair, reconstruction or removal of the consequences of an environmental accident under the apron may be several times greater than the increased costs during the construction. It may be appropriate to construct a hydrant system when the annual fuel throughput at a particular airport exceeds 200 million litres of fuel.

Even though a hydrant system of fuel distribution has been constructed in the airport, it is necessary to have several tankers available in case of failure or maintenance of a part of the hydrant system, so that fuel can still be supplied to the aircraft on aircraft parking stands, in hangars (for engine test), and for defuelling the aircraft if there is a need.

In summary, the main advantages and disadvantages of hydrant systems are:

Advantages:

- higher fire safety
- higher ecological protection (lower probability of fuel escape and spilling)
- smaller area required during technical servicing of aircraft
- smaller area required for parking of dispensers in the vicinity of the apron
- lower operating costs at high volumes of fuel supply
- lower demands for labour
- easier to ensure high quality of fuel
- aircraft may be very quickly refuelled with any quantity of fuel
- lower demands for maintenance
- reduction of traffic on the service roads.

Disadvantages:

- higher investment costs
- the aircraft should be accurately parked on the stand
- low flexibility
- high costs of repair in the event of a failure of the system
- in any case need for several tankers to be available.

10.4. SAFETY OF THE REFUELLING OPERATION

Aircraft kerosene has a high concentration of harmful substances and high degree of flammability (class II). During the construction, operation and maintenance of a

226   *Airport Design and Operation*

fuel system, adherence to the standards and set procedures shall be ensured so that water cannot be polluted with fuel and so that humans and property cannot be harmed in the event of a fire.

10.4.1. Ecological Damage

Even small leaks or spilling of fuel during refuelling may have the same serious consequences as a bigger escape of fuel after a failure of the equipment. Small escapes are more frequent, and more easily escape detection, but their effect is cumulative. Also, a small quantity of fuel contained in water can lead to significant levels of pollution, considerably lowering the quality of water and thus reducing its usability.

Fuel escapes may occur in four places:

- during transfer to or while stored in fuel stores
- from distribution piping
- during refuelling
- in hangars during aircraft maintenance.

Then fuel may get into water by one of the following ways:

- by rain drainage into surface waters
- by sewage conduit into a water treatment station
- by infiltration into underground waters.

Fuel most often gets into the rain drainage after an escape of fuel on the apron. Aircraft kerosene is a poisonous substance and its most noticeable effect is in watercourses. After an escape of fuel, fish and other aquatic animals and aquatic plants are weakened by direct toxic effects of aircraft fuel and perish as a result of a deficiency of oxygen which is consumed as the microbes degrade the fuel.

Installation of a separator of oil products in the drainage branch from the apron, or in other parts of the drainage system, is a constructional measure taken to limit the fuel escapes through rain drainage channels. In some cases, the sewer discharges out at a retention tank serving among other things for retention of storm waters. In the event of an escape of fuel, the drain from the tank may be plugged, the layer of fuel may be withdrawn from the tank surface, and water may be treated before being discharged into a watercourse.

The simplest operational measure is to clean up fuel leaks during refuelling *in situ* by sprinkling them with an absorptive substance. After absorption, the residue is carried away and the fuel is burnt off in a specified place or in a special furnace. Special absorbent textiles, which are able to absorb up to 20 times their mass, are most suitable for removing the spilled fuel. It is simpler to remove the fuel with them than with loose material.

When greater quantities of fuel are spilled, some airports use special suction sweepers for removing the fuel. At Manchester airport, fuel is washed away with a large quantity of water into the sewerage drain and later treated in a retention tank. The surface of the apron is thoroughly cleaned with a detergent. Detergent must not get into the sewage system, because the emulsion that would be created with water and fuel cannot be removed in the retention tank.

An effective way of reducing the number of cases of fuel spilling and to improve the discipline of handling agents providing aircraft refuelling is to invoice them with the costs connected with cleaning up.

Fuel escapes are often caused by the poor condition of the equipment. Therefore regular, visual inspection and scheduled maintenance are essential. Another preventative measure is to prohibit execution of aircraft maintenance and repairs on the apron, except cases of removal of failed aircraft, and permit them only in hangars, which are equipped with efficient separators.

Fuel may get into the sewage drains after leaks from aircraft in hangars and workshops. If collectors of oil products are not checked and cleaned on a regular basis, they cease to perform their function. If fuel gets into a water treatment station, it may disrupt the quality of the biological function of the station and disable it. Measures to prevent fuel escapes into a water treatment station are similar to those in the case of rain drainage.

When underground storage tanks or installations are damaged, or after a fuel escape on unsealed surfaces, fuel leaks into underground waters. Then it creates a layer on the water surface. New building regulations should prevent such an eventuality. However, there are still many installations that do not comply with the regulations. The solution then is for the affected area to be surrounded by drill holes from whence the polluted water is continuously pumped out and treated. In addition, monitoring drill holes are established, from whence samples of water are taken. There is a disadvantage that by pumping, the level of underground waters drops, having an unfavourable impact on vegetation. Another solution is separation of the affected area by insertion of a plastic membrane at a depth of 30 m, or by partitioning off by ramming in steel walls. Then cleaning is performed only within the defined section. However, in any case, the state of the underground water is worsened.

It is often difficult to detect a fuel leakage from the underground storage tanks or from the fuel piping. In spite of the fact that the majority of big airports have established a system of monitoring of fuel escapes by means of drill holes, it gives only an identification of the consequences of the fuel escape. According to the Technical Review of Leak Detection Technologies Crude Oil Transmission Pipelines Alaska Department of Environmental Conservation, methods used to detect product leaks along a pipeline can be divided into two categories, externally based (direct) or internally based (inferential). Externally based methods detect leaking product outside the pipeline and include traditional procedures such as right-of-way inspection by line patrols, as well as technologies like hydrocarbon sensing via fibre optic or dielectric cables. Internally based methods, also known as computational pipeline monitoring (CPM), use instruments to monitor internal pipeline parameters (i.e. pressure, flow, temperature), which are inputs for inferring a product release by manual or electronic computation. Marking substances may be utilised for an exact localisation of a failure and fuel escape. In the majority of cases, they are volatile materials which are added to the fuel. After several days, the place of damage location may be identified by means of a sensitive gas chromatograph. Since the marking substance has been used for locating the leak, the results are not distorted by any preceding pollution of land.

Depending on particular state regulations fuel hydrant systems are periodically subjected to tests at most airports. Historically two testing methods, both reflecting temperature changes in the pipeline, were used. During the first — *the volume test*, a constant pressure is maintained during testing and the volume of liquid to be delivered or removed to/from the system is recorded while the pressure remains constant. Results of two independent tests using high and low pressure are compared to eliminate the effect of temperature changes. The second — *the pressure test* is similar to the first one as it is also carried out under two different pressures. During this test pressure changes are measured. This test takes about 45 minutes for each section of the hydrant system.

Both methods are able to detect a minimum leakage of 0.002% of the fuel volume in a particular pipe section or minimum 4.3 ml/min. Even though both tests could detect system leakage they did not allow to the determination of the position of a fault. It was also necessary to put different sections of the hydrant system out of service.

Modern, acoustic or sonic leak detection systems, so-called rarefaction wave, also called an acoustic, negative pressure, or expansion wave is becoming a preferred leak detection methodology in many pipeline applications. This method allows system testing during regular operation. Leak detection is based on the analysis of pipeline pressure variations. When fuel breaches the pipeline wall there is a sudden drop in pressure at the location of the leak followed by rapid line re-pressurisation a few milliseconds later. The resulting low-pressure expansion wave travels at the speed of sound through the liquid away from the leak in both directions. Instruments placed at intervals along the pipeline respond as the wave passes. Another way of assessing pipe leakage is used at the stage of performance tests.

Very accurate flow metres are installed in each section of the piping to measure the quantity of fuel at inlet and outlet from the respective section. The results are compared, and if there is a difference, an alarm is put out.

At each airport a plan should be elaborated for the event of an oil accident which all service personnel must be acquainted with. The plan shall contain not only procedures of catchment and removal of escaped material and removal of polluted earth, but also preventive measures to be taken to prevent an oil accident.

In some countries (e.g. the United States), stringent limits of atmospheric emissions have been set by the regulations for environment protection. Among them, the contents of unburnt hydrocarbons have been regulated. The unburnt hydrocarbons may get into the atmosphere even during aircraft refuelling. Therefore on some airports, the vapours from aircraft tanks are drawn off and treated.

10.4.2. Fire Safety

A danger of fire during refuelling results from the combustibility of fuels. Aircraft gasoline is ranked among the Class I inflammables (the highest rank) and aircraft kerosene into the Class II inflammables. The prescribed provisions to protect against fire are related to that ranking. Among the passive ones is the extent of safety protective areas adjacent to the stores of aircraft fuels. Their size differs depending on the volume of storage tanks, their construction and the kind of stored product. The active measures prescribe the fire equipment, kind and quantity of extinguishing agents, procedures for fire fighting in compliance with the fire order in force (Figure 10.5).

Ignition of fuels may occur not only on contact with fire or with a material at high temperature, but also from a static electricity discharge created either from the movement of the fuel in the aircraft tank during the fuelling process or its accumulation on the surface of aircraft or vehicles. Therefore all parts of the fuel system must be provided with bonding and must be grounded. During aircraft refuelling, bonding should be provided also between the aircraft and the tanker or hydrant.

To stop aircraft refuelling from hydrant and prevent fuel leaking, shut-down in three ways is possible: (1) by Dead Man type of valve on the dispenser; (2) valve on the hydrant in the hydrant pit and (3) by an Emergency Fuel Shutoff System — EFSO button.

There are also various safety systems installed on truck tankers which are usually fitted with: (a) an automatic braking system which blocks truck movement unless the hoses are properly coiled and valves closed; (b) system preventing refuelling commencement unless the truck parking brake is applied and (c) Dead Man type

230   Airport Design and Operation

valves. Both truck tankers and dispensers are equipped with aircraft tanks overfilling protection, which is based on the maximum amount of fuel that can by filled in particular aircraft type and amount of fuel delivered. There must always be a minimum 2% of aircraft tank volume which must be left empty for fuel thermal expansion.

Figure 10.5: Fire safety of overground tanks consists of passive (that is fire breaks) and active measures (that is fire fighting technology). *Photo*: A. Kazda.

The fire protection of vertical overground storage tanks consists of water sprinklers for cooling a burning tank. The sprinklers are supplied from dedicated water reservoirs. For example the volume of the special water tanks at Hong Kong airport is 750 m^3. Aqueous Film Forming Foam (AFFF) (see also Chapter 19) is often used for extinguishing fuel fires. The foam is produced in a generator and transported by distribution system to the points of injection into the tank. Automatic or remote controlled discharge monitors are usually mounted on the railing.

It is essential that the aircraft be fully restrained and that there should be a clear path for the bowsers to move away if an incident was to occur. Fuel zones should be established 6 m around the fuel vents. There should be emergency stop buttons on hydrant stands and no vehicle should be parked in front of it. There should also be an emergency phone at the head of the stand, and a person nominated as responsible for the fuelling operation. If fuelling is to occur while passengers are on board, the passengers should be informed, the emergency exits should be kept available, and non-essential vehicles should be kept off the apron. If there is no airbridge, steps should be positioned both at front and rear doors, and the doors should be constantly manned. Some aircraft have fuel systems which allow fuelling the tanks through more than one fuelling point simultaneously. This usually requires fuel equipment to be positioned on both sides of the aircraft which increases the risk in case of fire.

Organising an intervention when a fuel escape occurs is usually within the competencies of fire brigades. They have both the required equipment and the necessary experience, they are able to assess the situation also from the viewpoint of the risks of possible fire.

10.4.3. Fuel Farms Security

Fuel farms at large airports can have a capacity of hundreds of millions of litres and could be attractive target for potential terrorists. They are always considered as high security objects. As they are usually located at airport remote areas they may appear as an easy target to potential attackers. To guarantee the required level of protection fuel farm double fencing is required, from which the outer fence could be an 'intelligent type', fitted out by microphone cable or other sensors. The fence system is often complemented by an automatic monitoring system or eventually even patrolled by armed commandos (see also Chapter 13).

10.5. AIRCRAFT FUEL — FUTURE TRENDS

It may be contemplated that aircraft kerosene will continue to be the most used aircraft fuel for at least the next 30 years. The assumptions that the crude oil reserves would soon be exhausted, which were made during the first oil crisis, proved to be unfounded. In the event that the crude oil reserves do eventually become exhausted, aircraft kerosene may be produced synthetically from other sources without a need to change the aircraft design concept.

The use of biofuels for air transport at this stage of knowledge is questionable. Under the Directive 2009/28/EC on the promotion of the use of energy from renewable sources this share rises to a minimum 10% in every Member State in 2020. The Directive targets the use of sustainable biofuels only, which generate a clear and net greenhouse gas saving without negative impact on biodiversity and land use. Many biofuels do not result in greenhouse gas emissions savings compared with conventional oil and they have serious additional environmental and social consequences from their production.

There were a number of test flights with biofuels conducted since 2008 which were followed by a few commercial flights with passengers after the American Society for Testing and Materials (ASTM) approval in 2011.

However, there are still many technological, technical and operational problems which must be researched before introduction into normal service. For example the current practice is that kerosene from different producers/suppliers is mixed at the airport fuel farm or upstream in the supply chain and supply streams are handled interchangeably. The same blend is than loaded to all aircraft. If an airline would purchase biofuels but then loads it into the airport fuel store it would not be possible to use it on a specific flight. In order to monitor rate of biofuel usage so as to keep track of CO_2 inventory for an airline's record but also because of different price of fuel, a parallel infrastructure would have to be set up.

The debate on global warming may also accelerate a change away from kerosene to hydrogen. Because the same amount of energy requires four litres of volume in the case of liquid hydrogen, compared with just a single litre of kerosene, the larger volumes of liquid hydrogen would require changes in the configuration of the aircraft. Also tanks for liquid hydrogen have to be either spherical or cylindrical in shape. Tanks on top of the fuselage are a possible solution for big passenger aircraft in the future. The use of cryogenic fuels might be limited to supersonic aircraft with a speed over twice the speed of sound.

Thanks to incremental improvements in engine design, the consumption of aircraft kerosene will continue to increase more slowly than air transport traffic. Such growth will alleviate problems in providing the quantities and rates of fuel supplies into aircraft.

In reconstruction or construction of large airports, there will be a tendency prevailing to install modern hydrant systems if sufficient throughput over 200 million litres of kerosene a year could be guaranteed. In small commercial airports, aircraft refuelling from tankers will continue to be the most frequent solution.

A system of collection, transmission and processing of data on aircraft refuelling is used at some airports. The monitoring system ensures a data collection and evaluation so that after refuelling a basis for invoicing may be available, or an invoice directly issued. Besides the fuel quantity, the modern computer equipment monitors also other parameters, such as fuel temperature, and automatically determines the total weight of fuel supplied into the aircraft. It could be expected that such systems will be increasingly used also at the smaller airports.

11

CARGO [*]

Bob Caves

11.1. INTRODUCTION

Cargo, and particularly mail, has been moved by air for as long as passengers. The first official mail flight was from Frankfurt to Darmstadt by Lufthansa on 10 June 1912.

By 1933, mail was being flown across the North Atlantic by flying boat with a mid-ocean stop for fuel, the aircraft being hauled on deck and catapulted off. A few years later, a more ingenious method of a multistage vehicle, the Short-Mayo composite aircraft, was tested. The mother flying boat took the piggy-back aircraft part of the way across, then launched it to complete the trip with the mail while the mother ship turned back. Many of the major air routes, domestically and internationally, were cemented by mail subsidies between the two world wars.

The freight element of air cargo started even earlier. A sheep, a cock and a duck were flown in a Montgolfiere balloon in 1783 at Versailles. The first flight in a fixed wing aircraft to carry freight occurred when a 30 kg bolt of silk was flown the 105 km from Dayton to Columbus, Ohio, in 66 minutes in 1919.

There followed a phase when IATA set freight rates on international routes. The freight was carried mostly on the basis of available capacity, passengers always being given precedence. There were many aircraft designed specifically for the freight market, often with an eye to the military market, both prior to and following the Second World War (e.g. Bristol 170, A W Argosy with front loading, CL 44 with tail loading). However, it was difficult to make enough money with all-freight aircraft to cover the initial cost of new aircraft.

These days, mail and freight has to pay market rates, but the market has continued to grow strongly as shippers take advantage of speed, cheaper packaging, higher safety and security of a shipment and lower insurance premiums to counter higher rates. The available capacity has grown as wide-body aircraft have appeared with

---

[*] This chapter is based in part on an MSc thesis by Stewart Walsh, late of TPS Consultants. The authors would like to thank him, and also Bill Blanchard of East Midlands airport and Darren Maynard and Charles Allen of DHL for their help and advice.

surplus belly hold capacity after loading passengers' bags (e.g. Boeing 747, 767, 777; Douglas DC 10; Lockheed 1011; Airbus A300), and as all-freight versions of passenger aircraft have been developed (e.g. A 300; Boeing 757, 767, 777; Douglas DC 8), usually with side-loading, but also the A300-600 and Boeing 747F with front loading (Figure 11.1), reducing the launch cost to manageable proportions. A list of representative freight aircraft is given in Table 11.1. A further option is to use Quick Change versions of for example the Boeing 737, the seats being taken out at night or at weekends and the aircraft used to carry containers which are rather shorter than the conventional containers for narrow-bodied aircraft.

Figure 11.1: Front loading reduces the handling costs. *Photo*: Courtesy CSA Cargo.

Even though there is this impressive range of available freighters, more than 60% of world air freight is taken on passenger aircraft. The advent of the A380 with its relatively small belly hold capacity was expected to increase the demand for all-freight aircraft, particularly as freight traffic is growing more quickly than passenger traffic in the developed world. However, the popularity of the 777-300 with 20–30 tonnes of belly freight has actually helped belly uplift to increase and pure freighter movements to fall. The future introduction of the A350 and the Boeing 777-X will also provide good belly hold capacity. Rates are no longer controlled by IATA and are determined by route and demand characteristics as well as by competition. Often demand is strong in one direction, allowing good revenue and the possibility of reduced rates for the opposite direction. There is a trend for the legacy carriers to sell off their all-freight aircraft and concentrate on specialised belly cargo like pharmaceuticals and express parcels, using the latest electronic communication developments.

Table 11.1: Freighter aircraft.

Aircraft	MTOM (kg)	Max payload (kg)	Main deck volume (m^3)	Wing span (m)	Doors
Boeing B737-300QC	56,470	12,270	120	28.88	Side
Lockheed L-100-30	70,310	23,440	127	40.4	Rear
Boeing B737-700C	77,560	18,780	120	34.32	Side
Boeing 727-200Adv F	83,830	21,180	137	33	Side
Boeing B757-200F	113,400	39,690	187	38.05	Side
Airbus Beluga	155,000	47,000	1400	45	Nose
Airbus A300-600F	165,110	54,750	265	45	Side
Boeing B767-300F	185,070	54,890	336	47.5	Side
DC-10-30F	251,750	69,850	371	50.4	Side
Boeing MD-11	273,070	93,170	443	51.5	Side
Boeing B777F	347,450	103,900	518	64.8	Side
Boeing B747-200C	356,080	100,010	503.1	59.64	Nose, side
Boeing B747-400F	362,880	112,900	604.1	64.44	Nose, side
Antonov 124	392,000	120,000	750	73	Nose, rear
Boeing B747-800F	435,456	140,000	689	68.5	Nose, side
Airbus A380F	590,000	150,000	945.8[a]	79.8	Side
Antonov 225	600,000	250,000	950	88.4	Nose

Sources: Manufacturers' web sites and www.freightersonline.com, adapted by the authors.
[a]Main deck, upper deck and lower deck.

236   Airport Design and Operation

Some major international (legacy) carriers have strong freight contributions to revenue and to profits, while others in different markets have only weak contributions. There is no agreed standard for cost allocation between passengers and freight. Those allocation methods that work on payload weight are unlikely to be as helpful as a revenue management system that takes account of passenger bags, no-shows, surcharges and actual rather than published rates.

Some low-cost carriers realise significant revenues by emphasising express shipments and/or by contracting with other airlines to handle their lower-hold space. However, the focus on passengers and operational efficiency (e.g. quick aircraft turnrounds) reduce the priority of cargo. Most LCC ignore the cargo market completely and focus fully on the passenger transportation (Figure 11.2).

Figure 11.2:   Most of low-cost carriers focus wholly on passenger market segment.
Photo: A. Kazda.

A growing proportion of world trade is now taken by air, and an increasing proportion of that is performed by the integrated, or express, carriers. These carriers offer an integrated door-to-door service with a guaranteed pickup and delivery time, relieving the customer of any direct involvement with customs and providing a tracking capability that allows the customer and the government inspection services to keep track of the shipment throughout the process. They mostly take shipments of less than 100 kg, but each of the four major players offers truly global coverage through intercontinental gateways, sub-hubs and fleets of delivery vans. Fedex, for example, in mid-2014 delivered approximately 4 million packages per day through 50,000 drop-off locations in more than 220 countries, using 650 aircraft, 48,000 vans and 160,000 employees.

The combination carriers (i.e. those catering for both passengers and freight) generally do not have such weight restrictions. They also offer a global product through their alliances and are also moving more towards an integrated product themselves. Shippers and freight forwarders now have a plethora of options from which to choose. It is quite common for integrators to use space on combination carriers and vice-versa. There are also airlines that specialise in heavy lift, using small fleets of unique aircraft like the AN 124 or the Mil 10 helicopter.

The freighter 'wet-lease' airlines (an arrangement that includes all facets of operating an aircraft on a carrier's behalf, including the airframe, crew and most, if not all, of the aircraft-related costs), or an aircraft, crew, maintenance, and insurance (ACMI) airline, satisfies important market requirements. It offers airlines flexibility to contract for air transportation services on a trial basis where demand is uncertain, or to augment existing markets, and/or to provide service in markets that are highly seasonal without an investment in dedicated equipment. ACMI services are offered for all freighter sizes, but most growth since the late 1990s has been in long haul, intercontinental wide-body freighter markets.

11.2. THE FREIGHT INDUSTRY'S CHARACTERISTICS

Most of the products shipped by air fall into one of the four categories:

- economically perishable goods
- physically perishable goods
- items for emergency maintenance
- strategic inventory management.

Efficient inventory management calls for 'just in time' lean manufacturing or quick response to consumer needs as in the fashion industry. There is also increased use of global outsourcing. Their supply chains need to be agile to change the type, the volume and the mix of product in short time frames.

These business trends have driven the growth of air freight, particularly the integrated carriers, and are expected to do so in the future, because they require overnight delivery whereas the combination carriers' schedules are more determined by the needs of passengers to travel by day. The trends also tend to reduce the ratio of value to weight, but the aircraft loads are still generally more limited by volumetric capacity than by weight limits. However, there are increasing risks to the supply chain in adopting these techniques from disruption in the logistics chain, in addition to problems at the plants, natural disasters, terrorism, etc. Carriers need to have a strategy in place for physical and electronic network redundancy and to work together with their customers on appropriate contingency plans.

Worldwide there were 1,738 freighter aircraft in 2011, 36% being standard body, 33% medium wide-body and 31% large. Over the decade to 2013, Freight Tonne Kilometres (FTK) managed to grow by only 2.5% per annum worldwide, due to 9/11, SARS, the second Gulf war, high fuel prices, the 2008/2009 financial crisis, the Euro/US fiscal crisis, natural disasters and political uncertainty Boeing predicts that worldwide air cargo Revenue Tonne Kilometres (RTK) will grow at 4.7% per annum between 2014 and 2033 while Revenue Passenger Kilometres (RPK) will continue to grow at the same 5.0% as it has for the previous 20 years. This will require some 1,460 additional freighters, more of which will be in the standard body and large categories. Much of this requirement will be met by converting older passenger aircraft, but there will also be a need for 935 new wide-body freighters.

The transport is provided by the four main alliances of combination carriers and four main integrators (DHL, UPS, TNT and Fedex), together with several independent smaller specialist carriers like Cargolux. The freight is generated either directly by frequent shippers or through forwarders who consolidate loads. There were some 15 global logistics firms (e.g. AEI, MSAS, Geologistics, Emery) and a large number of small ones, but many have been bought out by the integrators. The distinction between these logistics companies and the airlines is becoming blurred as the latter develop their own logistic capabilities and offer value-added elements as part of their customers' supply chain. The traditional logistics companies are tending to merge in response to these threats. The challenge of logistics is getting the right product to the right place in the right quantity at the right time, in the best condition and at a good price.

An essential element in the logistics of air freight is the use of Electronic Data Interchange (EDI). This allows the communication, tracking and tracing of documents by the shippers, the integrator and government agencies. A shipment may go through up to 40 controlling processes and need 12–15 documents. EDI helps to cut out much of this paper trail. Most consignments can be cleared through customs before arrival.

Security has become a crucial area in air freight (Figure 11.3). Frequent shippers/consigners are awarded 'known' status. The validation in the United Kingdom is done by independent assessors appointed by the Department for Transport. They check for secure premises, staff that have been through reference checks and are properly trained, access control, preparation, packing and storage of cargo, and transport of secure cargo to approved agent or airline. Similarly agents are given 'Regulated Agent Status'. In the United States, the Transportation Security

Administration has issued new security rules. All off-airport employees of approved forwarders must have background checks and all employees of cargo airlines must have a full criminal history check. Overseas departure points must have similar provision. All 4,000 'known' shippers are to go onto a central database. Secure areas have been extended to include ramps and cargo facilities. Airlines have the right to examine any cargo and its packaging except Post Office mail to check the correctness or sufficiency of information or documents. With some exceptions, all cargo from unknown shippers must be screened, as must all containers whose seals have been broken.

Figure 11.3: Importance of cargo operation security is constantly growing. *Photo*: Courtesy of CSA Cargo.

Most of the freight is transported through a few large hubs. This is mostly due to economies of scale and scope, giving keen rates. At airports in South East England, including Heathrow, each worker moves about three times as many tonnes per year than in the Midlands airports. Historically, 28% of the air freight flown from Asia to North America has gone direct, almost all of the rest went through Anchorage (29%), Tokyo (23%) and Seoul (16%). In the case of conventional freight carried by the combination carriers this hub pattern is generated by the historic location of the carriers, by geography and by the facilities at the airports. Time in flight and in transit is most important, a saving of 1 hour perhaps being worth $1,000 in airport fees. The availability of the entire freight infrastructure (freight forwarders, equipment, customs, etc.) is also essential. For the integrators' hubs, geography determines the catchment area within which pickup and drop-off deadlines can be met by trucking. The airport needs, in addition to the freight infrastructure, to be free of night restrictions, have available labour nearby at reasonable cost, good road access and, increasingly, the possibility of rail/air connections.

The performance of the top 10 cargo airports in the world in April 2014 is given in Table 11.2 in terms of tonnes loaded and unloaded.

Table 11.2: The top 10 cargo airports.

Airport	Throughput (tonnes in April 2014)	Annual Growth (%)	Leading carrier
Hong Kong	365,000	5.8	Cathay
Memphis	357,410	9.3	Fedex
Shanghai	264,336	10.3	China Eastern
Incheon	215,030	2.8	Korean
Dubai	207,317	3.7	Emirates
Anchorage	203,213	1.7	Fedex
Louisville	185,676	7.0	UPS
Miami	175,791	5.2	N/A
Tokyo Narita	175,580	7.3	JAL
Frankfurt	169,283	1.1	Lufthansa

Source: Airports Council International.
N/A — information not available.

In the light of the industry's characteristics and growth expectations, as described above, it is clear that the facilities at airports need to be developed to provide for efficient, effective and economic operations. At the same time, they must meet the required standards in terms of the environment and sustainability. Carriers are constantly searching for locations where the environmental and operating constraints may be minimised: DHL moved its main hub to Wilmington only a year after making a $220 million investment at Cincinatti partly for better runway capacity, only to move to Louisville while contracting UPS for its domestic operations but then back to a remodelled major intercontinental hub at Cincinatti in 2013. It has moved its main European hub from Brussels to Leipzig partly due to environmental restrictions. The design of airside, terminal and landside facilities will now be examined, bearing in mind these industry needs.

Cargo **241**

11.3. AIRSIDE DESIGN CONSIDERATIONS

The taxiway system should be free of capacity bottlenecks and minimise the in and out distances. The requirement is more important for integrator hubs, where many aircraft arrive almost simultaneously at night. This also requires runways to be aligned so that the arrival and departure routes avoid major centres of population and for taxiway routes to avoid directing engine exhausts towards the community. There should be unimpeded access to the freight apron.

Figure 11.4: Contoured ULDs on roller-bed dollys. *Photo*: A. Kazda.

The apron needs to have the capability to park all the all-freight aircraft likely to be on the ground at the same time, unless the maintenance or passenger aprons are close by and are likely to have spare capacity at the time of the freight operations. It is often approximately five times the plan area of the terminal. Freighters operating conventional cargo service tend to spend longer on the ground than integrator or passenger aircraft, up to 7 hours. Considerable flexibility should be built-in for growth in the size of aircraft: once a pattern of freight routes has developed, it is more likely that traffic growth will be by frequency and size of aircraft rather than by additional routes, particularly in the case of the integrated carriers.

242   Airport Design and Operation

The aircraft should be parked as near as possible to the freight terminal in order to reduce the amount of ground traffic movement. This is essential if loading bridges are to be employed, though this normally only happens with nose-loading B 747 operations. The freight transfer devices require a well-marshalled airside road between the terminal and the apron. The greater the percentage of contact stands, the less the need for other airside roads and the potential for accidents.

There should also be sufficient space for storing all the handling equipment and Unit Load Devices (ULDs). ULDs come in a large range of sizes and types, from 606 × 244 × 244 cm to 156 × 153 × 163 cm, some are rectangular, some are contoured to the shape of the fuselage, some are rigid, some are netted pallets and some have environmentally controlled interiors (see www.freightersonline.com and Figure 11.4). At the same time, it should be borne in mind that aprons are expensive to build because they are designed to support the standing loads from aircraft at their maximum weight, while most of the ancillary equipment imposes considerably lighter loads.

Figure 11.5:   Bulk loading cargo into a belly hold. *Photo*: A. Kazda.

The ULDs are taken from the terminal on flat roller-bed dollys and loaded onto the aircraft main deck or belly holds using high-loader (Hi-Lo) vehicles fitted with driven roller beds. Once inside the aircraft, they are transferred to roller beds on the floor of the aircraft, having been loaded in the correct order to achieve the necessary balance. Bulk cargo is loaded manually into the belly holds, having been brought in carts to the apron by tug and transferred to the aircraft door by self-powered conveyors (see Figure 11.5).

Cargo **243**

There is a lot of movement of cargo between the passenger and cargo aprons at most airports. It is therefore good practice for the distance to be kept as short as possible, within the framework of the airport's master plan. An airside road connection of sufficient capacity should be provided, with the capability to support 10 tonnes per axle, 12 m wide, a maximum of 4% gradient and a minimum turn radius of 20 m.

11.4. TERMINAL DESIGN AND OPERATING CONSIDERATIONS

11.4.1. Location

The air cargo terminal is a critical part in the air cargo supply chain. An inadequate air cargo terminal concept that is unable to accommodate the peak volumes of cargo may result in delays, while a concept that is not flexible enough to meet the changing demand may soon become obsolete during its service life.

A conventional freight terminal should be located as close to the passenger terminal as possible, commensurate with master planning indications to extend the facilities, and with geotechnical site constraints: earth-moving, drainage, utilities, etc. A terminal for an integrator should, in contrast, be separated as far as possible from other facilities unless there is likely to be substantial cross transfer with the combination carriers. Many integrators prefer to be on the opposite side of the runway, with their own taxiway system. The freight terminal should also be as close to the runway as possible, without infringing on any of the runway transitional surfaces, either from the building or from the tails of parked aircraft. Nor should it give rise to reflections that might threaten the integrity of the electronic guidance systems.

11.4.2. Design Parameters

The design of the terminal itself and the equipment with which it is fitted depends on:

✈ the type of operator and their service standards
✈ the expected rate of growth of demand and the ultimate capacity required
✈ the political and economic setting
✈ the airport and local authority planning constraints.

Conventional freight terminals may be used by a single airline (usually the based operator) or forwarder, or by several different companies each with their own sections of the building. This latter option clearly gives a more competitive market and creates more revenue for the airport as landlord. However, it causes duplication of the materials handling equipment, takes more space and increases the initial cost. Integrators usually construct and operate their own terminal. Their traffic usually consists of packages less than 30 kg and courier mail. Activity is very peaky, and the dwell time is less than with conventional freight. Target service standards also help to determine the staffing and facility requirements. Typical standards may be:

- consignments available for collection, examination or transhipment 3 hours after arrival
- cleared consignments available within 15 minutes of consignee arriving at import collection point
- customers to wait not more than 30 minutes after arrival for collection at truck dock
- cargo reception to be complete within 30 minutes of arrival at truck dock.

The level of demand for conventional freight depends on the freight tariff, the time spent in transit, the frequency and timing of services, as well as the economic characteristics of the region. Further, the future type of traffic is important: the domestic/ international split, the in/out/transfer split, the freight/mail split, seasonal fluctuations, daily and weekly peaking, special handling needs for heavy items, outsize items, perishables, dangerous goods, livestock, urgent goods. Most importantly, it is necessary to know whether the freight is to arrive and depart in bulk or on pallets or in small or large ULDs, and if the ULDs can be loaded direct from trucks that will be given access to the apron.

The political and economic setting will largely determine the availability of labour of differing skill levels and their rates of remuneration, as well as the life cycle costs, the availability of spares, the performance level of equipment and staff. There will be policies on airport ownership that will determine the degree of commercial practice on the airport, so influencing the dwell time of freight in the building. The border control and security authorities will determine the regulations governing security and truck access to aprons, though aviation's governing bodies are attempting to harmonise matters of security and safety worldwide. They will also determine the extent to which bonded goods may be allowed to move outside the airport, which in turn will bear on the process of 'known' shippers and the possibility of allowing some handling off airport, for example stuffing of containers.

Planning guidance from the airport or the local authorities might influence the location, the size and the height of the building, and the extent to which the airport activity can grow. It can require compliance with sustainability objectives for the building and its power consumption. It can also influence the mode of ground access through road building and modal shift policies.

Cargo 245

11.4.3. Mechanisation

All the above factors contribute to determining the level of mechanisation to be provided for handling the freight. The choice is essentially between:

- manual: manpower plus fork lift trucks
- semi-mechanised: roller beds or conveyors
- fully mechanised: Elevating Transfer Vehicles (ETV), Automatic Storage and Retrieval Systems, Transfer Vehicles.

Figure 11.6: Storage mechanisation reduces the labour costs. *Photo*: Courtesy CSA Cargo.

A high labour content may mean that costs rise over time, but it is more possible to react flexibly to demand peaks. The fully mechanised approach only really works with high volumes of containerised freight and in a setting that can guarantee good maintenance skills. Even so, the whole terminal can come to a halt if an ETV breaks down. ETVs work up to 7 m high with hydraulic or electric chain drive. They are only applicable if there is a need to store a large number of ULDs over three decks. British Airways, at its multi-level World Cargo Centre at Heathrow, prefers lifts and lowered roller conveyors so that there are more options in case of breakdown. Transfer Vehicles are powered roller conveyors mounted on a carriage with bogies which are electrically driven along rails in response to a command system. Mechanisation is much more expensive in terms of first cost, but reduces the labour requirement, causes less handling damage, less pilferage and there is less risk of mishandling (Figure 11.6).

A semi-mechanised terminal may have belt conveyor systems and powered flat roller conveyors where the rollers are chain-driven from the previous one. They will also have reorienting and transfer dock beds: some have wheels at right angles that rise up between the rollers, or powered ball decks, or heliroll rotation tables where the different quadrants are powered with a joystick.

11.4.4. Terminal Functions and Operations

There are five main functions to be performed in the terminal:

- conversion between modes of transport
- sorting, including breaking down loads from originators and consolidating for destinations
- storage, and facilitating government inspection
- movement of goods from landside to airside and vice-versa, or from aircraft to aircraft
- documentation: submission, completion, transmission.

A good terminal will have systems that will allow efficient movement, effective storage, easy sortation, accurate and timely inventory control, tight security and effective use of manpower. Getting everything right can make a very significant difference to the mishandling rate.

Most freight comes from the shipper by road and leaves the airport by air, but it is not always so. Airlines use trucks with flight numbers instead of aircraft on relatively short haul sectors, and rail is also beginning to be used instead of road as a way of meeting sustainability targets. Also, there is an increasing trend to integrated air/road/rail/sea ports as at Dubai, Sharjah and Seattle. The freight on the trucks may be loose, on pallets, or already containerised. If it is from a 'known' shipper, the truck will have been sealed before leaving the shipper's premises.

Unless there is a full container for one specific destination, the freight is taken out of the ULD or off the pallets and sorted by aircraft flight destination. All the freight for each destination is gathered together and stuffed into the container(s) appropriate for the aircraft being used. The same process applies to freight arriving by air for transfer to another aircraft, except that it comes into the terminal from airside rather than landside. The breakdown and stuffing is an entirely manual process, regardless of the degree of mechanisation in the terminal. It is good ergonomic practice to have height-adjustable platforms at the work stations. The platforms may also indicate the weight and sometimes the stability of the ULD. This information is vital for the correct loading of the aircraft.

Storage is needed mostly for incoming freight which is awaiting clearance or collection following inbound clearance, but also for inbound freight before breakdown, outbound freight awaiting consolidation or stuffing or awaiting departure, and transhipments. Collection can be a matter of an hour or two, but may, in some countries that have no strict policy for charging for it, extend to weeks as companies use the terminal as free storage. Typical times in the developed world are 20 hours for export, 40 hours for import and 32 hours for transhipment. Traditionally, goods take 6 days from shipper to receiver. Approximately 90% of this time is spent on the ground. Transport accounts for 12% of this time. The rest can be considered as delay, largely due to waiting for documentation to be completed due to lack of resources or information, or inaccurate delivery instructions, or problems with customs clearance. ULDs vary in height from 1.7 m for lower deck, through 2.4 m for the main deck, to 3.0 m for fully contoured, these dimensions determining the rack spacing required for storing them. Consignment storage is usually in 1 m^3 of racking fed by fork lift or by automatic storage/retrieval systems as employed in supermarket warehouses. They may extend to 20 units high by 30 long by 14 deep, to give 1,000 m^3. Both inbound and outbound freight may be subject to inspection by government agencies for contraband, drugs, illegal immigration, weapons, etc. Storage is much less necessary in integrators' terminals, where the freight arrives only just in time for onward shipment and where almost all of it will have been pre-cleared though the EDI. Empty ULDs are often stored outside.

Goods are usually moved from trucks into the terminal by trains of carts carrying bulk freight, pallets or containers pulled by tugs, preferably electrically powered. They are then manually off-loaded onto conveyers or taken by fork lift to be sorted by destination. Either the sort process deposits the goods directly at the stuffing platforms or they are again taken by conveyor or fork lift to the platform. The conveyor system for moving the goods inside the terminal can run to many kilometres in length. It is usually mounted high up to allow free movement of fork lifts and ULDs at ground level. Packages up to a maximum of 30 kg are put into trays on the conveyers. The filled containers are then moved to the airside dock designated for the aircraft by a system of roller beds, or omni-directional ball-top beds if there are changes in direction involved. In the reverse direction, a similar process of movement occurs, except that the trucks back up to doors in the terminal so that they can be loaded directly under a canopy rather than having to be fed by tugs or fork lifts.

11.4.5. Documentation

The consignments and their associated documents flow through parallel channels, closely tied together. The relationship is indicated in Table 11.3. The paperwork has traditionally been passed through four separate systems:

- Cargo Movement Control (CMC)
- Data and Documentation Flow (DDF)
- Cargo Community System (CCS)
- Management Information System (MIS).

CMC basically carries information about the status of the cargo within the terminal: what form, where, stage of processing. The information comes from manual input to a PC or from reading bar codes. DDF generates the Air Waybill and enters it in the airline's reservation system. The CCS for the terminal integrates airport data with that from carriers, forwarders and the control authorities, with a link to customs. MIS maintains overall integration, stratified so that, for example, senior management can review commercial results and access all data, while the operations staff can access short term data like expected schedules and load balance status.

The industry is now moving towards an Electronic Data Interchange System (EDI), shown schematically in Figure 11.7, which links all these separate systems together. The development of the Internet allowed the introduction of Logistics Management Systems (LMS) to replace the previous SITA (Societe Internationale de Telecommunication System) and the later Cargo Community System (CCS) which first allowed direct exchange of data between PCs. In 2009, IATA launched the e-Freight Project to replace all paper documents.

It has taken some time for many countries to become compliant, but by 2010, 80% of air freight volume used it, involving 44 countries and 32 airlines.

Apart from cutting out a great deal of paper, it should allow:

- electronic acceptance of all Air Waybills
- utilisation of bar codes to identify and track all freight through its journey
- electronic capture and transfer of Air Waybills through customs for faster clearance
- full availability of status for all customers
- 'carriage ready' presentation of all freight and documentation.

Table 11.3: The flow of consignments and documents.

Documents	Action	Goods
Inbound		
AWB/Manifest to cargo office Check for completeness of documents	Airline	To bonded reception Numerical check
Sorting and distribution of AWBs and Manifest copies (import/transfer)	Airline	Breakdown and segregation of import and transfer cargo
Issuing of Transfer Manifest/customs documents (in-bond document, etc.)	Airline/Customs	Transfer cargo conveyance to proper place (export section/other airline)
Completion of AWBs with Customs/Airline (rotation number, flight number and date)	Airline/Customs	Physical check (identification of import goods against doc). (Manifest/AWB/copy Load Sheet/Customs list)
Sorting of AWBs for processing (Brokers/Special cargo, etc.) Notification to Consignee/delivery of AWB to Broker	Airline	Storage/packing in bonded area according to Consignee/Broker and nature (special cargo, perishables, cold storage, etc.)
Preparation and filling of Entry forms with Customs examination (sampling system)	Broker/Airline/Customs	Conveyance of selected items to Examination area (sampling system)
Release check/stamp Preparation of documents for delivery	Airline/Customs	Customs release check. Withdrawal of goods from bonded into Cleared Cargo area/delivery to Consignee, reforwarding
Outbound		
Letter of Instruction/Export Declaration (licence, etc.)	Shipper	To Agent or to reception area of airline's warehouse
Issuing of AWB/completion of Export Declaration	Agent/Airline	Checking of weight/packing/labelling
Clearance Processing of Export Declaration/examination (sampling system)	Customs/Agent/Airline	Customs Check Point for area Airline's Acceptance Check Point (if received from agent) Examination (sampling system)
Check and final completion of AWB (Routing/bookings/special cargo) Distribution of AWB copies	Airline	Assembly in warehouse according to flight/destination
Final release to appropriate Flight/completion of Manifest/running off copies	Airline	Loading area tally against Manifest/Load List and assembly on carts/bins/pallets
Preparation of flight documents AWB/Manifest copies: (a) for customs file, (b) for ship's bag on board	Airline/Customs	Conveyance to aircraft Customs supervision (load check)

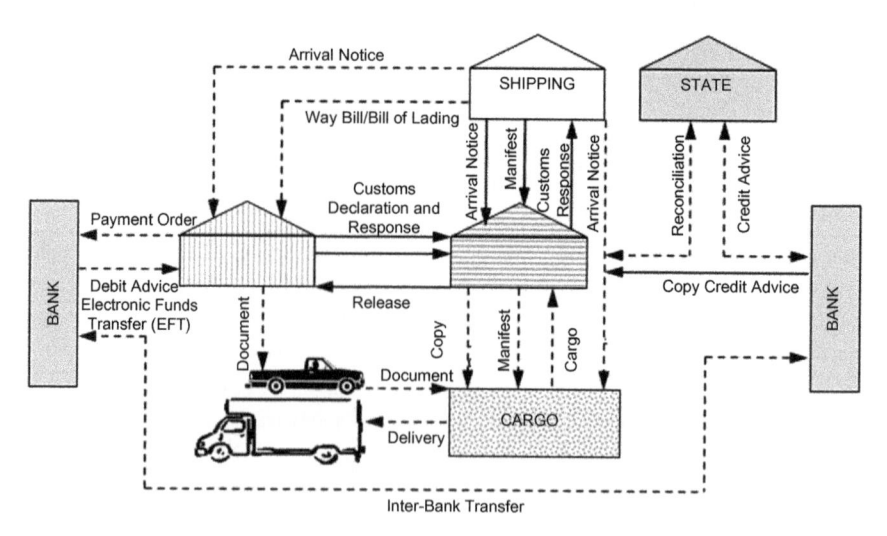

Figure 11.7: Electronic Data Interchange System. *Source*: Courtesy Stuart Walsh.

11.4.6. Utilities

The terminal design should take account of the need for a large number of facilities:

- potable water
- electric power, and a standby system, clean for computers
- lighting (200–300 lux at floor level, allowing effective colour reading and the use of PCs)
- fire detection and protection
- heating and cooling
- sewage disposal
- garbage disposal
- surface water drains
- lightning protection
- aircraft fuelling and ground power
- communications (phones, computer cables, antenna sites)
- ground to air radio
- two-way personal VHF radio
- public address system

↦ closed circuit CCTV

↦ document and message conveyors.

11.4.7. Security

A security system is necessary to protect against theft and, increasingly, to thwart acts of terrorism. ICAO passed responsibility to the freight agents to X-ray screen, search, and hold for 12 hours or depressurise all cargo that is not from a 'known' shipper. New regulations have been brought in, requiring screening of all international cargo inbound as well as outbound. Cargo bound for the United States has to be put through a portal on the ramp to detect explosives. In the United Kingdom, a machine which can thwart the threat from attempts to disguise or hide radioactive materials is being installed under an initiative called Programme Cyclamen.

A minimum requirement of a cargo terminal security system will have:

↦ control of access to staff and visitors

↦ a CCTV system

↦ perimeter fencing and lighting, and a road to physically inspect it

↦ a barrier control system

↦ background checks on all employees

↦ explosive detection system.

In addition, it should follow all the guidelines discussed with respect to passenger terminals and the airside (Chapters 12 and 13).

11.5. CARGO TERMINAL LAYOUT AND SIZING

11.5.1. Layout

A notional typical layout is shown in Figure 11.8. Most terminals have only a single work floor, the inbound and outbound freight being processed side by side. Essentially, the flow moves towards the airside from the truck bays through de-stuffing, sorting, transfer, re-stuffing, weighing, screening and out to the airside dock, and vice-versa.

There are separate channels for international, domestic and transfer traffic. The minimum depth required from the aircraft nose to the landside boundary is:

252   Airport Design and Operation

- ✈ apron staging area — 18 m
- ✈ airside road — 12 m
- ✈ unit loading area — 6 m
- ✈ terminal — 90 m
- ✈ truck docks — 25 m
- ✈ landside cargo road — 10 m.

Resulting overall terminal depth is in total of 190 m, plus another 10 m if aircraft are to be loaded through the nose. Columns should be kept to a minimum, subject to cost control. Some of the latest terminals have multi-level processing.

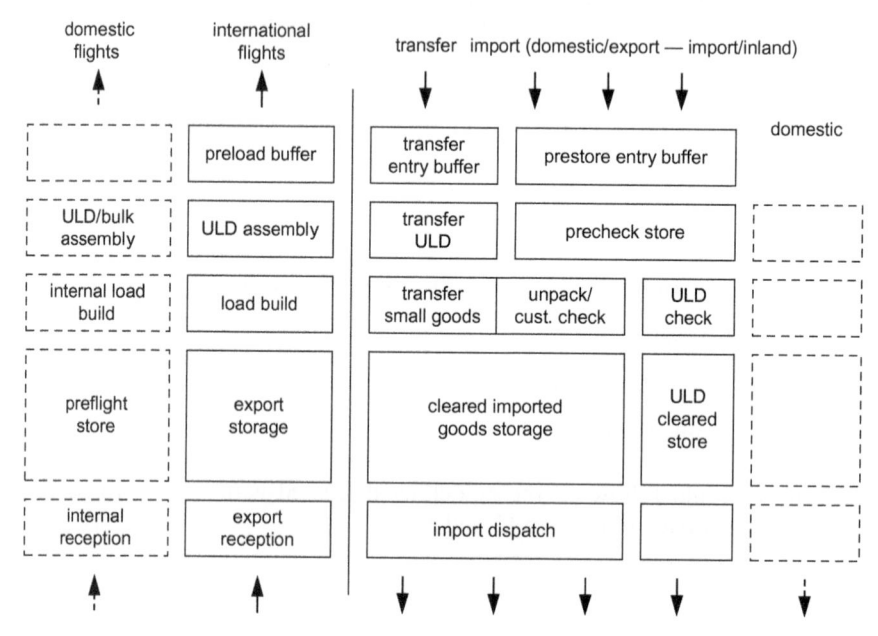

Figure 11.8: Notional layout of a cargo terminal floor. *Source*: Courtesy Stuart Walsh.

A choice has to be made between handling ULDs either with the short side or the long side parallel with the terminal frontage. The former is good for storage, there are more positions for 20 ft containers, and the build-up positions can be accessed from two or three sides. However, there are fewer build-up positions and a deeper building is required to get an adequate staging area.

IATA recommends that offices requiring direct access to the physical goods should be on the same ground level as the operations, while accounting and other offices should take advantage of the natural height of the building by being on a mezzanine. However, the physical separation between offices can result in dysfunctions. Canopies should be provided over the truck and airside docks.

11.5.2. Functions and Facilities

In addition to providing for the actual storage and flow of cargo, the terminal building has to house facilities for a large number of ancillary support activities:

- supervising and control offices
- airline offices
- freight forwarder offices
- customs inspection area and offices
- agricultural inspection area and offices
- security
- operations offices
- administration and support areas
- customer service
- battery charge area
- maintenance and spare parts
- plant room
- computer suite
- staff support
- aircraft spares and tools
- refuse collection.

It will also be necessary to provide special accommodation for a number of special cargos, all with their own requirements:

- vaccines, live organs, medicines — refrigeration
- bullion, documentation — strong room, dedicated 'airlock' type vehicle docks or hopper systems
- human remains — temperature control, mourning area
- livestock — climate control, quarantine, partitioning for species, racking for small cages, veterinary support

254 Airport Design and Operation

- dangerous goods — IATA regulations
- radioactive — shielding or spacing to meet exposure standards
- airmail — secure
- oversize or heavy cargo
- meat, fish, diary, flowers, plants — all need to be cooled.

Each type of special product needs its own separate work area for ULD build-up, sorting repackaging and quality control. Flowers and fruit need 4–6 °C and 80% humidity; fish need to be near 0° and near 100% humidity.

11.5.3. Sizing

The size of the building is determined partly by the above factors, and also by the amount of freight delivered at one time, the expected dwell time, the density of the freight, the size of equipment, and the role of the terminal. If most of the freight comes in belly holds of passenger aircraft, the flow through the building is less peaky than if it arrives on a B 747 freighter. The size of the aircraft determines the maximum unit load size, hence the batch processing size and the size of the load assembly area. The role of the terminal may be transhipment, international entry/exit, originating domestic or international flights, or a mixture of all these functions. The more originating and international freight there is, the more functions need to be performed and hence the larger the building. The cubic capacity of a terminal depends on the design year throughput and the expected seasonal, weekly and daily peaking. It is also influenced by the factors mentioned in the discussion above. The minimum height for processing is 5 m, more for automated terminals with high storage racking. It is customary to install offices on a mezzanine floor above the work floor. Also, there is an increasing trend to use multi-level buildings for the main functions, as space on airports becomes increasingly scarce and expensive (e.g. British Airways at Heathrow, Hong Kong, SATS at Singapore). A survey of some 50 terminals revealed averages of 0.5 tonnes/year/m^3 for throughputs less than 400,000 tonnes/year to 1.0 tonnes/year/m^3 for throughputs of more than 800,000 tonnes/year.

As long ago as 1964, the US Federal Aviation Administration issued an Advisory Circular with a space chart that suggested that 800 m^2 is needed for processing and a further 150 m^2 administration for a throughput of 150 tonnes/day, that is a total of 40 tonnes/year/m^2. or even 47 tonnes/m^2 if the offices are not counted, assuming 250 working days per year. This high level of productivity is hard to achieve. IATA suggests that the figure should vary between 5 and 17 tonnes/m^2 depending if the terminal is equipped for low or high automation.

The survey suggests that the average spatial productivity is 10 tonnes/year/m^2 of total area for terminals handling more than 250,000 tonnes/year or 5 tonnes/year/m^2 for small terminals with 50,000 tonnes/year. The spatial productivity varies from 1.87 to 52 tonnes/year/m^2, depending on the country, the degree of automation, the type of cargo and the quality of service. Even a manual terminal can be made to produce 40 tonnes/year/m^2, but only at a low quality of service and with health and safety being compromised. In terms of operating area, the throughput varies from 6.66 to 20.48 tonnes/year/m^2. This implies a productivity of 16 to 77 tonnes/year/m^2 of administration area. It is usual for offices to take 20–25% of the total floor area.

Other results of the survey, in terms of tonnes per year, were:

- tonnes per break station: 3,641–18,604; average 10,000
- tonnes per bulk storage bin: 38–643
- tonnes per ULD storage position: 167–783; average 550
- storage positions per ETV: 70–600; average 300
- bulk/ULD storage ratio: 0.52–13.25; average 3
- tonnes per landside dock: 2,520–22,727; average 4,000 for old, 10,000 for new
- tonnes per employee: 307–909; average 375 for medium automation
- tonnes per airside dock: 2,778–47,273; average 20,000.

The data obtained by the survey are only approximate. It would be preferable for the areas to be broken down by function, and account to be taken of the amount of bypass cargo and apron storage of ULDs, and some of the processing being done off-site. However, the above results do give some approximate indications of the space required. In general, the higher the degree of automation and the higher the throughput, the greater will be the spatial productivity.

The terminals for the integrated carriers tend to operate at the lower end of the spatial productivity ranges. They do not deal with all the specialised types of cargo, but the cargo is less dense and the daily utilisation is lower due to the peakiness of the demand. IATA suggests that the main integrator hubs manage 7 tonnes/year/m^2, while the smaller facilities can handle about 5 tonnes/year/m^2.

Airside doors should be 5 m wide and up to 5 m high, while landside doors are normally 4 m high by 3 m wide. They should be damage-resistant and secure. The presentation of ULDs to the aircraft depends on the aircraft's floor restraint system and its loading doors, and so affects the way the ULDs are organised for transfer to the ground handling equipment. This makes it convenient to have a ball-bed airside dock. The width of terminal doors depends on the presentation angle, as do the widths of the conveyor system's pallet dollys and the truck beds.

256 Airport Design and Operation

The space for each activity within the terminal is dictated by the maximum size of ULD or package handled. Typically, the truck dock loading bay will have scissor lifts or a vehicle that moves along the dock interface capable of handling 5 tonnes, with a width of 3.4 m and a lift from 0.5 to 1.6 m. Work stations are usually 3.6 m long, 2.7 m wide, lift from 0.5 to 1.6 m and can support 7 tonnes (Figure 11.9). Reorientation beds and weighing scales will have the same dimensions and weight limitations as work stations. Sorting chutes may be 50 m long and extend 4 m either side of a central conveyor belt.

Figure 11.9: Cargo palette work station. *Photo*: Courtesy CSA Cargo.

11.6. LANDSIDE DESIGN AND OPERATIONS

The landside traffic delivering or collecting freight consignments includes private cars, flatbed trucks, vans of different sizes, lorries (some with their own handling systems) and special container vehicles. The vehicles need an efficient road access system, with two lanes and a total width of 10 m, preferably dedicated to freight with its own access to the public road system. The access roads should, where possible, avoid centres of population, as both the size of the trucks and the frequent need to move at night can cause problems with local communities. The vehicles also need parking, security checks, docks on the landside face of the terminal, and support facilities such as documentation services, driver rest areas and fuel stations.

The trucks will often be backed up to individual doors in the terminal for collecting consignments, but may need to be parked adjacent to the terminal when delivering, with the consignments being taken to the terminal by fork lift. The secure area behind the terminal needs to be deep enough to allow reversing to the doors and also for the parking. The angled parking should be 22 m deep and 35 m from the loading bay.

Arrangements must be made for the employees to get to work. With the emphasis on minimising the environmental footprint, efforts should be made to ensure there is public transport available. This is a particular problem when most of the freight activity is at night, as with the integrated carriers. An adequately sized car park also must be provided for those who cannot take advantage of bus services.

11.7. FUTURE TRENDS

There is every expectation that cargo will continue to grow at least as fast as passengers, and that the integrated carriers will obtain an even larger share of the market: industry will press for ever shorter order cycle times and solid state technology will continue to increase. Trends already under way include paperless transactions (EDI), short haul trips by truck, larger aircraft and multi-level terminals. Other probable changes are as follows.

Conventional combination carriers will improve their systems to take advantage of their high frequency services. As logistics companies offer more added value to the shippers, more free zones will be located on airports. The trend to integrated rail/air/sea is likely to continue despite the problem of compatible containers. Sea may take 35 days, air 4 days and sea/air 16 days, and there are many customers for whom 16 days is a comfortable time to tie up inventory in return for rates that are approximately half the air rates.

As land on airports becomes harder to find and more expensive, more aspects of the cargo handling will find its way off airport, subject to customs allowing bonded transfer, as in Taiwan and Delhi.

11.8. DHL CASE STUDY

DHL is one of the four largest integrated carriers. It is now wholly owned by Deutsche Post World Net, the world's seventh largest employer. DHL was launched in the United States after deregulation. The US hub is at Cincinnati, where it has recently invested $105 million. It now has a building of 95,000 m^2 and 48 aircraft stands. The throughput is up to 90,000 pieces per hour with 1600 employees.

Worldwide, DHL uses more than 260 aircraft, 179 being jet-powered, 110 of which meet ICAO Chapter 4 noise and emissions regulations. It employs 77,000 people, has 32,500 vehicles, 4000 offices and 120,000 destinations in 220 countries, and serves 2.7 million customers each day through 500 airports and 31,000 service points. DHL generated revenue of 12.7 billion Euros in 2013 from its Express business.

In Europe, DHL operates 12 hubs. DHL moved from Brussels to Leipzig in 2008 after a new runway had been completed, partly because of strong environmental constraints at Brussels. The terminal has a strong emphasis on sustainability: with solar panels, combined heat and power plant, rain water being used for washing aircraft and with access to the rail system. It handles 2000 tonnes per night.

The main UK base is at East Midlands Airport (EMA), second only to Leipzig in size. EMA encouraged freight operations after it was privatised, building a large apron and warehouses at the East end of the airport, together with a cargo village to house 25 forwarders. Almost all of the freight is by integrated carriers (UPS and DHL), but there is also a strong Royal Mail hub, a fleet of 9 Boeing 737-300QCs carrying 20% of all Royal Mail flown post and small parcels in containers sorted in a 4000 m^2 warehouse on the airport. Mail peaked in 2011 and has since fallen off by 15% due to the competition for first class mail from electronic mail. EMA handled 293,000 tonnes in 2005, making it the main pure freight airport in the United Kingdom. Though UPS has a cut-off for pickup in London 1 hour earlier than with its Stansted operation, EMA's location is very central, its management style is friendly and helpful, and there is a relatively small number of people annoyed by noise. It has been proactive in using its noise monitoring system to police the aircraft movements. It is the first airport in Europe to adopt the Webtrack system to allow anyone to view the movements on a pseudo radar screen after a delay of 24 hours and interrogate it to find the details of each flight. An independent noise forum meets quarterly. There is over 90% compliance with Continuous Descent Approach (CDA) procedures. The airport has organised 24 hour scheduled bus services from the three cities within 20 km so that freight workers can get to work. It has a 3083 m CAT IIIB runway, 24/7 opening hours, good weather, maintenance facilities, a large Cargo Village and can handle all aircraft types, including the An 225.

The location allows DHL to meet tight pickup and delivery times to a large part of the United Kingdom by truck. DHL invested £35 million in a dedicated terminal in the year 2000, while the airport invested £15 million in the associated 160,000 m^2 apron on a site at the West end of the airport that allows DHL a maximum degree of independence in its operations. The terminal location required that the ILS beam for Runway 09 be narrow, so Category III equipment is in use for all precision landings.

The terminal has a 30,000 m^2 warehouse and 10,000 m^2 of offices. Also, there is a 10,000 m^2 truck canopy for the 35 bays and parking for a further 70 trucks. There are 450 truck movements each night. There is a workforce of over 750 people, and a capacity to handle 18,000 packages per hour or 1200 tonnes per night, using 4 km of conveyor belts.

The 18 stands serve 22 domestic and European destinations directly, together with flights to Wilmington, New York and Hong Kong, each night of the working week. The fleet consists of B 767Fs, B 777Fs, A 300s, B 757SFs and B 737s. All but the A 300s are equipped for Category III landings. EMA does not work as a pure hub where all aircraft are on the ground at the same time. Many of the aircraft are turned round within an hour. A tight schedule may result in a decision not to use the belly hold capacity because it takes longer to load. Up to 30% of the cargo is O-D, that is it is imported or exported. The rest is transferred between flights, mostly entering the terminal for sorting, though some is transferred directly on the apron. The target turnround time from unloading a package to reloading it onto an outbound connection is 90 minutes.

The Flight Control Room has a controller for each of three operations: fuelling, air crew communications and loading/unloading. They have a view of all stands on the apron, though not all of each stand can be seen when the aircraft are parked.

Most of the documentation travels ahead of the cargo electronically, allowing real time tracking of the cargo and virtually all the cargo to be cleared through Customs without physical inspection. It also allows the Network Control Group to match the expected cargo to the aircraft, to switch types to change capacity, contract space on commercial flights and accept cargo from the conventional airlines. They are able to call on a back-up aircraft when necessary for technical or capacity reasons. The group also performs the load and balance calculations for all the aircraft, using a computerised system developed in-house.

Security is maintained with a variety of systems. There are 101 CCTV cameras of very high quality viewing all aspects of the operation. All staff go through a screening process in the administration building attached to the terminal, similar to that performed on passengers. All trucks pass through a security lock where the 'known' shipper seals are inspected and there is X-ray equipment for screening any loads that do not conform.

Cargo flows from airside or landside into one area of the terminal. Much of the landside cargo is brought from the trucks in trains of cages. Airside cargo arrives in ULDs. The de-stuffing, sorting and stuffing of ULDs are all manual processes, though some pallets are loaded directly into containers by fork lift. All parcels are electronically reconciled with their proper container, after which the ULDs and pallets are moved on fork lifts or roller beds to the airside or landside doors for loading. Courier traffic is dealt with by a separate belt and automated sort system.

Almost all of the DHL flights are during the night, so the terminal has been opened to other freighter operators during the day, thus making better use of all the handling equipment. EMA has reduced rates during the daytime and has advantages in clearance time over the London airports, but the number of forwarders has not increased over time. DHL is now planning to invest £90 million to expand the sorting capability at East Midlands, in anticipation of substantial growth in the volume of traffic handled, mostly by increasing loads rather than the number of flights. The company is also developing drones, with the objective of taking the goods much closer to the customer.

12

PASSENGER TERMINALS[☆]

Tony Kazda and Bob Caves

Kauffman's airport law:

The distance from the entrance into the terminal building to the airplane is inversely proportionate to how much time you have to catch the plane.

(A. Bloch: Murphy's Law)

12.1. AIRPORT TERMINAL DESIGN PRINCIPLES

The airport terminal is often the first point of contact with the country for the arriving passenger. It is a shop window of the country and makes the first and, on departure, also the last impression on the passenger. From the architectural point of view terminals have always been and some still are a show piece representing the best of a particular country. It is, however, necessary to give priority to the functionality of the building by a suitable layout of the terminal and the way it is operated if the passenger is to go away or enter with a good impression. A passenger survey[1] shows that even the largest airport terminals can fulfil the passengers' needs and expectations and were designated as The World's Best Airport Terminals in 2014. In order of satisfaction they were Heathrow Airport Terminal 5, Singapore Changi Airport Terminal 3, Haneda Airport Int'l Terminal, Shanghai Hongqiao Airport Terminal 2 and Munich Airport Terminal 2. They consistently manage to give passengers a good experience of moving through an airport terminal. The design of the building depends not only on the number of the checked-in passengers but it must also have regard for the type of airline operation, in particular whether the airport is predominantly an airline hub or is serving mostly local point-to-point traffic. The basic design characteristics of terminals can be found in the literature (Ashford, Stanton, Moore, Coutu, & Beasley, 2013; Blow, 1996; de Neufville & Odoni, 2013; Horonjeff & Mckelvey, 2010; IATA, 2004). This chapter

---

[☆]The authors would like to thank Keith Polkey, Declan Rajasingam, Paul Boynton and Dave Milner of Heathrow Airport and George Hohlacov for their help and advice.
1. World Airport Awards (2014), SKYTRAX.

extends, rather than repeating, the literature by considering specifically the operational implications on design.

The main function of a terminal is to provide a convenient facility for the mode transfer from ground to air transport, and vice-versa. It is also the national frontier for international passengers, so needing to provide all the necessary facilities for this as well as those for the processing required by the airlines. The demarcation between the airside and the landside is usually within the terminal, marked by a security screen, the airside of which is subject to strict control of access prior to the boarding process. In all cases, only passengers are allowed airside after passing security screening, though at some US airports the baggage carousels are installed landside to allow greeters to help their relatives with baggage after arrival (e.g. Indianapolis airport). The landside is usually open access from where the passengers are dropped off their ground transport, through ticketing and check-in and up to the airside barrier. However, sometimes the local security situation requires the boundary to be drawn at the terminal entrance.

On an airport with a great number of transit and transfer passengers the airside transit part of the terminal has to be dimensioned sufficiently and must be equipped with the systems for transportation of the passengers so that it allows them to circulate rapidly and thus ensure short declared times between the connecting flights. The departure and arrival concourses can in this case be smaller. On airports with mainly origin/destination traffic increased attention has to be paid to the design of the departure concourse of the terminal.

Passengers and those accompanying them have many tasks to perform within the terminal, the more so with the trend to self-help rather than the use of porters and escorts. They have to find their way through a series of processes, often encumbered by coats, bags, dependants and disabilities. They also have to assist in being processed through check-in, security screening, passport control, the boarding gate, customs, immigration and picking up ground transport, often in a language which is not their first choice. The ambience the passengers experience influences them in three major ways: the stress experienced in accomplishing their group or personal goals, the form and nature of their social contacts and their feelings of identity and self-worth. Passengers are most stressed at check-in, security and boarding.

Travellers, airlines and other users of the terminal have their own ideas about the comfort, convenience, costs and ambience that should accompany the movement of passengers and their bags between aircraft and ground transportation. There are inevitable compromises to be made between capital and operating cost, between cost and level of service and between form and function, all of which influence a designer's ability to satisfy all the needs of all users. Only when passengers feel safe, know where they are going and have enough time to spare do they pay much attention to the levels of comfort and the attractiveness of their surroundings. The time needed to pass from the kerb to the aircraft, and its variability, is affected by

Passenger Terminals 263

the efficiency of the processing and the speed of moving through the building, the latter in turn depending on the crowding and the ability to find the correct route. The processing and crowding depend on the staffing levels, the number of stations at each processing point and the amount of space provided for a given throughput.

Efficient processing and movement requires good anatomical and physiological ergonomics for the passenger. Some of the common problems are:

Manoeuvring heavy trolleys:

- to check-in
- along extended 'crocodile' (or snake) check-in queues
- through customs to ground transport
- changing levels (except at Frankfurt and Kansai where the trolleys can be taken on escalators).

Check-in:

- height of the counter
- trolleys moving as passengers try to lift off their heavy bags
- impossibility of keeping weight of bags close to passenger's body when placing bag on scales
- hearing the operator's questions
- service 'justice' — in terms of being processed in order of arrival, which is ensured by 'crocodile' (or snake) queues at the expense of the appearance of a great queue length (Figure 12.1).

Movement through the terminal:

- slippery floors
- lack of assistance with coats and hand baggage — there is a trend to carry more bags on board but few airports provide trolleys in the airside of the terminal.

Despite these problems, the primary factors controlling passenger acceptance are the space provided, the minimum and maximum times required and the wayfinding ability.

The design of an airport terminal is affected by the composition of type of flights/ airlines (i.e. traditional carriers, low cost, long haul, charter flights, etc.) the mix of passengers and their requirements. For example, in airports with a great proportion of charter flights, the concourse in front of the check-in counters must be sufficiently deep and must offer enough space to cope with the inevitable long queues of passengers arriving in large batches and being processed through a small number of check-in desks. They are normally either migrant workers or package holidaymakers, frequently with over-sized baggage, and often impede free movement in the departure concourse. Airports with a high proportion of business travellers must offer them fast check-in and the shortest possible distance to the aircraft. It is therefore important, before starting the design of the terminal building, to know the type of airlines, type of airport operation, type of passengers, their composition and their requirements. The design is also significantly influenced by global trends in mobile technologies resulting in rapid growth of off-airport check-in but also changes in security rules.

Figure 12.1: The 'crocodile' queues should guarantee service 'justice'. *Photo*: A. Kazda.

However, the type of operation and the passengers' characteristics in the airport can change, either gradually or suddenly. Because of these changes, caused not only by economic but also by political factors, the terminal buildings must be designed very flexibly. This is particularly true for utilities: gas, electricity, water, telecommunications and security management systems (see Figure 12.2). Examples of such changes might be the introduction of a special channel for checking in Schengen passengers at the airports of the new European Union member states or changing concessions layout following the abolition of duty-free sales. Therefore the terminal buildings must be highly flexible to provide fast, simple and cheap reconstruction. Changes in the layout of the building must be executed so as not to disturb substantially the check-in or bag claim processes, given the difficulties in changing conveyor routings. In the past, designs have often been too inflexible. Instead of masonry partitions in the terminal building it is more suitable to use light assembled or wood partitions or even 'moving walls', which can be moved within one night or a few hours. It is then necessary to make the service installations in the floor or, even better, in the ceiling of the terminal building. The terminal must be capable of further and simple extension as the number of passengers increase without the necessity of further significant modification of its critical elements such as foundations, staircases and services. The terminal building layout must in particular provide for:

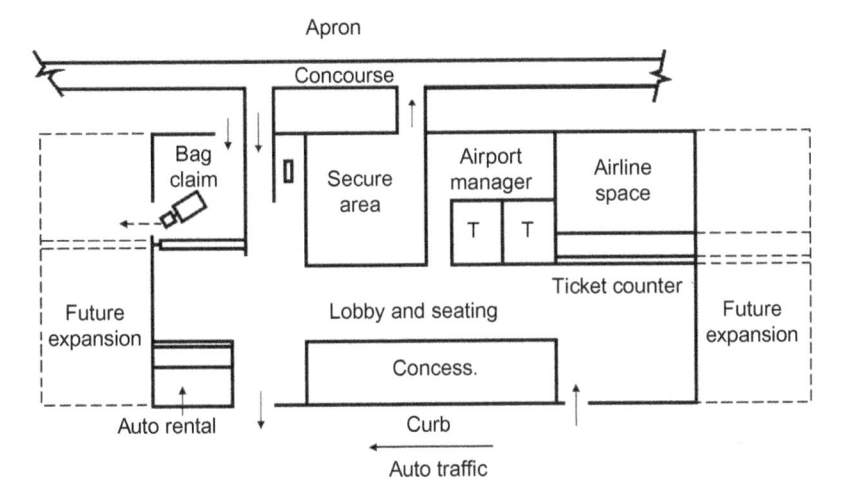

Figure 12.2: Example of a functional terminal layout. *Source*: FAA Advisory Circular, Planning and Design of Airport Terminal Facilities at Nonhub Locations.

- ✈ terminal airside connection
- ✈ terminal connection to the landside surface transport system
- ✈ as short as possible walking distances of arriving and departing passengers
- ✈ information for the passengers during the whole series of processes
- ✈ convenient connection for transfer and transit of passengers
- ✈ baggage handling system for local and transfer bags
- ✈ airside/landside security screen
- ✈ government controls
- ✈ appropriate sizing of all areas

✈ required mix and quantity of aeronautical and non-aeronautical services

✈ space for offices and all support services.

However well a terminal is designed, it will only achieve its purpose if it is operated efficiently. It is all too easy for standards that could be met to fall short because of poor management. An essential tool for maintaining quality is to make Service Level Agreements (SLA) between the airlines, the government services and the airport. The SLAs should be monitored by surveys, questionnaires and an easy to use complaints system. The agreements should cover, among other things:

✈ fixed ground power serviceability

✈ aircraft parking stand availability

✈ remote coaching performance

✈ people mover serviceability

✈ bag systems serviceability

✈ queuing standards

✈ security service standards

✈ ease of wayfinding

✈ availability of seats

✈ cleanliness.

12.2. AIRPORT TERMINAL LAYOUT

The terminal building is a connecting link between the airside and the landside. It is a connection between 'the sky and the earth'. First and foremost, it must provide basic aeronautical services; that is activities directly connected to provision of the processing of passengers. The building must provide a fast and shortest possible transition of the passengers from the surface transport through the outbound processes to the aircraft on departure and in the opposite direction on arrival. The arriving and departing passengers must be physically separated, not only for fast and fluent movement of the passenger flows but also in order to ensure security (see Chapter 13). The flows of passengers can be separated by fixed or moveable obstacles on a single level (horizontally) or on several levels (vertically). Separation of departing and arriving passengers on Schengen flights is not mandatory as the level of security is considered to be similar (i.e. single security stop).

For small airports, a single level concept is the most suitable (Figure 12.3a). In this concept the departing and arriving passengers and their baggage are separated horizontally, usually on the same level as the apron. The arriving passengers on small airports with domestic/Schengen flights often do not even pass through the terminal building and the baggage is directly handed to them from the baggage cart under a shelter or even near the aircraft. In this single level concept passenger loading bridges are not used, though the concept of a telescopic cover to protect passengers from the weather was introduced at Oakland, California, by 1930 and installed at the original beehive terminal at Gatwick in 1936.

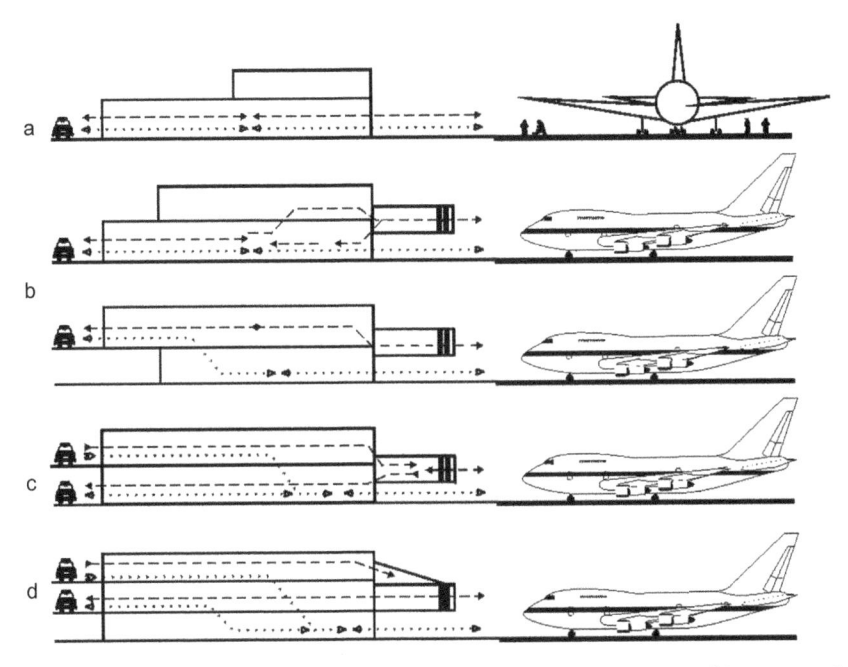

Figure 12.3: Airport terminal layout. (a) Single level concept, (b) one and half level concept, (c) double level concept, (d) three level concept.

If terminal buildings at larger airports were designed on a single level, the terminal would require a large area of land. It then becomes more convenient to separate the passengers vertically. The simplest type of vertical separation is a concept of one and a half levels in several variants according to local conditions (Figure 12.3b). Normally the departure and arrival landside concourses are on the same level side-by-side. The division of flows of departing and arriving passengers and baggage can be done at any point after the check-in process or alternatively immediately after the entrance into the terminal building. Both levels meet again on the apron if the passengers are to walk or be bussed to the aircraft. However, if passenger loading bridges are regarded as necessary for passenger safety or as economically viable, the flows meet again at the entrance to the airbridge. In either case, departing passengers are held behind a barrier until all arriving passengers have been cleared into the terminal.

268 Airport Design and Operation

The double level concept provides separation of the passenger flows even on the landside by vertical stacking of the road access system, though with the capability to move between the levels inside the building (Figure 12.3c). The double level concept is usually used for terminals with traffic volumes of above 5 million passengers a year.

The FAA suggests that it may be worth considering a two level terminal when the traffic exceeds 1 mppa, a two level kerb when it exceeds 2 mppa, and multiple terminals by the time traffic has reached 6 mppa, but, in most parts of the world, traffic would have to be considerably higher to justify these solutions.

In addition to the flows of arriving and departing passengers the three level concept also separates baggage vertically (Figure 12.3d). Also if transfers between Schengen and non-Schengen flights are expected, horizontal separation by means of the double level concept between flights is questionable. For those cases some airports use a three level concept (e.g. Vienna Terminal 3) for passengers' separation. It is particularly advantageous to use the three level concept at airports where the baggage transport system and also other systems for technical handling have been designed below the level of the apron.

The flow of passengers through the departing process must be direct, logical, limit the changes in the vertical level and be as short as possible. Maximum walking distances for the passengers are recommended by the International Air Transport Association (IATA) as shown in Figure 12.4 and described below:

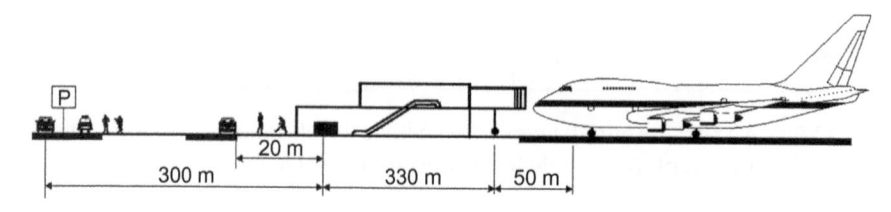

Figure 12.4: Maximum IATA recommended walking distances.

✈ from the departure kerbside in front of the
  terminal building to the check-in counter — 20 m
✈ from the farthest car park to the check-in counter — 300 m
✈ from the check-in counter to the farthest gate — 330 m
✈ from the gate to the aircraft — 50 m.

The walking distances are similarly specified also in the case of passenger transfer between aircraft. If the distances are longer than specified, it is advised to provide some kind of people movers for the passenger (see Section 12.7). ICAO suggests that the time from disembarking the aircraft to exiting the terminal should not exceed 45 minutes for international passengers. However, this limit seems to be the maximum time for the long haul flights and does not correspond with passengers' expectations on short haul point-to-point connections today.

12.3. AIRPORT TERMINAL CONCEPTS

The terminals at small airports have mostly been designed as centralised buildings, that is, where the processing of the passengers is done in one location rather than being distributed through several points in the terminal. The concept of the centralised terminal in combination with piers, fingers or satellites (refer to Chapter 7 for a discussion of these various terminal/apron layout options) is also used at large airports. It provides easy orientation for the passengers through check-in and security, optimum utilisation of space and concentration of services in the terminal building. However, as the number of stands increases, the distance to the outlying stands exceeds the recommended walking distances and therefore it is recommended to provide transportation for the passengers from the central-processing building to the gates, together with an effective information system. A central terminal building with a system of several parallel satellite piers interconnected by a transportation system makes an almost ideal solution for large airports if the space is available midfield, that is between parallel runways. It has a large capacity of both stands and peak hour passengers. It enables transfer of passengers to and from common travel areas without using the central building, which is then not required to handle these passengers. Therefore this design is convenient for the hub-and-spoke type of operation. This design was first used at Atlanta, and has been successfully adopted at Oslo — Gardermoen, Denver, London Heathrow Terminal 5, Washington Dulles or Madrid Barajas among others. It seems that it is possible to use central-processing terminals up to about 90 mppa and more than 150 or so gates, whether they have piers and moving walkways or have satellites connected with automatic people movers.

In the past, large airports preferred the concept of several decentralised buildings, with the traffic segregated between terminals by airline, by domestic/international or by some other categorisation of passengers. This was particularly typical for airports in the United States, where the construction of terminals is often financed by airlines. It is still preferred by, for example, Air France at Paris Charles de Gaulle Terminals 2A, 2B, 2C, 2D, while other decentralised terminals, that is 2E, 2F 2G serve different airlines and/or type of flight. It can offer very late check-in, there is no need for passenger moving or complex bag sort technology inside the terminal, the Flight Information Display System (FIDS) can be simple, there is plenty of kerb, and expansion simply means building another unit. However, when transfer between buildings is unavoidable, it is often necessary to provide transportation of the passengers by some form of people mover between the terminal buildings in order to minimise the connecting time (see Section 12.7).

Now that the Low Cost Carriers (LCC) are demanding dedicated low-cost airport facilities, a new era of decentralised terminals is being launched. Singapore Changi airport has a 25,000 m^2 single storey building at a cost of EUR 12.5 million with basic facilities and no airbridges to handle 2.7 mppa, that is 9 m^2/1000 ppa. The airlines will pay only 50% of the rent in the other terminals. Jet Blue built a 59,000 m^2 terminal on two levels at New York's Kennedy airport with 26 gates and a short-term capacity of 10 mppa, that is 5.9 m^2/1000 ppa. Kuala Lumpur opened a single storey low-cost terminal in 2006 for Air Asia, costing $US 30 million for 10 mppa with 72 desks in 35,290 m^2, that is 3.5 m^2/1000 ppa (Airline Business, March 2006). Lyon St. Exupery airport has refurbished an old charter terminal, and the passenger handling charge is only 16% of the normal charge. Some of these terminals are being built at only a quarter of the price of a traditional terminal. The LCCs have very few passengers transferring onto the legacy carriers, and do not even encourage hubbing on their own network, so there is little need for people movers in their case. If the need arose, bussing would be the appropriate option. However, some passengers do transfer between flights at large LCC bases (e.g. London Stansted, London Gatwick, Frankfurt-Hahn, Milano-Bergamo, Dublin). As the LCCs do not support interline services — baggage transfers and single ticketing, passengers have to buy two separate tickets. To transfer between flights they have to pick up their registered luggage and check it again for the next flight. This type of transfer produces additional load on parts of the terminal (baggage handling systems, check-in, security, etc.) and could be critical for 'pseudo-hubs' with limited capacity in the future. Another concept is the linear type of terminal building. This is particularly suitable for 'point to point' operations and for a simple passenger handling process, for example domestic operation. The linear terminal building is formed by multiple individual modules independent of each other. It needs a near-perfect information system for access so that departing passengers obtain precise instructions as to the terminal location they require. The main advantages of the concept of a linear building is minimum walking distance from the departure kerbside in front of the terminal building into the aircraft, and simple baggage handling. The main disadvantage is relatively high staffing levels, as it is necessary to have a security check in each module, though some airports and governments require a security check at each gate even in centralised terminals. Also, it is difficult to provide shopping and catering facilities, which are becoming an increasingly significant source of airport revenues, and it requires dedicated sets of gates for specific flights. In its curvilinear form (e.g. Dallas-Fort Worth) it is somewhat incompatible with moving walkways.

This concept hardly still exists in its classical form, except in very small scale terminals, though it can be seen in operation on the landside at Dallas-Fort Worth. It is best used for domestic point-to-point operations.

12.4. TERMINAL DESIGN

12.4.1. Design Methods

The standard approach to calculating appropriate space is to choose a representative busy hour, accepting that, for the few hours per year for which there will be more traffic, there will be unacceptable levels of crowding. The terminal capacity is never designed for just 'tomorrow's' needs but for future market evolution to a specific time horizon for example 10 years ahead. Sometimes the terminal building is designed and built much larger than required immediately, but technology, that is check-in counters, is installed only in a part of the building (e.g. Prague Terminal 2). This allows gradual distribution of investment costs. The crowding will be absent for most of the time, perhaps 15 years, when the traffic is growing towards the design busy hour, though it has to be said that most terminals then go on to operate for years after the design conditions are reached before expanded. Capacity of many terminals has been gradually expanded by interior layout changes and investments in technologies over the originally designed levels during the terminal life cycle. For example the London Heathrow Terminal 2 which was opened in 1955 was originally designed for 1.2 mppa. It was closed after 55 years of operation on 23 November 2009, having handled about 8 mppa during its last years of operation.

The US Federal Aviation Administration (FAA) uses a concept of the peak hour of the average day of the peak month to derive the representative peak hour. An approximate rule of thumb is for the average day in the peak month to be 1.05 times the annual average day, and for the peak month to be 10% of the annual traffic. The typical peak hour passengers (TPHP) may be as much as 20% of the day's traffic for small airports, falling through 11% for 2 mppa to 8.6% at very large airports. The FAA also uses a concept of Equivalent Gates (EQA) for the analysis of space requirements based on a nominal 100 seats per aircraft, so that a B747 gate is rated as 4.5, a B-747SP as 2.7 and an F28 as 0.5. The air transport movements are not usually as peaky as the passenger flows, that is the load factors are higher in the peaks.

IATA's concept of a busy hour is derived by taking the second busiest day of the average week of the peak month and selecting its busiest hour. Other peak hour concepts are the Standard Busy Rate (SBR) which refers to the 30th busiest hour in the year, the 20th busiest hour (at Amsterdam), the 40th busiest hour (Aeroport de Paris) and the 5 percentile hour in the year (BAA). The SBR can often be approximated by dividing the mppa by 3000.

A further option for identifying a design hour, appropriate for airports where there is a stable daily pattern through the week due to the operations being dominated by scheduled carriers or for airports with only a few commercial movements per day, is to construct a notional schedule and select the busiest hour. The problem with this approach is that the world is becoming increasingly uncertain, and it is nearly impossible to anticipate airline behaviour sufficiently far ahead.

The FAA (1988) and IATA (2004) have developed procedures, or models, for calculating required areas for each activity within the terminal, based in the busy hour flow and acceptable standards for queuing time and crowding. Note that the FAA often uses 'enplanements' rather than 'passengers', the latter implying the total number of passengers handled in a given time, whether arriving, departing or transferring. It is necessary to be specific about the flow rate for whichever facility or activity is being designed. The FAA models are appropriate for the US system and deal comprehensively with all activities in the terminal. The IATA models are aimed at terminals worldwide, but they are not so comprehensive and tend to be difficult to apply, leading to errors if not used with care. Both models tend to assume that the traffic is constant in the busy hour. There are also models for small airports developed by the FAA (1980) and by Transport Canada (1992). The latter, called STEP, is appropriate for airports up to 3 mppa.

Table 12.1: Level of service — space to be provided for passengers in different functions.

Activity	Situation	Level of service standard (LOS) [m^2/PAX]					
		A	B	C	D	E	F
Waiting and circulating	Moving about freely	2.7	2.3	1.9	1.5	1.0	Less
Bag claim area (outside claim devices)	Moving with bags	2.0	1.8	1.6	1.4	1.2	Less
Check-in queues	Queued, with bags	1.8	1.6	1.4	1.2	1.0	Less
Hold room; government inspection area	Queued, without bags	1.4	1.2	1.0	0.8	0.6	Less

Source: de Neufville and Odoni (2003).

Guidelines are available for the amount of space per passenger in the busy hour which will give a required level of service (LOS) (Table 12.1). The IATA LOS standards for space requirements range from 2.7 to 1.0 m^2 per occupant for LOS A to F respectively for long-term waiting space, reducing to 1.4 to 0.6 m^2 for gate holding rooms. The A level of LOS refers to free flow and an excellent level of comfort, C to a good level of service with stable flow, acceptable delays and a good level of comfort; D is adequate LOS, with acceptable delays for short periods, but the flow is unstable; F is unacceptable for delays and comfort, delays are unacceptable, there is cross-flow and the system breaks down. The crowding should not fall below level D for the TPHP, and most new terminals aim for C in the busy hour. Guidelines also exist, which are different from one airport to another, for acceptable queuing times and movement rates.

For the terminal building as a whole, the FAA gives a rule of thumb of 14 m^2 per one way peak hour passenger for the gross floor area of domestic terminals. IATA suggests 25 m^2 to 30 m^2 for domestic and charter terminals. Within the total space, 5% will be unusable, rather more than 50% will be rentable and rather less than 50% non-rentable. In the US domestic terminals, some 38% will be used by the airlines, 17% by other concessions, 30% by public circulation areas and 15% by airport administration. Adequate space must be also allowed for facilities directly connected with airport operations (e.g. baggage handling/sorting and security systems, operational staff changing rooms/lockers) and terminal technical operation facilities (e.g. fire equipment, electrical systems backup, heating/air condition systems, waste management systems, security screening for staff and goods). Space for those 'technical' areas can make up 40% of the total terminal area (e.g. Bratislava Airport).

Care has to be taken in the application of these analytical models. They are based on handbook formulas insensitive to the realities of each situation, and the formulas are easy to misunderstand and thus frequently misapplied for example the common design hour LOS standards should apply to simultaneous occupants (except the FAA guidelines per TPHP), but are often used per design hour passenger even if their dwell time is much less than an hour. Also, the TPHP differs in different parts of the terminal. They should be improved by a realistic appreciation of the dynamics and behaviour of sequences of queues, the psychology of crowds in such situations, and the ways airport users truly allocate the time they spend in passenger terminals — their slack time is often much greater than their processing time. People do not spread out evenly like gasses, but tend to congregate in specific places, for example the mouth of the baggage chute in the baggage reclaim hall. LOS standards should increase with exposure time. The requirement is also strongly influenced by whether the activity is to be joint use or dedicated to an airline, a route or a type of operation. Territoriality in public spaces leads to a wasteful use of public facilities.

'Intensity of use' and 'service time' are less value-laden terms than 'crowding' and 'time delay', and may therefore be more appropriate as general measures of performance. Because passengers are often accompanied by visitors, and because the occupancy may turn over several times in the peak hour, total numbers of people accommodated simultaneously may be a more appropriate measure than passengers per peak period. Current measures of performance based in the space requirements for a standard 'busy period' are poorly suited to terminals serving intensive hub-and-spoke flight schedules, aircraft fleets with large fractions of high capacity wide-bodied and commuter aircraft, and growing numbers of price-conscious passengers. In fact, there is a large variety in passenger perceptions of adequate space. The length to which an individual will go in accepting cheerfully what most would consider to be extreme inconvenience is remarkable, so long as it is his/her own choice. This is reflected in the recent calls by the LCCs for the minimum of facilities in new terminals in order to keep charges down.

The accuracy of the model results depends also on the availability and suitability of the input data. If there is insufficient data there can be only approximate solutions. For example, in the design of the kerb length in front of the terminal building it is necessary to know the split of the passengers arriving to the airport by car, taxi or by bus, the average number of the passengers in each of the means of transport, average time of stay for each vehicle and the length of waiting space necessary for each of the means of transport. Often, the length of the kerb has to be designed using the default values for all kinds of vehicles. For small airports, under 0.5 mppa, it is advisable to compare 'rules of thumb' values with expert judgement.

For airports with passenger volumes from 0.5 to 5 mppa, the use of fixed ratio models is problematic. The volumes of the passengers are already too big to assess the results obtained from static modelling by expert judgement, but too small for the system of the airport to be relatively stable. Therefore it is more appropriate to use simulation models for the dynamic modelling of such airports.

Simulation models allow individual parts of the building to be considered operating under different conditions. Thus, for check-in counters, it is possible to set different lengths of the queues for the different categories of the passengers: three passengers in the queue for the passengers of the first class, five passengers in the queue for the business class and a waiting time less than 15 minutes for 90% of all passengers, etc., to determine the number of counters required, and also to carry out sensitivity analysis around the central specifications. The models of each individual part of the building can be combined to see how the parts influence each other, taking account of the way passengers spend their spare time. For example, it is possible to investigate the effect of immigration processing on the crowding in the bag reclaim hall, depending on the relative productivity of the two activities, thus avoiding the need to provide duplicated waiting space. Again, the models are only as good as the data that is fed into them. The collection of the behavioural data is very time-consuming and costly. It is necessary to know the behaviour of each type of passenger (e.g. domestic/international/transfer, business/leisure, group size, disability, with/without bags, mode of arrival) and visitor for each decision point throughout the terminal.

There are several simulation models of terminals available such as ARCportALT (Aviation Research Corporation), PBFM (Passenger and Bag Flow Model) and CAST. These models tend to have to be tailored to each specific terminal floor plan, being constructed from CAD drawings. More generic models like ARENA and WITNESS do not need to know the shape and size of the facilities in advance of the modelling. Given the data, these models can generate facility size requirements, assess queue lengths, the degree and duration of crowding, the processing times, waiting times and dwell times throughout the day for each type of passenger. Some of them also assign gates and estimate the number of passengers and bags failing to catch flights. Once the model is calibrated and running, it is easy to carry out sensitivity analysis and 'what if' analysis to compare multiple options.

A major strength of these simulation models is their ability to present real-time graphic demonstrations of the flow using three-dimensional animation, and it is possible to watch queues and crowds build and decay relative to one another throughout the terminal. They also present the results of queue length, level of service, etc., in easily digestible graphics. In fact, the user has to continually remember the quality of input data in order to keep a sense of proportion as to the validity and accuracy of the results. Perhaps the most valuable aspect of using terminal simulations is that it asks difficult questions of the operations and management staff, and gives them a better understanding of the way their terminal works. It can act as a focus for discussions between all parties interested in the design and functioning of the terminal.

12.4.2. Component Design

The landside kerb is used by cars, taxis, courtesy busses and public busses for drop-off and pickup. The various vehicles are often segregated to separate kerbs, with the consequent need for most people to cross lanes of traffic. They are also segregated between arrivals and departures, either spatially in single level terminals or vertically in two level buildings. A typical total kerb length is 1 m per 10 PAX/TPHP, but this depends very much on the mix of traffic and the discipline at the kerb. It is often the case that vehicles stay longer than permitted, and strong policing is required. There is also a problem with security when vehicles are left empty while meeting or waving. Some airports have now banned the use of close-in kerbs, and force people to use a park and ride system or sometimes charges have been introduced for passenger drop off or pickup.

The ticketing lobby requires 4 m of queuing space, a counter and space behind the counter for staff and facilities; perhaps some 2500 m^2 for 50 EQA. The lobby is usually on the back wall of the check-in concourse.

The check-in concourse needs, in addition to the check-in or drop-off counters, some basic facilities like flight counter information, telephones, toilets and a small café for people who wish to linger with those who are seeing them off, but the general aim is to move passengers through the security screen quickly after check-in so as to avoid delays and crowding. It is becoming increasingly common to provide self-service kiosks, whether dedicated to an airline or Common User Self Service (CUSS) kiosks (www.iata.org/CUSS), together with fast bag-drop facilities for the kiosk users with bags. This is more common among the legacy carriers than with the LCCs and the charter carriers, whose reservation systems are not so compatible. An essential element of the terminal design is the communication system and their ducting to drive the reservation systems.

The area depends on the TPHP, type of traffic, counter configuration, per cent of e-ticket sales, level of service, average processing time and ratio of passengers to

276 *Airport Design and Operation*

accompanying people. It is preferable to keep the latter away from the counters and their queues. This is best done by having a crocodile queue with airline personnel checking the right to board at the entry to the queue. The concourse should be at least 20 m deep to allow for queues at right angles to the counters, and for circulation space behind the queues. LOS level B requires 2.3 m^2/PAX with a trolley while they move around the concourse. The normal speed can be presumed to be 0.9 m/s.

Check-in counters may be provided at train stations, car parks, downtown or at gate lounges, but it is most common to situate them immediately after the entry to the terminal. They may be arranged linearly facing the passengers when they enter the concourse (Heathrow Terminal 4, Manchester), end-on to the flow (Figure 12.5) (London Stansted, Hong Kong) or in islands (Heathrow Terminal 1, Dublin). The linear facing type may allow passengers to move on through the line of desks after processing, or require them to move across the queues to the departure gates. This latter arrangement can lead to confusion unless there is a lot of room behind the queues, but it is by far the most common layout.

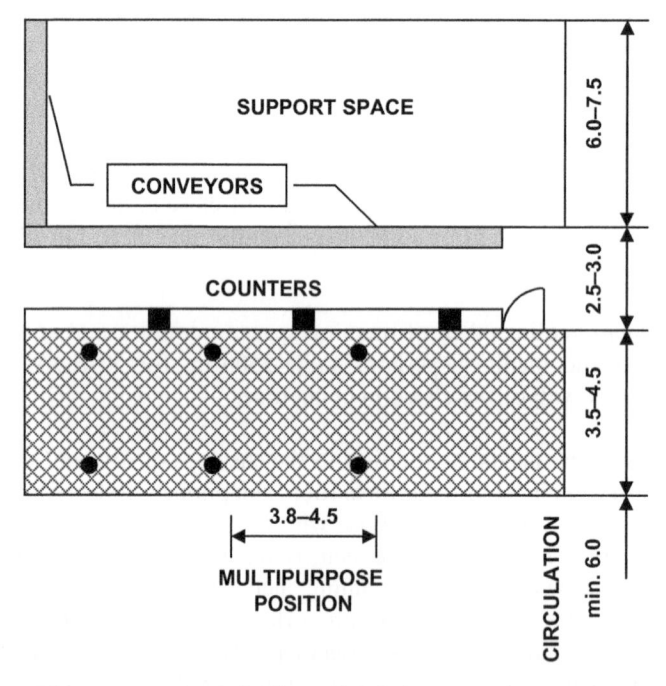

Figure 12.5: Linear counter (all dimensions in metres). *Source*: FAA AC 150/5360-13 Planning and design guidelines for airport terminal facilities.

The conveyors run along behind the counter staff, and are fed by a short feeder belt running back from each pair of counters. The flow-through types take more space and cost more to build and maintain (Figure 12.6). The island types have the counters arranged in a U shape around belts running in the direction of the passenger flow (Figure 12.7).

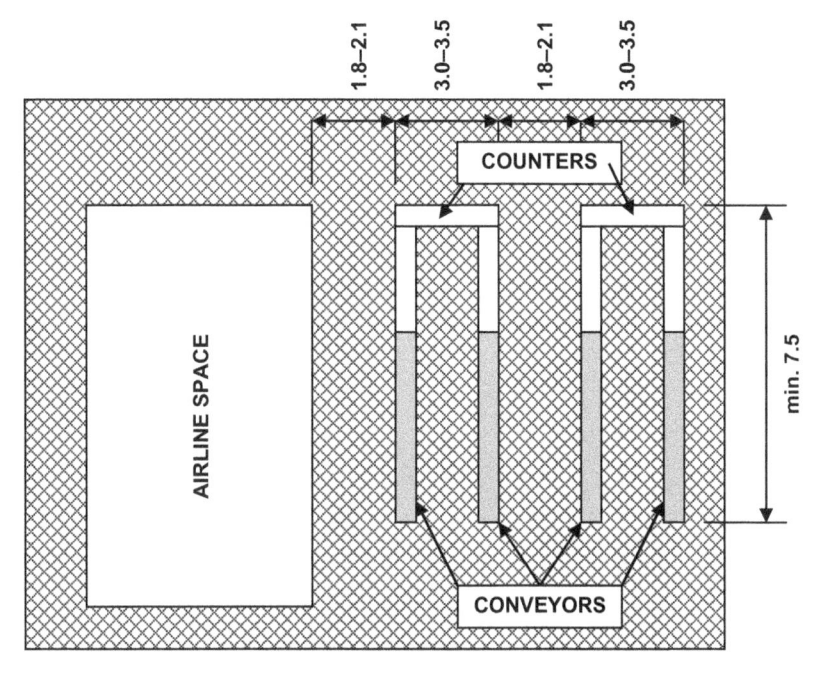

Figure 12.6: Flow-through counter (all dimensions in metres). *Source*: FAA AC 150/5360-13 Planning and design guidelines for airport terminal facilities.

They offer some element of flow-through and make efficient use of the conveyor belts because they can be fed from both sides. IATA has a preference for the island layout, with between 10 and 20 counters per side and with 20–30 m separation between islands. The average processing time is 2 minutes, depending on number of bags, size of group and procedures required by the authorities. The FAA suggests that, for 50 EQA, 86 m of counter frontage is needed for a hubbing terminal and 148 m for an O-D terminal. Each pair of counters takes approximately 3.5 m, so a total of 84 counters is needed. The area needed is 571 and 981 m^2 for hubbing or O-D respectively. It has to be remembered that these guidelines are for domestic operations in the United States. IATA's rule of thumb is that for 2500 departing passengers per hour (i.e. rather less than 50 EQA) and a 10 minute maximum queuing time, 70 economy desks and 14 business class desks are required. However, IATA also suggests that 5 economy, 2 business and 1 first class desks are needed for a B747-400, which seems to be rather less than the previous rule of thumb. Another rule of thumb is that nine counters are needed per 1 mppa. A B747-400 may present 80 passengers in the peak 15 minutes which would occur about 90 minutes before boarding. An A380 will therefore present perhaps 120 passengers in the same period, requiring 50% more desks if queues are to be kept under control. LOS level requires 1.2 m² per queuing passenger with few bags, rising to 2.0 m² for one with two or more bags. The acceptable queue time in economy is deemed to be 12 minutes, and 3 minutes in business class. However, there are no recommendations for new types of operation for example a high number of internet check-in, short haul passengers with hand baggage only or low-cost terminal operations.

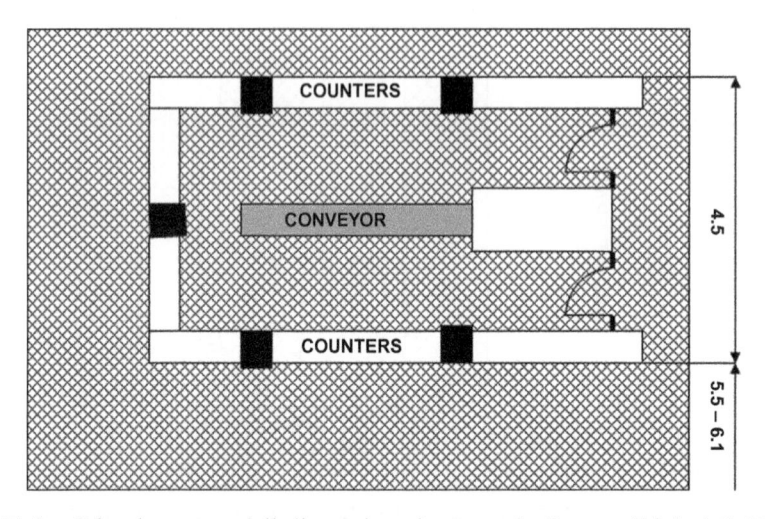

Figure 12.7: Island counter (all dimensions in metres). *Source*: FAA AC 150/5360-13 Planning and design guidelines for airport terminal facilities.

The out-going baggage handling system is usually installed under the floor of the departure level at large airports. It consists of:

✈ belts, or trains with tilting trays or destination-coded vehicles

✈ sorting devices that read the codes on the bags and either push the bags off the belts or tilt the trays to assign the bags to the flight make-up area

✈ a screening system

✈ a feed for transfer bags.

The system needs to be able to handle a large number of bags; between 0.8 and 2.2 bags per passenger, or between 1,600 and 2,200 bags per hour for 4 EQA and between 16,000 and 26,000 per hour for 50 EQA. IATA suggests that it should take no longer than 9 minutes for a bag to get from check-in to the furthest make-up area. Transfers from domestic to international should take less than 25 minutes, from international to international no more than 35 minutes. This may require fast track routes for rapid flight connections; it may be necessary to transfer directly on the apron.

The conveyor belts normally operate at between 0.4 and 0.8 m/s, giving 26–50 bags per minute, but can achieve 1.5 m/s if acceleration and deceleration belts are employed. Slopes should not exceed 22°. Chutes may be used for down-slopes, but there is a greater tendency to damage the bags. If there are many flights to be served, it is necessary to install belts that recirculate past the fixed make-up stations for each flight, where the carts for each flight are usually positioned parallel to the belt for ease of loading. Those belts with inclined beds can carry a double layer of bags. Semi-automated sorting can be installed, where a pneumatic arm pushes the bag onto the appropriate slide, actuated by a signal from the operator who reads the baggage tag. The system can sort about 30 bags per minute.

Tilt-tray systems have been used extensively for high volume terminals. They are complemented by conveyors onto which the trays discharge, so that the bags can accumulate before being loaded on to the carts. Destination-coded vehicles have been used recently instead of tilt-trays. These have an individual cart, or 'tub', for each bag, running on a static rail system and with the coding on the cart. They can be operated up to speeds of 12 m/s and cope with slopes of 33°. The carts can be moved by electric motors in the track which only power up when the cart approaches (e.g. TUB TRAX, working at 3,000 carts per hour at Leipzig), or by linear motors in the carts (e.g. BAG TRAX, working at 4,500 carts per hour at Amsterdam). The ALSTEC carts at Heathrow travel some 150,000 km/year. It is simple to pull a cart out of service without disrupting the service, whereas, if a belt fails in a more traditional system, the whole system is stopped. Whatever system is used, a backup system with 75% of the main system should be available. Experiments are currently being done on Radio Frequency Identification Technology (RFID) to track bags more efficiently. It should soon be possible to arrange home pickup of bags with RFID tags.

The size of the sorting and make-up hall depends on the bags per passenger and the type of technology employed. The beds are usually 42 m long. For manual sorting from belts, 170 m^2 is needed for 4 EQA and 2,100 m^2 for 50 EQA, with 1.3 bags per passenger (Figure 12.8). The area should be increased by 100% if destination-coded vehicles are used and by 150–200% with tilt-tray sorting. Aisles around sorting devices and carts should be 1 m wide, and traffic lanes for cart trains should be 3 m wide with 6.5 m turning radius. Great care should be taken in locating columns and in arranging for adequate ventilation.

Outbound passport control can achieve an average processing time of 20 seconds. As an example, if there are 7 desks and if the maximum desirable queuing time is 5 minutes, the queue could be 105 people long. The spacing in the queue is 0.8–0.9 m, 1 m^2 per passenger giving LOS level C. Space needs to be available in the check-in hall for this size of queue to develop. Clearly, if people present themselves at greater than the flow rate of 21 per minute for a sufficiently long time, for example if, say, 350 people leave the check-in counters in a 10 minute period, the queue could exceed the maximum desirable length. It should not be necessary to wait more than 5 minutes. It is often difficult to persuade government offices that more desks should be opened in such a case, so it is necessary to try to manage check-in so as to deliver a smoother flow, or provide more queuing space and accept a lower level of service. The process is also facilitated and speeded up with the introduction of Advanced Passenger Processing (APP), using biometrics and machine-readable passports.

Figure 12.8: For small airports a simple manual sorting system is sufficient — Brno Airport. *Photo*: A. Kazda.

Security screening of passengers requires a walk-through detection device, an X-ray machine for accompanied baggage and space for manual searches and recovery of the X-rayed items. The FAA suggests that each security station requires an area of 10–15 m^2, and that it can handle 500–600 passengers per hour (PAX/hour). However, the flow rate at Heathrow, until more severe requirements were put in place in late 2006, was 270 PAX/hour per facility. The rate fell to 210, but improved management of the process should bring the productivity back to 270 PAX/hour or 13.3 seconds per person. It is difficult to see how the FAA productivity estimates can be achieved with today's security regime, so the security queues should follow roughly the same calculations as for passport control. Meanwhile, the queues have sometimes been backing up to landside. Security checks at most airports are centralised, which is significantly more efficient both in terms of investment and use of personnel. Many airports changed security check configurations except where there is not much change or they alter the configuration of security controls to centralised (e.g. Vienna Airport) excluding parts where it is difficult because of terminal layout or operational issues (Vienna Airport Terminal 2 Schengen D gates).

Corridors have an effective width some 1 m less than the actual width, because of a reluctance to use the space in the 'boundary layer' near the walls. The effective width is also reduced by facilities such as telephones, and displays which encourage people to congregate in front of them. If the average walking speed is 74 m/min, a corridor with an effective width of 6m will allow a flow of 300–600 people per minute, depending on how wide the people are and how closely they follow one another.

The departure lounge should have, as a minimum, space for the essential activities of waiting for flights to be called and then queuing to board the flight, and the facilities required for processing the passengers. There should also be telephones, toilets and some minimum catering for those whose flights are delayed after they have passed through security. At today's load factors of around 75%, this requires approximately 1 m^2 per seat of the aircraft being boarded. Seats should be provided for some 50% of the expected number of passengers unless it is likely that they will be held for a long time, this being dependent to some extent on the boarding policy of individual airlines (Figure 12.9). The space required per seat is 1.5–2.0 m^2, and for circulating is 1.5–1.8 m^2/PAX for LOS level C.

Retail facilities (or concessions) were originally concentrated in the duty-free shop, initiated first by Aer Rianta at Shannon airport in the 1950s. This is changing within Europe with the international traffic between some 30 countries now being considered 'domestic' from the point of view of duty paid. This has caused a serious loss of revenue for the airports. Another near essential is the stationers or drug store. All the airside goods and employees have to be screened. It is necessary to have secure truck bays for delivery and waste removal, and to have back-of-house paths for food, waste, and stock so that such movements can be kept away from the passengers and air cargo.

Retail in all its forms is a growing activity within airports as they search for new ways of increasing revenue. At least some of this trend is to the passenger's advantage. Passengers generally find it useful to have shopping and catering facilities in a mix appropriate for the type and the volume of traffic. Some feel that too much retail gets in the way of the passenger flow towards the gates and causes delays. However, airports which put the retail facilities out of the main flow, like Washington National or Rio de Janeiro Galleao, lose revenue because the passengers cannot be bothered to search out the opportunities: the impulse buyer cannot

282   Airport Design and Operation

be easily captured. Also, if there are few outlets, the passenger does not get the price control that comes from competition.

Figure 12.9: There must be guaranteed minimum LOS for passengers who can wait a long time for their flights in the airport lounges. *Photo*: A. Kazda.

The passenger will tend to spend more when relaxed, so they need a comfortable base from which to visit the shops after having gone through the stressful check-in and security processes. The centralised concept of security screening, which is currently preferred by many airports, contributes to a quiet environment without further unexpected delays that encourages passenger purchases. Shopping at an airport is not like shopping in the high street. There is only a finite dwell time, the shopper cannot come back, and it is not the primary reason for being there. The spending will tend to be greater if the merchandise is visible, well lit and appealing, and if the shop is clean and neat. The solution developed by the BAA at, for example, Heathrow Terminal 4, is to combine the areas for the gate processing activities, retail and catering. The departure lounge is one very wide concourse with shared space for a wide range of highly visible shops. Revenues are high, the passengers have more to interest them and the flow is not impeded. The split of floorspace should be about 60% shops and 40% catering. The total concession space in a terminal will typically be between 700 and 1,000 m^2/mppa.

Catering is predominantly airside. The FAA sizing methods suggest around 350 m^2 is required per mppa, but it could be considerably more or less, depending on the type of traffic and thus on the degree of use. Some 20–40% of the seating for catering may be counted towards the total of required airside seats.

Gate rooms, where they are required or preferred, should have seating for 80% of the aircraft capacity at 1.7 m² per seat, and standing room for the remaining possible 20% at 1.2 m²/PAX. Then LOS level C is achieved with a 65% load factor; a 95% load factor results in LOS level E. There must also be room for the airline counter and for a queue to form for checking the boarding pass and bag reconciliation prior to boarding. There is a conflict between the airline's desire to process passengers early into the gate rooms, and the passengers' aversion to being kept corralled for what may be an indefinite period with few if any facilities. Toilets should be provided if there is security control on entry, which is not always the case. If there is a change of level from the gate room to the airbridge, escalators should not be used because the queue from the bridge could back up to the escalator. The IATA sizing methods suggest that, when gate rooms are built along the sides of piers for a Code F aircraft like the A 380, the total pier width needs to be at least 26 m. The gate room concept is still used at many airports, in particular those without airbridges and LCC operations where it enables short turnaround times. Passengers are checked when entering the gate room and released to board the flight when the aircraft is ready. This reduces boarding time. However, the shared gate-space concept with no enclosed space, so the concourse space could be shared by more gates or other services, is becoming increasingly popular. Automated reconciliation gate control system facilitates and speeds up gate control. The shared gate-space concept is further enhanced by the introduction of self-boarding gates. According to SITA: Air Transport Industry Insights; Airport IT Trends Survey (2013) just over 10% of airports are using self-service, that is self-boarding gates but by the end of 2016 the figure will jump to nearly 60%. The shared gate-space concept also supports airport policy of announcing a gate later to keep passengers in the shopping areas.

The gate-space facilities should satisfy new passenger's needs, including sockets for mobile phones and laptops charging. Passenger should also find good value for money snacks shops as most of short haul flights do not offer any refreshments on board, today.

Executive lounges are an important part of an airline's service to its Commercially Important Passengers (CIP). Business people need to be able to continue to conduct their business in suitable surroundings while they wait for their flights. Airlines sometimes share lounges when their business and first class market is small. At least four times the normal space should be used for sizing, to allow for the expected levels of comfort. The lounges should be near to the access to the gates. One operator of the A 380 has asked for a direct connection between the CIP lounge and the upper deck of the aircraft.

Inbound government controls cover immigration and naturalisation services; customs; public health; and lifestock and plant health control. Some countries install pre-clearance facilities at the point of flight origin, in which case provision needs to be made for those facilities at the origin airport rather than on entry. The areas downstream of inspection for passengers and for their baggage should be sterile. At the entry airport, strict segregation is required of international passengers from domestic passengers and the public by means of a physical barrier and with strict controls on entry and exit. There should be no public phones. The areas should be well lit and without glare. Baggage from international flights should also be segregated. Transit passengers should have their own secure waiting room. The total space required for the support of all these activities is approximately 750 m^2 per 1,000 peak hour international passengers. The processing time is longer than on departure, and may average 30 seconds, depending on the origin of the passengers. In completely smooth flow, the 1,000 passengers per hour could be served comfortably by 9 desks. However, passengers usually arrive in large batches from international flights, typically 300 within 10 minutes, during which time only 50% would have been processed by the 9 desks by the end of the 10 minutes, and people at the end of the queue of 150 passengers would have to wait a further 10 minutes. The largest international terminals receive more than 5,000 arriving passengers per hour, or perhaps 16 aircraft, so the flow into passport control is smoother and 40 desks might be sufficient. According to the IATA Manual an acceptable time for waiting in the inbound passport queue is 7 minutes. However, it could probably take longer to clear passengers at many low-costs or charter/holiday airports. Customs require inspection tables, interview rooms, payment facilities, kennels, a bonded warehouse and space for administration and accommodation for staff.

A separate *reception lounge* is required at gateway airports, where visiting dignitaries may be entertained while the formal processes are completed on their behalf.

The baggage claim hall needs to be close to the airside road system for ease of transferring bags from the carts to the reclaim belt, and also close to the landside access points so as to avoid long walking distances with baggage. Airside it contains tugs/dollys/containers pending flight make-up, tug charging (if electric), staff restroom, early bag store. Landside it normally contains one or more bag reclaim units (carousels), though a simple roller bed fed from airside by gravity may be used at smaller airports. It is important to provide currency exchange and toilets in this area, as passengers wish to use the time while waiting for bags to make themselves comfortable after the flight and before being met. The reclaim units may be either flat-bed or sloping bed devices, the latter being able to carry a double row of bags. They may circulate round from airside to landside on the same level, or a conveyor may provide a remote feed to an island carousel on a different level to the airside dock.

The length of claim frontage depends on the number of aircraft arriving in a peak 20 minute period. The arrival of 16 EQA in 20 minutes might need 150 m of flat-bed frontage. It is unusual for beds to have more than a 50 m frontage, in which case three carousels would be necessary, but this might increase to 6 if the number of bags per passenger and the percentage of passengers terminating locally were both high. Equally, if most flights were short haul domestic, two carousels might be sufficient. If the 16 EQA were 4 widebody aircraft, it would be most convenient to have one sloping bed carousel for each of the aircraft. The reclaim hall area should be approximately 9 m^2 per metre of claim frontage, to include storage of bag trolleys, lost bag facilities and airline offices. LOS level C needs 1.7 m^2/PAX.

Figure 12.10: A waiting time for bags of 12 minutes is deemed to be acceptable.
Photo: A. Kazda.

There should be approximately a 12 m space between the claim units. The airside input area should be approximately 3 m^2 per metre of claim frontage for a U-shaped flat bed, rising to 6 m^2 per metre for a remote sloping bed carousel. According to the IATA Manual a waiting time for bags of 12 minutes is deemed to be acceptable (Figure 12.10). However, it is difficult to generalise on this. The manuals do not specify how the time should be measured. The difference between arriving of the first and the last passenger could be significant. Waiting time will also considerably differ between short haul flights and long haul flights served by widebody jets. The introduction of the A 380 is causing airports to increase the size of their belt reclaim lengths from the existing 50–60 m to 75–90 m, usually by adopting W-shaped sloping beds, in preference to using two belts per flight. Airports should aim to have bags on the carousels before passengers arrive from immigration controls. Airports sometimes install baggage carousels landside (e.g. Indianapolis Airport) to allow meeters/greeters help arriving passengers with bags.

Some airports offer duty-free on arrival (e.g., BAA Heathrow, Luton, Rio Galleao). The new security measures may make this a more attractive addition to the pre-customs facilities.

The arrivals hall should allow unimpeded exit from customs. The essential facilities are: meeting point, currency conversion, hotel and tourism booking, car hire, ticketing for onward transport and car parks (remote and isolated payment stations are to be avoided), café, flight information, phones. The sizing depends on the same factors as the check-in hall, but the only long queues are likely to be formed by meeters wanting to be close to the entry point.

Airline offices are required for cabin service and personnel, aircraft line maintenance, managerial offices, flight operations, flight crew and cabin staff, secure and volatile storage. These activities might take 25 m^2 per EQA movement.

Depending on airport terminal size each terminal, as any other public building, has technical equipment. In the case of international airports those include fire equipment, heating and air conditioning, backup electric systems. It also has space to house waste management rooms, security checks of goods and personnel, detention rooms, locker rooms, offices and space for police, customs and other state authorities. These areas can take up to 50% of the entire terminal surface area.

It doesn't matter which method for a terminal capacity assessment or facility sizing is used. The result is always floor area for the different functional parts of the terminal. However, the terminal building is always a kind of 'block' a 'cuboid' where all parts must fit in, but they never do. Final design is therefore a compromise between the sizing of the different facilities and services and the terminal layout, depending on many preferences.

12.5. THE HANDLING PROCESS

12.5.1. Passenger Handling

The handling processes for passengers and their baggage can differ according to the characteristics of the passengers and the rules of the country. The most basic difference is the handling of domestic and international passengers due to differing border controls. There are three kinds of borders in Europe at present:

- ✈ Non-Schengen type of borders between the countries of the European Union and the third countries — they have the same character as in the past
- ✈ Schengen type of borders between the countries within the European Union that have signed the Schengen Agreement — these are treated as domestic movements

✈ Borders between pairs of countries of the European Union — the movement of goods is free, the movement of persons is inspected.

Other differences result from the customs, passport, safety and health regulations for individual countries.

Two basic check-in concepts for departing passengers, namely **check-in for individual flights** and **common check-in** are increasingly being replaced by **internet check-in** and **mobile check-in**. Passengers who checked-in off-airport and have hand baggage only can proceed directly to the gate and passengers who have to check-in their bags use staffed or automatic drop-off counters. At some airports passengers who did not check-in by internet or mobile must get their boarding passes printed from self-service kiosks (Automatic Ticket and Boarding — ATB system). The number of passengers using internet or mobile check-in is constantly growing but depends strongly on the type of airline. It is the highest for the LCCs as some of the LCC carriers have an extra charge for those who did not check-in by internet and print their boarding pass. At Prague Airport (2014) the share of internet/mobile check-in was 50–60% for LLC flights, 30–40% for full service carriers and 0% for charter flights. The share of the internet/mobile check-in is continuously rising.

With the **individual flight check-in** concept the passenger for a particular flight can be handled at one or several counters reserved for that flight. The data on the passengers and on the flight booking status, which are necessary for the bag loading, can be collected manually or taken from the reservation system. Charter flights and LCCs particularly tend to use this type of check-in, but with few desks. The large number of bags and the use of few desks can result in long queues. In addition, some of the LCCs tend to board without seat assignment but by order of check-in, so the queue forms some time before the opening of the counter two hours before the flight. However, recently some LCC found that boarding without seat assignment could take longer and they are switching to a seat assignment strategy (Ryanair). For some European LCC it is usual for the flight to close at check-in 40 minutes before departure, so that some 180 passengers, all with bags, are required to be processed within 80 minutes using only 2 desks. Some LCCs are charging for checked bags in an attempt to reduce the processing time and handling costs, and some charter airlines are asking passengers to arrive three hours before flight time. This means that some people will aim to arrive four hours before so as to be first in the queue, so adding to the congestion landside. From the passengers' point of view it would be appropriate if they could use more check-in desks, even it would incur extra costs. But many airlines are definitely not willing to do this.

With the **common check-in** concept, the passenger can check-in at any counter of the given airline, handling agent or even in any part of the terminal building which is suitably equipped. It requires computer technology to support this method of handling. Information on the passenger (flight booking status) must be available at each check-in counter. Computer technology allows comprehensive data processing in the course of the check-in process: seat allocation, data for aircraft loading, catering special requirements, etc. The advantage of the common check-in is an equal load on all check-in counters. Sometimes there is a relatively longer handling time of those flights which would have had short queues, and, if the handling agent provides the check-in, there may be a loss of identification of the passenger with the airline. The same counter is often used during the day by several airlines or handling agents. This is made possible by the CUTE system (Common Use Terminal Equipment), which provides access into the airline's host computer and to the Airport Operations Database (AODB). It needs a Local Area Network (LAN) infrastructure to allow data, video and voice transmission in both the public and the administration areas. The logo of the company can be displayed above the counter during the time it is actively using it. It may also be used by a single company with many flights, in which case the queue may take the form of a crocodile, particularly where the available depth for queuing is small. Passengers avoid being held up in a slow queue, but have to move their luggage forward many times.

The traditional passenger check-in process on departure is normally for the passengers to submit their air tickets and their passports/ID cards at the check-in counter, but in current reservation systems it is sufficient for passengers only to submit their passport or ID. This allows their booking status and identity to be checked. The checked baggage (the baggage transported in the aircraft cargo holds) will be weighed and, if the free weight limit has been exceeded, the appropriate fee will be charged. The passengers will get their boarding passes (if not having printed it already). The weight of the baggage will be registered and baggage receipts stuck on the boarding pass. On some flights the airlines use a 'piece' concept. In this case passengers are allowed to take one or more pieces of baggage of specified dimensions free of charge. All baggage consigned to the hold will undergo a security check (e.g. X-ray). The security screening of baggage can be made before the check-in or directly at the check-in so that the passenger can participate in it. However, automated security checks can be done at any part of the baggage handling process. Some countries also require outbound customs inspection. After checking in bags, in the case of international flights, passengers then move through to the emigration passport control and security checks of the passenger and any carry-on baggage and their personal items. They are then able to proceed to the departure concourse to wait until their flight is called. They may be called directly to the gate but, now that further security checks may be made before boarding and the number of passengers per flight is so high, they will sometimes be called to a gate room where the final processing can be done in advance of boarding.

In order to speed up the check-in process and decrease the requirements on the number of check-in counters and employees, an Automatic Ticket and Boarding (ATB) system has been introduced for passengers at most airports with regular flights. By means of a touch screen the passengers can find out whether there are free seats on the particular flight. The ATB machine will print out air tickets/boarding passes for the passengers, for which they will pay by credit card. ATB provides positive identification of the passengers and their baggage, the latter being taken to a drop point nearby. The CUSS system is a common user version of ATB (Figure 12.11). There is not much evidence that they save space in the concourse, but they can be located in car parks, in hotels or downtown. The sequence of individual stages of the check-in process can be different at various airports. In some airports the central security check has been placed at the entrance to the check-in area and only those holding an air ticket are eligible to enter the check-in area. The system of check-in immediately after getting out of the car on the walkway in front of the terminal building, *curb check-in*, is common in the airports of North America for certain categories of passengers. In this way, passengers are relieved of their baggage and may complete the check-in at the counter in the building. Many other combinations are also possible, including allowing check-in at the departure gate.

Figure 12.11: For scheduled non-LCC flights use of the self-service kiosks is compulsory before baggage drop-off at some airports. *Photo*: A. Kazda.

On arrival, the passengers on international flights first go through the passport control. For passengers with a biometric passport (also known as an e-passport), this contains biometric information that can be used to authenticate the identity of travellers. Automated border gates and e-passports also significantly reduce waiting times at the passenger's immigration control. The passengers can be divided into several channels, according to their nationality and visa regulations. For business or frequent travellers, the system of positive identification automated control is being introduced in some airports. On the first entry into the country and upon request, passengers are issued with an identification card, which, in addition to the basic passport data, contains also a biometrics description (e.g. dactyloscopic marks of the hand, face geometrics or iris characteristics). On later entries, a reading facility checks the biometrics of the passenger with the data on the card. Some types of flight require also a health control which follows the passport control if necessary. After the baggage has been collected by the passengers on international flights they go through the customs control. In order to speed up the customs inspection a concept of red and green exit is commonly used. The red exit is used by the passengers who wish to declare goods at the customs. The green one is used by the others, only spot checks being made. A blue exit can also be found at the EU airports. The EU nationals, to whom special customs regulations apply, use it. On exit from customs, a clear path is needed through the metres and greeters to the transport options. There should be as many options as possible: urban busses, coaches, metro, heavy rail, as well as pickup zones in the short term car parks and taxis. It is helpful to have ticket facilities in the arrivals concourse, including fixed price vouchers for taxis to save the passenger in a strange country having to haggle with the driver over the fare.

It is, of course, necessary to find one's way between these processing points. Perfect wayfinding will result in a minimum time for the journey. The wayfinding process involves using two different functions; information-processing and decision-making. Passengers are also influenced by their previous experiences. It could be expected that inexperienced travellers tend not to have an adequate understanding of how to navigate at airports and would require further information. Any lack of wayfinding ability will increase the uncertainty in predicting the time needed, as well as increasing the average time taken. This is particularly important because a prime psychological concern is to eliminate the unknowns. Some of the uncertainty comes from the difficulty in estimating the effect of the possible barriers to processing and movement, and some from the difficulty in actually navigating though the terminal. It is recognised that the spatial logic of a natural line-of-sight progression in a straight path from ground to the aircraft and vice-versa is the ideal design. This is easy to provide when traffic levels are low, or if a decentralised 'gate-arrival' design is used to minimise walking distances. It is, however, very difficult to provide natural wayfinding in terminals for millions of passengers per year without the depth of terminal becoming prohibitive. This was the concept used for the design of the terminal at Stansted, but it is imperfect because all the ancillary activities which have to be fitted in on the single floor level impede the view and the original concept had to be adapted to cope with much higher numbers of passenger (it was originally designed for 8 mppa but served more than 19 mppa in 2014) and also because the view of the aircraft is illusory in the sense that they are not accessible directly from the terminal but only from the remote satellite terminals via a people mover. It is also used in principle at the Hong Kong Chep Lap Kok airport which has more than 60 million passengers per annum (2014). The eye is, in fact, drawn to the airside end of the pier by a descending roof rather than a direct view of the aircraft. However, the forward motion is impeded by having to cross long check-in queues which form at right angles to the flow from kerb to aircraft, and by the need to change levels to access the people mover.

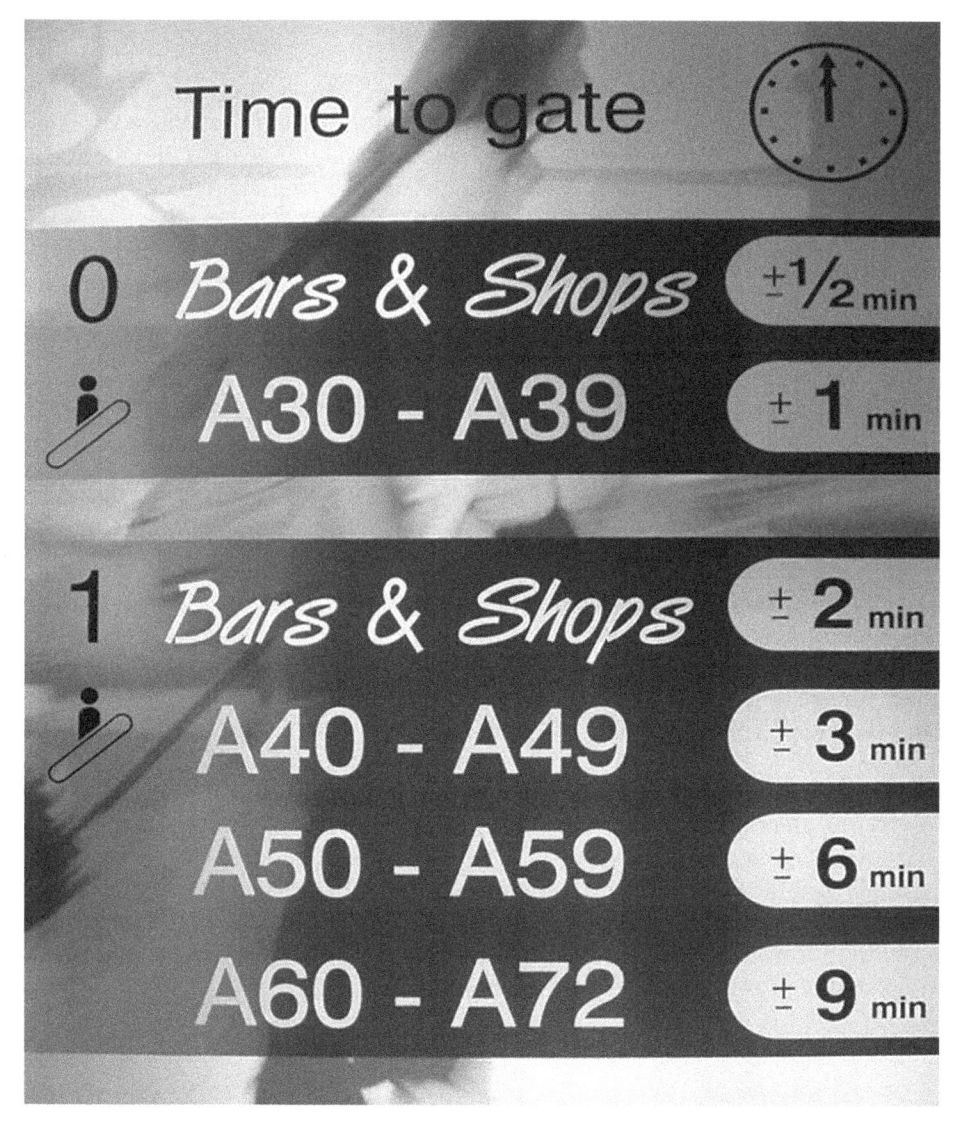

Figure 12.12: Some airports inform passengers of the time required to reach a gate. *Photo*: A. Kazda.

When complex routes through a terminal become necessary, an efficient signing system becomes essential. Considerable effort has been put into understanding the best way of communicating information about flights to passengers, and into assessing the advantages of flat screens or large flip board displays relative to verbal announcements. IATA and ICAO have set recommendations for symbols to try to cope with the language problem, and most major airport groups have their own brands of shapes, colours and fonts, each specific to a type of sign: aeronautical, concessions, amenities, emergencies.

The signs can be static, for example for emergency instructions, or dynamic for check-in, flight, gate, bag claim and car parking information. Popular colours are black on yellow, red on white, and dark blue on white. The signs should not be more than 10° from the natural line of vision and the lettering should be at least 1 cm high for every 3 m viewing distance. Despite these aids to design, some people still have to ask the way. The words 'departures' and 'arrivals' do not mean the same for everyone, nor is an upward-angled arrow or aircraft symbol always interpreted as a need to change level. Cultural differences (e.g. language, functional illiteracy) are other contributors to wayfinding difficulties, while the emotional state of individuals also influences their ability to decode and use the information to assist in the wayfinding process. There are also the perpetual problems of avoiding clutter with too many signs, differentiating between wayfinding and other signs, preserving visibility in crowded conditions with low ceilings, and assisting passengers who need to back-track.

While it is reasonable for airports to gain revenue from advertising, it should not take attention away from the essential information. The Vancouver airport terminal uses lighting graduated in brightness towards the nodes for subliminal wayfinding, and also uses the light fittings and carpet markings as pointers. The BAA and some other airports provide helpful indications of the time required to reach batches of gates (Figure 12.12).

Age negatively affects wayfinding performance and overall decline could be observed in all orientation skills in a healthy aging population sample. Aging individuals experience a decline in sensory, cognitive and motor skills, leading to greater wayfinding difficulties. Statistics show that the 10% of the population are over 60 years old, to rise to 50% by 2050.[2] Asia will host the largest number of elderly, followed by Europe and North America. In Europe, the percentage of people over 65 is expected to grow from 17.4% in 2010 to 29.5% in 2060.[3]

---

2. HelpAge International: Global ageing data 2012; Accessed 6.1.2015 on: http://www.help-age.org/global-agewatch/population-ageing-data/global-ageing-data/
3. European Commission, Eurostat (2014): Proportion of population aged 65 and over % of total population; Accessed 6.1.2015 on: http://ec.europa.eu/eurostat/tgm/table.do?tab=table&init=1&language=en&pcode=tps00028.

Wayfinding and being processed are both more difficult for the disabled. The EU legislation requires airports to differentiate as little as possible between the disabled and the able-bodied. The ICAO definition of a person with disabilities is 'any person whose mobility is reduced due to a physical incapacity (locomotory and/or sensory), an intellectual deficiency, age, illness or any other cause of disability when using transport, and whose situation needs special attention and the adaptation to the person's needs of the services made available to all passengers'. Up to 20% of the first world population has a significant disability, plus many more who do not report their disabilities. According to the Papworth Trust disability facts and figures (2010), disability is strongly related to age: 2.1% of 16–19 year olds, 31% of 50–59 years and 78% of people aged 85 or over. However, we can expect the number of disabled air travellers will be smaller than the share of the relevant age group as not all of them are willing to travel. Each type and grade of disability has its own set of ergonomic challenges.

Figure 12.13: Airport must designate points at which disabled persons or persons with reduced mobility can announce their arrival at the airport and request assistance. *Photo*: B. Badánik.

The Regulation (EC) No. 1107/2006 of the Parliament and of the Council of 5 July 2006 concerning the rights of disabled persons and persons with reduced mobility when travelling by air establishes rules to protect disabled persons and persons with reduced mobility against discrimination and to ensure that they receive appropriate assistance. Under the regulation a reservation or boarding can only be refused for justified safety reasons or if, due to the size of the aircraft or its doors, the embarkation or carriage of a disabled person or person with reduced mobility is physically impossible.

The rules make airports responsible for the assistance of disabled and elderly passengers from the point of arrival at the terminal through the whole process to the aircraft boarding without additional charges. Airport must designate points of arrival and departure both inside and outside terminal buildings, at which disabled persons or persons with reduced mobility can announce their arrival at the airport and request assistance (Figure 12.13). Air carriers are responsible for providing assistance in aircraft, including the carriage of recognised guidance dogs and up to two pieces of medical equipment per person. Air carriers and airport managing bodies must ensure that their personnel has appropriate knowledge of how to meet the needs of disabled persons and persons with reduced mobility, and should provide necessary training for their staff.

If it is at all possible, respect for the dignity of the disabled should be preserved, and this is best done by making it possible for them to use the same facilities and routes as other passengers by providing the same information and accessibility. Considerable progress has already been made in making physical provision by ramps (though these need a lot of space), large lifts and minimising changes of level, and now with aids to wayfinding. Some recent helpful initiatives are:

For the blind:

- synthesised voice calling out the floor and direction in all elevators
- tactile maps of the terminal.

For the visually impaired:

- high contrast flight information displays
- high contrast wayfinding information embedded in the flooring.
- audible warnings on travelators.

For the deaf:

- visual paging systems displaying text versions of audible announcements
- TDD service on public pay phones and at airport customer service points.

For the hard of hearing:

- amplified handsets on all counters

✈ induction loops

✈ public phones with sound boosters.

For all requiring carers:

✈ unisex washrooms in every cluster of washrooms.

Despite this progress, much more needs to be done if the disabled are really to be able to be as independent as the fully mobile.

12.5.2. Baggage Handling

Baggage handling is becoming a critical activity. The airlines are trying to shorten the turnround time between individual flights and at the same time, the average load factors are increasing. On big airports a long haul terminal may have an average of 210 passengers per aircraft. If the free baggage allowance of 20 kg of hold baggage per passenger is taken up, in the course of 40 minutes approximately 8 tonnes of baggage must be loaded and unloaded. One of the criteria for quality of service of the airport is the time passengers have to wait for baggage after disembarking from the aircraft and passing through immigration.

At small airports most of the activities connected with baggage handling can be carried out manually. The baggage is loaded on the baggage cart either directly or from the belt conveyor and dispatched to the aircraft, where they are loaded in the aircraft cargo holds. The baggage can even be handed directly from the baggage carts to the passengers on arrival. On medium size airports with the departure of several flights in the same time, sorting of baggage must be provided. Sorting can be done manually, when the baggage is picked up from the carousel by the handlers and loaded on the carts, or semi-automatic, in which case the operator will preselect an appropriate branch to which the baggage will be dispatched, according to the baggage tag. In both systems, because of the human factor mishandling can occur, leading to a certain number of lost or mislaid bags, which damages the reputation of the airport.

According to SITA 2014 Air Transport Industry Insights the Baggage Report, the performance of the industry with respect to mishandled baggage improved considerably between 2007 and 2013. Total mishandled bags dropped from 46.9 in 2007 to 21.8 million in 2013, mishandled bags per 1,000/PAX from 18.88 to 6.96 and costs of mishandled bags per passenger from 1.89 to 0.67 USD. According to the SITA 2014 report the number of mishandled bags per thousand passengers is the smallest in Asia 1.96; North America 3.22 and it is the highest in Europe 9.0 in 2013. The low number of mishandled bags in Asia is due to some unique factors. Mishandling is not entirely captured through the WorldTracer system hence the SITA Baggage report is not an accurate reflection of the mishandling in Asia. Also the hubs in the

EU that show a high percentage of mishandling deal with a lot of transfer bags, and it is inherent in bag transfers that they are prone to being mishandled. Further, at airport in Japan, the mentality is more towards zero tolerance for mishandling and their mishandling rate is indeed close to zero.

At large airports it is necessary to provide automated sorting of baggage. Automation can increase the capacity of a terminal building significantly and also improve the service standard. Automated sorting uses baggage tags with bar codes, QR codes, magnetic cards or electronic chips for baggage and destination identification. Bar coding is the most frequently used today. In this case the tags are automatically printed when passengers check-in. Reading of the code is carried out automatically by laser sensors, alternatively semi-automatically with manual laser sensors, and the data are stored in the database. The data are used for the baggage routing within the baggage sorting system. Usually each bag is also scanned before it is loaded in an aircraft hold so as to have thorough information on each bag position. These data are also compared with the data from the database on passengers. This provides positive identification, called baggage reconciliation, to ensure that passengers have boarded the same aircraft into which their baggage has been loaded. If any passenger does not board the aircraft it is possible to determine exactly where the baggage is and unload it for security. On airports with a great number of transfer passengers it is moreover necessary to provide sorting and redistribution of baggage in a very short time in order to guarantee the minimum transfer times of the passengers. It may be best to provide sorting of transfer baggage outside the main terminal building, particularly for online transfers within an airline's hub.

Arrival bags are taken from the aircraft and placed airside on a conveyor belt which carries the bags to the landside bag reclaim hall, either directly through the airside/landside screen or via a transfer belt which feeds a free-standing carousel with a sloping bed. Each carousel is assigned to one or more flights. The belt tracks are made long enough to present the bags sufficiently spaced to allow passengers to select their own bag without interference from other bags, that is the reclaim length is related to the size of the aircraft.

The baggage handling process still has a considerable human content, leading to many health and safety concerns.

12.6. NON-AERONAUTICAL SERVICES

If we consider the needs of the passengers, it is necessary to realise that the main reason why most people come to the airport is not the airport itself, but because they want to get somewhere else. Most of the passengers are in a condition of emotive stress, even though they may be experienced travellers. Therefore it is necessary to create a familiar and pleasant environment for the passengers, where they could relax and spend time before boarding. The airlines will provide this for premium passengers in their own lounges, but airports have seen the need to provide for all passenger types, and the opportunities that this brings.

At present many airport administrations achieve higher revenue from the non-aeronautical services than for the aeronautical ones. This is caused by several factors. One of the main factors is the need to increase revenue. The possibility of increasing revenue from landing and handling fees is in many countries strictly limited by the regulations of antimonopoly authorities and also by the strong lobbying of air carriers. At the same time the advantage of providing non-aeronautical services at the airports is that on the airports there is a high concentration of the more affluent members of society, and few of them are able to completely avoid some dwell time in the terminal before boarding the aircraft. Therefore during this time they can use shops and services, if they are some available and interesting, thus raising revenue for the airport and the concessionaires (the shop operators).

The minimum provision of aeronautical services is explicitly determined by technological requirements and by the scope of the airport operation. The extent of non-aeronautical activities depends more on the space available at the airport, on type of airport (hub, charter, low-cost, regional), on the management philosophy and the passenger volumes but also on particular country traditions and practices. Sometimes non-aeronautical services can reach such a level that they impede the handling process or passenger circulation (Figure 12.14). For example, they can limit the capacity of check-in counters, interrupt the flow of departing passengers or their view of the location of toilets or departure gates. In spite of the fact that diversification and variety of offered services is required, non-aeronautical activities should never disturb the smooth running and basic functions of the airport. But the reality is often different. At many airports areas originally designed for passenger circulation or waiting are taken up by shops. Passengers must make their way to their aircraft through a warren of retail facilities.

It is, however, possible to take advantage of the increased scope of services to make the airport operation easier. The offered activities can divert the passengers from the main circulation streams in the airport and help to distract the attention of the passengers from the problems and irregularities in the air transport system.

On many airports the traditional layout of the check-in process and non-aeronautical services is changing. It also depends on the overall terminal concept including security (centralised/decentralised). Within the new concept of location of services in the terminal building the passenger, passing through the terminal, actually goes through the shops. The needs and wants of the passengers gradually change. Typical passengers today expect to be offered more than last minute duty-free shopping at the gate. The passengers will require a wider range of services: hotels business services, and various catering possibilities of the same standard, quality and price to which they are accustomed in the High Street, simply value for

298   Airport Design and Operation

money. Customers can choose from the same range of goods, brands and fashion goods. Airports have also provided additional services: fitness centres, super markets, cinemas, casinos, etc., despite the additional floor area taken by these facilities. Security and flight delays have caused people to dwell longer in airports, causing problems for bored children, so many airports have installed play areas for children.

Figure 12.14: Public corridor effective width obstructed by the addition of shopping facilities. *Photo*: A. Kazda.

Though the airports' characters gradually change and become similar to the city centres, it is apparent that they will always differ by the scope of offered services, depending on the specific role within the catchment area (transit airport, origin/destination or hub airport).

12.7. Passenger Transportation — People Movers

The use of widebody jets at the end of the sixties necessitated extension of airports, particularly terminal buildings at hub airports. Greater wing spans required greater distances between the stands and longer walking distances for the passengers.

On the other hand, economic pressures forced the airlines to increase the daily utilisation of aircraft and to shorten their turnround times. At large terminals this necessitated the introduction of passenger transportation. Most of the means for transportation of passengers which are used in the airports were originally developed for transportation in cities. Some of them were improved or modified for use in the airports. As mentioned above, there are internationally recommended maximum walking distances for the airport buildings. Usually this distance is not supposed to exceed 300 m. If that distance is exceeded, assistance for the passengers should be provided. However, the practice is often quite different. In low-cost airport terminals passengers must often walk hundreds of metres. The second task of transportation facilities for the passengers in the airports is to minimise the connecting times between flights. The generic term used to describe all facilities used for passenger transportation in the airports is people movers.

Terminals with a flow of up to 5 mppa, depending on the terminal concept, often do not require installation of people movers within the building. Though it might be desirable to facilitate movement of passengers on the airport, each people mover means an increase of investment costs in the construction of the airport. It means also an increase in the operational costs, particularly to guarantee the availability and reliability of a facility. Therefore, if it is acceptable for the passengers, there is a tendency to avoid the use of people movers in the airport, except for escalators and elevators to change level.

The choice of a suitable people mover system depends on:

- speed
- capacity
- safety and security.

Other criteria for the choice are usually these factors:

- transportation distance and elevation difference
- required frequency
- reliability
- ease of use by the handicapped
- ease of use with accompanied baggage
- maintenance requirements
- design characteristics
- procurement and operational costs.

For transportation of the passengers between the terminal building and remote stands on the apron or between individual buildings on the airport it is possible to use buses or mobile lounges. The special airport buses used for the operation on the apron usually have a bigger capacity than regular buses. Because they have not been designed for regular operation on public services, they can be wider and have lower clearance. This makes it easier for the passengers to get in and out. In spite of the fact that special buses have higher capacity, sometimes the capacity of one bus is not sufficient. Regular buses are still used at airports with turboprops and small regional jet operations as their capacity is sufficient for these types of aircraft and also their acquisition and operational costs are lower than those of special airport buses.

A few airports use mobile lounges to transport passengers between aircraft or a remote terminal and the central-processing terminal. The advantage of the mobile lounge is simplification of the passengers' movements. The passengers do not have to change level as when using buses. If all the stands on the apron have been designed as remote and mobile lounges are used for transportation to them, there are advantages of a quieter and less polluted environment in the terminal building as aircraft are usually parked on remote stand positions. The vehicle can be used as a holding lounge while it is docked at the terminal frontage. They do, however, cost more than busses. All these special vehicles are usually wider and have implications for the design of the airside roads. In fact, whatever means of surface transport is used to access remote stands, there are serious implications for space, safety, pollution and minimum connecting times. One further option for connecting a central terminal with a remote pier, which does not involve people movers or vehicles, is to bridge the taxiways as has been done at Gatwick with Pier 6 and also at Denver. The Pier 6 bridge spans 128 m, is 32 m high, and saves 50,000 coach trips per year in serving 3.4 mppa off 11 stands.

The simplest and also the most widely used types of people movers within the terminal building are escalators for overcoming changes in level and moving walkways for near-horizontal transport. Moving walkways are mostly used for distances up to 200 m. They usually do not significantly shorten the time taken to reach the aircraft. The usable distance is limited by the walkway speed, which usually must not exceed 1.25 m/s. The length of walkway is also limited by the fact that it is only possible to get off of the walkway at its ends. Therefore several sections of walkways following each other have to be installed in the corridors to the gates. This is an incidental benefit when one section is out of action, in that the quality of service is not too seriously affected. Width of escalator or walkway should always allow bypassing passengers with handbags.

For transportation of passengers, flight crew, employees and visitors between individual buildings over longer distances at the airport, for example between terminal buildings (Gatwick, Kuala Lumpur), between a central terminal and satellites (Atlanta) or between the terminal building and a railway station (Birmingham) or parking lots, the use of walkways is not appropriate, because of their slow speed.

Shuttle buses are sometimes the solution but, as the demand becomes greater, it becomes appropriate to use an automated shuttle type of people mover either on one track or on parallel tracks (Figures 12.15 and 12.16).

Figure 12.15: People mover system, shuttle type, connecting airport buildings at Roma Fiumicino. *Photo*: A. Kazda.

They have been used for some time in the United States, at Atlanta, Seattle, Washington Dulles and Miami among others but also in Europe at Heathrow Terminal 5 or Madrid Barajas airports. The first shuttle type people mover in Europe was installed at Gatwick airport. The part connecting the new satellite with the existing main terminal building and the railway station was built first. Later another route was built, which connected the south terminal with the new north terminal (Figure 12.16).

A people mover with the highest passengers throughput is the system at Atlanta Hartsfield International, which transports more than 100,000 passengers a day (Figure 12.17). The system connects the four parallel satellite piers with the main terminal building. The people mover is located in a 1,600 m long underground tunnel together with the baggage transportation system and a walkway.

The reliability of these types of people movers is usually higher than escalators or movable walkways, and they are less vulnerable than buses to labour disputes. However, most failures bring the whole system to a halt, while a bus can easily be replaced. In the transportation peak hours the people mover operates at set intervals. Outside the peak it may be possible to call the vehicle by a pushbutton as with an elevator. There are many different types of construction and power systems. The Westinghouse company uses fully automated carriages on tyres with electrical drive, while the OTIS company uses its own technology, which is used for elevators, the carriages being driven by a steel rope.

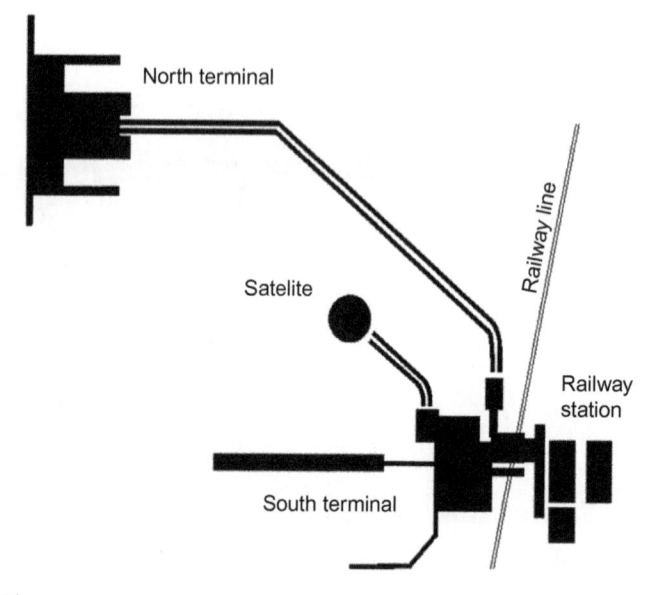

Figure 12.16:   Scheme of people movers at Gatwick airport.

On large airports, providing surface transportation of the passengers is just as important as the air transport. Often the surface transportation of passengers at the airport is a bottleneck, limiting the airport capacity. This is considered further in Chapter 14.

12.8. MOBILE AND IT TECHNOLOGIES

Terminal design and operation is and will be increasingly affected by the mobile and Information Technologies (IT) development. Airports are investing in IT not only to improve the passenger experience, but also to increase capacity while reducing costs of operation and to meet constantly growing demand for air travel.

While the Internet, in particular, is already a well-established travel tool, both for travel planning activities, such as flight search, reservations and check-in, new technologies like smartphones and tablets are getting more popular for real-time travel information, commercial applications and proactive solutions offering generic or personal passenger services during disruptions. Most passengers expect personalised automatic rebooking, to resolve the impact of flight disruption.

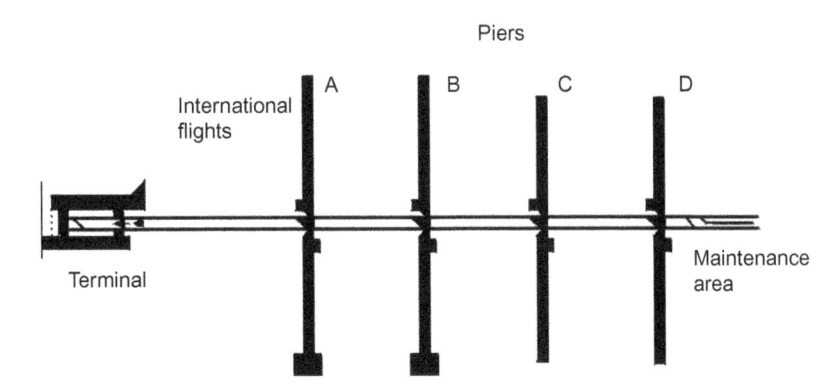

Figure 12.17: People mover system at Atlanta Harsfield airport.

Not surprisingly most airports offer free Wi-Fi, sometimes time or capacity limited, as according to SITA: Air Transport Industry Insights; The Passenger IT Trends Survey (2014) 97% of passengers carry at least one electronic device with them when flying and 18% carry three devices — smartphone, tablet and laptop. A strong migration of travel interactions to the smartphones could be expected in near future.

According to Air Transport Industry Insights Airport IT Trends Survey 2013 it is expected that passenger processing will globally remain the highest priority of IT investments in the forthcoming three years for the majority of airports (51% of surveyed airports), while investments in improving airport operations is the priority for airports in China (46%). By 2016 98% of airports plan to implement check-in kiosks; 90% plan to implement assisted bag-drop, 82% plan to implement bag-tag printing, 70% to increase common use check-in kiosks and 80% of airports expect self-service to become the most popular check-in method by 2016.

Most airports (95%) also focus their programmes on future mobile applications (apps) for passenger services. The highest value of mobile technologies seems to be in real-time information delivered to passengers but also to other airport users. Among the top mobile apps provided by 2016 which passengers expect to be able to exploit are the airport status, flight status updates, baggage status and the wayfinding, while about one third of airports will focus on mobile commercial opportunities. This is in line with user expectations but also with the scope of mobile services that airports are offering. Just over half of airports (51%) already provide real flight status updates to passengers via mobile apps, while only 18% are currently exploiting the retail opportunities. Among the commercial applications airports plan to offer are the possibility to pay for services and products which will be further enhanced by Near Field Communications (NFC) technologies. Airports are also evaluating social media applications for delivering passenger commercial services.

Most passengers want personalised services and about 40% of them are willing to provide personal data and to share their location information to improve the travel process, such as passenger flow monitoring, but only 29% of them are willing to share data for commercial purposes.[4]

Wayfinding support or Location Based Services (LBS) are important especially at large airports and during transfers where passengers are time constrained. Most big airports, in particular in Europe, were built up piecemeal and this is frequently reflected in the complicated transfer flows with several direction or level changes. Precise position and real-time/distance information to the next critical point for example security check, together with queue length information are then very important. Tailoring comprehensive wayfinding services to each individual is particularly beneficial for aging and disabled travellers because of their specific needs. Location Based Services combined with public transport information could eliminate information barriers in an unknown environment and thus increase public transportation usage by passengers with a direct impact on the environment. Improving information on passengers in real time could be used to improve level of service by better staff planning but also, in the future, for airlines to make a decision to hold a flight for a few minutes or directly re-book the passenger for the next flight.

However, it should be stressed that all IT and mobile technologies must be only a complementary, addition to traditional static signage, not a replacement for it.

Mobile and IT technologies could have a dramatic impact on airport design and operation. For example off-airport — internet/mobile check-in can reduce the space needed for the public departure lounge and number of check-in counters in particular at low-cost terminals. However, if the terminal function changes to take charter flights, the space in front of check-ins will not be sufficient to cater with long queues. The requirement for high flexibility in terminal design is therefore still important.

---

4. SITA: Air Transport Industry Insights; The Passenger IT Trends Survey (2014).

13

SECURITY

Tony Kazda and Bob Caves

13.1. UNLAWFUL ACTS AND AIR TRANSPORT

The expression 'security' concerns all unlawful acts connected with civil air transport. The character and danger of unlawful acts change over time not only with the change of exterior, particularly political, conditions, but the terrorists react also to the adopted measures for safeguarding the security of air transport.

The level of security precautions is one of the important factors by which the airlines assess an airport for operation in line with IATA Operational Safety Audit (IOSA) requirements. Airline reaction to an inadequate level of security could be illustrated by the Athens airport case where a B 727-200 of the American airline TWA Flight 847 was hijacked in June 1985. During the high-jacking one American national perished. Five months later on 23 November, terrorists hijacked EgyptAir B 737-200-Flight 648 on its Athens-to-Cairo route and landed at Malta. In order to achieve their aims, they started systematically to murder the passengers at regular time intervals. To rescue those who remained alive, a commando-style rescue operation was carried out. The result was 61 dead and 21 injured. The aircraft was completely burned out. After this event, American citizens were warned not to fly to Greece, since it was not capable of ensuring their security. The number of tourists from the United States to Greece decreased by half during the following tourist season.

With unlawful acts, the actual number of casualties is less significant than the threat that everybody who uses air transport could become the casualty. For example between 2004 and 2014 a total of 435 people died as a result of civil aviation unlawful acts (this excludes casualties of 8 March 2014 Malaysia Airlines Flight 370 B 777 whose disappearance remains unexplained). Is it possible to compare this with 33,561 people killed in car accidents in the USA just in one year — 2012?[1] It is the violence and insanity of terrorist acts that is so macabre, together with the fact that the victims do not get the chance to defend themselves against these acts.

---

1. National Highway Traffic Safety Administration Data Resource website. Retrived from http://www-fars.nhtsa.dot.gov/Main/index.aspx. Accessed on 6 December 2014.

It is necessary to realise that air transport is not the real target of terrorists. The targets are 'enemy' countries and their governments, upon which the terrorists want to enforce a change in their politics. To attack the 'enemy' country, the terrorist does not have to risk performing sabotage activities on the territory of a foreign country. On the contrary, the terrorist can wait for the 'country', represented by an aircraft of one of its carriers, to come to meet him. The terrorists themselves can choose the time, place and weapons for the attack. The aircraft itself may be worth several hundred million dollars. The unlawful act will become a central theme of the news bulletins of all television and radio stations of the world for several days, and in this way the terrorist will get the required publicity.

Terrorists are not the only source of threat. Others are: criminals, refugees, mentally unstable people, disgruntled employees and unruly or disruptive passengers.

The first recorded unlawful act in the history of aviation was the hijacking of an aircraft in Peru on the 21 February 1930. The hijackers seized the Pan American Airways Fokker F7 and used the aircraft to drop political leaflets. The first casualty of an unlawful act was the pilot of a Romanian aircraft which was hijacked in July 1947 during a flight from Romania to Turkey. The third case happened in April 1948. Seventeen hijackers from Czechoslovakia, among them two crew members of the aircraft, forced the aircraft to land in the American occupied zone of Germany.

Hijacking of aircraft was practically the only type of unlawful act until 1969. The hijacker needed the aircraft as a 'means of transport'. The highest number of hijacks was from the United States to Cuba when Fidel Castro's regime used it as a special combat tactic against the 'US enemy'. It is evident from the analysis of hijackings that there was a close dependence between the state policy limiting the free movement of citizens and the number of hijacks. After opening the frontiers the probability of 'traditional' hijacking decreased.

Where the politics cannot be changed, measures have to be introduced to counter the terrorists. The number of hijacks in the United States decreased substantially after the introduction of security measures. The measures included manual inspections and checks of passengers by metal detectors and inspection of their cabin baggage. The checks were supposed to prevent the would-be hijackers from carrying weapons onto the aircraft.

After 1968 the character of unlawful acts started to change. Terrorists began to use hijacking of aircraft to enforce their demands rather than for transport. Hijacking was a suitable tactic for terrorists because:

- ✈ the passengers and crew were suitable hostages
- ✈ the aircraft was a suitable, safe and temporary mobile prison for the hostages

✈ it was possible to place demands upon authorities through the media that monitored the case.

The countries in the Middle East and in Europe were not able to react quickly and effectively to hijacking and they were slow to recognise the problem represented by transit and transfer passengers. In Jordan at Dawson's Field three aircraft were destroyed on the ground by Palestinian terrorists on 6 September 1970. This led to the introduction of X-ray screening.

Terrorist tactics changed again after 1986. Increased security measures at the airports made it more difficult to carry weapons onto aircraft. Therefore the terrorists adopted methods which were easier and safer for them. Before 1986 sabotage as a coercive measure was seldom used. The difference from hijacking was that sabotage did not make it possible to keep hostages, to state conditions and to blackmail authorities. Those sympathising with the terrorists realised that they could also become victims. The responsibility for the committed crime could also be claimed by other 'rival' terrorist groups. In this period, sabotage, as distinct from hijacking, was considered to be a condemnable act also in the countries where the terrorists were based.

From the beginning of the 1980s the number of unlawful acts decreased but their character and the consequences of the terrorist actions became more severe. The formation of fundamentalist groups supported by extremist countries changed the situation. Terrorism became an integral part of political and ideological struggles. Terrorists began to use sabotage and sabotage threats against aircraft and airports as coercive measures. Aircraft were destroyed in the air and on the ground. In Sri Lanka on Air Lanka Flight 512 3 May 1986 a bomb was placed in the tail of the aircraft during refuelling by the Liberation Tigers of Tamil Eelam militants; there were massacres in the check-in concourse at the airports of Rome and Vienna on 27 December 1985 for which Abu Nidal Organization claimed responsibility; the Air India B 747 Flight 182 was destroyed in flight on 23 June 1985 over the Atlantic Ocean by a bomb planted by a Sikh extremist group; a Korean Air Flight 858 B 707 exploded in mid-air on 27 November 1987 above the Andaman Sea upon the detonation of a bomb planted inside an overhead storage bin in the aircraft's passenger cabin by North Korean agents; a Pan Am Flight 103 B 747 exploded on 21 December 1988 over Lockerbie in the United Kingdom for which the Libyan leader Muammar Gaddafi admitted Libya's responsibility; and a UTA Flight 772 DC-10-30 aircraft was destroyed over the desert between Chad and Niger on 19 September 1989 by bomb which had been loaded on the aircraft at Brazzaville airport by a terrorist group centred upon Islamic Jihad.

The original measures used against hijackers proved to be insufficient against terrorists. The terrorists used more sophisticated means including highly effective plastic explosives, firearms made of plastics and composites and masked bombs. The measures for ensuring security were therefore complemented by inspections of all checked and hand baggage, air cargo, galley equipment, mail etc. Airport employees were included in security inspections and also all other persons that entered the airside area of the airport. Computer-controlled systems were used to an increased extent for inspection of authorisation of entry into specific areas. Also hitherto accessible parts of terminals, workshops and maintenance areas on the airport were hereafter included in the security cordon. The checked luggage had to be reconciled with the passenger before the aircraft could leave the stand.

A threat of a different type is by mortar bombs on airports from beyond the exterior perimeter of an airport. The first mortar attack on an airport was launched in January 1975 on Paris Orly airport. After the Irish Republican Army mortar attacks on London Heathrow airport in March 1994, it seems that increased attention must be also paid to the land around the airport, particularly to the car parks or unattended areas.

A new type of attack was used by terrorists on 11 September 2001. Four aircraft were hijacked, two B 767s and two B 757s. Two of the aircraft were flown into the World Trade Centre Towers. Another aircraft was flown into the Pentagon complex collapsing part of the structure. The last one crashed outside Pittsburgh, the hijackers having been challenged by the passengers. During these unprecedented acts, where al-Qaida extremists used aircraft as weapons, 33 crew members, 214 passengers and 19 hijackers were killed, as were thousands of people on the ground. After 11 September, new security measures were adopted to enhance security including air marshals (sky marshals) presence on board some airlines. US Department of Transportation recommended cockpit doors be reinforced in order to deny access from intruders. At US airports, only passengers with a valid air ticket were allowed beyond the security points (except with special medical or parental needs), each passenger could take one cabin bag only, all electronic items, for example laptops and mobile phones were subjected to additional screening, there was a limit on the amount of jewellery or other metal objects that passenger could wear and travellers had to remove all metal objects prior to passing through the metal detectors to facilitate the screening process.

Another type of threat to commercial airliners are shoulder-fired Surface-to-Air Missiles (SAMs) sometimes designated as Man-Portable Air Defence Systems (MANPADs). The SAMs were developed at the end of 1950s for protection of military ground forces against combat aircraft and at present there are about 350,000–500,000 in the military arsenals worldwide. There are no accurate statistics on the SAMs attacks on civilian airliners as most of the cases were from war zones where civil registered aircraft operated as military transport aircraft. It seems that about 12 civil airliners were shot down since 1978, most of them in Africa.

There have been a few attempts to shoot down an airliner since 2002. The first one was on the Boeing 767 of Arkia Israeli Airlines in 28 November 2002. The aircraft was departing from Mombasa Moi International Airport (Kenya) when two SAMs were launched by al-Qaida terrorists in the airport vicinity, but fortunately both missed the aircraft.

The second attack was conducted on 22 November 2003 during a DHL operation from Baghdad airport (Iraq) where the Airbus 300 DHL express courier service was struck by a shoulder-fired missile which damaged the left wing. The pilot was able to fly and land the aircraft only by differential engine thrust control.

On 9 January 2007, an AerianTur-M Antonov An-26 crashed near Balad Air Base (Iraq) while attempting a landing. The official cause of the crash was poor weather conditions, however, witnesses claim the aircraft was shot down by a missile and the Islamic Army in Iraq claimed responsibility. Thirty-four of the thirty-five civilian passengers on board died.

In Somalia on 23 March 2007 an Il-76 aircraft operated by TransAVIAexport Airlines crashed in the outskirts of Mogadishu having been hit by a SAM, killing 11 Belarusian civilians on board; one passenger who initially survived, died hours later.

On 17 July 2014, Malaysia Airlines B 777-200, Flight 17, flying from Amsterdam to Kuala Lumpur, crashed near Donetsk in the eastern part of Ukraine. It is assumed it was shot down by a missile. All 283 passengers and 15 crew are reported dead.

There are different types of SAMs on the black market from Soviet production like SA-7, SA-14, SA-18, as well as the US Stinger, French Mistral or Chinese FN-6, and plenty of other types. Typically they have a target detection range of about 6 miles and an engagement range of about 4 miles, so aircraft flying at 20,000 ft (3.8 miles) or higher are relatively safe. There are different systems of SAMs guidance and control. The Infrared (IR) missiles use passive guidance and depending on the year of production they could be from the first to fourth generation. The IR missiles are most popular among terrorists and sometimes are referred to as 'fire and forget'. The Command Line-of-Sight (CLOS) missiles must be flown by an operator who visually aims at the target. They require a properly trained and skilled gunner. Laser Beam Riders use laser for missile guidance to the target. These missiles are resistant to all currently known countermeasures, such as flares which are designed primarily to defeat IR missiles.

Unfortunately there is no single solution to mitigate the SAMs threat to civil aviation aircraft. To enhance aircraft protection and security, a combination of on-board countermeasure systems, air traffic control procedures, pilots training and strengthening of airport and local security measures might be used.

The advantage of the terrorists is that they themselves choose the manner and time of fighting. It is difficult to presume what the future weapons or type of an attack of terrorists will look like. State sponsored terrorists have access to technically sophisticated weapons and they go through professional training, which enables them to overcome more and more sophisticated countermeasures.

During the summer of 2006 a terrorist plot to detonate liquid explosives on board at least 10 aircraft over the Atlantic was discovered and foiled by British police. As a response to this new identified threat it was necessary to further tighten security rules at airports concerning the items in carry-on baggage. For a short time after a security alert in August 2006 in the United Kingdom, severe limitations were put on the size of bag that could be carried through to the aircraft, together with a total ban on liquids. This caused huge delays particularly at major airports. A further requirement was to take off shoes and have them passed through the X-ray machines. Some US airports, like Dallas/Fort Worth, provided disposable slippers to encourage passengers to take their shoes off early, as well as bins to dispose of banned substances. These restrictions caused big bottlenecks to occur before the X-ray screening, at the magnetometers as boarding passes were taken through and after the screening while shoes and coats were replaced.

In November 2006 new rules were introduced in the United Kingdom and a few days later also in other states raising the limit from restricted size of hand baggage of 16 cm × 35 cm × 45 cm to aviation industry standard of 25 cm × 45 cm × 56 cm and easing the amounts and types of liquids in the cabin baggage. Most liquids were banned from cabin baggage and they could not be taken through security checkpoints. The ban included all drinks (apart from baby milk), toothpastes, perfumes, shampoos, hair gels and other similar items. These items were permitted through the security screen only in small quantities within separate containers, each of a capacity not greater than 100 ml. Containers had to be brought to the airport in a single, transparent, re-sealable plastic bag, with capacity not more than 1 litre. Each passenger was allowed to carry only one bag of liquids. Passengers were also prohibited from taking aerosols through security checkpoints. Fortunately, passengers could purchase most of these items at the airport shops after the security points.

Tightening of security measures forced some attackers to modify their tactics and aims. Some of them decided to focus on ground targets. Those attacks are difficult to prevent as they are aiming at public areas with no specific security precautions. However, at some airports there is a security check on the airport access road and all vehicles are searched for arms or explosives (Entebbe).

At Glasgow Airport terminal, two men tried to drive in with a Jeep Cherokee loaded with several petrol containers and propane gas canisters on 30 June 2007. Fortunately security bollards in front of the entrance stopped the car from ramming its way inside the terminal and the gas canisters did not explode. A 20-year-old suicide bomber from the North Caucasus committed a suicide attack in the international arrival hall of Moscow Domodedovo airport on 24 January 2011.

The bombing left at least 37 casualties and about 180 injured. At a car park outside the airport terminal of Burgas airport a suicide bomber attacked on 18 July 2012 the passenger bus transporting Israeli tourists from the airport to their hotels, after arriving on a flight from Tel Aviv. The explosion killed the Bulgarian bus driver and five Israelis.

Fortunately the number of the most dangerous offences, that is acts of sabotage, hijacks or bomb threats in comparison with minor incidents or theft, is relatively small. All unlawful acts create a 'pyramid'. Most of the problems both at the airport and on board the aircraft are caused by unruly and disruptive passengers (Figure 13.1). According to IATA statistics the number of incidents with unruly passengers per 1,000 flights increased from 0.57 in 2009 to almost 0.8 in 2010, a 30% increase. In other words, in 2009, there was 1 unruly passenger incident for every 1,760 flights which increased to one unruly passenger incident for every 1,256 flights in 2010.[2] The same increase could be expected at the airports.

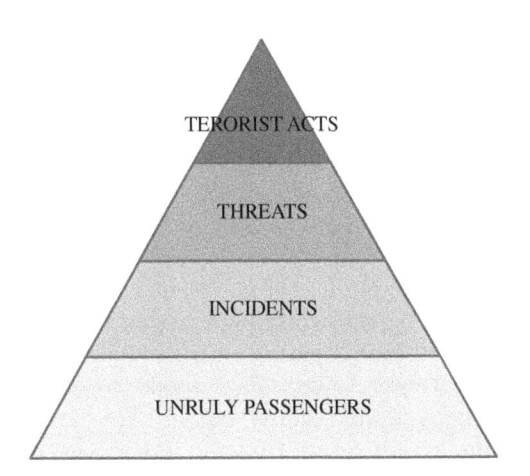

Figure 13.1: A 'pyramid' of unlawful acts.

In May 2012 IATA submitted a Working Paper to ICAO special sub-committee of the Legal Committee to include the issue of unruly passengers on the modernisation of the Tokyo Convention 1963. The severity of some unruly passenger incidents, along with their operational consequences, is a growing cause for concern and can in the long run influence the overall level of crime in aviation. Airports and airlines should therefore develop a zero tolerance policy to unruly passengers.

---

2. IATA submission to ICAO AVSEC Panel, AVSECP/21-WP/21 (2010).

13.2. Legal Framework of International Aviation Security

As has been mentioned, the purpose of most of the attacks against civil aviation is to create political pressure on individual governments. The government politics directly condition the level of threat to civil aviation and a single unwary declaration of a high government representative can mean an immediate threat of terrorist attack with extensive negative economic consequences for airports and civil aviation in general. The protection of civil aviation must therefore be an integral part of the national security plan of each country. It is illogical that the protection against unlawful acts be paid by airports, airlines or passengers, who themselves are jeopardised as a consequence of the political attitude of the country. Moreover, under international law the governments must ensure protection for all companies and individuals that are on its sovereign territory without any discrimination. In spite of this, according to the EC regulation No 300/2008, each Member State may determine in which circumstances, and the extent to which, the costs of security measures taken under this Regulation to protect civil aviation against acts of unlawful interference should be borne by the State, the airport entities, air carriers, other responsible agencies or users. In the long run, the costs are transferred onto the final users — passengers, or will be borne by the taxpayers. Some airports directly specify the security charges.

The threat of unlawful acts is at the same time a very effective way of waging war against another country. It is estimated that the total damages to US civil aviation in connection with the war in the Persian Gulf exceeded the costs spent on military actions.

International cooperation and standardisation is important as well as co-ordination of procedures between individual countries in order to ensure the security of civil aviation. Specific legal acts were always in response to violent unlawful actions against civil aviation. The issues of security were for the first time addressed at the international level in the agreement in **Tokyo** in **1963 Convention** *on Offences and Certain Other Acts Committed On Board Aircraft* (Tokyo Convention) which deals with the issues of acts 'disturbing the security of aircraft or travellers'. As of 2013, the Tokyo Convention has been ratified by 185 states.

The growth of politically motivated terrorism and unlawful acts in the 1960s was addressed by the **Hague Hijacking Convention** of **1970**. *Convention for the Suppression of Unlawful Seizure of Aircraft* is a multilateral agreement by which the contracting states are committed to prohibit and punish aircraft hijacking. As of 2013, the convention has 185 state parties.

The Montreal Convention of 1971 — *Convention for the Suppression of Unlawful Acts against the Safety of Civil Aviation* dealt with acts of sabotage. The Convention sets out the principle that a signatory to the treaty must either prosecute a person who commits one of the offences or send the individual to another state that requests his or her extradition for prosecution of the crime. As of 2013, the Convention has 188 state parties.

The Montreal Supplementary Protocol of 1988 — *Protocol for the Suppression of Unlawful Acts of Violence at Airports Serving International Civil Aviation*, Supplementary to the Convention for the Suppression of Unlawful Acts against the Safety of Civil Aviation. The Protocol enhances aviation security by including airports serving international civil aviation in the scope of protection of the Montreal Convention of 1971.

Montreal Convention 1991 — *Convention on the Marking of Plastic Explosives for the Purpose of Detection* is a multilateral anti-terrorism treaty that aims to prohibit and prevent the manufacture or storage of unmarked plastic explosives.

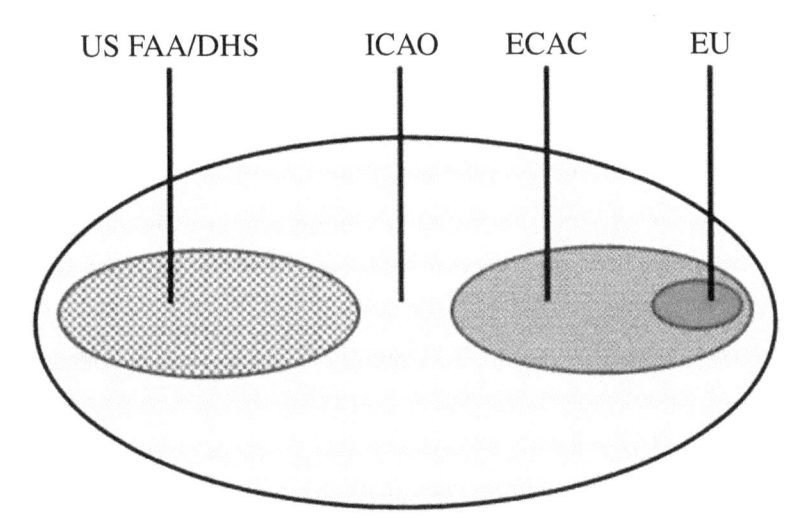

Figure 13.2: Security international legal framework.

ICAO standards for security issues are dealt with in Annex 17. **Annex 17 Security**: *Safeguarding International Civil Aviation Against Acts of Unlawful Interference* was published for the first time in August 1974. Since then it has been amended when necessary. The latest Amendment to Annex 17 — Amendment 14 — became applicable on 14 November 2014 (Figure 13.2). It requires countries to have a National Aviation Security Programme (NASP). The host country has the responsibility for:

✈ the airside/landside barrier
✈ control and use of identification (ID)
✈ central screening of passengers and baggage
✈ control of sterile area
✈ ramp security
✈ armed reactive force

314   Airport Design and Operation

- effective communication
- control of area overlooking the airport.

The method of implementation of the security standards of civil aviation is different from country to country. For example, the responsibility for security of airports could be divided among three parties:

- the government being in charge of inspection of passengers and their cabin baggage
- the airport being in charge of airport areas behind the security check and the whole premises of the airport
- airlines assuring compatibility of individual registered baggage with passengers and security of their own ground equipment.

ICAO Doc 8973 (Restricted) **Aviation Security Manual**, 8th ed., 2011, assists Contracting States to implement Annex 17 by providing guidance on how to apply its Standards and Recommended Practices (SARPs). Annex 17 and Doc 8973 are constantly being reviewed and amended in the light of new threats and technological developments that have a bearing on the effectiveness of measures designed to prevent acts of unlawful interference. The latest updated version on areas such as unpredictability, behaviour detection techniques, landside security and screening of persons other than passengers has been incorporated in this guidance material.

In the European Union (EU) the aviation security is ruled by the Regulation (EC) No 300/2008 of the European Parliament and of the Council of 11 March 2008 on common rules in the field of civil aviation security and repeals Regulation (EC) No 2320/2002. The regulation establishes common rules in the EU to protect civil aviation against acts of unlawful interference. The regulation's provisions apply to all airports or parts of airports located in an EU country that are not used exclusively for military purposes. The provisions also apply to all operators, including air carriers, providing services at the airports. It also applies to all entities located inside or outside airport premises providing services to airports. In line with the regulation every Member State shall draw up, apply and maintain a national civil aviation security programme and also national quality control programme.

Detailed measures for the *Implementation of the Common Basic Standards on Aviation Security* are laid down by the *Commission Regulation (EU)* **No 185/2010** of 4 March 2010. The rules are further complemented by a 'security manual' European Civil Aviation Conference — **ECAC Doc 30, Part II** (restricted) *ECAC Policy Statement in the Field of Civil Aviation Security*, 13th ed., May 2010.

In the United Kingdom an integrated approach for aviation and border security was adopted which involves the Home Office, the Department for Transport and

UK Visas and Immigration. This contributes to the system's close co-ordination, and processes improvements to be more effective across the range of organisations involved in aviation and border security checks. A new focused risk-based approach to aviation security is adopted. This gives operators the flexibility and responsibility to design their security processes. A Security Management System (SeMS) was also introduced in line with EASA requirements.

Following the 9/11 terrorist attacks in the United States, Congress passed the Aviation and Transportation Security Act (ATSA) on 19 November 2001, which established the Transportation Security Administration (TSA). The Act also transferred the responsibility for civil aviation security from the FAA to the TSA. On 22 February 2002, FAA and TSA published a final rule transferring most of the FAA's aviation security rules, including FAA's Sensitive Security Information (SSI) regulation and protected vulnerability assessments for all modes of transportation. The Homeland Security Act of 2002 (HSA), signed on 25 November 2002, established the Department of Homeland Security (DHS) and transferred TSA from DOT to DHS.

13.3. THE AIRPORT SYSTEM AND ITS SECURITY

To be able to guarantee the required level of security and to achieve low operational costs it is necessary to meet some principal requirements of terminal design. Brian Edwards mentions in The Modern Airport Terminal three distinct approaches to effective security design: surveillance, space syntax and territoriality.

An open and spacious terminal interior and its surroundings allow effective *surveillance* (Figure 13.3). Common open space can be easily supervised from a single point. Good design is a prerequisite of crime prevention and a required level of security. Monitoring of enclosed parts of terminal such as shops, corridors and lounges could be performed indirectly by CCTV cameras.

All parts of the airport terminal are designed for the optimum number of passengers and this guarantees terminal space occupation at optimum passenger loads. *Space syntax* is a measure describing the space occupation. High levels of passenger density make direct surveillance difficult, while empty large spaces may cause a feeling of discomfort or even fear. The space syntax could be improved by closing parts of terminals off-peak, improving direct surveillance by guards and improving terminal lighting.

In the complex terminal design process we should also target the issue of *territoriality* — that is identification by the user of a space to the particular area they are occupying at that moment. This could be difficult in public buildings.

316   Airport Design and Operation

However, if a sense of territoriality could be created, the users will take care about the 'designated space' and could challenge criminal or anti-social behaviour themselves. For passengers, the 'territory' which they could take care of could be a group of seats or for retailers the space could be demarcated by a specific floor type. A person, who, for example leaves an unattended bag or rubbish there could be challenged by the 'territory owner'. The space could be defined by colour, type floor or furniture or a signage. The staff and passenger could not only recognise 'their' territory but also have 'the responsibility' to control it.

Figure 13.3: Spacious, clear and open space design allows direct and simple terminal surveillance. *Photo*: A. Kazda.

According to EC regulation No 185/2010, at every airport landside, airside and security restricted areas (SRA) must be defined. The SRA is a part of an airport to which screened departing passengers have access, through which screened departing hold baggage may pass or in which it may be held, unless it concerns secured baggage; and a part of an airport designated for the parking of aircraft to be boarded or loaded.

At airports where more than 40 persons hold airport identification cards giving access to security restricted areas, critical parts of security restricted areas must be established. Critical parts include at least all parts of an airport to which screened departing passengers have access; and all parts of an airport through which screened departing hold baggage may pass or in which it may be held, unless it concerns secured baggage. Whenever a critical part could have been contaminated by for example an unscreened person, a security search of the parts must be carried out to ensure that it does not contain prohibited articles.

Fast changing security threats require modifications in the design and construction of terminal buildings and other areas of the airport. The earlier terminals seldom comply with today's requirements. Old terminal buildings mostly do not allow separation of departing and arriving passengers, so that temporary solutions have to be adopted in order to separate the flows of passengers. It is also problematic to estimate which security requirements will have to be met in the future. It is therefore necessary to design the terminal buildings with the maximum flexibility.

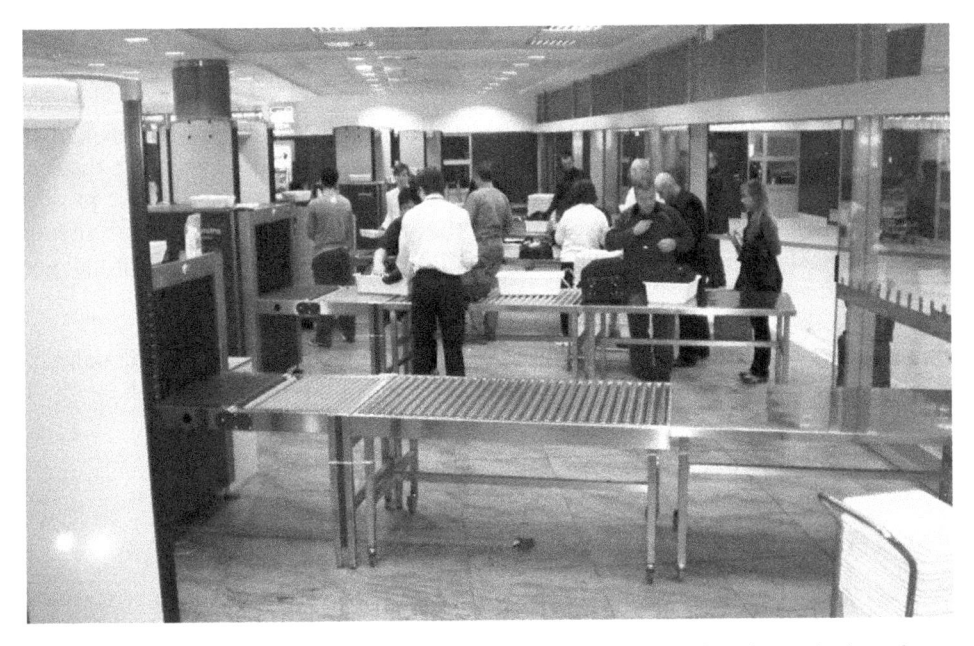

Figure 13.4: A centralised security system is currently used for the majority of new terminals. *Photo*: P. Kusý.

Measures to combat unlawful acts required substantial intervention in the check-in process and in the design of the airport terminal, because the threat originally had not been taken into account. The first measures consisted in setting up security control of passengers and their cabin baggage in the gate, immediately before boarding the aircraft, or alternatively directly before boarding the aircraft upon the exit from the gate. For flights with special security requirements, the inspection of checked baggage was carried out bag by bag. Sometimes, the baggage was reconciled with the passenger on the apron before boarding. It was possible to hold the air cargo in the warehouses for some time to eliminate the possibility that a consignment with a concealed explosive and timing device would be loaded onto an aircraft. However, these measures turned out to be impractical. When the inspection of the passengers took place at the gates, there was often a delay to flights.

Decentralised security check locations had to be established at the entrance to the holding room for each gate. This increased demands on personnel at the checkpoints and on their technical equipment requirements. The queues of passengers before the security control could prevent other passengers from moving freely. After screening the passengers are held in a sterile gate holding room. Some operators still prefer decentralised screening as they believe that it provides a higher level of security (i.e. Singapore Airport), some airports use a decentralised system as it would be operationally difficult to make changes for the centralised concept (Vienna, non-Schengen D gates).

*A **centralised security*** system of inspections tends to be given preference in the design of new terminal buildings (Figure 13.4) and some old terminals changed to the centralised security concept in recent years (e.g. Vienna). Thus, by the time the passengers reach the entrance to the gate (or to the transit concourse) they will already have passed through one or several security filters. The centralised system has several advantages:

- passenger inspection is carried out before the entrance into the airside circulation area, so the passengers do not have to wait until their flights have been announced
- the flow of passengers through the security inspection is substantially more stable with smaller peaks than at individual gates
- there is higher utilisation of technical equipment and personnel
- passengers are, after passing the security filter, more relaxed, there is no additional risk of delay so they tend to spend more in retail and catering facilities.

The centralised security inspections have also their disadvantages:

- corridors or concourses to the gates must be kept security-sterile
- the security-sterile area must be equipped with the required services
- before entering the sterile part all employees and goods also have to be checked

✈ people accompanying the passengers are prevented from having free access to the whole terminal building; this decreases the commercial use in that part of the terminal. Exceptions need to be made for elderly and handicapped passengers, who were used, at least in the United States, to being accompanied by the dependants up to the entrance of the aircraft

✈ separation of departing and arriving passengers is difficult.

If centralised screening is carried out at the airport terminal it could be complemented by an additional security check at the gate or pier entrance for high risk flights. In some airports, where it is necessary to introduce a special security mode, the preliminary security inspection of all persons is carried out immediately on their entrance into the check-in hall. Sometimes the security checks are even at the entrance of terminal (e.g. Antalya; Entebbe) or in front of the terminal and only passengers with valid air tickets are allowed to enter the building (Entebbe).

Security requirements are often contrary to the architectural intentions of the design of the terminal building. Balconies, terraces and entresols, which divide the internal area of the building in a suitable way, could be convenient observation points for the terrorists, or a place from where shooting could take place. In the event of a bomb attack, large glass areas, providing natural light, can be very dangerous. Glass shards are a source of extensive injuries. All glass in these areas should be toughened, foil coated or laminated blast-resistance security glass and secured firmly to a robust structure. A blast which would do only superficial damage to a modern-framed construction would cause moderate damage to load-bearing masonry. In the check-in hall and in other passenger areas, controls should ensure that it is difficult to leave baggage which might contain a bomb. A particularly easy place to leave a bomb is inside a rubbish bin. They should therefore be transparent or, as is the case in Copenhagen, bombproof so that the horizontal blast wave is negated.

Some older airport terminals have been combined with car parks (e.g. Toronto, Charles de Gaulle — Aérogare 1). If a car were to be placed there, and the explosives it contained were detonated, there would be extensive casualties and damage. In a period of increased danger of terrorist attacks, the floors of the garages directly adjacent to the areas for the passengers must be closed to the users and modified so as to limit the spread of the detonation wave. In the design of new airport premises, the parking places should principally be located with no direct contact to the terminal building, whether they have been designed as parking lots in the open or in multi-storey garages. However, it is necessary to find a compromise, so that the walking distance does not exceed the IATA recommended limits. At the same time it is necessary to provide inspections of parking places, not only to provide security, but also for prevention against theft.

It might well be advantageous for a terrorist to place an explosive in the left-luggage rooms of a terminal building. Therefore, the baggage must be security screened before accepting it into custody. If the airport uses lockers for baggage, they should be placed outside the terminal building.

For flights with special security requirements, it is appropriate to reserve one part of the check-in counters with a separate entrance for the given flight. The counters must be physically and visually separated from other parts of the building. The formation of a queue in front of the entrance to the counters must be eliminated, so that no attack could be made against it. In some airports, a whole terminal has been assigned to flights with special security requirements. This gives the possibility to provide increasingly complex security standards, but there is a clear demarcation of the premises which makes the target obvious to the potential terrorist.

It is necessary to pay special attention to maintaining security during construction or reconstruction of the airport premises. The workers of construction companies are sometimes not too willing to undergo the security measures. It is very difficult to find out whether they are criminally unimpeachable. Because of a generally high turnover of staff, it may not be difficult for a terrorist organisation to infiltrate their members into a construction company. Inspection of the movement of materials is made difficult by a great number of cars, lorries and other delivery vehicles. The only solution is a consistent separation of the areas where the work takes place from other premises of the airport. Even then, it is not possible to exclude the possibility that weapons or explosives might be smuggled into the airport and concealed there for later use.

An increased effort to maintain security must also be made during emergency situations in the airport (fire in the terminal building, emergency landing of an aircraft, etc.). Emergencies can be evoked intentionally in order to distract the attention of security units from a terrorist attack.

13.4. SAFEGUARDING OF AIRPORT SECURITY

13.4.1. Security as a Service

In civil aviation, safeguarding security has to be considered as a service to passengers. A feeling of safety is one of the basic needs that must be satisfied. However, what one category of passengers considers as adequate safeguarding of security, for example families with children, can be considered as a useless nuisance by other groups, for example business passengers. Therefore, a compromise has to be searched for, between the level of safeguarding security, the time required for security inspections of the passengers, and the costs. The required security standard has to be guaranteed in the airport. On the other hand, the airport operation must not be paralysed, nor must the high quality of service of air transport be affected by disproportionate prolongation of the check-in process with the introduction of strict and time-consuming security inspections.

The system of safeguarding security is unique in each airport. It will be different on large international airports such as London Heathrow, New York J.F. Kennedy or Amsterdam Schiphol from small airports serving perhaps two or three holiday charter flights daily.

At small airports, the application of basic standards might be inconsistent and their implementation might be practically impossible. In such case the state has the option of applying alternative rules for providing an adequate level of security protection. According to the (EC) No 300/2008 Regulation[3] the various types of civil aviation do not necessarily present the same level of threat. In setting common basic standards on aviation security, the size of the aircraft, the nature of the operation and the frequency of operations at airports should be taken into account when considering the grant of derogation. However, based on a risk assessment, states could also apply more stringent measures than specified in the regulations. However, the requirement has always to be the same, namely to prevent the possibility of locating a bomb on board the aircraft or to prevent the airport premises being penetrated or aircraft being boarded by a group of terrorists.

The manner of safeguarding security will depend on the scope of the airport operation, individual destinations, airlines, kind of operation (scheduled, chartered or general aviation) and the airport size. This must be specified in the airport security programme and the quality control programme. In designing the airport protection it is necessary to consider the following:

+ protection of the airport perimeter, aircraft on the ground whether in hangars or on movement areas of the airport, operational facilities, stores and terminals
+ limitation of movement of persons and vehicles into security sensitive areas on the airside
+ security checks of passengers and employees
+ control of movement of passengers and separation of arriving and departing passengers.

13.4.2. Airport Perimeter Security and Staff Identification

The standard of safeguarding security on the manoeuvring area and other areas of the airport depends on the reliability of monitoring access to the airport and identification of the employees. It is necessary to emphasise that there is no security system which could not be overcome. Modern protection of an airport perimeter rests

---

3. Regulation (EC) No 300/2008 of the European Parliament and of the Council of 11 March 2008; on Common Rules in the field of civil aviation security and repealing Regulation (EC) No 2320/2002.

322   Airport Design and Operation

above all in technical installations, which can identify the place and manner of breaching the perimeter with high accuracy and reliability. At the same time, overcoming an obstacle requires a finite time during which it is possible to send a security commando to the place of violation. On small airports in the third world countries, the outside perimeter of the airport covers several kilometres, and it is very seldom protected by high-quality fencing. On some airports in Africa this has even allowed attacks of organised groups on the cargo holds of aircraft while they were standing in the queue and waiting for permission to take off.

Figure 13.5:   DTR 2000 — fencing for areas with high security needs. *Photo*: Courtesy of Magal Security Systems.

However, the large open space on the airside of airports allows the threat to be visible, at least in daylight, and provides in this way enough time to identify the intruder and to respond before they can threaten the terminal area. It is possible only to hope that the security commando has been properly trained and motivated in order to be able to control and defend the apron and the terminal building.

Since the number of workers is in most cases low on such an airport, their identification by means of simple cards, possibly with colour differentiation for authorisation of entrance into the specified areas, will be sufficient.

The protection of the airport perimeter is decisive not only for safeguarding the security but also for safety. Perfect fencing prevents penetration of animals to the movement areas and their possible collision with the aircraft. Airport perimeter fencing fulfilling both security and safety functions is required by Annex 14, Volume 1, Chapter 9, and also by the Doc 30 — ECAC Policy Statement in the Field of Civil Aviation Security. A higher security standard is required in large international airports, or at least of high security, sensitive or vital premises in such airports (Figure 13.5). The required standard cannot be usually achieved by common fencing. In order to secure the more important facilities like fuel farms, transformer stations or control towers, barrier type protective systems have been used, as well as other types of systems which function on different physical principles. The quality of security of a system is measured by the probability of intrusion detection, threat classification and low False and Nuisance Alarm Rates (FAR/NAR). The quality of security is one of the most important parameters for judging the level of safeguarding a facility and can be expressed mathematically.

The probability of system intrusion is in particular affected by:

- number of false alarms in relation to the unit of length and time (e.g. to 1 km of length in one month)
- reliability of the whole system (depends on the reliability of individual parts)
- monitoring of the condition of the security system
- raising the alarm upon failure of basic functions of the security system.

The intruder can be successful in overcoming the security system, if at the same time four conditions have been fulfilled:

- perfect knowledge of the system (mechanical construction, physical principles and software) by the intruder
- sufficient time and ideal working conditions
- availability of special tools and facilities, which are not normally carried by people
- after an unsuccessful first attempt there is a possibility of an unlimited number of other attempts.

324   Airport Design and Operation

When selecting a security system, the structure has to be assessed also from the point of view of the effect of the environment on its function. Again, there is no perfect system. A system suitable for one locality can be, from the point of view of physical principles on which the system functions, inapplicable in another locality.

- A *microwave system* is unreliable on uneven terrain or if there is vegetation (e.g. grass, shrubs) in the detection zone. It can be easily penetrated in the vicinity of the transmitter and receiver — the intruder will crawl through under the microwave beam without being discovered.
- In the *electric field system* the shape of the electric field will change if a large body comes close to it. This change will be recorded by wire sensors. The basic requirements of faultless operation are perfect suspension of the system, ensuring the correct distances between the top wire and the earth and the prescribed stretching of wires. If these conditions are not met, false alarms take place.
- *Systems sensitive to pressure* or deformation record the changes in mechanical load on the soil when a person walks above the sensor. False alarms arise by the movements of surrounding trees and poles (due to the wind) and precipitation phenomena (rain, hailstorm).
- *Infrared radiation* (the so-called infragates) systems are sensitive to fog, which changes and disperses the infrared beam. Similarly their functioning is affected by dust and by rapid changes of temperature which can condense water on the sensors.
- *Vibration* or *deformation detectors* can be installed on the finished fence and record vibrations of the fence evoked by intruders. The system is sensitive to false alarms evoked by wind, hailstorm etc.
- *Radar* is a high-tech system used for security surveillance at medium and large airports for automatic detection of intruders in particular in combination with automatic infrared cameras. The radar type is usually an E-scan Frequency Modulated Continuous Wave (FMCW) Doppler Ground Surveillance Radar. Surveillance radars operate on 15.7–17.2 GHz frequency. The range of radar is typically up to 3.2 km for a crawling person and 7.4 km for a walking person. The problem of radar is that it cannot 'see' behind an obstacle (so-called 'radar shadow'). This could be overcome by a combination of signals from more radars.

It would theoretically be possible to create a perfect, insuperable system by combining various security elements. However, this is practically not possible. By integration of several components into one unit the number of elements increases and the reliability of the system decreases. The number of false alarms increases and the integrity of the overall security system decreases. Moreover, by integrating several systems the price of the security provision increases.

In order to prevent the crossing of the fencing by means of a 'high bridge', it is possible to increase the level of detection by installing an independent camera system monitoring the protected area parallel to and within the guarded line.

Whatever systems are used, there should be a physical inspection and search of the perimeter by guards at least once a day. It should not be possible to cross the airside/landside barrier via plumbing ducts, air vents, drains, storm water facilities or utility tunnels which must be protected by physical grids.

The sorts of place which might be identified as particularly vulnerable are gate posts, rivers, power plants, control tower, centralised air conditioning, the approach lighting system, emergency access routes and the drinking water reservoir. These should all be monitored by CCTV in a central security control centre.

The security network should link the following systems: check-in questions; biometric access control; production of passenger ID bag screening, status, reconciliation; intruder detection; CCTV infrared; car park number plate recognition; passenger and staff security displays. The communication network should be able to survive the failure of any single cable or support system, terrorism interference at one or two points, extra high demand and an aircraft accident.

In the event of an incident with an aircraft, there needs to be an isolated parking position and a blast-containment area. The airport design should minimise places where devices may be concealed. Lighting should be such as to allow guards to see intruders but conceal guards from them.

13.4.3. Employee Security Procedures

On all bigger airports it is necessary to ensure computer control of authorisation of entrance of employees into specified areas of the airport. The most common manner of inspection is the use of magnetic or chip cards. At the entrance point the employee pulls the card through the slot of the reading device. The authorisation of his or her entry into the given area will be checked and entrance will be permitted or refused. At the same time it is possible to register the entrance or leaving the given area and the movement of the employee on the airport. If security is violated it is possible to evaluate retroactively the presence of the employee on the airport. The system should also monitor forced intrusion into the guarded areas for example through the fire door. It must also enable the inspection of validity of the cards. Cards with limited validity, for example issued to part-time workers, building workers or concessionaires, must be refused entrance into the protected area after expiration of the validity.

Inspection of the entrance into sensitive areas from the security point of view can be tightened by the combination of the security card with the personal code of the employee, the Personal Identification Number (PIN), which has to be typed on the keyboard after the card has been inserted into the reading device. This will limit the abuse of the card if it is stolen.

Identification of an employee in highly sensitive areas can be further enhanced by the implementation of biometric signs (dactyloscopic signs of the hand, voice identification — voice spectrum upon pronouncing of the password, scanning the iris or face geometrics) or by comparison of the portrait of the worker scanned by CD camera with the picture stored in the database. Some firms offer chip identification cards which can be read at some distance from the scanner. Such systems or cards are still more expensive.

Acquisition and operational costs can be reduced by decreasing the number of inspection points into the building. Then it is possible to combine the computer systems for the identification of employees with the control systems of the buildings, which fulfil other functions such as fire alarms, control of the heating, automatic switching off of the lights in case of absence of the worker in the workplace, that is, the building management system.

The computer systems for identification of the airport employees and securing the airport are becoming much cheaper and so allow the numbers of security service workers to be reduced. At the same time the costs of maintaining the system increase, while reducing the number of workers can be the source of conflicts with the unions in some countries.

In EU countries, airport identification cards can be issued only after a background check has been conducted which includes, among others, criminal and employment/ education records in all states of residence during at least the preceding five years. In some countries a 20-year background check is required when employing airside workers, whether by the airport or by the concessionaires.

Any possible contact of the airport employees with terrorists when, for example, cleaning aircraft, can be made more difficult by random rostering of workers into work groups with the use of a computer programme.

13.4.4. Measures in Relation to Passengers

One of the important measures for increasing security is complete separation of arriving and departing passengers. In the Schengen countries' airports a system of one-stop security screening is already introduced. Passengers screened at one Schengen EU airport can mix after arrival with screened departing passengers and during transfer/transit are not security re-checked. One-stop security results from recognition of equivalence in security processes and quality control. The main benefits involve removing unnecessary duplications and inefficiencies, facilitation of passenger movement and cost reduction. At many airports in the third world countries, the security measures are insufficient and smuggling of a weapon or explosive on board the aircraft could be easier than in the countries with a controlled level of security measures. After arriving from such a country the terrorist could hand over a weapon to his accomplice, who had already gone through the security inspection. Separation of flows of arriving and departing passengers makes it radically more difficult to hand over a weapon or any other inadmissible material to another passenger.

It is also common to generate profiles of passengers to differentiate the scope of the security inspection. It is obvious that it is necessary to pay more attention to two men coming from a country sponsoring terrorism than to a group of pupils from let's say Scandinavia on an excursion. Large established airlines have much information in the reservation systems which can also be used for data input for security inspection. If a suspicious passenger or a group of passengers who cannot be assigned to common categories appear on the flight, increased attention can be paid to them. There are several databases on the most wanted terrorists in the world.

From the beginning of 2004, as one of the security measures after the 11 September, an agreement came into force allowing US Customs and Border Protection (CBP) to access European airline reservation databases to use up to 34 items of information about each passenger. In the new scheme airlines will supply the data to the US counterparts. Advance Passenger Information (API) refers to passenger data which are required by many governments for security, immigration and customs purposes today. Passenger data format was standardised through the cooperation of World Customs Organization (WCO), the ICAO and IATA. However, the format is not accepted by all states yet.

As with the US Government, many other states, based on bilateral agreements, require access to Passenger Name Record (PNR) reservation data today.

The data structure could be divided into five basic categories:

Information about the passenger: name; address; date of birth; passport number; citizenship; sex; country of residence; US visa number (plus date and place issued); address while in the United States; telephone numbers; e-mail address; frequent flyer miles flown; address on frequent flyer account; the passenger's history of not showing up for flights.

Information about the booking of the ticket: date of reservation; date of intended travel; date ticket was issued; travel agency; travel agent; billing address; how the ticket was paid for (including the credit card number); the ticket number; which organisation issued the ticket; whether the passenger bought the ticket at the airport just before the flight; whether the passenger has a definite booking or is on a waiting list; pricing information; a locator number on the computer reservation system; history of changes to the booking.

Information about the flight itself: seat number; seat information (e.g. aisle or window); bag tag numbers; one-way or return flight; special requests, such as requests for special meals, for a wheelchair, or help for an unaccompanied minor.

Information about the passenger's itinerary: other flights ticketed separately, or data on accommodation, car rental, rail reservations or tours.

Information about other people: the group the passenger is travelling with; the person who booked the ticket.

However, there are also protected data like information on political orientation, racial or ethnic data, religious or philosophical opinions, trade union membership, and data concerning the health or sex life of the individual. Further security enhancement will be accomplished after full implementation of new passports with biometric data.

In the United States the government's Terrorist Screening Center (TSC) created and maintains the No Fly List, a list of people who are not permitted to board a commercial aircraft for travel in or out of the United States. In August 2013, a leak revealed that more than 47,000 people were on the list.[4] Future work on the passenger data issues will according to IATA information concentrate on new and innovative methods of data transmission, the concept of a single window into governments for data and innovative and broader-reaching use of data to maximise its benefit.

Security inspection of the registered baggage was introduced as a countermeasure after the sabotage on Pan Am Flight 103 on 21 December 1988 over Lockerbie in Scotland. The aim of the inspection is the discovery of explosive or other banned substance (combustible materials, acids, drugs) in the baggage of the passenger. Performing security inspections of registered baggage meant a substantial change in the common checking process. The majority of airports with commercial flight operations have automatic systems of registered baggage screening which usually trace suspicious baggage before the passenger passes the security check and could be retained there to be present during the subsequent manual baggage inspection. All airports must ensure 100% inspection of all registered baggage by technical devices with specified parameters or in combination with manual inspections.

---

4. 'In First, Government Officially Tells ACLU Clients Their No Fly List Status'. ACLU, 10 October 2014.

Another requirement of baggage inspection is to ensure reconciliation of the registered baggage with the passengers. It has to be ensured that all passengers that have handed in their luggage during the check-in process have also boarded the appropriate aircraft. There are some exceptions for when flight schedules are disrupted. The system of baggage reconciliation should also ensure a permanent monitoring of the movement of passengers and baggage in the areas of the airport and the information as to which part of the aircraft hold space the baggage was loaded, in case it is necessary to search out and unload the baggage. It also has to be ensured that after inspection there is not a possibility to replace it or to put anything into the baggage.

The system of baggage reconciliation has two principal shortcomings. In fact there are suicide bombers, who try to bring the bomb on board themselves and detonate it, and also there could be persons that had a bomb planted into their baggage without their knowing it.

The baggage reconciliation at most airports is usually provided by computer comparison of databases of baggage tags with the boarding card of the passenger, which were generated at check-in. The baggage tag can have a record with a bar code or QR Code which could be gradually complemented by the RFID technology. Some systems use a baggage tag record as well as a boarding card record. The code is at the same time used as information for sorting the baggage. Before boarding the aircraft or on passing chosen specific locations, the boarding card could be inserted into a reading device and in this way the movement of the passenger can be monitored. However, in some countries there is a legal problem where a system of passenger movement monitoring is considered as an infringement of civil liberties.

Baggage inspection can be performed manually or with the use of technical facilities, at present particularly with the use of X-ray units. Manual inspection of baggage has several advantages over the X-ray inspection, for example identification of the majority of objects in the baggage is easier compared to the interpreted display on the screen, it has lower initial costs, and thoroughness of inspection depends primarily on the time spent and so it is possible to react promptly to the external conditions (Figure 13.6).

On the other hand, inspection with the use of X-ray units has several advantages over the manual inspection. It is easier for the X-ray unit to discover secret partitions in the baggage or objects hidden inside other things, and it enables higher productivity and hence decreases labour requirements. Also, the passenger's privacy is less disturbed when his or her personal belongings are being inspected and it is difficult for the potential terrorist to predict how successful he or she can be in hiding the weapon against the X-ray inspection.

Automation of security inspection with the use of technical devices has a number of advantages in comparison with the inspections that are executed personally, by security screeners. The machine is different from humans in that it does not decrease

330   Airport Design and Operation

in efficiency and reliability when repeating monotonous activities. On the other hand, the technical facilities are only the tools that speed up the security inspection and make it easier. In order to ensure effective inspection the screener must be properly trained and acquainted with the equipment in order to be able to use it effectively.

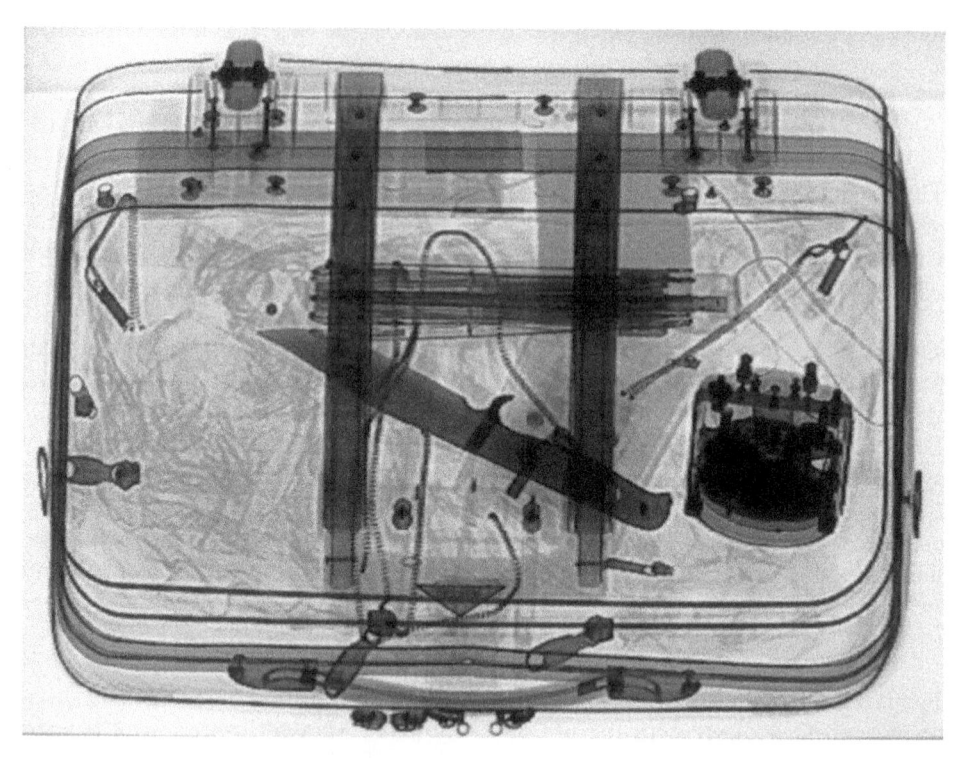

Figure 13.6:   X-ray black and white image of a baggage. *Photo*: A. Kazda.

It is compulsory for all international airports to have 100% hold baggage screening. If automated, the baggage inspection may be designed as a five-stage process. At the beginning of the process there is a computer-controlled X-ray unit that automatically sorts out the baggage. The suspicious baggage is identified according to the weight of articles in the baggage, affinity of shapes of the objects with suspicious objects, and the atomic weight of substances. During this test, the 80% of baggage, which is not suspicious, is sorted out and can be immediately loaded into the aircraft.

Approximately 20% of baggage which does not pass the automatic inspection is subject to further inspection by an X-ray unit which automatically marks the possible explosives with a red colour. The security inspection screeners will inspect the baggage on the X-ray unit monitor. Suspicious objects in the baggage can be enlarged on the monitor. Sometimes it is difficult to identify plastic explosives because they can be shaped into the form of common objects. It is more difficult to hide timing devices or primers. Therefore, the operator looks particularly for devices connected by wires. In this stage of inspection another 19% of baggage is sorted out and sent off to the aircraft.

Each suspicious piece of baggage is subject to a third stage of inspection. In this stage a gas analyser is used, which can, by means of gas chromatography, discover the traces of some explosives. On the basis of this test it is possible to eliminate approximately a further 0.9% of baggage.

The baggage that does not pass the gas test will be subject to manual inspection. In a separate room and in the presence of the passenger, the baggage is opened and properly inspected by an explosives expert. According to airport experience the manual inspection is necessary for only one of 5,000–7,000 pieces of baggage.

A different approach has been chosen by some operators, which in addition to technical facilities relies on ascertaining the passenger's profile in a conversation with the security inspection worker, together with proper manual inspections. In the course of the inspection the passenger is asked questions separately by two workers, the answers being compared and evaluated. Answers are of the same importance as the reactions of the passengers. A precondition of their correct interpretation is high professionalism and experience of the security service workers. The precondition of this approach is not only extensive support of the state but also top-level professional training and high motivation of the workers.

To speed up the security screening process and facilitate movement of frequent passengers, crew members or airport staff, various systems of positive identification using biometric identification patterns were introduced recently at some international airports. Frequent travellers could usually sign up for a programme, have their biometric scan registered and pay a fee. Airports use this type of service not for arriving, but departing passengers to expedite the check-in and security screening. Iris recognition is increasingly used as a convenient way of positive identification. Passengers do not need to put down their baggage and the biometric recognition is usually accomplished within less than a second.

13.5. Detection of Dangerous Objects

As has already been mentioned, inspection with the use of technical equipment has a number of advantages, which make the security inspections easier and speed them up. At the same time they also limit the possibilities of human factor errors. When selecting the equipment it is particularly necessary to take into account the airport size, number of passengers in the peak hour, passenger and flight profiles, employees' level of income, equipment acquisition costs and average detection costs.

Detection devices for discovering explosives and other prohibited objects have to meet many criteria. They must be able to discover weapons and explosives, which are used in the military or in the civilian sector. The sensitivity of the device must not be affected by the location of the explosive or the type of container and, moreover, the device has to function with a minimum of subjective human input. The price of the device and its peak capacity are also important. Several devices have been developed, which meet most of these criteria. These include metal detectors, X-ray units, gas analysers, vacuum chambers and use of trained dogs.

13.5.1. Metal Detectors

Modern metal detectors can discover weapons made from ferrous and non-ferrous metals, some also some types of composites or metals with low electrical conductivity. Metal detectors are usually based on a pulse induction principle where wire coils which serve as transmitters and receivers are placed in the frame. Coils generate short pulses of current through the coil of wire and generate brief magnetic fields. When a metal object passes through the metal detector, the pulse creates an opposite magnetic field in the object which triggers an alarm. Most modern detectors also indicate the approximate height of the metal object enabling security personnel to rapidly locate the object. The walk-through metal detectors are most common and, being relatively cheap devices, are used even at small airports for passengers and employees screening. However, traditional walk-though detectors are not reliable in detecting metal in shoes or on the lower body extremities. New types of walk-through detectors are able to scan shoes or, alternatively, the passengers are asked to get a shoe scan by stepping on a special scanning device.

The problem of metal detectors is that they cannot detect non-metallic concealed objects made of ceramic, plastic or some types of composite material weapons. Traditional X-ray technologies used for luggage screening are unsuitable for checking people because of high ionising radiation charges and they are also unsafe for some passengers, for example those with pacemakers, who still have to be checked with a hand search.

13.5.2. Millimetre-Wave Scanners

Millimetre-wave scanning is one of the technologies used by full body scanners where the waves are reflected back from an obstacle and create a 3D image. The technology is based on millimetre-wave gigahertz technology that is completely harmless to people (Figure 13.7). It is designed for screening of people at airports and other high security sensitive areas. At this wavelength clothes become transparent, so the technology can detect metal or non-metallic objects under clothing, including the person's body. Even though the technology can detect hidden weapons, it also raises questions of privacy and whether it should be used on the general public at the airports, as the body shape is visible on the screener's monitor.

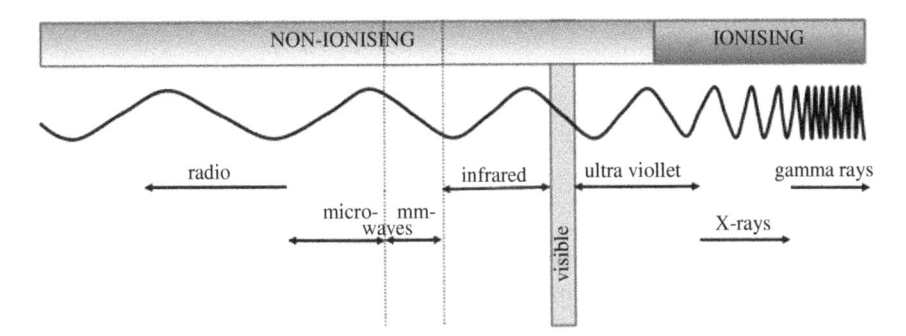

Figure 13.7: Millimetre-wave scanners are based on non-ionising technology which is harmless to people.

13.5.3. Backscatter Screening

Backscatter technology is based on the X-ray ionising radiation. Contrary to a traditional X-ray technology which is based on the transmission of X-rays through the object, backscatter detects the radiation that reflects from the object and creates an image.

The backscatter scanners will typically merely create a 2D image whose pattern is dependent on the material property. Because of privacy concerns the photo-quality images which could be seen by the person viewing the scan the new types of scanners display just a silhouette indicating places where a hidden object could be searched. The backscatter scanners can reveal all metallic, plastic, ceramic and composite objects but also they are good for imaging organic materials. X-ray radiation dose from backscatter scanners is extremely low. For example the dose for a six hour flight is 200–400 times larger than a backscatter scan, or a standard chest X-ray is almost 100 times higher.

13.5.4. X-ray Units

X-ray devices were first used at airports in the United States but also in Czechoslovakia at the beginning of the seventies. They have become an effective device for discovering potential hijackings, which at that time represented the greatest danger. The first X-ray units were not sufficiently sensitive and the display of the baggage on the monitor in real time did not provide enough time for the operators to recognise dangerous articles in the baggage. The picture was not sufficiently clear. On the other hand, at that time the screeners were looking just for weapons.

334 Airport Design and Operation

In addition to other things the X-ray capabilities are characterised by two, to a certain degree different parameters:

- resolution, indicated by the thinnest steel wire that an X-ray unit is able to display on the light background
- penetration rate, measured as the thickest steel plate that an X-ray unit is able to radiate through without having the plate displayed as a completely black object.

However these criteria are not accepted as a standard worldwide.

X-ray units have to meet health standards concerning radiation leakage. According to the standards valid in the United States, radiation leakage must not exceed 0.5 milli Roentgen per hour (mR/h). In modern devices this leakage usually does not exceed 0.1 mR/h. The X-ray system must be safe for photographic film, magnetic carriers of data or data in electronic notebooks.

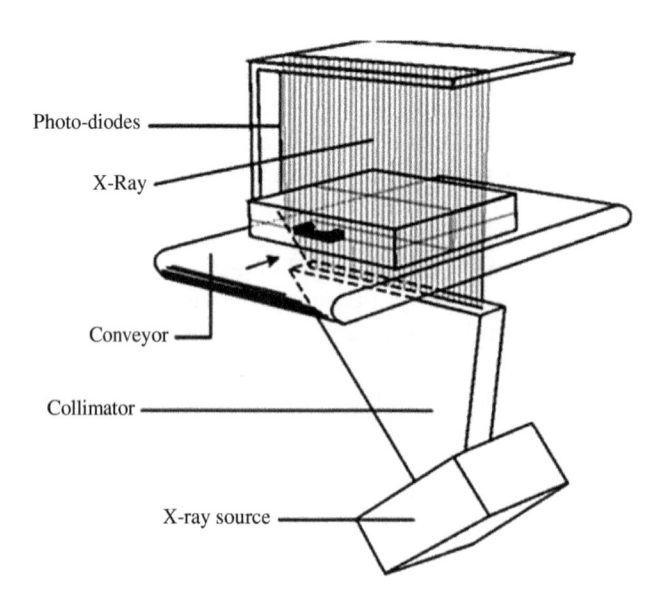

Figure 13.8: Principle of X-ray unit.

In 1980 the firm EG&G Astrophysics introduced Linescan System I, the first digital system, which used as sensors a series of silicon diodes, producing black and white images. The system gave a high quality of display and high reliability. Sensitive sensors enabled the radiated output to be decreased so that the device was safe for a photographic film. The device has been further improved, individual objects in the baggage being displayed in pseudo-colours depending on the material's thickness and density. The principle of an X-ray unit is shown in Figure 13.8.

In 1988 the first X-ray unit functioning on the principle of radiation of two different energy outputs (Dual Energy Systems) was introduced. This type of X-ray unit is known by the abbreviation E-Scan. E-scan is capable of distinguishing organic and inorganic substances by their atomic numbers. Objects with atomic number below 10 are classified as organic and on the monitor they are displayed in orange/brown shades. Explosives and narcotics have these low atomic numbers. Plastics and explosives such as Semtex or C-4, which also have a low atomic number, are displayed by an orange warning colour on the monitor, which attracts the attention of the operators. Objects with atomic number higher than 10 include metal objects (e.g. weapons). These are displayed in shades of blue. Very thick objects that cannot be penetrated by low energy radiation, are displayed in green. In this way the attention of the well-trained operator is drawn to the suspicious objects, the baggage then being manually searched.

Another type of X-ray unit uses Computer-Aided Tracing to search for suspicious objects. Shapes of dangerous objects (guns, knives, etc.) have been saved in the computer memory, these being compared with the objects in the baggage. The attention of the operator will be drawn to the object by the computer.

An X-ray unit which gives a three-dimensional display of baggage is being offered under the name Z-Scan. The three-dimensional display is achieved by associating two emitters in one X-ray unit. The first radiates the baggage from the side and, with a time delay, the second cross-ways from the bottom. The results of both X-ray units are processed by a computer, which assigns a division of weights in vertical sections through the baggage. From individual sections a three-dimensional picture of the baggage is generated. By means of an algorithm the location of atomic weight of plastic explosives is identified and the attention of the operator of the device is automatically drawn to them.

There are also X-ray units using the technology of computer tomography, which was originally used in health care, for identification of dangerous objects in the baggage. As in the case of E-Scan, the baggage will be initially radiated by 70 kV energy and then by 140 kV. The results of both tests are processed. By comparison with the database, the system is automatically able to determine suspicious objects. They will be displayed in red and the attention of the operator will be drawn to them. The advantage of this device is that it can also work fully automatically and suspicious baggage will be sorted out so that the operators can concentrate on the suspicious baggage.

The majority of modern X-ray units have functions that substantially simplify the work of operators and remove the shortcomings of the first X-ray units with imperfect television monitor displays. Digitalised video recording allows the object to be searched for any period of time with further processing of the picture, for example enlarging of one part of the baggage. Emphasising the edges of individual objects and wires (Edge Enhancement) makes it easier to discover electronic primers.

Identification of objects behind very thick objects (lead/steel plates) is made easier by gamma emphasising (Gamma Enhancement), which makes the picture lighter or darker.

Replacement of traditional photodiodes by a combination of optical fibres and a CCD camera has lowered interference and substantially improved the picture quality even with a high-speed conveyor belt. While with traditional X-ray units the speed of the conveyor belt can be around 0.25 m/s^1, with the new technology a high-quality picture is provided with conveyor belt speeds above 1 m/s^1. This substantially increases the capacity of the system of baggage inspection.

Modern X-ray units equipped with a microprocessor are more reliable and require less maintenance. They allow automatic calibration and inspection of the device with identification of faults.

Modern detection units must satisfy specific tests and are classified as either Explosive Detection System (EDS), Explosive Device Detection System (EDDS), Primary Explosive Detection System (PEDS) or Threat Image Projection (TIP). The equipment can be used only in a strictly specified way and in line with the equipment characteristics and performance. The definition of the equipment is as follows:

Explosive Detection System: A system or combination of different technologies which has the ability to detect, and so to indicate by means of an alarm, explosive material contained in baggage, irrespective of the material from which the bag is made.

Explosive Device Detection System: A system or combination of different technologies which has the ability to detect, and so to indicate by means of an alarm, an explosive device by detecting one or more components of such a device contained in baggage, irrespective of the material from which the bag is made.

Primary Explosive Detection System: A system or combination of different technologies which has the ability to detect, and so to indicate by means of an alarm, explosive material contained in baggage, irrespective of the material from which the bag is made. Capabilities of PEDS are lower than those of EDS; their use should be limited in time and operating procedures should be adapted to detection capabilities.

Threat Image Projection: A software programme approved by the Appropriate Authority which can be installed on certain X-ray machines. The programme projects virtual images of threat articles (e.g. guns, knives, improvised explosive device) within the X-ray image of a real bag under examination or complete virtual images of bags containing threat articles, and provides immediate feedback to the X-ray machine operators of their ability to detect such images.

13.5.5. Gas Analysers

Terrorists prefer plastic explosives. Some of them, such as Semtex and C-4, are more effective by one third than TNT and this is twice as effective as dynamite. The basic components of all plastic explosives are either RDX (cyclotrimethylentrinithramin) or PETN (pentaerytrytol tetranitrate). Highly explosive substances pass almost immediately into a gaseous state after explosion, the speed of detonation of C-4 being 8,052 m/s.

The plastic explosives were developed during World War II because they are more effective and also resistant to the environment, as they can be detonated only by a primer, and also because they can be formed. By precise dosing of additives and shaping, it is possible to give them almost any appearance, from bricks, leather shoes to children's toys. Another advantage of plastic explosives, for example Semtex, is that they have practically no smell and their emissions can be identified only with difficulty.

Therefore, the member states ICAO adopted in March 1991 the convention on marking plastic explosives to aid their detection. In the course of production the plastic explosives have to be marked by one of the four agreed substances. These are ethylene glycol dinitrate; 2,3-dimethyl-2,3-dinitro-buthane; para-mono-nitrotoluene or ortho-mono-nitrotoluene. The aim is to increase the level of emissions of material so that they can be more easily identified by common gas analysers and by dogs. Unmarked explosives are limited to use for military and police purposes.

The devices for identification of explosives are used as a supplement to other technical facilities, for example to X-ray units or as independent devices. In order to identify explosives on the passengers, the gas analysers can be built into, for example, a revolving door or special chamber. The device can further be equipped with a magnetic card sensor for inspection of the entry of employees into especially high risk areas (e.g. air traffic control centre). Manual detectors can be used for inspection of baggage.

Some of the detection devices work on the principle of gas chromatography (Figure 13.9). Emissions from the investigated object are sucked by a pump into the first chamber of the detector, where they will hit the membrane of the silicone rubber. The emissions of the explosives pass through the membrane into another chamber, where they and also the air molecules are exposed to a weak source of radioactive radiation, which ionises them and the molecules acquire an electrical charge. The ionised molecules are then driven through a tube with an electrical field pulsating with the frequency of 50 Hz, after which they hit a scanning electrode, which records them in the form of a weak electrical current.

If the emissions of explosives are really present in the sample that is being investigated, after ionisation they get bigger and heavier than air ions and therefore they

338   *Airport Design and Operation*

move more slowly through the tube. At that time the device records one more, later current impulse. From the position and size of the second current peak, the microprocessor will specify the composition and concentration of emissions of the investigated substance. The relative concentration of emissions is usually shown on the display of the device.

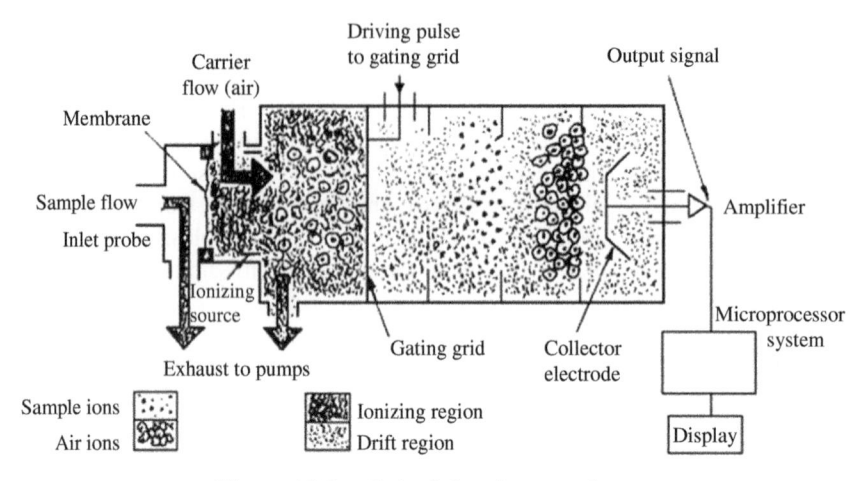

Figure 13.9:   Principle of gas analyser.

A certified gas analyser must satisfy requirements on Explosive Trace Detection Equipment (ETDS). ETDS is a system or combination of different technologies which has the ability to collect and analyse particles on, or vapour from, contaminated surfaces or contents of a luggage, and so to indicate by means of an alarm the presence of traces of explosives.

13.5.6.   Vacuum Chambers

At some airports the baggage and air cargo are subject to additional inspection in a vacuum chamber. The chamber simulates the low pressure which would be achieved in the aircraft in which the explosive might have been vacuum primed. The possible charge may then be fired in a safe environment. The possibility of exploding any primer by pressure vibrations, ultrasound or by electrical impulses is also being used.

13.5.7.   Dogs

Using trained dogs is still a very common way of discovering explosives (Figure 13.10). Dogs are very fast, work effectively and are able to sniff out what most of the devices cannot discover, such as clear plastic explosives and trinitrotoluene. Another advantage of dogs is that they can penetrate places where the human would never get. The disadvantage of dogs is that they cannot work by themselves and independently.

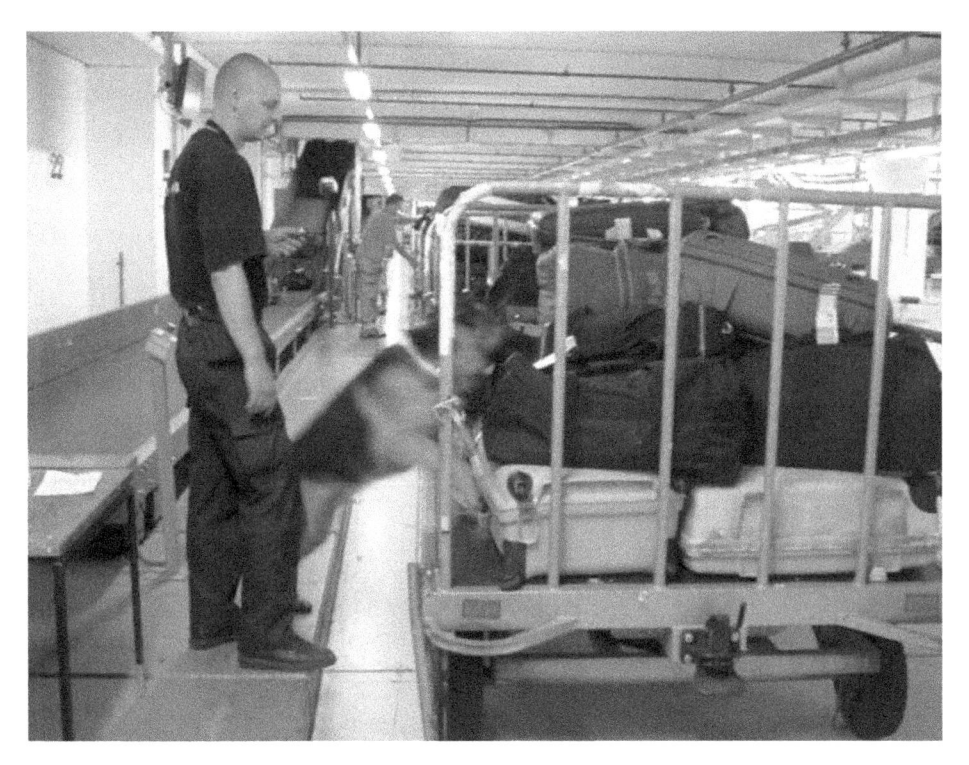

Figure 13.10: Dogs are indispensable when searching for explosives and drugs. *Photo*: T. Kazda.

Also, depending on the conditions applying in a particular country, there can be high costs for taking care of them. They must be handled and trained by an experienced dog handler. A dog may lose concentration, lose interest or its attention can be distracted by another object. In spite of that, it carries on performing the trained activity. Another problem is that the smell of the sought object can be covered, either intentionally or not, by other strong smells for example by solvents, ammonia or fuel.

Therefore, it is best to use the dogs more as an additional procedure, for example for inspection of aircraft, baggage and objects when there is a danger of bomb attack. The dogs are less suitable for regular inspection of baggage.

13.5.8. Liquid Scanners

After a plot to blow up trans-Atlantic flights in 2006 a liquid/gel ban for all liquids in the carry-on baggage (except baby food) in containers over 100 ml was introduced. To ease the regulation, speed up security checks and to facilitate the air transportation new scanners are tested for their ability to determine the chemical composition of liquids. Most of the scanners can analyse the liquid contents of bottles in non-metallic containers up to 3 l in size without opening them by use of laser technology.

13.6. Conclusion

Each airport must have an airport security programme and the quality control programme which are part of an airport's Security Management Systems (SeMS), the extent of which corresponds to the airport size and its importance. The SeMS provides airports with a risk-based framework to create a security culture. In addition to other things it must specify not only the duties of security departments but also of individual employees for different kinds of emergency situations. Examples of such situations can be fire, attacks on the airport and its premises (stores, fuel farms, air traffic control, apron etc.), bomb threat or hijacking of aircraft or landing of an aircraft with the terrorists on board. Those situations should be evaluated in the security risks assessment matrix (Table 13.1) where a probability of each situation, its impact on the airport operation and severity of action should be assessed.

Table 13.1: Security risks assessment matrix — severity/probability of event occurrence.

Severity / Probability	Catastrophic	Hazardous	Major	Minor	Negligible
Frequent	Critical risk	Critical risk	Critical risk	High risk	Moderate risk
Occasional	Critical risk	Critical risk	High risk	Moderate risk	Moderate risk
Remote	Critical risk	High risk	Moderate risk	Moderate risk	Low risk
Improbable	High risk	Moderate risk	Moderate risk	Low risk	Low risk
Extremely improbable	Moderate risk	Moderate risk	Low risk	Low risk	Low risk

As already mentioned, the probability of occurrence of high risk events is (fortunately) very low and it may happen that employees at small- or medium-sized airports would be never confronted with such a situation. In spite of regular training they might not be able to recognise a critical situation (Black Swan Theory). Some airports are reviewing experience from different sectors (nuclear industry) where even small/negligible incidents are registered and evaluated to increase the overall security level and company culture.

A higher quality of technical facilities decreases the probability that the saboteurs can get the explosives or weapons on board aircraft by traditional means. However, this does not mean that they will give up their activities. On the contrary, it is possible to assume that they will look for other ways to achieve their objectives. Therefore, it must be ensured that all routes to the aircraft are protected. It is no use installing facilities for detection of explosives for millions of dollars if the security measures are such that it is easy to get to the apron and to the aircraft without being detected. Equally it is a waste of money if the operating personnel are not sufficiently trained and motivated. In other words, high-quality technical facilities are welcome; however it must not be forgotten that they can be effective only when fully integrated into a sophisticated security system.

To improve the security system and also to facilitate air transportation Airports Council International (ACI) and the IATA have signed a Memorandum of Understanding (MoU) jointly to develop Smart Security (SmartS) in December 2013. The aim of SmartS is to improve the passenger journey from curb to boarding. Passengers would be able to proceed through security checkpoints with minimal inconvenience, while maintaining security resources are allocated based on risk, and airport facilities are optimised.

14

LANDSIDE ACCESS

Tony Kazda and Bob Caves

14.1. ACCESS AND THE AIRPORT SYSTEM

The primary advantage of air transport is speed, particularly for long haul where it has completely supplanted shipping, but also for those short-haul trips where it is in competition with surface transport, in particular high-speed trains. However, the average speed is reduced by the ground portion of the trip. The trip does not start or finish in the airport, but at home, at the hotel, at the workplace etc. The passenger is just as concerned to reduce time on the ground as in the air part of the trip, and just as annoyed by any delay, whether in the air, the terminal or on the way to and from the airport. The total time of transportation 'from door to door' is decisive for the passenger. The attractiveness of an airport markedly decreases if the time of access by surface transport exceeds a certain maximum time. For short-haul trips this might be as short as 30 minutes, while 2 hours or more might be acceptable for long haul, inclusive tour or low-cost carrier trips. By increasing speed of surface transport, it is possible to increase the catchment area of the airport.

Transportation time to the airport is just one of the three most important factors affecting the decision process of the passenger for a particular airport, the other two factors being price of the flight ticket and number of flights (frequencies) offered but it is also influenced by the quality of services offered by the airport itself. The quality of surface transport affects the size of the catchment area of the airport in the competitive market between several airports. Surface transport quality and quantity should not need to become a limiting factor of the development of air transport at an airport if plans have been properly developed. Yet, paradoxically, it may be the very success of increased access traffic that limits an airport's growth, if the environmental impacts of the ground transport are too severe. The problem remains that in some cases there is often no co-ordinated approach to the planning for an airport and that for the town and the region that it serves, particularly with respect to ground transportation (Table 14.1).

The decision of which mode of transport is selected by a passenger is a personal choice, however, an airport can considerably influence the decisions by enhancing the offer, improving the quality of service or decreasing the price of public services. For example Boston Logan and Geneva airports offer free transportation to downtown in order to promote public transport. The responsibility of ground transport system planning rests on the local/regional authority but the quality of the ground infrastructure and transportation significantly influences the airport market. When planning a new airport or airport expansion, in depth analysis of the airport access system must be carried out to match the ground transportation system capacity with the terminal and airside capacity. The ground transport system characteristics differ considerably in relation to the airport size and type of operation, for example point to point/hub and spoke; traditional or low-cost carriers; short-/long-haul routes. Airports with large shares of holiday travellers usually have a high percentage of the ground transportation provided by coach, for example Faro. The transport system should also take into account other airport activities, for example aircraft maintenance organisations, parcel services or cargo operations which could generate significant number of users from the employees segment but also freight/goods transportation. The peaks of traffic to an airport often mirror the peaks in the local traffic. Sometimes the responsibility for planning and operation of the various modes of transport has been divided among several departments.

Table 14.1: Distance of some airports from city centres and connection times by public transport.

City	Airport	Distance [km]	Travel times (public transport) [min]
London	Gatwick	43	31
Hong Kong	Chek Lap Kok	40	23
Chicago	O'Hare	35	29
Paris	Roissy	28	35
New York	J.F. Kennedy	27	38
New York	Newark	25	32
London	Heathrow	24	55 (15[a])
Tokyo	Haneda	19	22
New York	La Guardia	13	18
Frankfurt	Rhein Main	10	12
Paris	Orly	10	31

[a]By Heathrow Express.

From the beginnings of air transport the airline companies laid great emphasis on the provision of transportation of the passengers to and from the airport. In the early days of aviation many passengers had no other means of transportation to the airport except the airline coach. The airports have been, contrary to the railway stations, located outside the towns they served. Therefore the airlines found it quite natural to provide transport for airport access, mostly by bus. The first connections of the airport and the town by rail were built in Berlin Tempelhof Airport where the old terminal, originally constructed in 1927, became the world's first with an underground railway, and a new railway station was opened at London Gatwick in September 1935 and was initially served by two Southern Railway trains an hour, on the Victoria to Brighton line.

After the Second World War, as a result of mass development of passenger car travel, there was an increase in the use of the car for transportation to/from the airport, and the need for improvements to the road system to allow convenient access. Now there is an emphasis on increasing the public transport share of airport access trips, both for the travelling public and for airport employees, in the face of road congestion and environmental concerns. The share of passenger public transport use at London airports between 2003 and 2013 is shown in Figure 14.1.

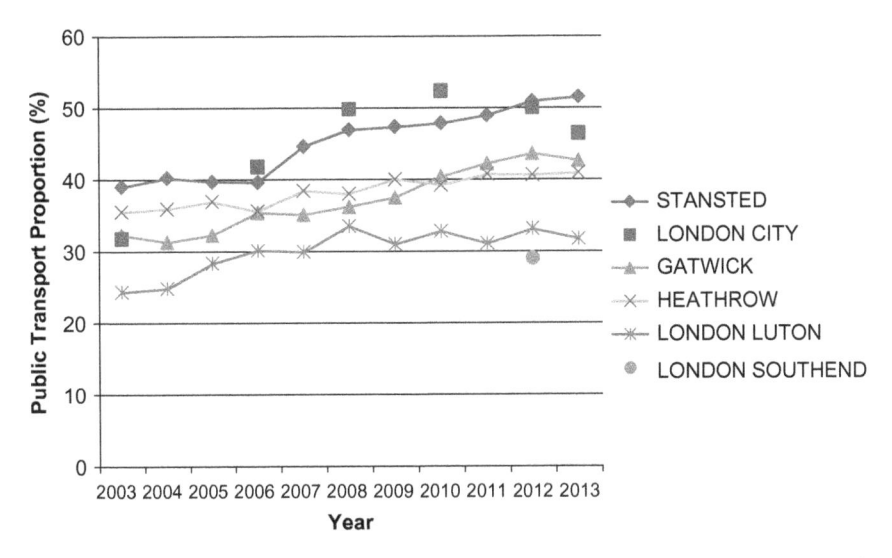

Figure 14.1: Passenger public transport use at London airports 2003–2013. *Author*: R. Moxon, Cranfield University; *Source*: United Kingdom Civil Aviation Authority, Southend Airport (2014).

In designing the airport and planning its development, the transportation to and from the airport has to be considered as an integrated system, including transportation by passenger cars, taxis, car rentals, buses, coaches and railway transport, not forgetting also the use of marine transport and helicopters where appropriate.

346   Airport Design and Operation

14.2. SELECTION OF THE ACCESS MODES

A gradual increase in the share of the available high occupancy modes of transportation normally occurs as the airport grows, and this change should be encouraged on the grounds of environmental impact and balanced capacity. Car trips will always predominate at small airports. Growth in the share of high occupancy vehicles as the traffic increases will normally take the form of public (mass) road transportation at moderately sized airports, while high-capacity rail transportation should have a substantial share in the large airports. Public transportation should have some role by the time an airport reaches two million passengers per year (mppa), including airport employees who do not always have access to a car. The mix between public and private transport significantly depends on the region, the airport type, the standard of living, and is influenced by the airport size and distance of the airport from the city. Bigger airports are able to invest more resources and are also usually located near large cities which are able to invest into public transportation (Figure 14.2). It is, however, difficult to achieve the 50% share to public transport that is often stated as an environmental goal. Definitely airports need a certain size to support public transportation services; on the other hand size alone does not explain high ridership. Usage of public transport also depends on the airport distance from the city centre, quality of public transport and geographic specifics (remote airports, airports located on islands).

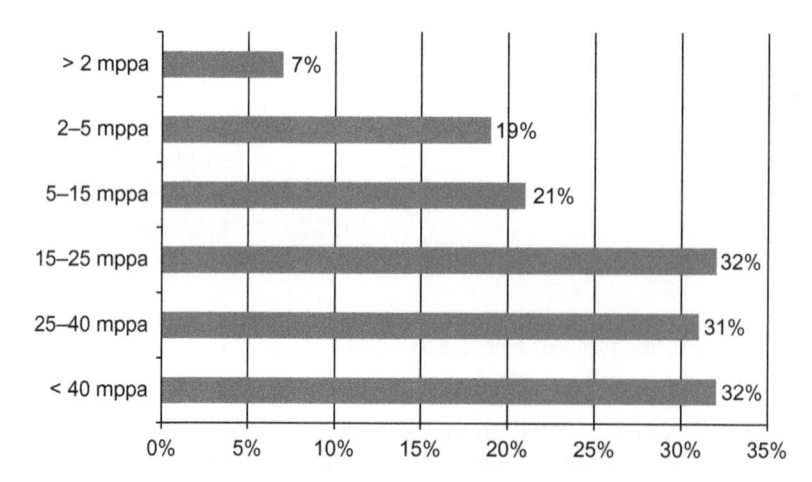

Airport size	Public transport share
> 2 mppa	7%
2–5 mppa	19%
5–15 mppa	21%
15–25 mppa	32%
25–40 mppa	31%
< 40 mppa	32%

Figure 14.2: Use of public transport increases with the airport size. *Source*: ACI, DKMA: Ground Transportation (2012).

Many factors make the car the preferred mode of access for passengers, including the low operational cost, the convenience for carrying bags and family groups, and the instant availability. It is therefore not easy to get people to move to high occupancy modes even if they are competitively priced, frequent, reliable and form part of a transport network that allows access to the complete catchment area. The airport administration should support the aim ensuring the passenger can get to and from the airport quickly, easily and simply so that they do not miss their flight. If the management does not do this, some potential air passengers will decide to use other airports or other modes of transport. From the management point of view, a consideration will often be the large revenues which come from the car parks. Surface transport must be considered as a part of the 'product' of the airport. Managers have to work to fulfil the airport's access needs by closely cooperating and encouraging the local authorities and transport operators to respond by investing in roads and operating services.

Theoretically it would be possible to ensure the change in the share of individual modes by making the mass transportation more attractive or by making the private transportation less attractive by imposing road tolls, high parking fees etc. However, individual groups of people accessing the airport, who will mostly be passengers and those accompanying them, employees or visitors to the airport, will all rank those factors differently. The following factors can be identified as affecting the selection of the mode of transport:

- The availability of the mode (ownership/availability of private car usage/ transportation on one end or both ends of journey, public transport modes availability)
- The distance of the airport from the home or workplace
- Duration of the individual segments of the transportation process (waiting, time to access the mode of transport, transportation time, time from the mode/ parking to the airport check-in)
- Frequency in the case of public transport
- Standard of comfort and quality of transport, which includes ease of use, number and quality of seats, handling of baggage, number and difficulty of transfers en route, possibility of secure parking
- Reliability of transport, including transfer reliability (for public modes), (how much time reserve is it necessary to allow in order not to miss the flight?)
- Total generalised cost of transportation (parking fees, value of time etc. must be included as well as fares or marginal cost of using private transport)
- Other factors such as personal safety, privacy, flexibility.

Research shows that the decisive factors for passengers are price, transportation time, number of changes and baggage handling. For airport employees particular concerns are price, flexibility, availability and personal safety for shift patterns out of normal hours.

14.3. CATEGORIES OF SURFACE TRANSPORT USERS

Airport access is needed not only by the airline passengers but also by other groups of users. As a rule of thumb the number of employees who daily commute to the airport represents one quarter to one half of the daily number of airline passengers at medium size and large airports. In addition there are the accompanying persons (meeters and greeters) who in some countries outnumber the passengers, and the visitors to the airport, this latter category amounting to 5% or 10% of the total. The roads and public transport networks also have to cater for the needs of the local non-airport traffic.

It is necessary to emphasise that there are also other factors which distinguish the airline passengers from other groups of surface transport users that have often a decisive effect on the selection of the kind of transport provided, and these can be different in each case. It is, for instance, necessary for business people to have fast reliable transport, leading to the development of the Heathrow Express. Therefore it is necessary to distinguish categories of passengers using surface transport, characterised by the factors such as:

- the reason of the trip to the airport,
- type of flight (scheduled — charter, short haul — long-distance flight),
- duration of the stay and
- social and economic factors (income, age, occupation, size of household, car ownership).

Each category will have different requirements. For example for the air transport passenger the prime concern is to get to the airport in time, so as not to miss the flight. They might not know the public transport system well which limits their flexibility.

On the other hand employees, because they commute every day, are much more price sensitive. In their case the cumulative cost of transportation must be considered in the family budget. A problem of public transport for airport employees is that airports are located out of the metropolis area and in particular in case of small and midsize airports they mismatch with the main streams of public transportation. Most airports have 24 hours operation, and many work shifts do not coincide with typical schedules. There is also a large variety of constraints and regulations with respect to the shift timing, parking reimbursement, overtime work etc.

14.4. ACCESS AND TERMINAL OPERATIONS

The characteristics of the various categories that determine access requirements also affect the requirements for the terminal. Those who want a fast and reliable access trip also want to move quickly through the terminal with as few surprises and delays as possible. They will perhaps only expect to check-in at the departure gate 10 minutes before the flight closes, having been assigned their seat when making the online reservation. In contrast, those with no time pressure will accept a relatively slow coach journey to the airport and a long wait in the terminal in exchange for the assurance of not missing their flight. For example the tour operators encourage these passengers to arrive early, and schedule their coaches similarly. It will be particularly important to them if they are travelling on inclusive tour or on discount fares not to miss their flight, since their ticket will not be interchangeable. They may be in the terminal for 3 hours or more, and will require many more facilities for food and relaxation.

A similar disparity exists between long and short-haul passengers. The airline will need the long-haul passengers to be checked in early because there are more formalities to be completed and there is more baggage to be checked in. Also, since fewer airports offer long-haul flights, the average access time will be greater and the passengers will need more time to organise themselves before flying. The proportion of leisure travel on long haul is high, so a high percentage of the passengers will have discounted non-transferable tickets and will have the same desire not to miss their flight as the inclusive tour passengers. So, depending on the type of passenger, each will have their own punctuality targets for the access trip and will adjust their start time to meet it. If they are inexperienced travellers, they may well experiment with the access trip in order to get some idea of how long it will take and how to navigate along it.

The actual dwell time in the terminal depends on the actual time of the complete access trip, including finding a parking space and transferring to the terminal, and the planned margin of safety together with the reliability of making the trip in a given time. The actual time for a trip will depend on the distance, the mode used and the traffic conditions. For each mode, it is likely to be normally distributed with a variance that is strongly related to the average travel time. If passengers have a good sense of the variance, they will leave a margin of safety that will get them to check-in at the prescribed time with a given level of confidence, say 99%. Those with longer journey times will then, on average, spend longer in the terminal even though holding the same ticket type for the same flight as passengers with shorter access times.

Similarly, access trips on modes or routes with low reliability will have a high variance. This will tend to be known to passengers, and they will again adjust their behaviour to achieve the target punctuality. Thus passengers on low reliability access modes or routes will again spend longer in the terminal, since they will, on average, have arrived earlier. If passengers do behave in this way, and if they value their time, they will switch to a more reliable mode according to weather conditions and time of day.

The terminal capacity is influenced by the number of people and time spent in the terminal; therefore it also depends on the access ground transportation characteristics. However, this has also other implications. For example public transport reliability improvements will cause passengers to arrive at the terminal later, so their dwell times will be shorter which could negatively influence retail sales. The terminal capacity could be also influenced by a change of operator type, for example traditional carrier will be replaced by low-cost airline, even though the numbers of passenger will stay the same, because of different passenger access travel behaviour.

In any case, the terminal should be designed in terms of space and facilities to meet their needs. The correct provision of space will depend on how and where the passengers wish to spend the slack time. Shopping has now become an accepted use of slack time, and in some cases is a deliberately planned activity. However, the processing may also be affected by the knock-on effects of access considerations, particularly if the design uses simulation of representative peak aircraft schedules and distributes the passenger behaviour around them.

14.5. ACCESS MODES

14.5.1. Passenger Car

The level of car availability, whether for the drop-off mode or the drive-and-park mode, is considerably higher at the home end than at the other end of the journey. The high standard of living in developed countries means that cars are widely used for transport, and particularly by air passengers who tend to be more wealthy than average. The car may be owned or rented, the latter being used by passengers whose trip started at the other end of the route and whose local trip origin is a workplace or hotel. Passengers prefer cars because of their flexibility and comfort. The additional factor for the business person is the short door-to-door time. The less well-off also prefer cars because they are perceived to have the lowest operational cost. The real operational cost of transportation by one's own car can be relatively low, but the airport parking fees can be high. In some cases tour operators or hotels will reimburse the parking fee to the passenger, or they may provide courtesy coach transport from dedicated car parks.

The high share of private car transport causes growing problems to airport administrations. It is necessary to build new parking places and access roads, and to increase the size of the drop-off zones in front of the terminal building. Constructing multi-level car parks in the vicinity of terminal buildings is costly and takes up space that could be better used for commercial purposes. If the parking places are provided in the remote areas of the airport, in most cases it is necessary to provide courtesy transportation to the terminal building for the passengers, which increases the operational costs. On the other hand, revenues from parking places represent an important part of the airport revenues, most particularly in most of the developed world. At the biggest airports the revenues from parking can be similar to the revenues from landing fees. Most of the airports solve the dilemma of the distance of the car park from the terminal building by dividing the car parks between short-term high price spaces in the vicinity of the terminal and the long-term remote car parks with a lower charge per day. The share of the number of places in the long-term and short-term car parks depends on the space available at the airport. A rough estimate of 1000 spaces per million passengers per year may be used, but the total and the split between long and short-term parking will vary depending on the characteristics of the traffic in terms of short to long-haul flights and the business to leisure passenger.

Figure 14.3: Kerb in front of the departure concourse for passenger drop-of. *Photo*: J. Hladký.

Space in the central part of Heathrow airport in the vicinity of the Terminals 1, 2 and 3 is very limited and can only be accessed through a tunnel, and the rate per day in the short-term car parks is approximately 2.2–2.6 times higher than in the long-term car parks. There are different car parks, with different services and pricing schemes depending on flight date, time and method of booking (e.g. internet) at Heathrow airport. The older airports with little space for further expansion have the biggest problems with building adequate access roads.

352 Airport Design and Operation

The second important parameter of airports with a high proportion of passenger cars for access is the length of the kerb in front of the departure concourse (Figure 14.3). Particularly at airports where the majority of passengers is driven there by another person, called 'kiss and ride', the length of the kerb in front of the departure concourse can be a critical point at peak times as the same central terminal area continues to handle the extra traffic by serving more aircraft gates. Airports try to limit the share of 'kiss and ride' transport not only because of the traffic load and road congestions but also for environmental reasons. The 'kiss and ride' traffic results in doubling the number of car trips and therefore about double the amount of emissions (Figure 14.4). Some airports introduced charges to discourage loading and unloading passengers by private cars. For example Luton airport charge for priority set down — drop-off and pick-up is £2 up to 15 minutes. However, it is also possible to drop-off and pick-up passengers for free at Mid Term Car Park and they could take a free shuttle bus to the terminal. In general, the 'kiss and ride' is less attractive the longer the ground access trip but its attractiveness also drops with public transport improvements.

Figure 14.4: 'Kiss and ride' transport doubles the number of car trips to the airport. *Photo*: A. Kazda.

The rule of thumb for initial design is 100 m of kerb per million passengers per year. Operational measures can be adopted, such as speeding up the unloading of passengers in front of the terminal with the aid of traffic wardens. It is also possible to organise the traffic in front of the terminal building in two or more lanes. However this increases the possibility of accidents or incidents when passengers are crossing the lanes to the terminal. Multi-level short-term parking, with direct access to the terminal from the garage, is perhaps the most effective solution, other than providing remote drop-off points and transfers by bus or rail. It is most important for the private car user that clearly visible signs should be positioned on the roads and on the terminal kerbside areas well in advance of desired destinations. The terminals, airlines and car parks should be colour-coded and easy to see at night. The information should be concise, quickly identifiable and easily understood.

A large percentage of private vehicle access trips is not only likely to lead to congestion; it also has a negative impact on the environment. At most airports the ground trips associated with the airport generate a greater share of air pollution than the aircraft movements. However, at small airports access almost always depends on private transport. These airports' flights consist in the main of short and medium distance routes by regional aircraft providing spoke connections to hub airports or direct regional point-to-point transportation. The surface transport must support this system by providing highly reliable and quick transport 'from door to door'. In these lower density situations this can only realistically be provided by the private car, so it is not surprising that it represents the major means of access for all segments of users. To support the local community, free parking is sometimes offered at small airports. Mass transportation to small airports is usually not economically feasible.

14.5.2. Taxi

The taxi has similar requirements for the design of the kerbside in front of the terminal as the private car. However, its kerb occupation time and parking requirements are considerably smaller, the latter consisting of a pool for immediate service and a longer term zone where taxis can wait their turn to be called forward to guarantee a steady flow of taxis to the arrivals kerb. It would be possible to consider the taxi as an almost ideal means of transport for an airline passenger. It is reliable, comfortable, ensures the direct transportation between the trip origin directly to the kerb in front of the terminal building and substantially reduces luggage handling problems. In addition, a taxi eliminates the problem of unfamiliarity with the public transportation system which is usually much higher at the non-home end of the trip. As per ACI Ground Transportation Report (2012) taxi is used on average by 20% of air passengers. However, there are large variations depending on the airport and it depends not only on type of passenger (business or leisure) and type of airline operator (LCC or traditional) but also on size of the airport, its distance from the city centre and the availability of other modes of transport.

Costs are sensitive issues for taxi usage. Costs of transport, and this includes also taxi, are usually linked to the costs of living. According to ACI Ground Transportation Report (2012) the cheapest taxis are in Africa and Middle East ($23 on average) and the most expensive in Europe ($53 on average). Obviously the price depends on the length of journey and for airports located far from the final destination may actually be very high. In general, the longer the ground access trip, the less competitive is the taxi. The high cost relative to using a private car is partly mitigated by avoiding the need to park, and, relative to public transport by the ability to share the cost among a group of passengers.

Although the airport administration is not directly responsible for the operation of taxis, the bad impression formed from low-quality or poor value taxi service has an impact on the overall image of the airport. Therefore many airport administrations regulate taxi fare policies and lay down criteria for the acceptable operation of taxis. Some airports display information on fixed taxi price for downtown or other destinations in the airport arrival hall. Important quality control factors are:

- to ensure the number of taxis meets the demand, particularly at night and in the time when the mass transportation is not available
- to ensure the high quality and fair price of services
- to deal with security issues in some countries and to discourage unofficial operators

Shortage of taxis can occur particularly at smaller airports when two flights are arriving close together. The problems of shortage of taxis and quality of service are mutually interdependent. This is mostly solved by awarding licences to serve the airport for a limited period of time. At most airports, any vehicle is allowed to drop-off passengers at the airport, but only authorised or permitted vehicles are allowed to pick-up passengers. The majority of airports issue licences against a fee and thus control which taxis are allowed to pick-up passengers at their airport. Sometimes the taxi licences are managed by the city. Other airport administrations issue licences free of charge on the basis of regular evaluation of the quality of service.

A disadvantage of taxis, and also of private cars, is access to the airport in peak times, where the journey could last longer than by coach which may use special priority lanes or by rail transport or underground if available (e.g. Charles de Gaulle). On the other hand at some cities taxis can use public transport fast lanes (e.g. Dublin).

At some airports, in particular in the United States and Europe, there are limousines operated as an alternative to taxis and providing upmarket competition. Limousines are specialized vehicles licensed for operation by pre-booking. They are usually the most expensive form of automobile ground transportation.

14.5.3. Minibus

In some countries transportation between the airport and the hotels or other main traffic generators in the town is provided by minibuses. The minibus, or van, may run according to a timetable or wait until it has been filled. It is usually cheaper than a taxi. Some hotels provide courtesy minibuses free of charge to and from the airport. Minibuses represent a blend of public and private transport. They offer some of the door-to-door advantages of a taxi and, at the same time, more security, comfort, flexibility and speed than a scheduled bus service. They have a high daily utilisation, though they often depart with few seats occupied because, in order to be attractive, they have to be frequent. This can result in a large environmental impact per passenger and congestion at the kerb. This is despite the fact that they take only slightly more space at the kerb than a taxi and substantially less than a bus. An additional problem with those minibuses operated by private companies, as opposed to sponsors like hotels, is that competition encourages them to occupy the kerb as long as possible. This needs strict policing.

14.5.4. Bus

Depending on the type of airport and availability of other modes of transport, in particular rail, bus could acquire a significant share of ground transportation at airports. This is particularly likely in touristic areas for inclusive tour travellers or in big Asian cities such as Nanjing or Incheon but also to get airport employees to and from work. Several types of bus service to airports can be distinguished.

Normal scheduled services of the local metropolitan authority, which operate as part of the regular bus system, are used more by employees than passengers because of their frequent stops and poor provision for luggage. However, in countries where the use of public transport is high municipal buses may be more used for airport access by passengers who are locally resident and are familiar with the route structure. Non-residents can find bus systems quite complex or confusing and may therefore prefer other modes. However, this is changing rapidly due to the availability of Internet based travel information and booking systems. Each airport website has a detailed description of the ground transportation services and there is usually a link to the transport provider's website for timetables, ticket booking and purchase but also smart phones real-time information (see also Section 12.8).

356   Airport Design and Operation

Airport express service, coaches dedicated to airport passengers are more costly, more comfortable, usually offer some amenities such as Wi-Fi, and they are faster than the scheduled buses. They usually link the airport with the municipal rail or bus terminals and the major hotels with only a few stops before the final destination. They work well for visitors, but are less convenient for locally based travellers who are more likely to want to start or finish their trip from their homes in the suburbs.

Figure 14.5:   East Midlands airport is connected to principal cities in the region by a bus system. *Photo*: A. Kazda.

Longer distance scheduled coach services compete with rail, taxi, private car or even air access to hub airports from more distant cities whose own airport does not offer the same range of air services. They overcome the need for the change of mode that is required on most rail systems to complete an airport access trip, but tend to be slow, infrequent and, as with the shuttle coaches, they may suffer from delays due to road congestion. They are mostly run commercially by a private company, but sometimes an airline operates the service primarily for its own passengers and may even check bags through to the trip destination. Some low-cost carriers flying from secondary airports use this system of buses to increase their load factors. Coaches for passengers on inclusive tour holidays, particularly those using charter flights, are organised by the tour operators. Some operators provide a pick-up service at the beginning of the holiday, but the main use is to convey passengers from the holiday airport to their accommodation. They are usually organised into a hub and spoke operation, where a series of flights arrives close together, and the passengers are assigned to whichever one of a bank of waiting coaches is to call at the required hotel. The coach parks for this type of operation at the major holiday airports like that at Palma in Majorca have spaces for hundreds of coaches. The theoretically high efficiency of this process is often not achieved because of flight delays.

Another example of interconnection of air and bus transport is the system created by National Express Group plc. They owned East Midlands airport in Great Britain (until it was bought by the Manchester Airport Group) and also the National Express interurban coach company. The airport had no direct connection to rail transport, so a network of coach services was created with good and reliable connections, which is an important marketing tool in the competition with the nearby airport in Birmingham (Figure 14.5).

Both the National Express and the tour operators' models for interlinking air and coach services are examples of private initiatives in line with policies in the US and Europe to provide 'joined-up', or 'integrated' public transport and thus create a more sustainable transport system.

14.5.5. Railway Transport

The recent political push for a sustainable transport system, combined with road congestion and frequent delays, has caused a renaissance of rail public transport in accessing airports. Many airports are supporting increased use of rail transportation, both to serve the nearest cities and also to increase their catchment, to the extent that airport expansion has been made conditional on achieving targets of up to 50% of passengers using public transport. The United Kingdom requires airports to compile access plans which will increase the public transport share. Good examples of the various rail access options already exist, for example:

- metros at Madrid, Stuttgart, London Heathrow
- high-speed networks at Frankfurt, Charles de Gaulle, Copenhagen, Lyon, Madrid
- light rail at Baltimore/Washington, Bremen, Vancouver
- high-speed dedicated at Heathrow, Oslo, Stockholm, Hong Kong, Kuala Lumpur
- rail connections at Geneva, Gatwick, Vienna, Prestwick, Fiumicino, Barcelona, Brussels, Munich, Birmingham United Kingdom.

358   Airport Design and Operation

It has to be emphasised that rail transportation is fundamentally a high-density mode. Its most effective use is to cater for high levels of traffic. Theoretically, rail transport allows a good connection to all parts of the catchment area and can substitute for other kinds of transport (Figure 14.6). The decisive factors are the extent of the network, and the feasibility and reliability of the connections. Dedicated rail access, of which there are many examples worldwide, usually connects the airport only with the town centre, which is convenient for visitors to the city. However, if the airport is used mainly for originating journeys, most of the passengers tend to start their trip from home and tend to have a car available. They would most likely either drive to the airport or be dropped off.

Figure 14.6:   High-speed rail connection can increase the airport catchment area.
Photo: J. González Manso.

Dedicated rail transport offers direct service between the airport and city centre. It is usually easier to use it as it is well advertised and described. It usually runs on the same rail track as regular trains and it is faster, but significantly more expensive than the latter at airports offering both services. This is the case when, for example comparing rail services at Vienna (CAT and S-bahn); Kuala Lumpur (KLIA Ekspres and KLIA Transit); Gatwick (Gatwick Express and Southern trains services). Dedicated rail is usually not convenient for business passengers as their destination is not the station in the city centre, but a hotel or a company in the catchment area. This would require them to catch a taxi from the city train station, so they usually prefer to take a taxi directly from the airport to avoid mode change. The rail access share depends also on the location of the station at the airport. It should be located either directly in or under the terminal building or it should have an easy connection to it, preferably a dedicated one rather than an urban bus that is exposed to the difficulties of the local road network. If a link is necessary, it adds another mode change to the trip. On the rail part of the airport access journey, the need to make one interchange, as at New York JFK, or Barcelona, can reduce demand to only 40% of the direct demand, a second interchange reducing it still further to 30% (Lythgoe and Wordman, 2002). If a spur can be built to a nearby main line, as has been done very effectively at Amsterdam Schiphol, Köln-Bonn, Geneva and Zürich, it may be possible to avoid an interchange and also expand the catchment area of the airport.

The decision to connect the town and the airport by rail transport depends not only on several factors, particularly:

✈ volume of airport passengers per a year

✈ possibilities of connection to the existing transport infrastructure

but also:

✈ split of traffic between scheduled, charter, business and leisure passengers.

There are several options for providing the link that should be compared. Building a special line is, depending on its length, economically justified only for large airports with a minimum of 7 mppa for a spur line and normally more than 10 mppa for a dedicated link. Provision of a rail link is now often a necessary condition of further growth of airports of this size. Most of the tracks have originally not been designed for connection to the airport and therefore the plans to do so are often compromised by the sunk investment and existing land uses. Conventional heavy railway systems are typically characterised by higher speed, lower frequencies and longer distances between the stations. Rail transport has been available at Gatwick airport ever since it opened and its high proportion of total surface transport results from the convenient and fast connection to London and the underground network, and to other parts of the rail network. Munich airport has been connected by means of two rapid transit links. The airport administration has set an ambitious target of achieving a 40% proportion of rail from the total surface transport. Frankfurt airport offers long-distance and regional train connections but also uses rail for cargo access, driven by the ban on lorries in Germany at the weekends. There is the threat of a similar ban, but only for older lorries that cannot meet the EU emission standards, in the London area.

The underground or metro systems are characterised by short distances between frequent stops. The underground with short headway times is mostly suitable for the employees and for people with business at the airport. Its advantages are high frequency and low cost. The disadvantages are that it is relatively slow and usually not suitable for the transportation of larger pieces of baggage. For instance the journey by underground from the centre of London on the Piccadilly Line to Heathrow takes 50 minutes on average and has 18 stops. Heathrow has three London Underground stations — one for Terminals 1, 2 and 3, one each at Terminals 4 and 5. The special heavy rail Heathrow Express connects the airport with Paddington railway station only. Non-stop trains run every 15 minutes. The journey to Terminals 1, 2 and 3 in the central area of the airport takes 15 minutes and a few minutes longer to Terminals 4 and 5.

The Oslo Airport Express Train connecting the centre of the capital with the Gardemoen airport has gained a very high market share. The Airport Express Train is responsible for approximately 32% of airport ground transportation (2012). One reason for this is the high proportion of passengers who originate from the city but also the relatively long distance from the city. Where the market is more geographically diverse, it is more difficult to capture a good share of the market.

> The first rail link to Gatwick on the Victoria to Brighton line was opened in September 1935 with a direct connection to the Beehive terminal (opened in 1936). The express service started in 1984. Gatwick Express charges £31.05 return between Gatwick and Victoria and takes 30 minutes. Southern Rail's stopping service costs from £30.60 return and takes 32–51 minutes. In January 2006, Gatwick Express was rated Britain's top train company by the government-funded National Passenger Survey, with 93% of passengers saying they were satisfied. The number of Gatwick Express passengers has increased by 30% since 1995 but its share fell more recently to 24% of air passengers arriving at Gatwick by rail in 2003. The number of Gatwick Express passengers fell because the rise of budget airlines at the Gatwick airport was attracting more 'price-sensitive passengers' who preferred to catch the slower Southern services to Gatwick because they were much cheaper.

The dedicated rail links are often thought to need seamless check-in downtown to make them attractive. However, BAA closed the check-in facility at Paddington station in the aftermath of the 2001 security problem on the grounds of cost and the check-in at Madrid Nuevos Ministerios station was closed in 2006 for security reasons. Also, the chance of lost bags doubles with downtown check-in. Another question is the relation between air transport and long distance high-speed railway transport. Some airport administrations consider rail to be an important competitor to air transport on short and medium distance routes. The construction of high-speed lines and an increase of the quality of railway transport should, according to the opinion of this group, support this view (Figure 14.7). New strengthened security measures and resulting delays at the airports can motivate more passengers to use high-speed rail on medium distances routes instead of air transport.

Figure 14.7: High-speed railway station situated at the airport. *Courtesy*: Lyon-Saint Exupéry Airport; *Photo*: M. Renzi.

Progress towards rail links has been slow in North America because of the car culture and the environmental impacts that would be incurred by building new lines through populated areas. Atlanta has the MARTA system to Hartsfield-Jackson airport, New York JFK links into the metro and heavy rail networks via its Air Train system and San Francisco has a BART link to Oakland airport via a shuttle bus and the BART system has been extended to San Francisco International airport. However, several airports have plans for rail access, driven by legislation to reduce emissions and by road congestion, including Chicago, Denver and Toronto (under construction is to be completed in time for the 2015 Pan American Games).

Others look on rail more as being an important partner of the airports, which allows a substantial reduction in the time of transportation to the airport and thereby increasing the catchment area. The stops of the high-speed railway cannot be designed so close together as with conventional rail so they can operate only in areas with high population density. Stations for them will therefore be built only in the largest airports. There are relatively few destinations in continental Europe to which it would be practical to travel and return by rail in a day. At the same time these airports have the biggest problems with the number of runway slots available, particularly in the peak periods. In these cases high-speed rail can really replace some regional flights and the capacity released can be used for long-distance flights. But even large airports will have to contribute to the investment in the high-speed railway, while it needs a convenient location close to the line of the rail track to make a direct connection to a smaller airport economically feasible. These smaller airports can be connected to the conventional network of fast trains. Those medium size airports with hub operations might absorb the regional flights displaced from the largest hubs but at the moment this does not really happen. From some smaller airports in Great Britain it is more advantageous to fly to New York through Amsterdam than through Heathrow. In some cases the airports themselves will be affected. After the Euro-tunnel opened under the English Channel, it became easier to access Brussels airport by rail from some parts of south-eastern England than London Heathrow.

Relative to rail, air transport has the advantage of flexibility, which is important particularly for small airports and small airlines. Changing routes and frequencies is much easier for air transport than for railways. However, it is difficult to justify the use of scarce runway slots at congested airports for these very short-haul routes.

The idea of combining air and rail services is attractive in some countries. There is a long-standing co-operation between Swiss and Swiss Railways. In 1982, Lufthansa and Deutsche Bundesbahn also initiated cooperation. Today the Rail&Fly programme allows the use of rail from more than 5,600 Deutsche Bahn train stations and tickets can be purchased from anywhere in Germany for as little as €29 in conjunction with an international Lufthansa flight ticket.

It is obvious that a direct rail connection substantially increases the airport's attraction and, in the near future, airports will be judged on whether they have such connection or not. Those without a station directly on the airport will risk being relegated to a minor role in the system.

14.5.6. Unconventional Means of Transport

Unconventional means of transport for transportation between the town and the airport may include different types of elevated railway systems, monorail systems, magnetic levitation (MAGLEV) trains, air-cushion vehicles, and helicopters, that is are all those modes which are not conventional.

Hundreds of schemes were proposed in the 1970s, but very few have been implemented. The main reasons are the high costs of building the infrastructure and for the operation of the facility, the slow development of the technology and the need for planning approval in the face of objections to the land use and visual intrusion.

Another problem is that most of these systems are not fully integrated in the surface transport net which usually means an increase of the total transportation costs. However, there are a few examples of successful unconventional lines connecting airports with other transport modes stations.

Figure 14.8: Orlyval transportation system. *Photo*: M. Filo.

Perhaps the best known is the Tokyo Monorail which connects Tokyo Haneda International Airport and Hamamatsucho Station in the town centre. The line was opened in 1964 to cope with the increased traffic during the 1964 Summer Olympics Games. The elevated line is 17.8 km long and it takes 22 minutes to reach the Haneda Airport Terminal 2 from the town centre. Currently, the line serves nine stations. It is often cited as the busiest and most profitable monorail line in the world.

Orlyval shuttle is a convenient connection in between Paris — Orly airport and RER station Antony on line B. The Orlyval was opened in October 1991 (Figure 14.8). The fully automatic line is unfortunately not integrated into the railway/metro network.

The fastest railway is the Transrapid system operating since November 2003 between the Pudong Shanghai International Airport and the Shanghai Lujiazui financial district (China). The 'train' covers 30 km, reaching in regular service a speed of 432 km/h during a journey time of 6 or 7 minutes. Maglev trains require special rail track where carriages are levitated, guided and propelled by electromagnetic forces. The construction of the infrastructure is very expensive.

Helicopters avoid most of the planning problems but generate too much noise at the city centre heliports and under the routes, and it is difficult to find slots in the air traffic at the airport unless special non-conflicting routes can be developed. The routes tend to be at low level, so there is noise nuisance under the flight path and people living there feel that they are being spied upon on their own property. Similar objections caused the closure of the Heathrow/Gatwick transfer link.

There are a few airports located on islands as Male or Hong Kong which use boats for public transport in large scale.

14.6. Airport Ground Access Improvements

The expected passenger growth in the 2030+ time horizon is tremendous and it is not only necessary to fly them, but also to get all passengers, meeters and greeters and staff to and from the airports. Air transport capacity enhancement usually considers air transport as an isolated and independent transportation system and the problems of airport ground access and its effect on airport terminals operation are underestimated.

There are traditional ground transportation management strategies for how the share of public transport could be increased. It is possible to encourage the use of public transportation by reducing or not charging fees; to support improvements of air quality by promoting the use of vehicles using alternative fuels or hybrid vehicles, or by requiring the consolidation of courtesy vehicle services; encouraging use of public transportation by providing fast lanes for buses etc.

Time is extremely important in air transportation. However, clock time is different from perceived time. Everyone knows that an hour spent with a loved one has the potential to pass quickly but in a traffic jam, when we are trying to catch our plane, the delay can seem to last much longer.[1] Many times we pay for taking a longer but

---

1. *Note*: Clock time is measured objectively, while perceived (or cognitive) time reflects users' experience. Paid travel time costs should be calculated based on clock time but personal travel time costs should be calculated based on perceived time.

faster route and we are willing to pay more for safe and fast transport to avoid a car journey at night or in a blizzard.

Conventional evaluation practices tend to ignore transport qualitative factors, assigning the same time value regardless of travel conditions, and so they undervalue service improvements that increase comfort and convenience. Discretionary passengers (people who have the option of driving) tend to be particularly sensitive to service quality. Increase of public transport quality often increases transit ridership and reduces automobile traffic. Numerous studies have quantified and monetized (measured in monetary units) travel time costs by evaluating how travellers respond when faced with a trade-off between time and money, for example when offered an option to pay extra for a faster trip. All factors of transport quality and comfort (level of service), waiting conditions, crowding, transfer, reliability, frequency, safety, security, real-time travel information, speed and even aesthetics could be evaluated as they affect the passenger perception of a particular mode of transport.

Transport qualitative improvements together with other comprehensive measures could influence decisions of some passengers to use public transport increasingly.

15

VISUAL AIDS FOR NAVIGATION

Tony Kazda and Bob Caves

15.1. MARKINGS

15.1.1. Markings Requirements

In the beginning of aviation, flights were performed only under Visual Meteorological Conditions (VMC). At low speeds, the natural view of the terrain provided the pilot of an aircraft with sufficient information for approach and landing under good visibility. As speeds increased, it was necessary to enhance the natural perception of the aerodrome by markings, and to provide the pilot of an aircraft with additional information by marking the runway and other movement areas and to standardise the markings.

The terrain picture is enhanced by picture texture with objects, trees and other details. The picture texture is important particularly in the last phase of the final approach, during the actual landing and during take-off, when the pilot's attention is concentrated into a narrow visual angle in the direction of the runway and most of the texture is acquired by peripheral vision.

In the initial and middle approach phases and during a circuit, when the aircraft is further from the aerodrome, the picture structure is important from the viewpoint of providing visual information. It is formed by the horizon and by natural features such as hills, valleys, rivers, fields and the like. The pilot of an aircraft may determine the aerodrome position from the picture structure.

Simple marking of ground movement areas is sufficient for conditions of impaired visibility by day, and in good meteorological conditions by night. Similarly, it is convenient to make use of the markings also for providing the pilot of an aircraft with additional information during approach and landing, such as runway width and length. Natural properties of a picture seen by day in good visibility are complemented with these markings to emphasise the crucial visual information on the location of the aerodrome and its runways. Only in extraordinary cases, when the aerodrome is located in a featureless landscape like desert, is it necessary to supplement the picture structure and texture more than this.

The requirements for visual information during a take-off are limited to emphasising the aircraft's position relative to the centre line and the start-of-roll end of the runway, and the ability to determine the speed and location of the aircraft relative to the take-off end of the runway during the take-off run, the rejection of a take-off and after landing. During the final phase of an approach, the most important requirement is to be able to assess the runway width and length, and the runway length remaining after crossing the threshold.

Directional guidance on the apron and on the taxiway system is given to the aircraft to ensure that it stays on the taxiway and that it remains at a suitable distance from obstacles. The information necessary to facilitate the pilot's orientation at taxiways and on the apron is again emphasised with markings.

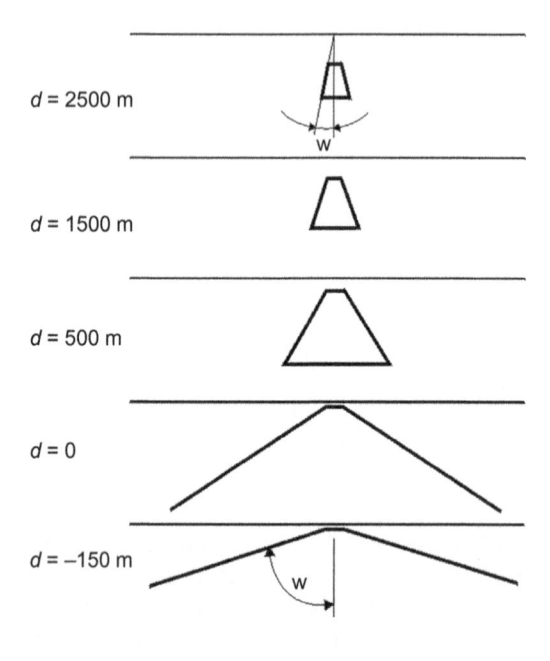

Figure 15.1: Change of perspective angle on approach with descent angle 2°30'. Source: Čihař (1973).

The runway perspective, as indicated by the converging angle of its edges, is particularly important in the final approach phase. The magnitude of the perspective angle is expressed by the relation (Figure 15.1):

$$tg\omega = \frac{d}{h}$$

where d is runway half-width and h is pilot's eye level above the runway.

During an approach, the perspective angle increases, and the pilot loses accurate information about the runway direction. Therefore on wider runways, it is necessary to mark not only the runway centre line, but to enhance also the runway edges by marking in order to improve the directional guidance of the aircraft.

The area of the runway system markings exceeds several thousands and even tens of thousands of square metres. The choice of a material for markings should aim to minimise the total costs. The specifications of coating compositions for aerodrome markings differ from those of coating compositions for road markings. Especially at aerodromes with moderate traffic, or side-strip markings, the coating will be more affected by meteorological effects or the accumulation of dirt than by mechanical wearing. The paint for taxiway and apron markings should be colourfast. There are several types of paints developed especially for aerodrome markings. The durability of cheaper paints, based usually on oil (alkyd) or water based emulsions, is approximately three years, depending on the airport traffic volume. Spraying machines may be used for these types of markings. A drying time of 30 minutes is usually acceptable for airport operation requirements.

Two-component paints are usually more expensive. They have a durability of 10–20 years, but, in general, it is more difficult to mechanise the marking. It is also necessary to consider that marking in a runway touchdown zone at aerodromes with a high traffic density will quickly become coated with rubber, and therefore it is not efficient to apply more expensive materials with longer durability.

The basic requirements for the markings are the colour, opaqueness, reflectance and roughness of the paint surface. The paint colours are determined by Annex 14, Aerodromes, Volume I, *Aerodrome Design and Operations*, Appendix 1 in the specified range of the colour spectrum. The reflectance of the marking surface may be improved by adding retro-reflective beads (glass spheres) into the paint. Glass spheres with refractive index of 1.9 or greater with size gradation between 0.4 and 1.3 mm diameter have been found to give the best results for aerodrome marking. The runway surface markings should not show any significantly different braking effect than that of the remaining runway surface.

If the antiskid characteristics are not sufficient, aluminium oxide and angular glass in sizes which will pass through a 150 µm sieve and less than 5% will be retained by sieves with 45 µm mesh were found to be effective. Adding the reflective component into the paint also improves the roughness of the marking surface.

15.1.2. Marking Types

The requirements for the physical characteristics of particular markings are given in Annex 14, Volume I, Chapter 5. The extent of the runway markings depends on the runway type and reference code. They apply to the following markings:

370   Airport Design and Operation

Runway markings (Figures 15.2–15.4):

✈ runway designation markings
✈ runway centre line markings
✈ threshold markings
✈ aiming point markings
✈ touch-down zone markings
✈ runway edge markings.

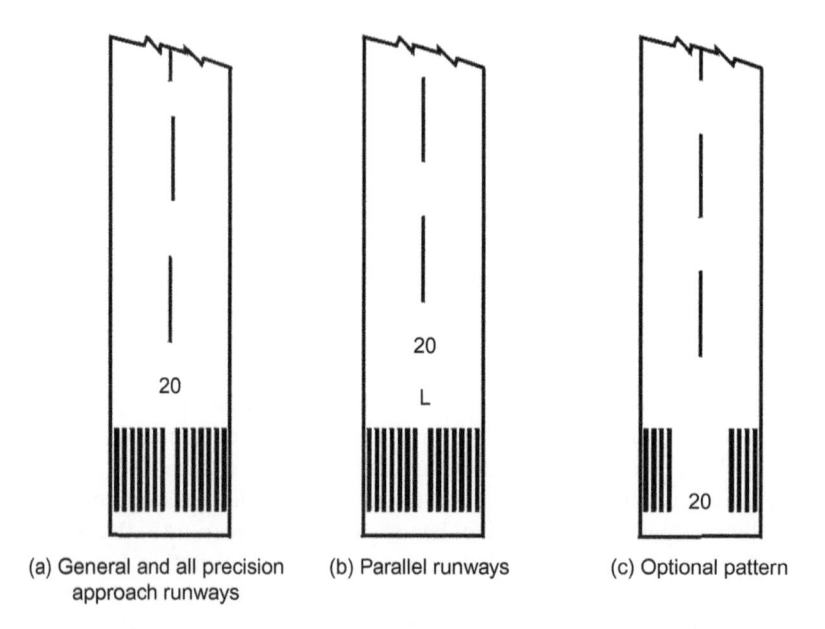

(a) General and all precision approach runways
(b) Parallel runways
(c) Optional pattern

Figure 15.2: Runway designation, centre line and threshold markings. *Source*: Annex 14, Aerodromes, Volume I, *Aerodrome Design and Operations*.

Taxiway markings (Figure 15.5):

✈ taxiway centre line markings
✈ taxi-holding position markings
✈ taxiway intersection markings.

Apron markings:

✈ aircraft stand marking

✈ stand identification

✈ lead-in line

✈ turn bar

✈ alignment bar

✈ stop line

✈ lead out line

✈ apron safety lines.

Other markings:

✈ aerodrome VOR check-point marking.

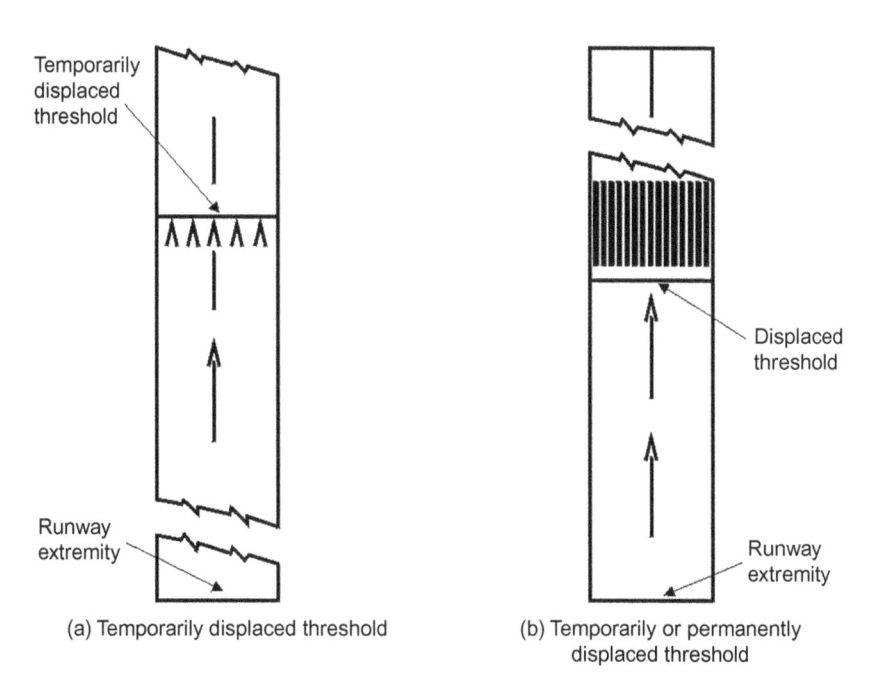

Figure 15.3: Displaced threshold markings. *Source*: Annex 14, Aerodromes, Volume I, *Aerodrome Design and Operations*.

15.1.3. Signs

Signs are installed at the airport to assist pilots' orientation on airport movement areas, to give them instructions and information. There are two basic types of signs: mandatory instruction and information signs (Figure 15.6).

Mandatory instruction signs specify the point or location beyond which any aircraft or vehicle is not allowed to proceed without control tower clearance. These signs include runway designation signs, holding point signs for Category I, II or III and NO ENTRY signs. Mandatory signs always have white letters on a red surface. The characteristics are given in Annex 14, Aerodromes, Volume I, *Aerodrome Design and Operations*, Chapter 5 and in the FAA AC 150/5345-44F.

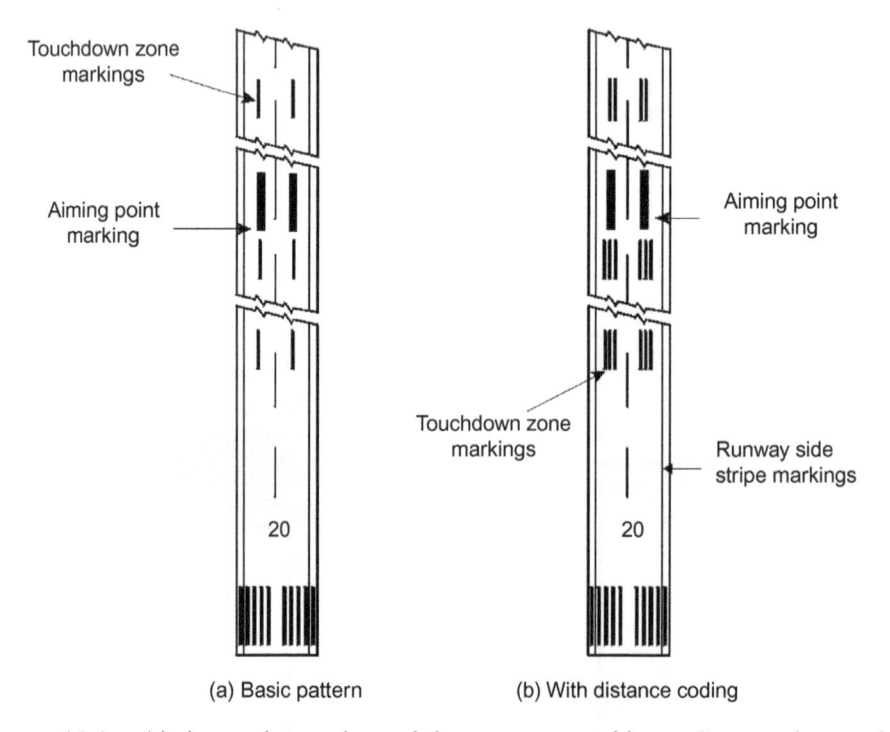

Figure 15.4: Aiming point and touchdown zone markings. *Source*: Annex 14, Aerodromes, Volume I, *Aerodrome Design and Operations*.

Information signs are used to specify location, direction or destination. They provide information on runway exits, directions to taxiways, aprons, stands etc. Information traffic guidance signs have black letters on a yellow surface. Location signs have yellow letters on a black surface.

Signs are usually constructed from lightweight metals or from light UV-resistant plastics. Plastic materials have better characteristics for reducing radar signal reflections, are practically maintenance free (except cleaning and changing of the light tubes) and have a long operational life-time. For night or low visibility operations all signs must be illuminated externally or internally — usually by 2 or 3 (depending on the size) horizontal fluorescent tubes which are placed inside to ensure an even distribution of light. The panels are mounted on columns with breakable couplings to satisfy frangibility requirements.

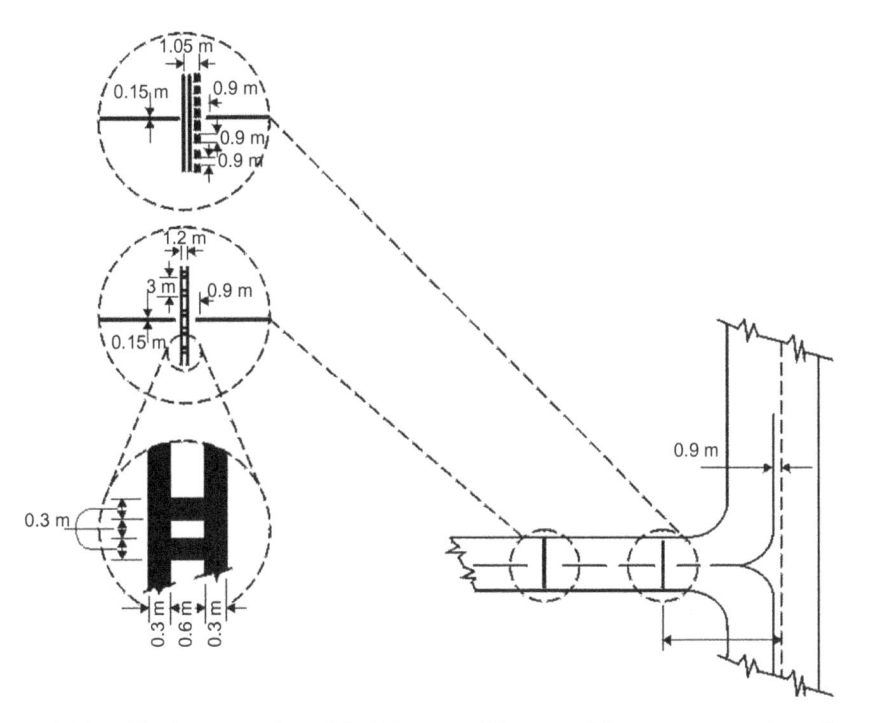

Figure 15.5: Taxiway and taxi-holding position marking — examples. *Source*: Annex 14, Aerodromes, Volume I, *Aerodrome Design and Operations*.

15.2. Airport Lights (*Author*: František Bělohradský, deceased, Consultant, Prague, CZ)

15.2.1. Characteristics and Components of Airport Lighting Systems

15.2.1.1. Introduction

When looking at a lighted runway from an aircraft cockpit during night approach, one can believe that there have been no changes in the airport lighting systems

374 Airport Design and Operation

during the last 50 years. In fact the light colours are certainly the same and there were only a few modifications in the lights' layout. The most important changes are not visible. In the new systems the latest technologies and materials are used to achieve high economic effectiveness, safety and reliability of the modern lighting systems.

Figure 15.6: Information sign (left) and mandatory instruction sign (right). *Photo*: Courtesy of Bratislava M.R. Štefánik Airport.

An important, and for the time being irreplaceable, role is played by the aerodrome lighting systems, in addition to the markings and the radio-navigation and radio-location aids. They were all made more important by increases in approach speed, and the need to improve safety. In addition, the need for better reliability called for a reduction of meteorological limits to operation. Under conditions of impaired visibility during the day and at night, the last phase of approach and landing has to be performed with a visual reference according to information created by a picture of lighting systems that augment the purely passive markings. As with other equipment, the aerodrome lighting systems are also standardised internationally by ICAO. The standards arose from the experience of Great Britain in particular, and later from the conclusions of work of the Visual Aids Panel (VAP) and All Weather Operation Panel (AWOP) teams.

The requirements for providing aerodromes with lighting systems vary with the type of runway equipment and reference code. The requirements are different for non-precision approach runways and for precision approach runways. Runways suitable for precision approaches are divided into three categories. Each category has a different meteorological limit for landing and take-off, and the requirements for lighting systems are also different. The ICAO categories are given in Table 15.1.

Table 15.1: ICAO precision approach categories.

Parameter	Category				
	I	II	III A	III B	III C
Decision height	60 m (200 ft)	30 m (100 ft)	N	N	N
Visibility/RVR	800/550 m	−/300 m	−/175 m	−/50 m	N

Source: ICAO: EUR Doc. 013 - European Guidance Material on All Weather Operations at Aerodromes, 4th ed.; 2012.
Notes: N — not defined. More information about the precision approach categories are in Chapter 3.

15.2.1.2. Light sources

The first generation of light sources were paraffin flares. These were supplanted by classical lamps designed for a maximum current value of 6.6 A with a bayonet cap. They are still used at many aerodromes for taxiway edge lights and to illuminate signs. They are made in a power input range of 30, 45, 65, 100, 200 and 300 W. Disadvantages are their large size and relatively small light flux.

First generation halogen lamps, of the so called double cap type, brought about a considerable reduction of the outer dimensions of lamps and an increase of light flux. Up to now they have been used in older types of inset lights. It was often necessary to refocus the lamp filament against the reflector, and the lamps have a relatively short service life because they were extremely heat stressed in the inset light fittings.

The second generation of lights, particularly the elevated ones, use halogen lamps with an exact position of the filament. They have extremely small outer dimensions, high light flux and a long service life of up to several thousand hours. They are made in an output range of 30, 45, 65, 100, 150 and 200 W.

Halogen lamps with an in-built miniature reflector in a single integrated unit are the most recent light sources to be applied in inset lights. These make it possible to design highly effective optical systems, and reduce the required input of the lights. The light sources made at present have inputs of 40, 45 and 100 W. They are outstanding for their long service life of up to 1,500 hours at 6.6 A., equivalent in normal use of up to 8,000 hours, depending on current level, of service before failure. An important feature of tungsten halogen lamps is the ability of the control circuit to modify the brightness of the lights in relation to operating conditions simply by varying the current flow through the lamp filament. Halogen lamps are robust enough to be used in airport operational conditions. They are able to withstand temporary excessive peak current five times the rated value for a limited time. They are technologically proven, mature and on the peak of their life-cycle. The new generation of light sources improves the operating reliability of the inset lights, and at the same time reduces the costs of electrical energy and maintenance of the lighting systems.

For low and medium intensity lights, LED sources are now being used. The main advantages of the LED lights are their high luminous flux, low energy consumption extremely long operational life of the order of 50,000 hours and better colour output without the need of filters. However, LEDs are delicate electronic devices. They fail to operate at excessive currents beyond their rated values or when subjected to high operating temperatures. They require DC stabilised power supply. LEDs are still a relatively new and rapidly evolving technology when compared with incandescent lamps. Typically the LED sources are used in the taxiway edge lights and low and middle intensity obstacle lights. A few companies have introduced taxiway inset lights designed for Category III operations.

15.2.1.3. Lights and fittings

According to Annex 14, Volume I, Chapter 5, the Aeronautical Ground Light is defined as: 'any light specially provided as an aid to air navigation, other than a light displayed on an aircraft'.

The ground lights may be classified according to the following criteria:

- place of installation
- type of mounting
- radiation characteristics
- character of the light.

They may be classified also by where they are located in the aerodrome lighting system as:

- obstacle lights
- approach lights
- visual approach slope lights
- threshold lights

- runway centre line lights
- runway touch-down zone lights
- runway edge lights
- runway end lights
- taxiway edge lights
- taxiway centre line lights
- stop bar lights
- safety lighting bars or 'wig-wags'
- lighting of information boards.

The types of mounting are:

- elevated
- inset.

The elevated lights are supported above ground level. The inset lights are firmly set in the pavement. Their cap protrudes only a few millimetres above the surface, and can bear the forces created when the wheels of an aircraft or a ground vehicle roll over them. The optical system of the light is fixed in the cap.

The radiation characteristic of the lights may be divided into:

- reflector: inset lights may be unidirectional lights or bi-directional lights.
- omnidirectional lights, which produce the light into all angles of the azimuth.

The radiation characteristic of the lights may be:

- symmetrical
- asymmetrical
- semi-asymmetrical.

The character of the lights may be classified as:

- constant light
- flashing
- flickering.

378 Airport Design and Operation

Using these characteristics, it is possible to create names for the lights and thus specify them further, for example:

✈ elevated reflector approach light

✈ unidirectional inset runway centre line light

✈ elevated omnidirectional taxiway edge light

✈ bi-directional inset taxiway centre line light.

Figure 15.7: Frangible safety masts installation — Amsterdam Schiphol.
Photo: A. Kazda.

15.2.1.4. Frangible safety masts

The approach light system must be designed in one plane which must be horizontal laterally and not be depressed longitudinally more than 1:66 below the horizontal plane for the first 300 m from the end of the runway strip and 1:40 thereafter.

It is therefore necessary to mount the lights on posts or masts when the ground is significantly below the runway elevation. Since 2005, the supports must be frangible; that is they must be of light weight and designed to break, distort or yield in the event of a collision. Supports less than 12 m should be frangible (Figure 15.7) from the base and those higher than 12 m need only be frangible 12 m down from the top. The challenge for the designer is to provide this frangibility while retaining sufficient stiffness and strength in the design-case winds. While simple steel or aluminium posts can be used for short supports, it is more efficient to use glass fibre for supports higher than 2 m: tubes for single light fittings or lattice masts if an array of lights is to be supported. Glass fibre is corrosion-resistant and almost maintenance-free. It is, though necessary to arrange for easy maintenance of the lights, usually by articulating the taller masts or using a pivoting base for shorter ones, unless access can be easily arranged by using a cherry-picker (an elevating platform which is adjustable for height and direction).

15.2.1.5. Requirements for aerodrome lights

The requirements for lights, their luminosity characteristics, mechanical properties and construction are determined not only by Annex 14, Aerodromes, Volume I, *Aerodrome Design and Operations* and other ICAO documents, but also by national standards. The requirements for lights may be divided into:

✈ lighting

✈ mechanical fitting

✈ electrical

✈ other requirements.

The lighting requirements are determined by Annex 14, Aerodromes, Volume I, *Aerodrome Design and Operations*, Appendix 2 by isocandela curves (Figure 15.8). Isocandela curves for the lights designed for Category I operations are wider horizontally. In contrast, the requirements for light intensity are higher vertically and lower horizontally for Category II. That requirement is even more conspicuous with the lights designed for Category III. All the manufacturers of lighting systems make efforts to provide lights that comply with the requirements for installation into systems for any of the Categories I to III.

Two basic factors are specified for evaluating the luminosity characteristics of the lights:

✈ mean light intensity, which is the integral of light intensity in a space beam (Istr)

✈ evenness of distribution of the light intensity: in a defined beam a minimum light intensity should be 0.5 Istr and a maximum light intensity of 1.5 Istr.

380   Airport Design and Operation

The requirements for the Thorn company's EL-AT threshold lights and their compliance are shown in Figure 15.9.

The mechanical fitting requirements are separately specified for elevated and for inset lights. The elevated lights should be light-weight, and easily frangible in the event of being hit by an aircraft or a vehicle. The resistance to wind and jet blast is shown in Table 15.2.

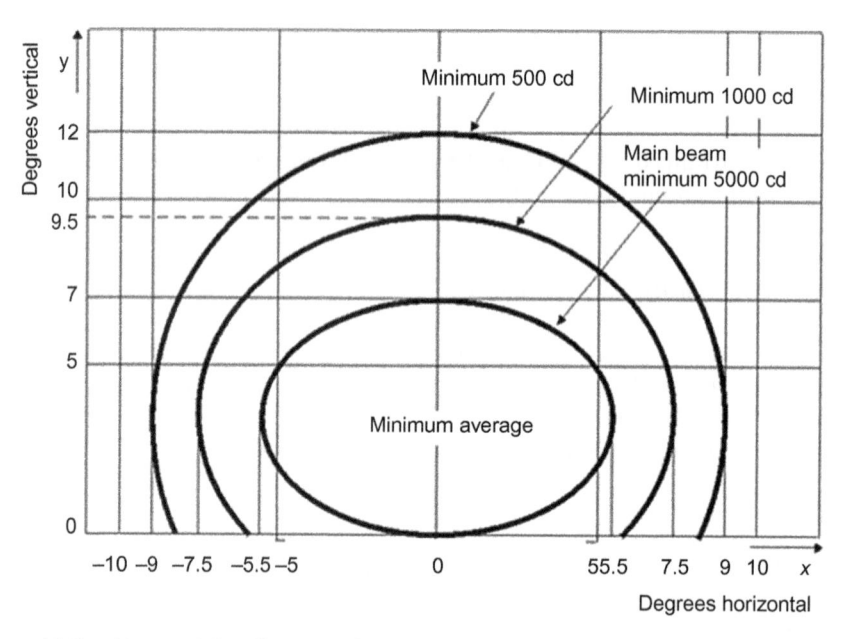

Figure 15.8:   Isocandela diagram for runway edge light where runway width is 45 m (white light).

The frangibility is defined as a property of an object to maintain its structure and resistance until a specified maximum force. If that force is exceeded, the mounting should deform or break in such a way as to create a minimum hazard for the aircraft.

In approach systems, if required due to descending terrain, the so-called safety masts are used. They are light-weight, usually lattice structures or towers made of aluminium or laminate, which will be frangible if hit by an aircraft.

The lights installed within approximately 2 m of the ground are supported by tubes or tripods attached to frangible fittings at ground level. The mechanical strength of the fitting should not be greater than 700 Nm for lights with the upper edge less than 360 mm over the level of the terrain. If the approach lights are higher than this, the strength of this light fixture may be correspondingly higher.

The inset lights should be resistant to dynamic effects caused by aircraft and ground vehicles passing over them. The basic mechanical requirement relates to the strength of the light cap expressed as a pressure applied on a surface unit. For a common inset light with a diameter of 300 mm, the required static strength of the cap shall be 230 kN. In addition, the temperature at the point of contact between the inset light and the tyre of an aircraft created by temperature transfer, or by heat radiation of the lamp, should not exceed 160 °C during a 10-minute exposure.

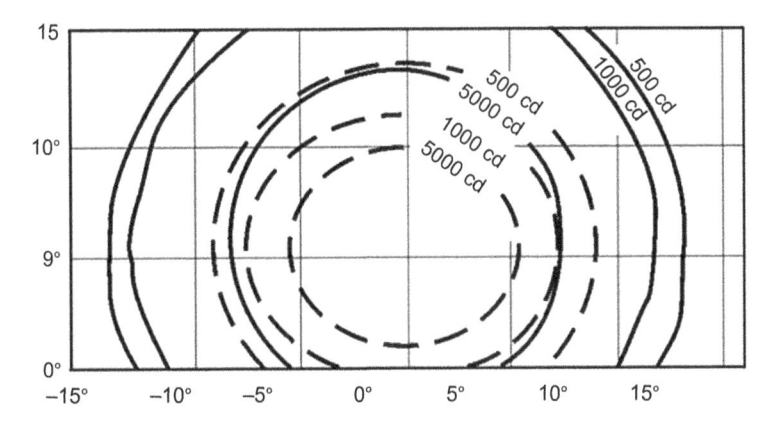

Figure 15.9: Isocandela diagram EL-AT (cd), threshold light (green light).

A low-voltage cable from the insulation transformer to the light is designed for a voltage of 450–750 V. The insulation resistance of the light should be at least 2 MΩ. They must operate within a range of temperature from −40 °C to +55 °C. The fittings of elevated lights must be constructed so that the exchange of a damaged or broken light can be quick and uncomplicated even in severe winter conditions.

With inset lights, the principal factors relate to the security of the fitting, the service life, the light source and speed of change of a cap with its optics *in situ*. The light bulb in the cap of an inset light should be changed in a workshop. It is also important that an inset light should resist damage by winter maintenance equipment like snow ploughs.

15.2.2. Characteristics and Components of Airport Lighting Systems

15.2.2.1. Approach and runway systems

15.2.2.1.1. Non-instrument and instrument runways. Lighting systems of moderate light intensity are used for non-instrument and instrument runways under VMC conditions at night. It is often sufficient to construct only runway edge lights, runway threshold and end lights for non-instrument runways.

If possible, it is advisable to construct a simple approach lighting system so as to improve the directional guidance. This is a mandatory requirement for instrument runways.

The simple approach lighting system consists of a single row of lights on the extended centre line of the runway, out to a distance of 420 m from the runway threshold with a crossbar at a distance of 300 m from the runway threshold (Figure 15.10). The simple approach system may consist of individual lights, or of short lighting bars, called barrettes. The effect of a barrette is created by at least three lights located next to each other. The runway edge lights and the threshold and runway end bars should be located at the runway edge, or at the most 3 m from its edge.

Table 15.2: The resistance of elevated lights to wind and jet blast.

Light	Wind/jet blast speed (m/s)
Approach	51
Visual approach slope	51
Threshold	103
Runway edge	155
Taxiway edge	103

Source: ICAO Aerodrome design manual, Part 4, Visual aids, 4th ed, Doc 9157-AN/901 Part 4, Montreal 2004 and producers data.

Omnidirectional lights are used for simple approach systems and for edge lights for instrument runways and for non-instrument runways that are to be used under VMC at night. The omnidirectional lights provide guidance not only in the direction of approach and take-off, but because they give light in all directions, they also provide informational guidance for circling flight. Some types of lights combine the properties of reflector and omnidirectional lights. The combination of one omnidirectional light and two reflector lights that have been used for a precision approach runway has now been substituted by a single light unit. It represents an important capital cost saving.

The Thorn company's EL-EAH light is an example of such a light. The ICAO requirements for lights on runways with a width of 60 m are fulfilled with a power of only 150 W, while for runways with a width of 45 m, 100 W is a sufficient input with reserves of 20–50% at a medium light intensity.

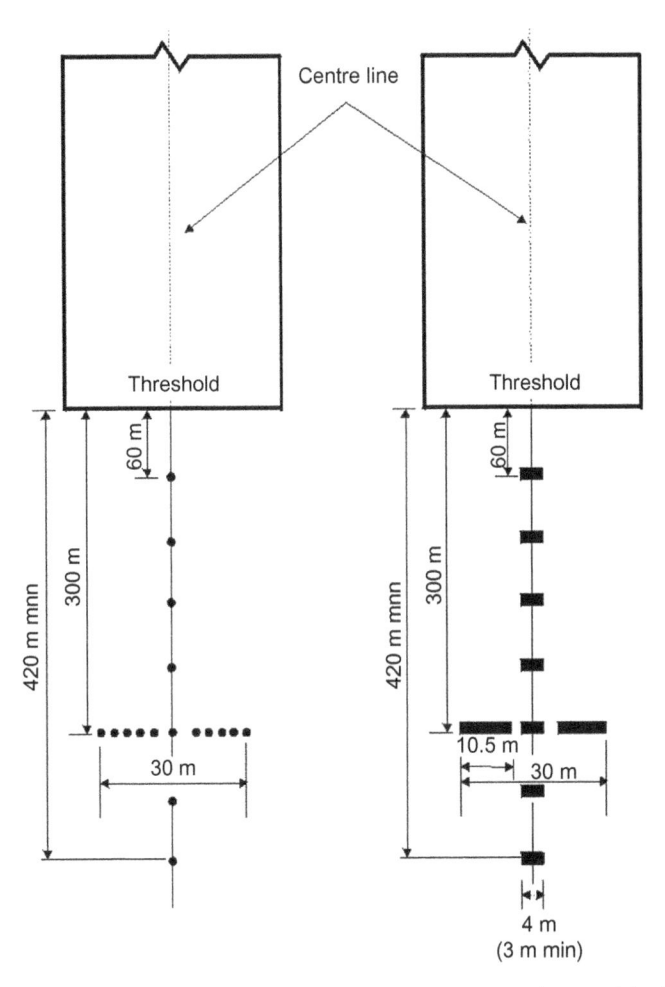

Figure 15.10: Simple approach lighting systems. *Source*: Annex 14, Aerodromes, Volume I, *Aerodrome Design and Operations*.

The light (see Figure 15.11) weighs only 1.9 kg and has excellent aerodynamic properties. Resistance to wind and jet engine blast was tested in a wind tunnel, and it is well above the required 560 km/h. Rust resistance is provided by the use of aluminium and stainless steel. The optical system is formed by a dioptre of heat resistant glass and two Fresnel lenses. The optical system bearer is formed by a solid aluminium casting, which in addition ensures good heat removal.

384   Airport Design and Operation

Changing the lamp bulb is very easy and quick, which is particularly important at aerodromes with a high traffic density or severe climatic conditions. The lamp bulb is equipped with a connector. The whole upper part of the light is usually changed for a new one, the lamp bulb being checked in the maintenance workshop. Changing the lamp bulb does not usually require more than 30 seconds. The mounting for the light is frangible (guidance on design for frangibility is contained in the *Aerodrome Design Manual*, Part 4, Visual aids).

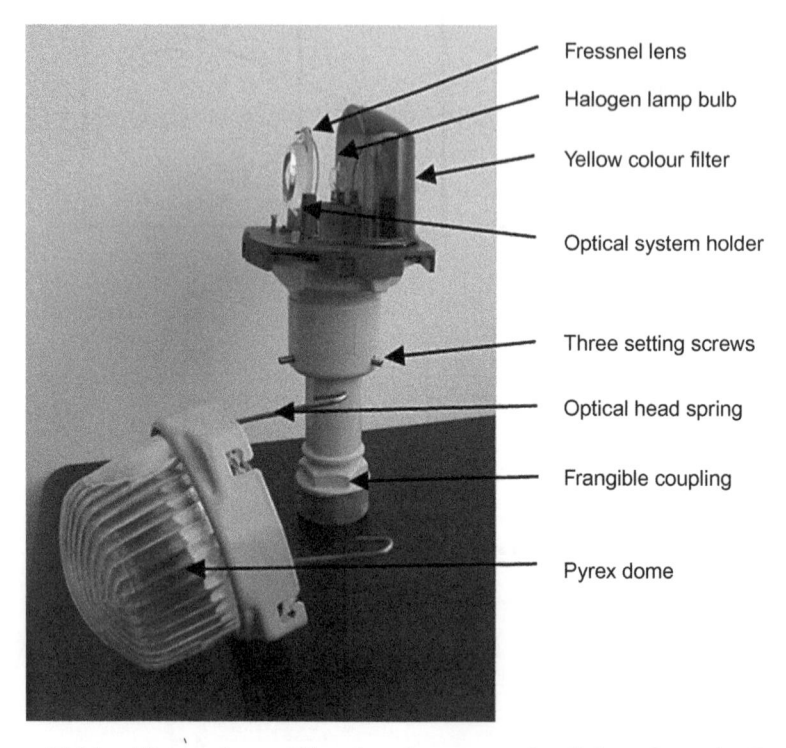

Figure 15.11:   Elevated omnidirectional runway edge light. *Photo*: A. Kazda.

15.2.2.1.2. Precision approach runway.   A Category I runway should be equipped with a Category I approach system. A diagram of the lighting system is shown in Figure 15.12. Two types of approach systems have been approved as a world-wide standard, an older type system, CALVERT and a newer type, ALPA ATA. The systems differ not only in the location of the lights, but also in the types used.

In addition, the ALPA ATA system has sequenced flashing lights, which more easily distinguishes the lighting system from other light sources in the aerodrome surroundings and gives positive identification of the runway. The sequenced lights flash twice a second. The flash moves along the system towards the runway threshold.

The time of each flash is so short that it does not dazzle the crew of an aircraft. The flash easily penetrates fog, and thus the pilot gets information about the runway centre line orientation several seconds earlier than from approach light systems with steady lighting.

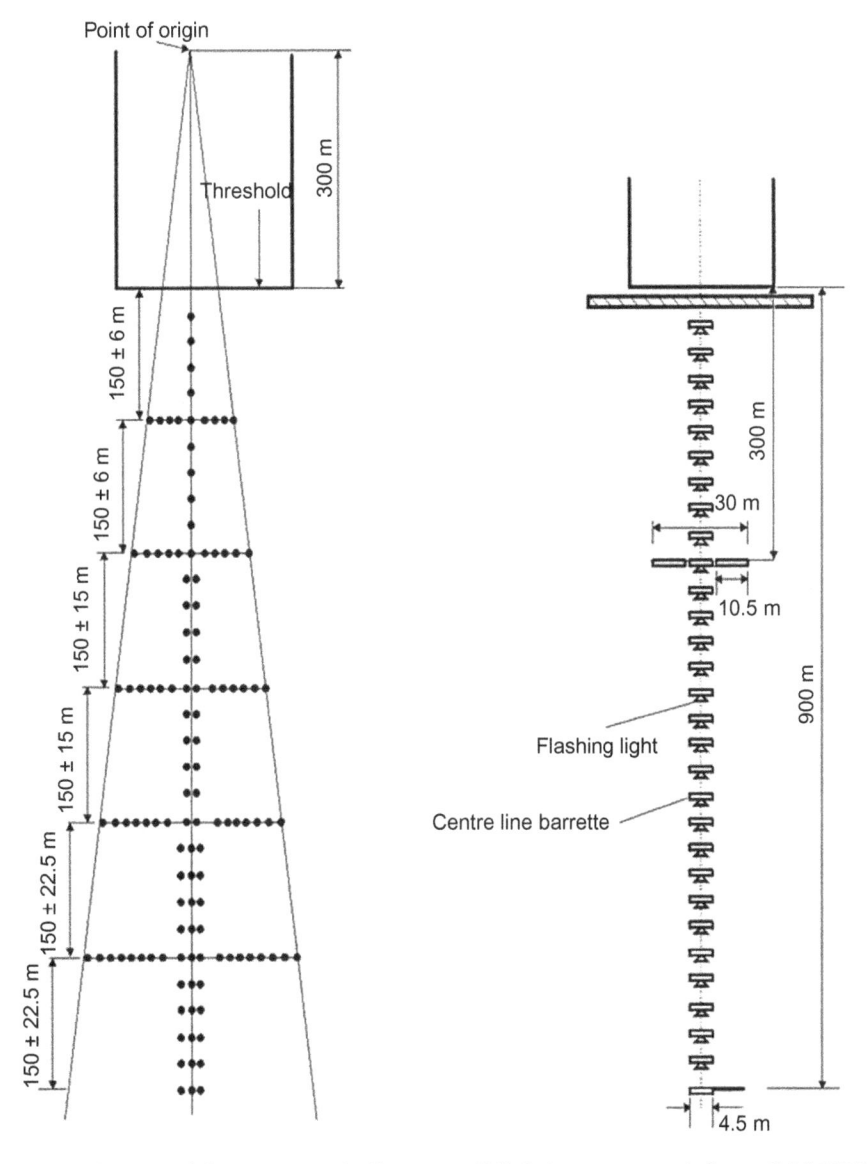

Figure 15.12: Precision approach Category I lighting systems left — CALVERT, right ALPA-ATA. *Source*: Annex 14, Aerodromes, Volume I, *Aerodrome Design and Operations*.

386   Airport Design and Operation

A high voltage supply for the discharge lamps for the sequenced flashing lights is provided from a control box, which is fixed to the light support or is located in the ground.

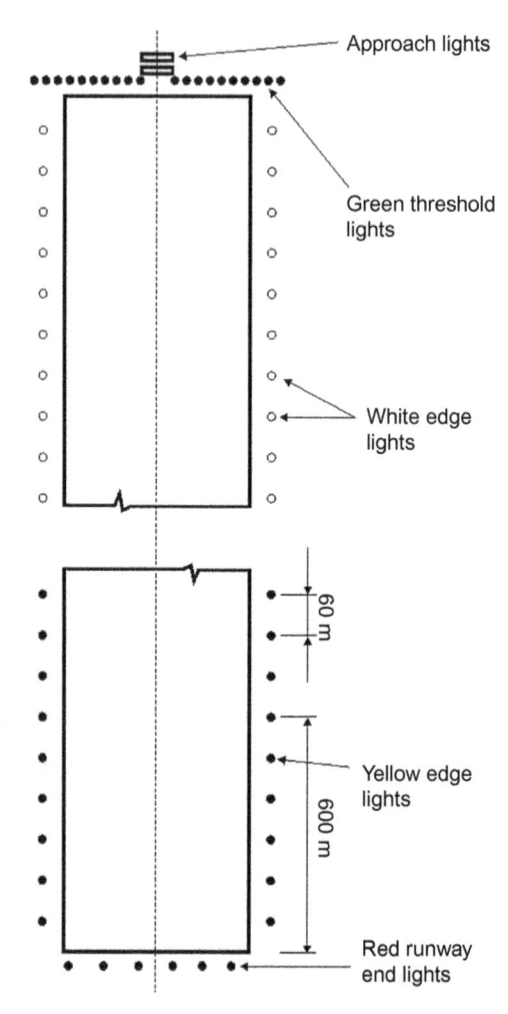

Figure 15.13:   Runway lighting — precision approach Category I runway.

ICAO states the requirements for the lights' properties, including the character of the white colour of the light and the mean light intensity in a defined beam of 20,000 candelas (cd). The requirements are usually complied with by reflector lights.

Threshold lights forming a crossbar should be located at the runway threshold (see Figure 15.13). The threshold lights produce a green light with a light intensity in a defined beam away from the runway of 10,000 candelas (cd). Runway end lights are also located at the runway end, facing the runway. The lights inform the pilot of an aircraft of the runway end position. They produce red light in a defined beam with a light intensity of 2,500 candelas (cd).

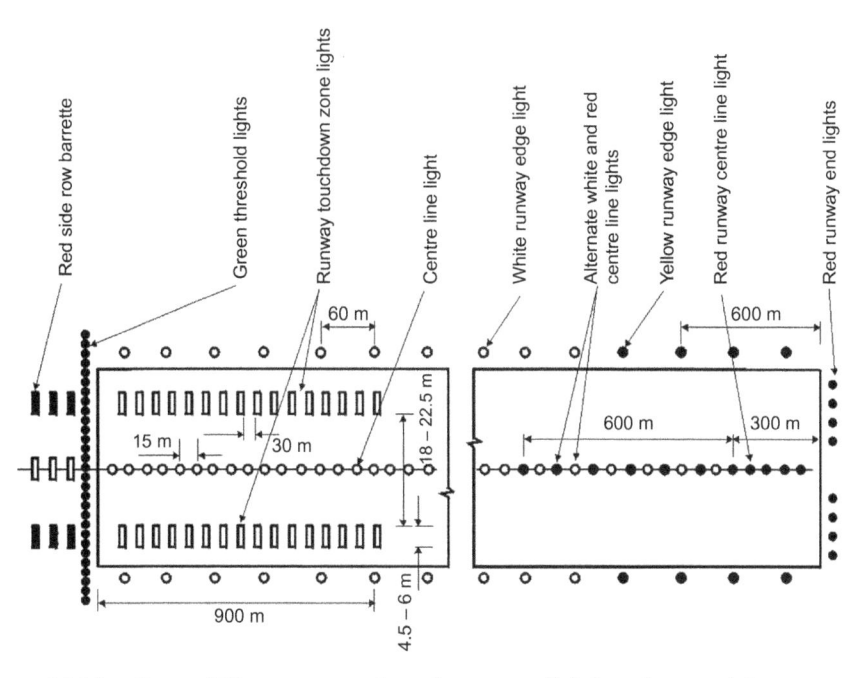

Figure 15.14: Inner 300 m approach and runway lighting for precision approach runway Categories II and III.

The runway edges are indicated with runway edge lights. They are either reflector or omnidirectional lights producing permanent white light with a mean light intensity of 10,000 candelas (cd) in a defined beam. In the last 600 m of the runway length (or half of the runway if the runway is shorter), the lights are yellow, and thus inform the pilot of the distance from the runway end.

The lights are supplied in the majority of cases by a serial distribution system and constant current regulator as described in Chapter 16. When the electrical power supply from the mains is interrupted, the supply is provided from a standby power source. For a Category I runway, the electric power supply should be restored within 15 seconds.

388   Airport Design and Operation

Category II of the ICAO meteorological limits is characterised by a runway visual range from 300 m to 550 m and by a decision height of 30 m (100 ft). These requirements are at present considered to be the limits for aircraft and crews to perform manual approaches and landings.

The Category I approach lighting system (CALVERT or ALPA ATA) is augmented for Category II operations with red side row barrettes in the last 270 m before the runway threshold. The reflector lights have a light intensity of 5,000 candelas (cd) in a defined beam. The approach system also contains a set of flashing lights.

A Category II runway has many more lights than a Category I runway (see Figures 15.14 and 15.15). Besides the runway threshold lights, two other important additional lighting systems are incorporated. Under Category II meteorological conditions, the efficiency of the runway edge lights is reduced, and the runway centre line marked with inset lights becomes the most important means of giving precision guidance to a pilot on the runway during landing, roll out, as well as during take-off.

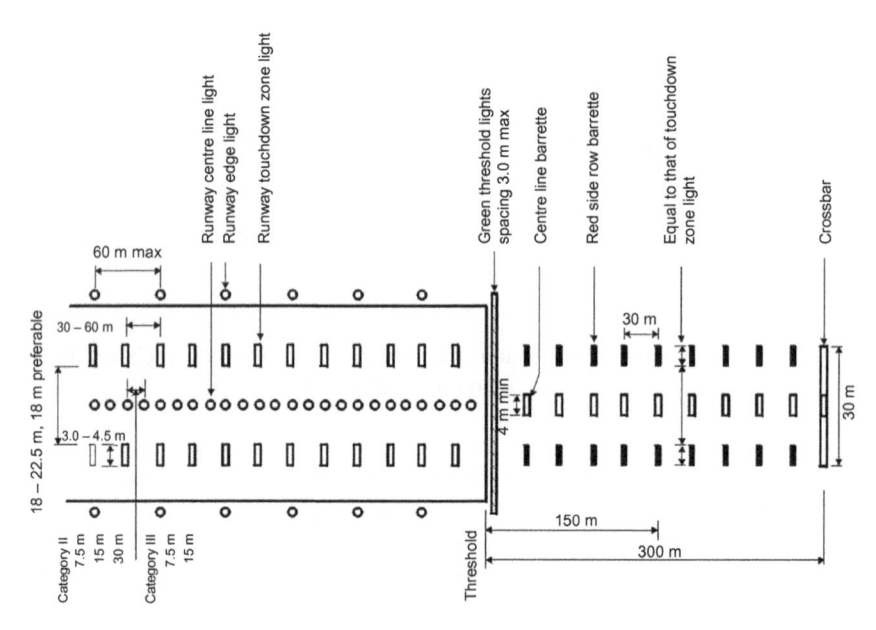

Figure 15.15:   Runway lighting for precision approach runway Categories II and III. *Source*: Annex 14, Aerodromes, Volume I, *Aerodrome Design and Operations*.

The centre line lights show steady white light of variable intensity. Inset white and red lights shall be placed alternately in the centre line from the last 900 m to 300 m from the runway end, while the centre line lights show only red in the last 300 m. Another system indicates the runway touchdown zone, again by means of inset lights. The lights of the touchdown zone shall be located with a longitudinal spacing equal to that of the runway edge lights, and form an extension of the series of approach barrettes. Contrary to the lights of the approach system barrettes, which are red, the touchdown zone lights are white.

Figure 15.16:   Runway inset touch-down zone light. *Photo*: A. Kazda.

The inset lights are often run over by the wheels of aircraft. Their construction needs to allow for that, and they should be fitted flush with the pavement surface. Their egg-shaped body is made either of cast steel or of aluminium alloys (Figure 15.16). The lid is provided with cut-outs for the light beams. The optical system inside the light consists of a light source and reflector, glass prism and colour filter, as the case may be. The lights are stressed not only dynamically as the aircraft runs over them, but also thermally, by the heat of the lamp. Even in heavy traffic conditions, in damp and in aggressive environments like winter maintenance, they need to be reliable. A white light with a light intensity of 5,000 candelas (cd) is required in a defined beam. The most advanced inset lights use a special light source, whose halogen lamp and a reflector form a single integrated unit. The halogen lamps have a minimum size and long service life. Effective optical systems allow the power requirements of the lights to be reduced, so reducing the heat inside the lights, and thus extending the service life of the lamps. The design of the lights is simple, the operation and maintenance is much cheaper than those of the earlier types of lights.

390 Airport Design and Operation

The IN-TO light made by the Thorn Company (Figure 15.17) is an example of a modern inset light for taxiways. Optical prisms, which were glued in the older types of lights, are fixed into the light cap by means of a silicon seal. The light is provided with a double seal, the optical system being closed with a cover with a seal, and the other seal being between the cap and the mounting. The light power requirement is only 40 W.

Figure 15.17: Taxiway inset light. *Photo*: A. Kazda.

The connection between the electric power supply from the isolating transformer and the light cap is waterproof. The colour filters for obtaining the required light colour are dichroic — a thin layer is applied on the surface of colourless glass. Such filters have much greater light transmittance than the coloured glass filters. In the old types of 'traditional' coloured glass filters the luminous intensity decreases rapidly up to 45% with a yellow filter 15% with a green 12% with a red and to only 2% of the original luminous intensity if a blue filter is used. The upper part with the optical system is mounted in the lower part of the light, which is itself fixed into the pavement.

For an ICAO Category II runway, a higher reliability of the electric power supply is required. In the event of a mains supply failure, a standby power source should be able to recover the electric power supply within one second.

Visual Aids for Navigation 391

The runway equipment for Category II, with the Runway Visual range (RVR) limit reduced approximately by half that for Category I, represents a large increase of cost in comparison with a Category I runway. The number of lights and other components and equipment, such as isolating transformers, light fixings, constant current regulators, is many times higher. The cabling is also longer. The inset lights, which must be used in the touchdown rows and the centre line lights, are much more expensive than the usual elevated ones, and also their installation and maintenance are more expensive. The price of equipment for Category II runways is increased also by other required facilities.

The standby source of the 'short break' type for Category II is also more expensive than the standby source with automatic start, commonly used for Category I lighting systems. In addition, meteorological equipment, in the form of RVR measurement, and more comprehensive ILS precision approach systems are required for Category II.

Category III approaches are divided into sub-categories A, B and C. The basic difference compared with other categories is that in Category III the landing itself is conducted automatically, and the lighting systems serve the pilot only for monitoring the very final phase of approach and landing. The practical limit of the usefulness of the lighting systems may be considered to be Category III B. When Category III C conditions exist, only non-visual approach methods can be applied. Landing and guidance of an aircraft to the runway is fully automated.

The location of lights of the Category III approach and runway systems does not differ from the Category II. The edge lights are not normally visible in Category III conditions, so the lighting systems of taxiways should be complemented with centre guidance from inset lights, which are green, or green and yellow for both directions between the runway centre line and the perimeter of the ILS/MLS critical or sensitive area.

For Category I, II and III approach systems, reflector type elevated lights are used (Figure 15.18). They should be light-weight, highly resistant against corrosion and jet blast. The reflector is usually made of aluminium, chemically polished and protected with a cover of light-weight aluminium alloy. The diffuser for green and red lights is made of heat resistant glass, and therefore no colour filter and its attachment are needed. The cooler in the rear part of the reflector lamp significantly extends its service life.

Changing the lamp is facilitated by easy access to the lamp after lifting the rear plastic cover. The lamp is equipped with a connector. The power requirement is very small, being only 150 W for a white approach light, and 100 or 150 W for a threshold light, the ICAO requirements for light intensity still being fulfilled with reserves of 20–40%.

392 Airport Design and Operation

The intensity of lighting systems must be adequate for the meteorological conditions — visibility, background brightness, and must fully satisfy pilot requirements. Light intensity control must allow adjustments of the light intensity to meet the prevailing conditions. Separate intensity controls must insure settings of compatible intensities for approach lighting system but also runway edge lights; runway threshold/end lights; runway centre line lights; runway touchdown zone lights and taxiway centre line lights.

Figure 15.18: Honeywell High Intensity Approach Light Type ATR 12. *Photo*: Courtesy Honeywell Airport Systems.

15.2.2.2. *Approach slope indicator systems*

Where an improved visual guidance of an aircraft should be provided on the approach, the runway must be equipped with a visual approach slope indicator system.

The approach slope system should be provided if:

↳ the runway is used by jet aircraft
↳ the approach is to be performed over a water surface, or over a featureless terrain (desert, snow-covered terrain) in the absence of sufficient incidental lights in the approach area by night

Visual Aids for Navigation **393**

✈ misleading information is produced by surrounding terrain or runway slopes (see Figure 15.19)
✈ dangerous objects are present under the approach path
✈ the terrain beyond the runway ends involves serious hazard for an aircraft if it touches down before the threshold or overruns beyond the runway end
✈ the aircraft may be subjected to turbulence, wind shear or micro bursts during approach.

Visual approach slope indicator systems are installed at virtually every civil aerodrome. An additional advantage is that they correct any tendency to fly low in heavy rain. The most used is the Precision Approach Path Indicator (PAPI) system, and some countries still use the Visual Approach Slope Indicator System, the Visual Approach Slope Indicator System (VASIS), or T-VASIS in Australia and New Zealand.

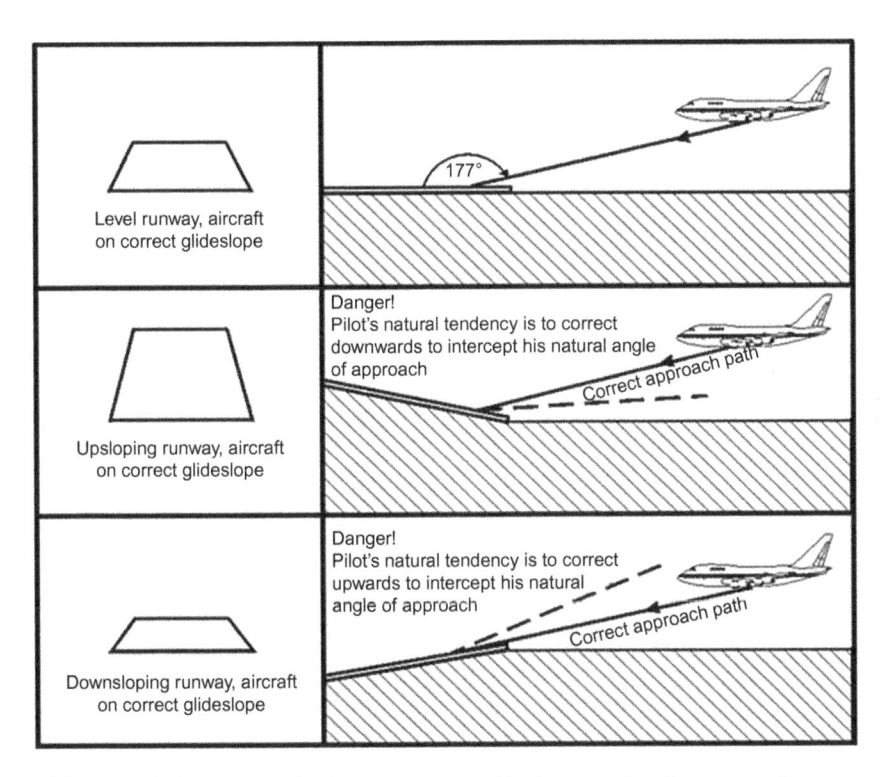

Figure 15.19: Possible illusions due to the longitudinal runway slopes.

The PAPI system is not only a more precise vertical guidance of the pilot during the approach down the glideslope, but it also has fewer lights in the system then the other two systems. PAPI gives a better indication of rates of change of angle above

394   Airport Design and Operation

or below the glideslope than VASIS. The costs of acquisition and construction are usually lower, and it is more economical to operate.

The PAPI visual slope light unit has a light beam horizontally divided into two sectors, red in the lower part and white in the upper part. The geometrical arrangement and horizontal adjustment of the lights ensure that the pilot of an aircraft can see red sectors of the two lights nearest to the runway and white sectors of the other two lights when on the correct approach slope. If the aircraft falls below the glideslope, the white colour of the other two lights changes to red at precise angular changes depending on the extent to which it is below the nominal slope. If the approach becomes higher than it should be, the red lights change to white. The geometrical arrangement of the approach slope indicator systems and the angles of adjustment are shown in Figure 15.20. The lights can be repeated on the other side of the runway in order to improve roll perspective, but both sets must give completely consistent indications.

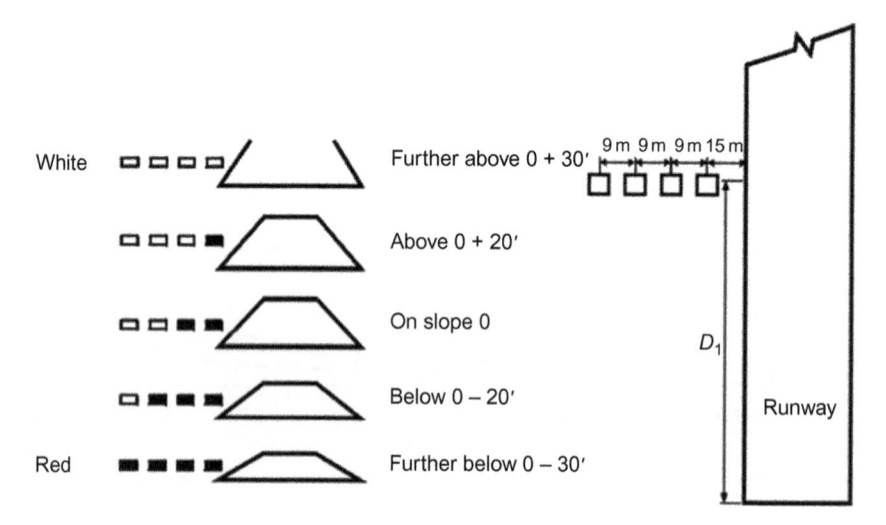

Figure 15.20:   PAPI light units settings and indications.

The construction of the visual slope light units does not resemble reflector or omnidirectional lights in any aspect. The light beam should be very sharply divided into a white and a red sector. The optical system contains one or two lenses, a red filter and a reflector. It is common to fit two optical systems in one light (Figure 15.21). The required power is then 2 × 200 W. In the APAPI systems — approved for aerodromes without international traffic, lamps of 45 and 100 W are also used, because the required range of the systems is lower. The required range of the PAPI system is 7.4 km on a clear day. The range is, in fact, usually 12–15 km, and, on a clear night, it can be 30 km and more.

Visual Aids for Navigation 395

The American Federal Aviation Administration (FAA) has approved another type of approach slope indicator system. It is the Pulse Light Approach Slope Indicator (PLASI) made by the De Vore Aviation Corporation. The equipment may be used on aerodromes as well as on heliports. The equipment consists of a single light fitting. As with the VASIS and PAPI systems, PLASI also uses white and red lights to indicate the required vertical position of an aircraft.

Figure 15.21: Visual slope light unit (uncovered). *Photo*: A. Kazda.

A constant white light indicates that the aircraft is correctly located relative to the glidepath. A pulsing white light indicates that the aircraft is above, and the pulsing red light that it is under the glidepath.

The transition between the position on the glidepath and off it is very sharp. The bigger the deviation of an aircraft from the glidepath, the higher is the frequency of flashing. Nearer to the axis, the light pulses more slowly. Thus the pilot of an aircraft easily acquires information on the extent of deviation of the aircraft from the glidepath axis. The PLASI may be simply adjusted to an arbitrary glidepath angle from 0 to 90° above the horizon. It requires only a small space for its installation. Therefore it is usable also for heliports on structures at sea.

396   Airport Design and Operation

The equipment has a battery with discharge lamps. When there is a failure of the light source, the battery is automatically switched on. The light output is controlled automatically by a photoelectric cell. Both the purchase and operating costs of the PLASI are less than those with the PAPI system.

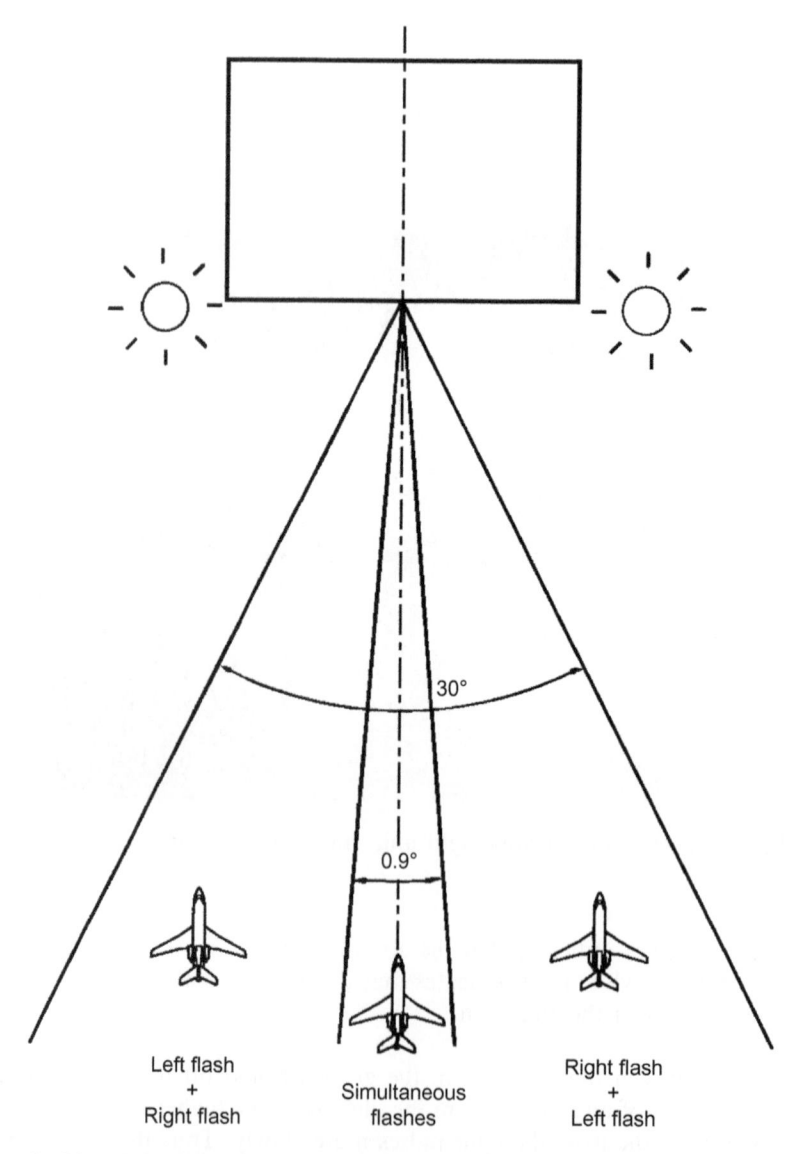

Figure 15.22:   Principle of system of approach azimuth visual guidance (SAGA).

For airports and heliports a new system of azimuth guidance was developed by the Thorn Company. It is the System of Azimuth Guidance for Approach (SAGA) (Figure 15.22). The system includes two lights with unidirectional rotating beams placed symmetrically on each side of the threshold.

The pilot receives, every second, a luminous message comprising two flashes issued in sequence by the two rotating beams of the system. When the aircraft flies inside a 0.9° wide angular sector, centred on the approach axis, the pilot sees two lights flashing simultaneously.

When the pilot flies inside a 30° wide angular sector, he sees the two lights flashing with a delay which varies from 60 to 330 ms according to the aircraft position in the sector. The further the plane is from the axis, the greater the delay, the sequence effect showing the direction to the axis.

15.2.3. Heliport Lighting Systems

Some visual systems designed for aerodromes are also used for lighting heliports. However, helicopter performance characteristics and the type of heliport operation impose special requirements on the heliport lighting systems. In general night and restricted visibility operations require lighting (optical navigation aids) and illumination (ground illumination) of the heliport and heliport equipment.

A heliport approach lighting system is usually installed when it is necessary to specify a particular approach direction. The approach system must be at least 210 m in length and should consist of a row of white lights placed at intervals of 30 m. There is a crossbar 18 m in length at a distance of 90 m from the perimeter of the final approach and take-off area. The lights beyond the crossbar may be steady or sequenced flashing to give a positive identification of a heliport and to distinguish the approach lights from the background lighting.

If the heliport is situated in an area where obstacle clearance or noise procedures are prescribed or it is essential to fly a precisely defined track towards the final approach and take-off area, a visual alignment guidance system (VAGS) should be used. A system of approach azimuth visual guidance (SAGA) (Figure 15.22) is one of the currently available systems fulfilling specifications for visual alignment guidance.

Where installation of VAGS is necessary or if a helicopter requires a stabilized approach, some type of visual approach slope indicator should be installed at the heliport. For this purpose a standard APAPI system could be used. There is also a special system designed for heliport operations only — Helicopter Approach Path Indicator (HAPI) (Figure 15.23). The signal format from HAPI is similar to PLASI.

HAPI signals determine four sectors by different colour and light formats in the horizontal plane between 1° and 12° above the horizon (Table 15.3).

Figure 15.23: Typical HAPI installation at elevated heliport. *Photo*: A. Kazda.

The advantage of HAPI against other standard systems is that the equipment consists of a single light unit and could also be used for helidecks or elevated heliports.

For night operations the Final Approach and Take-off Area (FATO) should be marked by white omnidirectional lights placed along the edges of the FATO. At either surface level or elevated heliports lighting systems, perimeter lights or Arrays of Segmented Point Source Lighting (ASPSL) or luminescent panel (LP) lighting could be used for the Touchdown and Lift-Off area (TLOF). Floodlighting is also allowed for TLOF identification at surface level heliports. The perimeter lights must be green fixed omnidirectional lights placed along the edge of the TLOF.

Where applicable the heliport lighting equipment can be further complemented by an aerodrome beacon, heliport beacon, winching area floodlighting and aiming point lights. At each heliport approved for night operations at least one illuminated wind direction indicator must be installed.

At aerodromes, the runway lengths have increased over time and so have the number of lights and lengths of cable, but the dimensions of heliports has remained practically the same. Heliport lighting systems are placed on a relatively small area. For this particular reason the traditional serial distribution system with constant current regulators used at aerodromes is more complicated and more expensive. A parallel supply system is typically used at heliports. Both elevated omnidirectional or inset lights are connected by a parallel feeder system to a simple distribution and control box.

For omni-directional lights and the illuminated wind direction indicator there is no need for variable light intensity setting. In the case of approach lights and visual approach slope indicators an intensity setting of usually 10%, 30% and 100% is used for steady lights and 3%, 10% and 100% for flashing lights.

Table 15.3: HAPI signal format.

Sector	Format
Above	Flashing green
On slope	Green
Slightly below	Red
Below	Flashing red

15.2.4. Lighting of Obstacles

Usually there are objects around an airport which penetrate the obstacle clearance limits (see Chapter 3) or may create a hazard to aircraft. If the object is immovable but could, according to aeronautical study, affect safety, the object should be marked. If the aerodrome is used at night, it should also be lighted. The way the obstacle is marked or lit depends on the object's character, its position and height, and the type of aerodrome operation. In general the obstacle lights could be divided into three groups:

Low-intensity obstacle lights

Low-intensity obstacle light Type A or B — red light with peak intensity between 10 and 32 cd/m^2. Low intensity lights could be used for marking of less extensive objects with height less than 45 m above the surrounding terrain (Figure 15.24).

400 Airport Design and Operation

Figure 15.24:   Low intensity neon obstacle light. *Photo*: M. Maťaš.

Medium-intensity obstacle lights

Medium-intensity obstacle light
    Type A — white flashing, with effective intensity of 20,000 cd/m².

    Type B — red flashing, with intensity of 2,000 cd/m².

    Type C — red fixed, with intensity of 2,000 cd/m².

Medium-intensity obstacle lights should be used where an early warning of danger is required or the object is extensive or its height above the surrounding terrain is greater than 45 m. When medium-intensity obstacle lights are used it is usually not required to use colour markings for an object.

High-intensity obstacle lights

High-intensity obstacle light Type A or B are white flashing with effective peak intensity of 100,000–200,000 cd/m². These lights should be used for objects with height of 150 m and more above the surrounding terrain or for recognition of towers, wires and cables by day. The light characteristic is not omni-directional as is the case with low or medium intensity lights, but the flash is directed by mirrors. For marking high towers or pylons usually 4–6 high intensity lights are necessary.

The popularity of an LED type (Light Emitting Diode) of low and medium intensity obstacle lights has grown recently. This is not only due to a very low energy consumption, that is about 12 W in the case of a low intensity light, but also its extremely long service life — more than 100,000 hours. When properly installed the LED type light is practically 'maintenance free'. Because of its low energy consumption it is possible to use photocells as energy source.

As a light source for medium intensity obstacle lights a xenon discharge lamp is usually used. To meet the light intensity criteria, some types of lights work in a dual regime with white flashes during the day and red flashes at night.

15.2.5. Light Control

15.2.5.1. Remote control equipment

The remote control equipment is a part of the aerodrome lighting system. It serves not only to regulate, but also to monitor the lighting system operation. Monitoring equipment that operates in real time is an essential requirement for approving the lighting systems for Category II and III conditions. It is necessary to control from a distance a large number of constant current regulators, and at the same time to assess a large amount of feedback data on the condition of the lighting systems. The remote control system rationalises the operation so that:

- ✈ the air traffic controllers are not overburdened by handling numerous control elements and redundant information on the condition of the lighting systems
- ✈ the maintenance personnel may receive detailed information on the condition of the lighting systems, as a base for their activity. That information is usually brought together in a central control board.

The systems at small aerodromes and heliports, where the number of constant current regulators is small, do not require complex control equipment. From each constant current regulator, a multi-cored control cable is led out, by means of which the orders and feedback signals are transmitted directly to a simple board in the control tower.

At aerodromes equipped with Category I, II and III precision approach systems, a whole range of constant current regulators operates simultaneously, often to different degrees of light intensity. Movement of aircraft on more complex systems of taxiways is controlled by switching on the lights for the appropriate standard taxiing paths. Pre-programmed control systems prevent the air traffic controllers from being overburdened by a complicated control situation in such cases. For example, in the 'take-off' programme, the approach system is not required to be on. The air traffic controller chooses a runway direction and the degree of light intensity, depending on the runway visual range or according to the pilot's request. The programme ensures the activation of the lighting systems necessary for take-off with the required degrees of light intensity. In addition to control panels, clear diagrams are also used, where the activation of the particular lighting systems is shown. Lamps were initially used to show this, later being replaced by light emitting diodes (LED). The air traffic controller was informed of a failure of a part of the lighting systems or constant current regulators by the flickering of a lamp or a diode, and by aural signalling that can be switched off if necessary.

Current practice makes full use of computer technology for the control of the more complicated Category I systems, but in particular for Category II and III systems. This has both operating and economic advantages. The expensive multi-cored cable is substituted by a cheap two- or four-cored one. Large lighting systems and sub-systems, safety lighting bars and stop lighting bars may be controlled and monitored, together with the power system of the aerodrome and meteorological and radio-navigation equipment.

The air traffic controllers regulate the lighting systems through several controls, and the information is presented in a summarised form in monitors. For example, in the event of a failure of an important system, the air traffic controller may receive only the information that the lighting system is not capable of operation under Category II conditions, but that it complies with the requirements for an operation under Category I conditions. For maintenance purposes, the monitoring system provides detailed data on the condition of the equipment failure in terms of reduction of insulation in a loop, the number of lamps not meeting the required output, and the like. In real time, the monitoring systems record the condition of the lighting systems, air traffic controllers' activity, failures and their repair.

Recently, the so called 'addressing system' of monitoring and control has been used for complex light systems of taxiways and stop bar lights. Two systems of communication are used, either a serial feeding cable, or a special cable network and optical cables.

Each light in a serial loop has its own precise address, and the system allows not only monitoring, for example it identifies two consecutive unlit lights, but also the control, individually or in groups of stop bar lights, sections of runway centre line lights, etc.

15.2.5.2. Single lamp control and monitoring

Current aerodrome lighting systems need separate series circuits when switching. In the case of lighting systems with a small number of lights like stop bars, sections of taxiway centre line or PAPI systems, every circuit needs a separate cable and a constant current regulator in the traditional solution (Figure 15.25). Such system design results in high installation and operation costs. In traditional systems the lamp failure monitoring function is able to detect the number of burnt out lamps only, but not their positions.

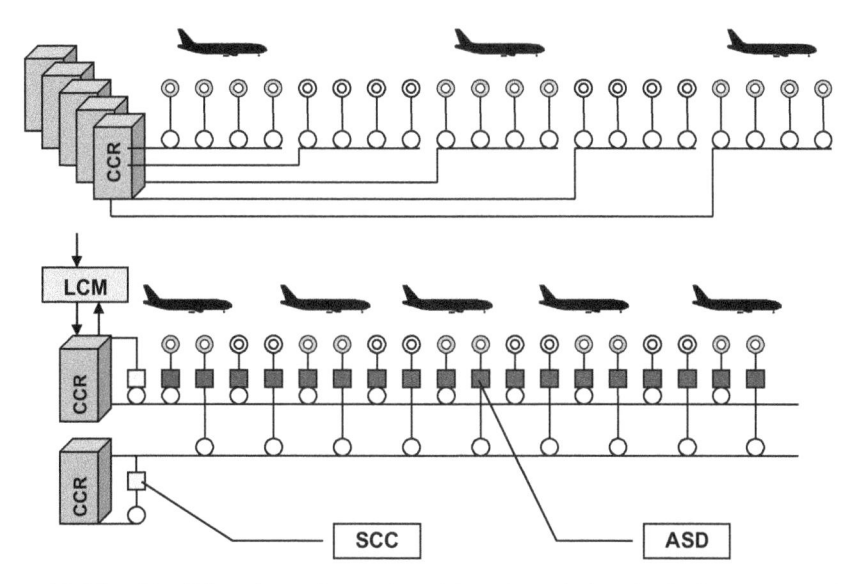

Figure 15.25: Traditional configuration (above) and the single lamp control and monitoring (below). *Source*: Honeywell Airport System brochure.

The single lamp control and monitoring system developed by Honeywell Airport Systems permits the control and monitoring of single lamps or groups of lamps within one series circuit. At large airports with complex systems of taxiways the single lamp control is a cheaper and safer alternative for aerodrome ground lighting systems. The control of a lamp, or group of lamps, is performed by an addressable switching device (ASD) which is plugged in between the isolating transformer and the lamp and is controlled via signals on the standard series circuit cable. Data transmission is achieved by a current modulation in the series cable. The constant current regulators (CCR) are controlled by the lamp control and monitoring (LCM) unit and further distributed through the series circuit coupler (SCC).

Another advantage of the single lamp control and monitoring is, depending on the weather conditions, the activation of a variable number of taxiway or runway centreline lights. During Category II or III conditions all lights must be active while during good visibility the distance between the lights can be doubled. This results not only in energy savings but it also increases the service life of lamps. Single lamp control and monitoring can also be used for a particular lamp's failure monitoring. This allows the introduction of a system of preventive maintenance and quality control for a precision approach runway Category II or III, and to reach standards of light serviceability in line with Annex 14, Volume I, Chapter 10 requirements.

There are also other companies offering similar solutions for taxiway lights control. Most of these applications could be used in the Surface Movement Guidance and Control Systems (SMGCS). The main purpose of providing an SMGCS or A-SMGCS (A — Advanced) is to enhance safety and capacity of aerodrome ground operations in particular during low visibility operations. In the 'visual' part of the system the pilot receives the information required for taxiing primarily from the taxiway centre line lighting, from stop bars and from taxiway guidance signs. The taxiway centre line lighting is only activated on the respective cleared areas. This lighting is switched on and off automatically by the taxiing guidance system after clearance or with monitoring from controllers. The stop bars established from red inset lights are installed right through the taxiway and they prohibit overruns when switched on. The stop bars are positioned in particular on the access taxiways to the runway in order to prevent the possibility of a runway incursion. Additionally, the stop bars are used at the taxiways to control taxiway traffic. More information on the A-SMGCS is given in Chapter 17.

15.2.6. Lighting Systems Construction and Operation

15.2.6.1. Lighting systems design and installation

Although ICAO and FAA have implemented a high degree of standardisation and uniformity into lighting systems in aerodromes, the installation of lighting systems is usually specific to each case. It is affected by the following factors:

- geometrical characteristics and arrangement of runways, taxiways and stands
- arrangement of power supply components (transformer substations) and the aerodrome control tower
- soil and climatic conditions
- previous habits and experience of the lighting systems operator, which are usually taken into account in the project
- anticipation of future additions to the system.

A renewal of the lighting systems is often connected with a reconstruction of the runway surface. The speed and quality of the works are influenced by the following factors:

- thorough preparation of the work, including the scheduling of work

✈ harmonisation of civil and electrical aspects of the work

✈ phasing the works to make the optimum use of climatic conditions.

Utilisation of an optimum season of the year is of principal importance, in particular in northern regions, where a favourable climate period is very short. Low temperatures and high air humidity have a negative impact in particular on installation of inset lights by gluing with epoxy resins into a concrete or asphalt pavement, and on insulation of cable couplings.

15.2.6.2. Maintenance of the lighting systems

The role of the lighting system maintenance personnel is to ensure that the equipment functions reliably during its service life. This is effected by:

✈ the technical quality of the lighting systems

✈ the quality of the project design

✈ the quality of construction works

✈ the maintenance organisation and equipment

✈ the level of knowledge and skills of the maintenance personnel

✈ the timeliness and availability of information on the real condition of the equipment in operation.

The most important parameter of the lighting systems to be monitored is the number of serviceable lights. The requirements for ensuring the serviceability are specified by Annex 14, Aerodromes, Volume I, *Aerodrome Design and Operations*, Chapter 9. A light is considered serviceable if its light intensity in the required direction has not dropped under 50% of the light intensity of a new one. Especially for in-pavement light fittings, reductions in light output well in excess of 50% can frequently occur unless a good maintenance regime is guaranteed. The reductions in light output are mainly due to contamination by dust, rubber deposits and de-icing fluids, misalignment of the optics within the light fitting and/or misalignment of the fitting. For a precision approach runway category II or III approach and runway lighting systems, measurements of intensity, beam spread and orientation of lights should be undertaken using a mobile measuring unit to analyse the characteristics of the individual lights at least twice a year for inset lights and once a year for other lights. Conformance with photometric standards may be demonstrated by the use of mobile measurement equipment which typically takes 10 to 15 minutes for runway lights measurement.

The requirement for the Category I systems is that at least 85% of lights are serviceable in:

- precision approach system
- threshold lights
- runway edge lights
- runway end lights.

Considerably more stringent requirements apply to the Category II and III systems. At least 95% of lights should be serviceable in the following sets:

- inner 450 m of the Category II and III approach system — runway centre line lights
- threshold lights
- runway end lights.

Furthermore, the following should be serviceable:

- at least 90% of touchdown zone lights
- at least 85% of lights of approach systems to a distance of 450 m from the runway threshold, and further
- at least 75% of runway end lights.

Besides the above requirements, there are supplementary conditions, such as the requirement that no two consecutive failed lights shall occur in the runway and taxiway centre line systems. In the stop bar lights, only one light failure is allowed.

With the standard 20-year service life, the initial cost of the lighting systems is a relatively small part of the total life-cycle costs. The quality of the equipment in its entirety is of prime importance.

The main effects that are directly connected with the costs on maintenance are:

- the price and service life of the light bulbs
- the resistance of the lights and fittings to corrosion
- the resistance of the lights and fittings to the effects of air traffic, such as the impact of the jet engine combustion products, snow and ice cleaning equipment
- the quality of the cables and connectors
- laying the cables into cable ducts and the isolating transformers into easily accessible shafts
- the reliability of the constant current regulators and other electric equipment

✈ the reliability of the remote control and monitoring equipment, and the timeliness and availability of information on the lighting system condition

✈ low power requirements of the lights and the related costs of electrical energy.

Figure 15.26: Solar LED light Model A702C has low power requirements, extremely long service life and is practically maintenance free. *Photo*: Courtesy Carmanah Technologies Corp.

These costs are all influenced by the costs of the materials and labour. It is best to use the same team of workers who also maintain other electric equipment. They should also participate in the installation of a new lighting system. In practice, the system of daily, weekly, monthly and half-yearly maintenance checks has been proved to be desirable. The maintenance activities are recorded in service logs. The organisation of the maintenance is designed in the light of previous practice and climatic conditions. Where the climatic differences between summer and winter are big, and inset lights are in the majority, it is convenient to carry out extensive maintenance before the arrival of the winter season, including the complete replacement of lamps in the inset lights, and another extensive maintenance intervention in spring. The maintenance in winter months may be made easier by laying the cables into cable ducts, putting the isolating transformers into shafts, and by a proper construction of inset lights.

The effectiveness and speed of reaction in the maintenance of the lighting system is influenced also by equipment as well as personnel. It is necessary to have a suitable vehicle with radio communication and radio telephone. In the vehicle, equipment should be available for inspecting the cable network, also a compressor and wrenches for maintaining the inset lights.

The light intensity of inset lights is often impaired by impurities in the optical system, such as the remains of rubber from aircraft tires. In order to maintain a permanent light intensity to the required limits, it is necessary to clean the inset lights. Cleaning may be carried out by compressed air containing crushed walnut shells or olive stone pips, compressed liquids, mechanical brushes or a combination of these. Full cleaning of the optical parts of inset lights is carried out in maintenance workshops when the light bulbs are replaced.

15.2.7. Trends in Lighting Systems Development

It may be expected that the future development will use the serial distribution of aerodrome lighting systems, except heliports, where a parallel supply system will be applied. The manufacturers will make an effort to produce unified equipment, including a wide variety of normative requirements. Reduction of power requirements of the lights and utilisation of light sources with long service life provides the manufacturers with a competitive advantage. Operationally undemanding inset lights of good quality will find their implementation in connection with upgrading the aerodromes to higher categories of meteorological limits.

A modern lighting system is characterised by small power requirements, long service life of components and light sources, and easy, quick and cheap maintenance (Figure 15.26). The control and monitoring systems will provide efficiency for the operators and diagnostics of the system condition for the needs of an effective maintenance operation.

16

ELECTRICAL ENERGY SUPPLY

Tony Kazda, Bob Caves and František Bělohradský (deceased)

16.1. BACKGROUND

The consumption of electrical energy of a large international airport may be compared with that of a town with 15,000–30,000 inhabitants. Besides electrical energy required for the aids to air traffic operation such as lighting systems, radar, communication and meteorological systems, it is necessary to take into consideration electricity consumption in hangars, buildings and other airport facilities. This chapter will deal only with the supply and distribution systems of electrical energy for lighting systems and other equipment for air traffic operation.

16.2. ELECTRICAL SYSTEMS RELIABILITY AND BACKUP

For lighting systems, telecommunication and radio-navigation aids, Annex 14, Volume I, Chapter 8 specifies the requirements for a standby electrical energy source in terms of the maximum switch — over time to a standby power supply. The requirement for a standby source may be fulfilled by constructing an independent supply of electrical energy to the aerodrome, connected with a 100 kV transformer station, or by a standby source complying with the required parameters. The decision to construct an independent supply line to the aerodrome should take account not only of investment costs, but also the electrical energy supplier's pricing policy and the use that may be made of surplus electrical energy.

On small aerodromes, the power to all the equipment may be covered by a central standby source. On large aerodromes, the number of installations that must be safeguarded increase, as do the requirements for power output of standby sources and the distances between the installations. The requirements for standby sources vary considerably in switching time and in the required power output of the standby source, according to the kind and category of the aerodrome. The greater the distance of the installation from the standby source, the lower is the reliability of the whole system as a consequence of possible failures in transmission lines. Therefore at larger aerodromes it is not practically possible to cover the entire consumption by a single, central standby source. For that reason, a decision on the number of standby sources and their arrangement should be made not only on the basis of an

410 Airport Design and Operation

economic assessment, but consideration should be given also to the reliability of the electrical energy supply. For Category II and III precision approach runways, the probability of failure of electrical energy should not by higher than 1×10^{-1} per year. The design and provision of electrical power systems for aerodrome visual and radio navigation aids must guarantee that an equipment failure will not leave the pilot with inadequate visual and non-visual guidance or misleading information. In practice we should guarantee the approach and landing of aircraft after passing the final approach fix (FAF).

Figure 16.1:   Short-break diesel generator. *Photo*: A. Kazda.

The standby sources, which are in the majority of cases diesel generators, are usually located directly in the transformer stations that feed the lighting systems (Figure 16.1). A special separate room is established for them, in which a constant temperature should be maintained, exhaust gases off-take and dispersion should be provided, and the fuel store reliably designed to prevent fuel accident/ incidents. There are also new independent standby generator systems in standard containers fulfilling these requirements. In the majority of cases, the radio-navigation and communication systems are backed up by battery sources with automatic switching.

Table 16.1: Secondary power supply requirements.

Runway	Lighting aids requiring power	Maximum switch-over time
Non-instrument	Visual approach slope indicators Runway edge Runway threshold Runway end Obstacle	1 minute 1 minute 1 minute 1 minute 1 minute
Non-precision approach	Approach lighting system Visual approach slope indicators Runway edge Runway threshold Runway end Obstacle	15 seconds 15 seconds 15 seconds 15 seconds 15 seconds 15 seconds
Precision approach Category I	Approach lighting system Runway edge Visual approach slope indicators Runway threshold Runway end Essential taxiway Obstacle	15 seconds 15 seconds 15 seconds 15 seconds 15 seconds 15 seconds 15 seconds
Precision approach Category II/III	Inner 300 m of the approach lighting system Other parts of the approach lighting system Obstacle Runway edge Runway threshold Runway end Runway centre line Runway touchdown zone All stop bars Essential taxiway	1 second 15 seconds 15 seconds 15 seconds 1 second 1 second 1 second 1 second 1 second 15 seconds
Take-off RWY for RVR <800 m	Runway edge Runway end Runway centre line All stop bars Essential taxiway Obstacle	15 seconds[a] 1 second 1 second 1 second 15 seconds 15 seconds

Source: Annex 14, Aerodromes.
[a]One second where no runway centre line lights are provided.

The requirements for a standby source of electrical energy and switching time are given in Table 16.1. According to Annex 14 Volume I, the switch-over time for lighting systems is defined as the time required for the actual intensity of a light measured in a given direction to fall below 50% and recover to 50% during a power supply changeover, when the light is being operated at intensities of 25% or above.

In order to ensure those requirements are met, a whole range of technical solutions may be adopted. Besides an independent supply of electrical energy, some aerodromes have heating stations equipped with low pressure steam turbines and generators which can supply electrical energy to selected facilities. This is termed co-generation. At some airports, the heating/power stations serve as the main source of electrical energy (e.g. Munich airport) and the connection to the 'public' high voltage line is the standby. The standby sources may be divided into the following groups:

Rotational standby sources — Diesel generators:

✈ with manual start-up — within 1 minute

✈ with automatic start — within 15 seconds

✈ short-break — within 1 second

✈ no-break — uninterrupted power supply.

Static standby sources:

✈ battery short-break — within 1 second

✈ battery no-break — uninterrupted power supply.

Diesel generators with manual start-up within 1 minute are used in smaller aerodromes with a non-instrument or an instrument runway. On larger aerodromes, they may be used for the provision of supplies of electrical energy to individual facilities such as the lighting of buildings, lifts, and the like. In general, they provide outputs up to 2.5 MVA.

Diesel generators with automatic start within 15 seconds are connected with an automatic control box which monitors the parameters of the electrical network. When the specified voltage and frequency parameters of the network drop under the determined limit, the automatics will start the diesel generator and switch the supply over to the standby source. If the first start is unsuccessful, the automatics will repeat the start procedure after several seconds. Diesel generators with automatic start are used not only as separate equipment, but also in combination with battery or other backup sources (see Figure 16.4), where a battery or different source of energy provides an immediate backup of electrical energy in the event of a failure of the network, and after starting, the diesel generator will recharge the system.

In the 'classic' systems a short-break diesel generator (Figure 16.2) has a heavy flywheel, which is driven at a constant speed through a gearbox by an electric motor supplied from the mains. An alternator, which in the normal state of the network runs with no load, is on one side of the flywheel shaft. On the other side of the shaft, there is a diesel generator connected by means of an electromagnetic clutch. When the monitored parameters of the electric network drop below the determined values, the diesel generator is automatically started and connected by means of the electromagnetic clutch to the flywheel and alternator. While the diesel generator reaches its full power, the alternator is driven by the turning flywheel. At the same time the automatics switch the appliances over to the alternator terminals. In the event of a failure of electrical energy, there is a short-term voltage drop of about 10%, and a frequency drop of about 4%.

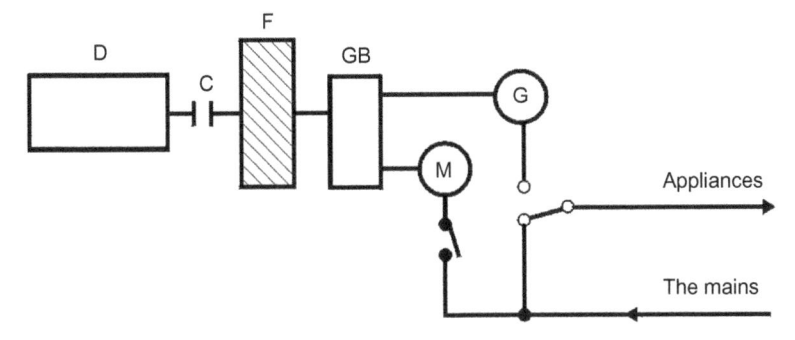

Figure 16.2: Short-break diesel generator. D, diesel generator; C, clutch; F, flywheel; GB, gearbox; M, motor and G, alternator.

The no-break diesel generator — or uninterruptible power supply (UPS) (Figure 16.3) is used when no short-time electric current failure is permitted, for example in a supply of current for any of computer systems or secondary radar. The no-break equipment differs from the previous type in that the electric motor constantly drives not only the flywheel, but also the alternator, from which electricity is supplied to appliances. In the event of an interruption of the mains supply, there will be only a very short-term voltage and frequency drop.

The traditional short-break and no-break equipment is expensive to purchase and to operate. This is not only due to requirements to provide a high reliability of the system, but also due to the need for high power inputs for the lighting systems of higher categories of landing aids. Some firms supply equipment that may be used differently according to meteorological conditions, normally as a diesel generator with automatic start, and when meteorological conditions drop under the determined limits, it may be switched over to short-break and no-break modes.

A different approach is represented by the NZ² system offered by the Caterpillar company. The immediate power backup is provided by a high energy-density integrated flywheel-motor-generator unit to store power (Figure 16.4). It is called Kinetic Power Cell Technology. The 273 kg heavy flywheel rotates vertically on a ceramic bearing in a vacuum chamber and is magnetically lifted to unload the flywheel bearings. When the flywheel reaches 4,000 RPM it is fully functional and capable of supporting the load during a power outage. The full charge of the flywheel is at 7,700 RPM. Total time to complete the start-up sequence to full charge status is less than 5 minutes. The UPS system is nearly maintenance free. It needs just basic maintenance, diagnostics and the bearing cartridge replacement approximately every three years.

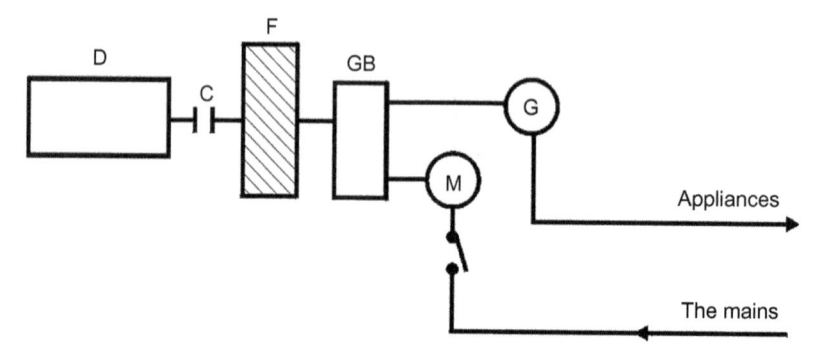

Figure 16.3: No-break diesel generator. D, diesel generator; C, clutch; F, flywheel; GB, gearbox; M, motor and G, alternator.

Power from the Caterpillar's UPS with Kinetic Power Cell Technology is immediately available up to a critical load. The UPS provides only 15 seconds of backup time, but this is enough for start-up and connecting a diesel-generator which can take the full load. The system has very high energy efficiency, between 96 and 97%. It is much more environmentally friendly compared to the battery-based systems. The NZ² system is delivered in a 6 m long standard container.

Static or battery-based UPS systems are more common than any other alternative UPS system. They work by storing power in a string of batteries. When power from the utility is abnormal, power is pulled from the battery string. After the energy supply returns to normal, the batteries are recharged in anticipation of the next event. Battery based UPSs are typically used for low capacities and immediate backup of computers, communication equipment or navigation aids with low energy consumption. The endurance required before the diesel generator starts (if necessary) may vary from several minutes up to several hours. Modern batteries with a closed cycle can have a high service life up to 10 years, and high capacity with small dimensions. Usually the small UPSs are installed as close as possible to the appliance but it is also possible to centralise supply for the systems with a fall-out not longer than 1 second to a single transformer station. The battery strings must be housed in a separate room that must be well ventilated. Most common valve-regulated lead acid (VRLA) battery-based systems have extensive maintenance needs. Regular inspection, terminal cleaning, tightening of connections and voltage measurements are crucial. In addition, several tests must be done to determine battery performance levels. Battery backup systems work with 86–93% energy efficiency.

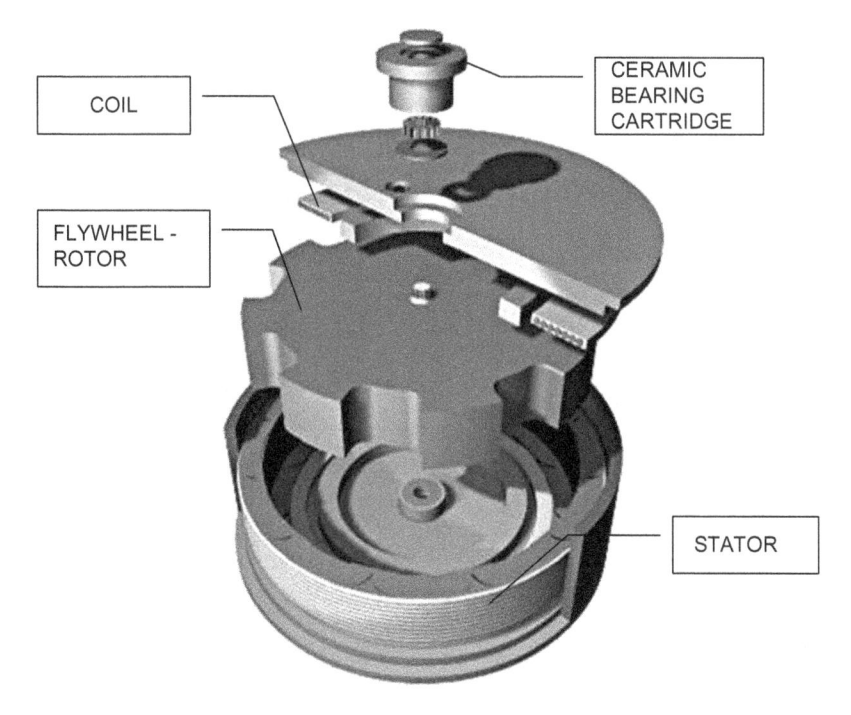

Figure 16.4: Integrated flywheel-motor-generator unit. *Source*: Courtesy Phoenix Zeppelin Ltd.

In spite of the fact that the static standby sources are still used mostly for appliances with low consumption, sometimes they are still used as sources with very high outputs for a short time. A static standby source may be designed as short-break but usually as no-break — UPS, according to how it is connected (see Figure 16.5). The capacity of the high output battery systems is typically designed to be able to work for 15 minutes under full load. After that it is supposed that the backup is provided by diesel generators which also recharge batteries.

Some types of batteries used in UPS systems are more reliable than others but they are still considered to be the weak link in backup power sources. In fact, 70% of all UPS problems are directly related to their batteries. The main problem is that even with repeated load testing, there is no satisfactory evidence that the battery system will survive next full load discharge.

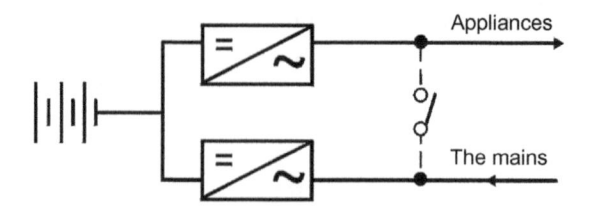

Figure 16.5: Static standby source; no-break — full line, short-break — dashed line.

16.3. SUPPLY SYSTEMS

The supply of lighting systems may be provided in two ways, namely by parallel or serial distribution systems.

16.3.1. Parallel System

The well-established parallel distribution system (Figure 16.6) is used at present for small lighting systems. It uses a constant voltage source. A three-core cable is connected to the switchboard of the transformer station equipped with a stepping control transformer for reducing voltage. Every individual light is connected to cable conductors by means of the isolating transformer. An advantage of the parallel distribution system is its simplicity.

The maintenance personnel need not be qualified for work with high voltage. The current in halogen lamps of the parallel distribution system is not regulated. As the tungsten gradually settles on the lamp filament, the lamp resistance drops and the current in the lamp increases, the tungsten melts, evaporates, and the lamp acquires its initial parameters. These lamps give a constant light during their whole service life. The main disadvantage of the parallel distribution system is that the voltage drops with distance from the transformer station, and thereby also the intensity of the more distant lights drop. For that reason, the parallel distribution system is only suitable for small areas, such as the lighting systems of heliports or aprons.

16.3.2. Serial System

16.3.2.1. Serial System — The Principle

The serial distribution system (Figure 16.7) is now almost universally adopted for runway lighting systems. Contrary to the parallel distribution system, which is supplied with constant voltage, the serial distribution system operates with constant current. It consists of a closed loop of single-core cable. Isolating transformers are put into the circuit, which ensure that there will not be a break of the primary circuit when the secondary circuit is interrupted (lamp bulb failure) and that the remaining lights continue to operate.

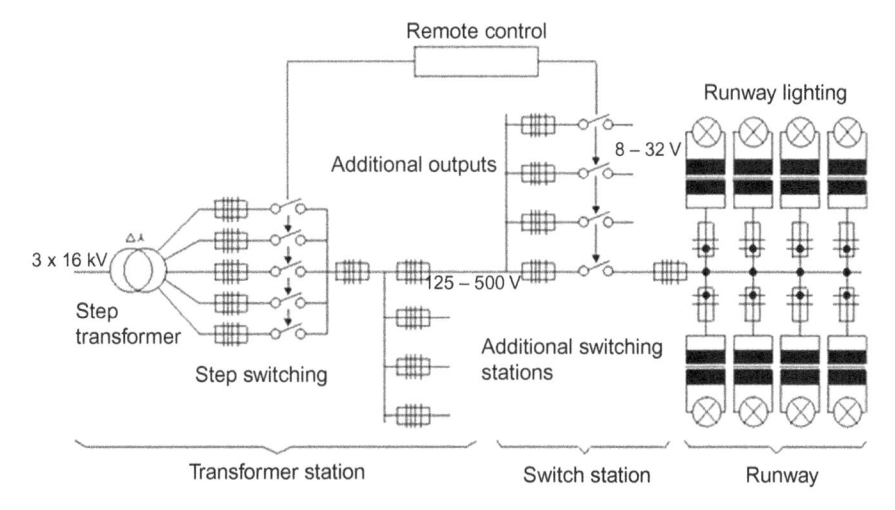

Figure 16.6: Parallel distribution system — principle.

16.3.2.2. Serial System — Components

The distribution cable, the quality and reliability of particular components, and the quality of installation all contribute to the reliability of the lighting systems. The components of a serial distribution system are:

+ single-core high voltage supply cables
+ high voltage connectors
+ two-core low voltage cable
+ low voltage connectors
+ isolation transformers
+ contact boxes
+ constant current regulators
+ system of control and monitoring of the lighting systems.

418   Airport Design and Operation

The number of elements in a lighting system increases with the airport category. Therefore for lighting systems of higher categories, components of higher quality should be used in order to avoid a loss of reliability of the whole system. The system of a Category II or III runway includes up to 100 km of single-core high voltage cables, approximately 1,000 isolation transformers, 2,000 high voltage connectors, and a large number of low voltage connectors.

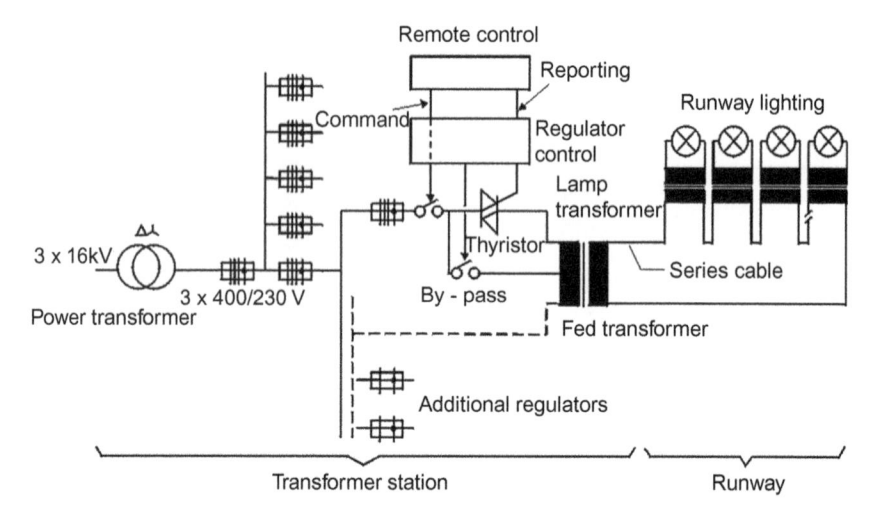

Figure 16.7:   Serial distribution system — principle.

High voltage cables are normally single-core cables with a copper core with a cross section of 6 mm^2. They are designed for an operating voltage from 500 to 5000 V. The usual range of operating voltages is 500, 600, 750, 1,200, 3,000 and 5,000 V. The cables are unshielded at the lower voltages. Shielded cables limit the interference caused by the thyristor regulation of the constant current regulators. Cables can be made in exact lengths, and equipped with standard connectors while in the production plant. Cable is more often supplied on drums, and connectors are mounted on site at the aerodrome. If the connectors are made in the production plant, the final mounting of the lighting system can be completed more quickly.

Advantages of a cable system with connectors mounted in the aerodrome are:

✈ simplification of design works

✈ simpler production

✈ savings of up to 10% of cable length

✈ easier installation when cable ducts and transformer shafts are used, since the required lengths may be provided with a centimetre precision.

Current is fed from the isolation transformer to the lights via low voltage (LV) cables. The LV cables are equipped with standard FAA connectors. In the event of the light failing, the connector is severed, and the light disconnected from the cable. If the light is distant from the isolation transformer, an extending LV cable (flexo) is used. This is a two-core LV cable equipped with a standard termination and FAA socket. LV cables are designed for a voltage of 250 or 600 V, while the operating voltages are usually between 6 and 45 V.

Figure 16.8: Cable installation in ducts and isolation transformers in shafts allows maintenance also in hard winter conditions. *Photo*: A. Kazda.

The isolation transformers are enclosed in rubber or plastic material, which ensures that they remain water-tight (Figure 16.8). The isolation transformer is provided with two high voltage (HV) leads for connecting two power cables and one LV lead for connecting to a light. The transformer output range is 30, 45, 65, 100, 150, 200 and 300 VA. The HV leads are designed for a voltage of 5,000 V, the LV lead for a voltage of 250 or 600 V.

The serial lighting systems are fed from constant current regulators. These are power supply transformers with control and monitoring circuits. The control of the intensity of the lights is most often in five steps that are 1%, 3%, 10%, 30%, 100% of light intensity. In some systems, such as PAPI and taxiway side lights, the light intensity is controlled in three steps of 10%, 30% and 100%. The seven steps control, with 80% light intensity, is used only occasionally. That setting ensures high light intensity, and does not reduce the service life of the light sources. The monitoring system provides information about the condition of the constant current regulators in terms of condition of the loop's insulation and the number of failed lights to indicate the need for maintenance of the lighting systems and to alert the air traffic controllers. The constant current regulators are placed in transformer stations and supplied from LV switchboards. Connector boxes, which allow several circuits of the lighting systems to be connected to a single constant current regulator, are also part of the supply system.

The constant current regulator maintains constant current of a specific value throughout the whole circuit despite a varying load, caused, for example, by burnt-out filaments in the lamps. The constant current regulators usually have an output range of 4, 10, 20, and 30 kVA. In more advanced lighting systems, which have a lower power demand, the power output range is usually 2.5, 7, 7.5, 10, 15, 20 and 25 kVA, though power outputs over 15 kVA are rarely used. In the power input loop of 15 kVA, a power cable with an operating voltage not higher than 3000 V may be used. Output current of constant current regulators is usually 6.6 A at the maximum intensity of the lights. At lower degrees of light intensity, the output current is proportionally lower.

A constant current regulator must indicate certain conditions:

- operation status
- failure
- insulation condition of the loop
- number of failed lamps
- degree of light intensity in the loop.

A clock records the number of hours the loop operates at a maximum light intensity and the total number of hours of operation. In simple systems, the constant current regulators usually indicate only operation, failure and insulation condition of the loop.

The constant current regulators are equipped with over-current and over-voltage protection, which automatically disconnect them so that the lights cannot burn out if the power supply cable or separation transformers are damaged. An electric charge may be induced in the long loops of the lighting systems when lightning strikes near the lights during a storm, which may damage the constant current regulators. Therefore the loops are connected with the constant current regulators by means of a lightning arrester. Another possible solution is the installation of a strong non-insulated and grounded conductor in the ground over the power cables.

In some cases it is convenient to supply several circuits of the lighting systems from a single constant current regulator, for example two PAPI systems, a few STOP lighting bars, various sections of taxiway inset lights, etc. Connector boxes placed beyond the constant current regulators, or in modern lighting systems, directly built into the constant current regulators, serve for switching of the specific loops. The boxes are equipped with HV connectors. On the basis of instructions from the control tower, the particular circuits are connected according to the requirements and pre-determined algorithms. For example, the runway centre line inset lights are connected to two circuits. Only one of them is always switched on, according to the runway that is in use. The runway centre line inset lights are lit only in the direction either of landing or of take-off. The PAPI systems are similarly supplied from constant current regulators for both directions of landing. In the case of taxiways, the connector boxes operate in a different way. According to the instructions from the control tower, one, two and even more circuits whose lights may be lit simultaneously are connected. A typical example may be supplying two stop bar lights from a single constant current regulator.

Modern constant current regulators are controlled by a microprocessor. The microprocessor regulates and monitors all the conditions and protections of a regulator. A more reliable and simpler communication system is also provided between a constant current regulator and the control and monitoring system, including an unambiguous identification of a failure of the regulator indicated in the office of the maintenance manager. The system of A-SMGCS is discussed in Chapters 15 and 17.

In order that the reliability may be improved, the lighting system supply is provided by a whole range of circuits. Depending on the number of insulation transformers in the loop, their power input, and HV cable length, the power input of the constant current regulator may be determined. The biggest load is usually in the loops of an approach system of high light intensity, and in the loops of runway edge lights. Every such loop supplies from 60 to 100 lights with a power input of 150–200 W, and the loop power input may reach up to 20 kVA. Small lighting systems, for example STOP-bar lights, PAPI and simple approach lighting systems require relatively small power input. The power input of the lights, including losses in the separation transformers, is given in Table 16.2.

The losses in the cable reach approximately 130 W per 1 km of the cable length. The calculation of the power input of the loop of an approach system of high light intensity is given in the following example.

Example — power input of the loop:

$$P = A \times B + L \times Z = 65 \times 170 + 3.2 \times 130 = 11,466 \text{ VA}$$

where P is power input of the loop; A is number of lights (65 pcs); B is power input of the light (177 VA); L is cable length (3.2 km); and Z is loss in the cable (130 W/km). In this case, the calculation shows that the load from the lights is decisive, while the losses in the cable are of little importance.

Table 16.2: Power inputs of the lights.

Light bulb (W)	Power input (VA)
45	55
65	77
100	115
150	170
200	222
300	325

16.4. Electrical Supply to Category I–III Lighting Systems

The lighting system of an aerodrome contains a whole range of systems which should be reliably supplied and controlled. Each aerodrome has its specific features. The electrical supply for a lighting system of low or middle light intensity of a non-categorised aerodrome is usually simpler, and contains from 2 to 8 constant current regulators. On the other hand, Category II and III systems are provided with a more extensive supply system containing a greater number of constant current regulators. For a Category II aerodrome, the usual number of constant current regulators is 15–25 in one transformer station. The electrical supply to the extensive taxiway systems in Category III is very complicated. The centre line guidance for a runway with inset lights is divided into a whole range of separately fed sections, requiring a large number of constant current regulators and connector boxes. For example, Munich airport, equipped with the lighting systems for Category III, is divided into 1,200 separately supplied circuits.

17

RADIO NAVIGATION AIDS

Tony Kazda and Bob Caves

17.1. BACKGROUND

The pioneering stage of radio navigation can be traced back to the 1930s when the radio direction finders (RDF) and the LORENZ systems (the name of the factory) ultra-short-wave landing equipment were developed and used in Germany in military operations for the first time. The LF/MF (low frequency/medium frequency) radio range system has served the US civil airways since about 1931, and a network of hundreds of stations covered all parts of the nation.

Today the vast majority of commercial flights are carried out under Instrument Flight Rules (IFR) even in good weather conditions. Many airline companies have strict rules that all their flights must be operated under IFR. Visual reference is, however necessary for the final approach and landing in most cases. Thus radio navigation aids create a complex integrated system with visual aids and electric supply systems to allow regular and safe air transport operations. Radio navigation aids are also referred to as non-visual aids. The required performance of the non-visual aids varies with the category of runway at which it is used. There is a wide range of equipment and facilities which could be divided into two main groups:

↣ Radio navigation aids
↣ Radars and radar based systems.

The system is complemented by voice communication systems of moving and fixed radio communication networks. In this chapter an outline of equipment and facilities related with the airport operation is presented.

The technical requirements and characteristics of non-visual aids are specified primarily by *ICAO Annex 10 Aeronautical Telecommunications, Volume I Radio Navigation Aids* and Volume IV *Surveillance Radar and Collision Avoidance Systems*.

The installations and locations of particular navigation aids and systems are usually specified on the basis of the regional air navigation agreements which are valid for a

424   Airport Design and Operation

period of at least 5 years. The non-visual aids used for approaches, landings and departures and for the terminal area and aerodrome surveillance and control are the following:

- Instrument Landing System (ILS)
- Microwave Landing System (MLS)
- Global Navigation Satellite System (GNSS)
- VHF Omnidirectional Radio Range (VOR)
- Non-Directional Radio Beacon (NDB)
- UHF Distance Measuring Equipment (DME)
- Transponder Landing System (TLS)
- Precision Approach Radar (PAR)
- Surveillance Radar Element (SRE)
- Surface Movement Radar (SMR)
- Advanced Surface Movement, Guidance and Control Systems (A-SMGCS).

GNSS, VOR, NDB, DME and SRE are used also as en-route navigation facilities.

17.2. RADIO NAVIGATION AIDS

17.2.1. Instrument Landing System (ILS)

The first ILS tests started in 1929 in the United Kingdom and six systems were installed and certified by the CAA in 1941. The first ICAO certified system using army standards was adopted in 1949. In the 1960s the first ILS equipment for 'blind' landings was approved. The system principles remained unchanged until today.

Each ILS system comprises the following components:

- VHF localizer (LLZ)
- UHF glide path (GP)
- VHF marker beacons.

All three parts are equipped with monitoring systems, remote control and indicator equipment.

The *localizer* antenna provides the directional guidance on approach and landing. It is a massive and large-scale construction to guarantee the emitting characteristics (Figure 17.1). It is located typically between 300 and 600 m behind the runway end. The distance must ensure that the antenna construction lies below the take-off/ approach obstruction clearance surface (see Chapter 3). The minimum distance behind the runway end is 300 m to satisfy the Runway End Safety Areas requirements (see Chapter 5). The housing for LLZ transmitters is usually located 100–120 m to the side of the aerial. The LLZ antenna must be positioned on the extension of the runway centreline.

Figure 17.1: The ILS LLZ 13 element antenna – Žilina Airport, fencing in the LLZ vicinity may not be metallic. *Photo*: A. Kazda.

The VHF localizer operates in the band from 108 to 111.975 MHz. The localizer and glide-slope frequencies are paired in accordance with Annex 10, Volume V specifications and only the localizer frequency needs to be tuned in the aircraft. For example, every localizer operating at 110.3 MHz has a glide slope operating at 335.0 MHz. If the localiser is equipped with a DME, it is also paired.

The LLZ antenna emits two horizontally polarised beams in the shape of a 'lobe'. When approaching the runway, the carrier frequency on the right side 'lobe' is modulated by a 150 Hz tone and in the 'lobe' on the left side 90 Hz frequency modulation is predominant (Figure 17.2).

426 Airport Design and Operation

The receiver in the aircraft determines the Difference in the Depth of Modulation (DDM) between the 90 and 150 Hz signals. On the course line, that is the extended runway centreline, there is a locus of points of equal 90–150 Hz modulation. The localizer signal coverage must be sufficient for a typical aircraft's avionics installation. It must extend from the antenna centre to the distance of:

- 46.3 km (25 NM) within ±10° from the centreline,
- 31.5 km (17 NM) between ±10° and ±35° from the centreline and
- 18.5 km (10 NM) outside ±35° if coverage is provided.

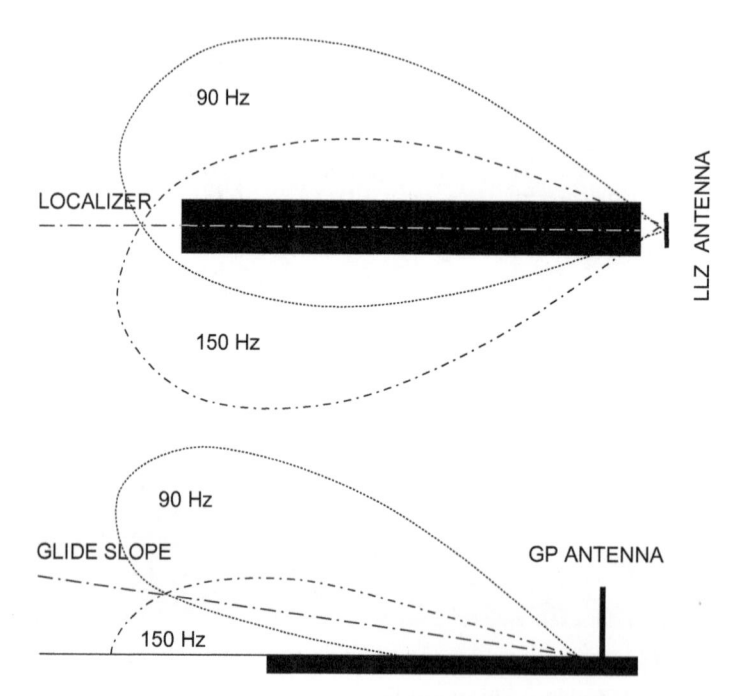

Figure 17.2: The radiation pattern of the localizer and glide-slope antennas. *Source*: Instrument Landing System Operational Notes, Dept. of Aviation, CAS Australia.

The localizer signals must be receivable vertically within the specified coverage at and above the 600 m height above the threshold elevation or 300 m above the elevation of the highest point within the final approach areas, whichever is higher. The upper limit is defined by a 7° inclined surface.

Depending on the ILS performance Category, there are strict minimum field strength signal requirements and other signal quality parameter specifications within the localizer coverage. To guarantee the signal quality in mountainous or in broken terrain, special two-frequency systems could be used. The coverage can be further improved by an LLZ antenna with a greater number of elements.

The *glide path* antenna is typically located 225–380 m from the runway threshold and 120–210 m to the side of the runway centreline, usually on the side where the interference from passing aircraft is smaller (Figure 17.3). The glide-slope signals are partially formed by reflection from a steel grid on the ground in front of the GP antenna. The UHF glide path operates in the band from 328.6 to 335.4 MHz. Also a GP transmitter emits modulated signals with the 150 Hz tone predominating below the path and the 90 Hz tone above the path. The glide path equipment should be adjustable by 1° above or below the standard approach angle of 3°. Within ± 8° from the centreline the signal coverage must be guaranteed to a distance of at least 18.5 km (10 NM).

Figure 17.3: The ILS GP antenna, view in the direction of approach. *Photo*: M. Hromádka.

The *marker beacons* operate on the frequency of 75 MHz and radiate a fan shaped field pattern vertically upwards. There are usually two marker beacons installed (middle marker and outer marker) or, if it is required by the approach procedures, a third – inner marker, can be added. The marker beacons are installed in defined positions on the approach where the pilot could check the correct height of the aircraft.

The outer marker (OM) is located typically 7.2 km (3.9 NM) from the threshold, except when it is not practicable, in which case it must be placed between 6.5 and 11.1 km (3.5 and 6 NM) from the threshold. The outer marker modulation is 400 Hz and it transmits two dashes per second continuously. The purpose of the OM is to provide height, distance and equipment functioning checks to aircraft on intermediate and final approach.

The middle marker (MM) should be located 1,050 m (3,500 ft) or ±150 m (500 ft) from the threshold. The middle marker modulation is 1,300 Hz and it transmits alternate dots and dashes. The purpose of the MM is to indicate the imminence, in low-visibility conditions, of visual approach guidance.

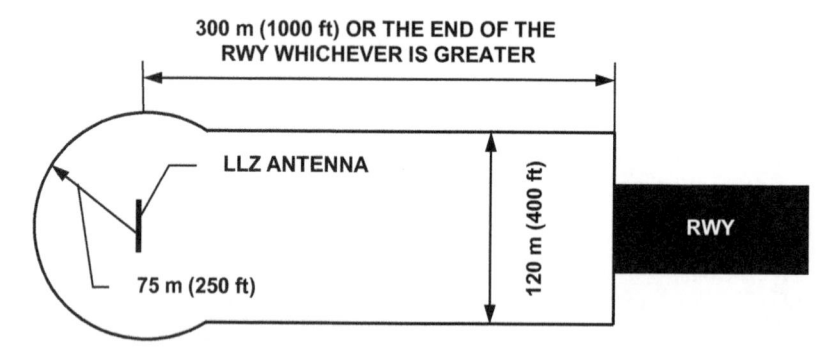

Figure 17.4: The LZZ critical area. *Source*: ICAO Annex 10 Aeronautical Telecommunications, Volume I.

The inner marker (IM), if installed, is located between 75 m (250 ft) and 450 m (1,500 ft) from the threshold. The inner marker modulation is 3,000 Hz and it transmits continuously at six dots per second. The purpose of the IM is to indicate in low-visibility conditions the imminence of arrival at the runway threshold. The IM location is at the point of the lowest decision height for the Category II minima.

To guarantee the quality of the ILS signals, the terrain around the localiser and glide path antennas must be perfectly level without depressions or uphill slopes. In winter the area must be cleared of high snow deposits which may cause signal distortions. During Category II or III operations, ILS critical and sensitive areas must be kept free and protected at the airport.

The localizer critical area dimensions depend on the LLZ position with respect to the runway end (Figure 17.4). The sensitive areas are larger and depend on the ILS/

runway Category, the type of LLZ antenna and the critical aircraft which uses the runway under the low-visibility conditions.

17.2.2. Microwave Landing System (MLS)

A Microwave Landing System (MLS) is a precision approach and landing guidance system. It was originally designed as a replacement for ILS after 2010. The first version of MLS was developed and tested in Australia at the end of 1970s. The version called Interscan was approved in the United States by the FAA in 1975 and three years later also by ICAO. MLS has many advantages compared to ILS.

MLS can operate on 200 channels in the frequency spectrum from 5,031.0 to 5,090.7 MHz, comparing with only 40 channels in the case of ILS. The antenna is typically located 300 m behind the runway end but it does not need to be installed in a specific position at an airport. MLS antennas are much smaller and lighter than ILS/LLZ due to the shorter wavelength of the frequency spectrum.

In its basic configuration the system has five functions:

- Approach azimuth
- Back azimuth
- Approach elevation
- Range
- Data communications.

All MLS signals, except the DME, are transmitted on a single frequency only through time sharing. The MLS beam scans across the coverage area at a fixed time rate and both azimuth and elevation data can be calculated by an airborne receiver which measures the time interval between scans.

The system is very resistant to any type of interference by fixed or moving objects or snow accumulations. The signals are more stable than in the case of ILS. MLS does not need protection of large sensitive and critical areas. It is hardly disturbed by aircraft in flight or when taxiing on the ground. Because of this the aircraft separation on approach during low-visibility operations can be smaller, resulting in higher runway capacity during these conditions. Also the accuracy of MLS is much higher than with ILS.

However, the system was not taken up quickly enough. By the 1980s MLS was installed experimentally in only a few locations. For the airline companies it would have been necessary to install MLS in addition to ILS avionics if flying to those few airports equipped with MLS systems. In 1994 the United States decided to terminate further MLS testing in favour of the GPS which should supplement or replace existing MLS systems. Phasing down of MLS systems in the United States was planned to begin in 2010. MLS systems, however, will still be used for military operations as the MLS can be utilised as a mobile system.

Because of specific weather conditions in the United Kingdom with a relatively high percentage of Category III operations and radio frequency congestion in the core London area, the United Kingdom decided to proceed to MLS implementation without relying on GNSS systems. MLS is operated at Heathrow airport in parallel with ILS. There are specific separation procedures to allow smooth operations for aircraft equipped with ILS or with MLS to increase the runway capacity during Category II and III conditions.

17.2.3. Global Navigation Satellite Systems (GNSS)

For the purposes of Civil Aviation a Global Navigation Satellite System (GNSS) is defined as a global system for positioning and timing which includes the constellation of one or more satellites, aircraft receivers and system integrity monitoring; extended if necessary, to support required navigation performance for the intended operation. Global navigation satellite system is a designation not only for the basic systems such as GPS, GLONASS, Galileo and BeiDou but also augmentation systems such as Aircraft-Based Augmentation System (ABAS), Ground-Based Augmentation System (GBAS) and Satellite-Based Augmentation System (SBAS).

There are several global navigation satellite systems under construction or in a development or testing phase. At the beginning of 2007 only one system was fully functional – the US Global Positioning System (GPS). The GPS was developed, funded and is controlled by the US Department of Defence. GLONASS development was initiated in 1976 in the Soviet Union and the system was completed in 1995. But its practical usage was delayed due to delays in receivers development and the full global coverage was achieved as late as 2011.

The basic operating principle of all satellite systems is simple and is the same. The crucial element is an atomic clock measuring time very accurately. Each satellite sends out personalised signals indicating the precise time when the signal left the satellite. The receiver station is able to distinguish a particular satellite and from the time delay between transmission and reception of each GPS radio signal it calculates the distance to each satellite. To fix the receiver's position simultaneous signals from at least four satellites must be received (Figure 17.5).

Even though there are many thousands of civil users of GPS world-wide, the system was designed for and is operated by the US military. GPS fulfils the civil aviation performance criteria specified in *ICAO Annex 10, Aeronautical Telecommunications, Volume I Radio Navigation Aids* for basic navigation, initial approach procedures and non-precision approaches. However, for the precision approach performance criteria specified for the system's accuracy, integrity and time to alert, continuity and availability, it is necessary to improve the accuracy parameters of the system. For example, in the ground-based systems like ILS or MLS, the emitted signals can be easily monitored and the response to the transmitter status is immediate. There is no such way of rapidly warning the GPS user of a 'standard' GPS failure. The most promising satellite-based navigation system, with high positional accuracy of at least 7.6 m, is the Wide-Area Augmentation System (WAAS) where the satellite constellation is complemented by a Ground Segment of multiple Wide-area Reference Stations (WRS) and the Spaces Segment with geosynchronous communication satellites to fulfil civil aviation user needs. The system was developed by the FAA and the US Department of Transportation. As an alternative to the US WAAS system, the European Geostationary Navigation Overlay Service (EGNOS) was developed in Europe. It is a Satellite-Based Augmentation System (SBAS) consisting of three geostationary satellites and ground reference stations located throughout the operational area. It provides reliability and accuracy of the positioning data for existing satellite navigation systems.

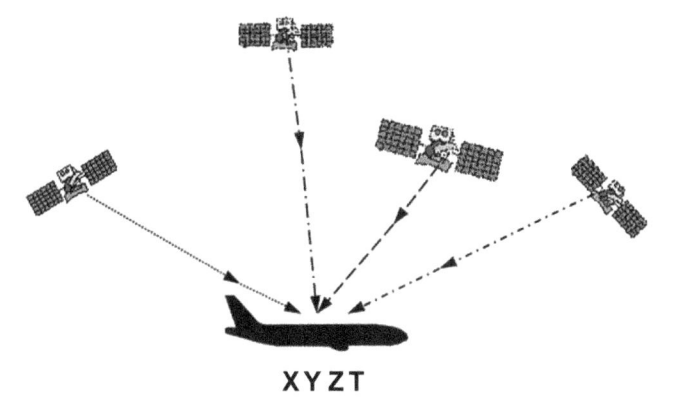

Figure 17.5: From the time delay the distance from satellites and the position is calculated.

The new Galileo satellite navigation system which is being built by the European Union should meet all requirements for the civil aviation systems. According to the Galileo Headquarters information,[1] full completion of the 30-satellite Galileo

---

1. http://ec.europa.eu/enterprise/policies/satnav/galileo/satellite-launches/index_en.htm

432 Airport Design and Operation

system is expected by 2019. Another navigation satellite system under development is the Chinese BeiDou, which plans to provide global coverage around 2020.

17.2.4. VHF Omnidirectional Radio Range (VOR)

VHF omnidirectional radio range (VOR) was developed and internationally standardised in 1949 to provide safe navigation coverage over the 360° azimuth. The frequency spectrum, between 108.0 and 117.95 MHz with 50 kHz channel separation, ensures that the signal is practically insensitive to atmospheric disturbances. VOR has much higher accuracy and reliability than, for example, NDB.

Figure 17.6:   D-VOR/DME ground station Nitra. *Photo*: A. Novák.

However, the range is limited to the line-of-sight, therefore the transmitters are usually located on elevated ground or mountain tops. The biggest advantage of the system is that VOR defines exactly each 'radial' from the facility. If it is coupled with DME it clearly and simply specifies the position fix.

The standard VOR transmitter antenna generates a composed radiation pattern consisting of a non-directional and a figure-of-eight pattern in the same manner as in the direction finders. The resulting cardioid pattern is electronically rotated thirty times a second. The second signal is transmitted by a non-directional antenna. It is amplitude modulated by a sub-carrier of 9,960 Hz, which is frequency modulated by 30 Hz. On top of that the signal is amplitude modulated by a Morse code for the station identification.

The receiver error is mainly the result of signal reflections close to the transmitter. With Doppler VOR the 30 Hz amplitude modulated signal is transmitted by a reference antenna and the reflections are greatly reduced. Doppler VORs are in principle more accurate than conventional VORs. They are more immune to reflections from buildings and terrain. The antenna is formed from a large number (40–50) of antennae mounted in a ring around a reference antenna. The amplitude modulated 9,960 Hz signal is sequentially switched to the elements to create a rotating signal (Figure 17.6).

17.2.5. Non-directional Radio Beacon (NDB)

A Non-Directional radio Beacon (NDB) is in principle a simple radio transmitter which radiates a signal equally in every direction (hence 'non-directional') in the band 190–1,750 kHz. The signal is modulated with a Morse code for identification. In most countries the NDB station is identified by a three-, two- or one-letter Morse code. NDB is a very old and inaccurate radio navigation equipment with many disadvantages. The NDB signals are affected by atmospheric conditions, rough terrain, mountain ridges, coast line refractions, night effects and electrical storms.

Figure 17.7: NDB identification NIT (Nitra). *Photo*: A Novák.

However, an NDB has at least three advantages over the more sophisticated VOR. It is cheap, it is very simple to install and, as a result of the frequency band used, the NDB signals can follow the earth's curvature, hence the signals can be received at greater distances and at lower altitudes. Practical range in the frequency band 200–400 kHz is about 300 km over the mainland and about 500 km over the sea.

The NDB antenna system consists of a horizontal antenna fixed on masts of 20–60 m height. The antenna length controls the transmitter's frequency. For reliable operation, it is necessary to locate the transmitter on terrain which is flat up to 150 m from the NDB. No obstacles, in particular electric lines or metal constructions, are allowed in the transmitter's vicinity (Figure 17.7).

At airports, NDBs are used for non-precision approaches. Two NDBs in the approach path determine the approach direction and also serve as a fix on the approach for the height check. In precision approaches they are usually co-located with the markers. After the certification of GPS procedures many NDBs at airports are being dismantled.

17.2.6. Distance Measuring Equipment (DME)

Distance measuring equipment (DME) provides pilots with distance information between the aircraft and the location of the DME ground equipment. The DME is a navigation aid which can measure the distance from an aircraft to a ground reference point. It can serve more than one aircraft simultaneously. Most DME installations are co-located with VOR equipment. VOR provides the directional information and together with distance from the DME they specify the exact position, that is they give the 'aircraft fix'.

The basic principle of DME is to measure the time of a pulse sent by the aircraft's DME interrogator, which is then detected by the DME transponder and sent back. The aircraft's equipment detects the pulse sent by the transponder and measures the time difference t (μs) between the first pulse sent and the received pulse.

The DME operates in a frequency band from 962 to 1,213 MHz. The channel spacing is 1 MHz. Each channel has its own number from 1 to 126 and the channels are subdivided to X and Y channels or to X, Y, Z a W in case of DME/P (Precision). The frequency of the DME equipment depends on the co-location with a particular VOR/DME station. The practical DME range varies from 10 to 20 NM at an altitude of 25,000 ft up to 100 NM or even more if the aircraft altitude is 50,000 ft.

Every aircraft which has DME equipment (interrogator) sends interrogation pulses at the receiving frequency of the DME station. These pulses are detected in the ground station and the DME transponder is activated. The coded reply pulses are sent back with a fixed 'Reply Delay' on a sending frequency separated by 63 MHz from the given interrogation frequency, except DME/P which are usually used in connection with ILS/MLS systems and where the delay is not applied to increase system precision. All pulses appear in the form of pulse pairs with a fixed Pulse Pair Spacing. The fixed Reply Delay is introduced to give an aircraft flying nearby a chance to transmit all pulse pairs before receiving the corresponding reply pulses.

The ground stations also transmit a Morse identification code on the 1,350 Hz modulated frequency which is heard in an aircraft as a dot-dash signal.

17.2.7. Transponder Landing System (TLS)

Transponder Landing System (TLS) is a new landing aid certified by the FAA up to Category I precision landing limits. It was developed by the US ANPC — Advanced Navigation & Positioning Corporation as an alternative system for airports with low IFR traffic volumes.

From the pilot's viewpoint, the TLS works like a conventional ILS and the same ILS instruments are used for the aircraft position indication. In the TLS system the ground stations interrogate the aircraft's transponder. The aircraft's position is determined by measuring the arrival time and angle of radio signals transmitted from the aircraft's secondary radar transponder. The range of the TLS is typically 22 NM in a sector of 45° to either side of the runway centreline but the system could be also used for 360° airfield surveillance.

During the approach the inbound aircraft is identified by the standard four-digit transponder code and a track is established for guidance purposes. At an update rate of 10 Hz, the system compares the aircraft's location with its 'correct' one and calculates the differences. The 'correct position' data are transmitted to the aircraft by VHF localizer and UHF glide-slope signals modulated with 90 and 150 Hz tones as in the ILS systems.

The system can track 25 aircraft (or equipped ground vehicles as a way to prevent runway incursion accidents) but only one aircraft could be 'on the beam' at a time, because the TLS generates a correction based on its position. It is difficult to estimate the TLS prospects relative to the FAA programme of Wide-Area Augmentation System (WAAS).

17.3. RADAR SYSTEMS

17.3.1. Precision Approach Radar (PAR)

Precision Approach Radar (PAR) is the only 'passive' system with precision approach capabilities. The pilot during approach must fully rely on the instructions from the air traffic controller who monitors the aircraft position on the two screens displaying the azimuth and elevation situation. The PAR specifications are still included in the *ICAO Annex 10, Aeronautical Telecommunications, Volume I Radio Navigation Aids*. However, because of the very high operational costs of the equipment and the requirements for air traffic controllers' and pilots' updating of qualifications, the system is practically never used in civil aviation today. The Ground Controlled Approach (GCA) is used mostly in military operations.

17.3.2. Surveillance Radar Element (SRE)

Initial development of radar systems can be traced back to as early as the beginning of the 20th century. However, the first functional systems were not brought into operation until World War II. The basic principle of primary radar is that the radar waves reflected back from objects in the air or on the ground are captured on the radar antenna, processed and shown on the radar screen as a target. The size of the target on the screen depends, besides the radar technical characteristics, on the aircraft size, type and location of the propulsion units, type of materials used in the aircraft construction, type of paint used and the target distance from the radar.

There are many different radar types and many different applications. At airports and in airport terminal areas, radars are used for monitoring aircraft and ground vehicles and controlling them to increase capacity, expedite traffic and increase safety especially during low-visibility operations.

Surveillance Radar Element (SRE) must be able to detect within an effective aircraft echoing area of 15 m^2 at a range of at least 46 km. For this purpose, radars with a wavelength of 3 cm are used. The radar installation point must guarantee perfect 'vision' over the full 360° of azimuth.

17.3.3. Surface Movement Radar (SMR)

Surface Movement Radars (SMR) are mainly used for aerodrome movement area surveillance and control especially during low-visibility conditions or at night. For this purpose, millimetre wave radars with aperture antenna are used. They are usually located on the highest point, for example on the aerodrome tower, as their vision is limited to the line-of-sight. To complement the SMR picture in the 'blind areas' behind buildings, automatic cameras or millimetre wave sensors are sometimes used.

17.3.4. Advanced Surface Movement and Guidance Control Systems (A-SMGCS)

It is difficult to know where it would be most appropriate to place a description of the Advanced Surface Movement, Guidance and Control Systems (A-SMGCS), because usually it is a combination of visual, electronic, radio navigation and radar aids.

All aerodromes have some form of Surface Movement, Guidance and Control Systems (SMGCS). The simplest ones are just painted guidelines and signs. The more advanced and complex include switched taxiway centre lines and stop bars (see Section 15.2.5.2).

The main benefits from the implementation of A-SMGCS are associated with, but not limited to, low-visibility surface operations. The significant differences between the functions of a SMGCS and an A-SMGCS are that the latter should provide more precise guidance, surveillance and control of aircraft and vehicles on the movement area, and should also be able to ensure spacing between all moving aircraft and vehicles especially in low-visibility conditions when spacing could not be maintained visually (Figure 17.8).

Typically the base functions of A-SMGCS should allow the following:

- Surveillance
- Control
- Routing
- Guidance
- Conflict alert.

There are different sub-systems used to provide A-SMGCS functions to aircraft and vehicles. They can be based on surface movement radars, mode S implementation, millimetre wave sensors, induction loops, optical sensors or a combination of all these devices.

An-SMGCS is usually used at airports with high throughput where it would be difficult or even impossible to maintain surveillance and control on the movement areas with conventional surface movement radar (SMR). A-SMGCS should provide situational awareness not only to air traffic controllers but also to those aircraft and vehicles that can come into proximity with each other. A-SMGCS also reduces voice communication, improves the usage of surface guidance aids and increases reliance on avionics in the cockpit to help guide the pilot to and from the runway.

The technical standards for an A-SMGCS are recognised to be the most demanding for the most critical conditions in terms of traffic density, visibility conditions and aerodrome layout. Implementation of facilities and procedures to these levels will, therefore, be not appropriate at all aerodromes.

438   Airport Design and Operation

An A-SMGCS needs to be related to the operational conditions under which it is intended that the aerodrome should operate. A failure to provide a system appropriate to the demands placed on an aerodrome will lead to a reduced movement rate or may affect safety.

Figure 17.8: During low-visibility conditions it is not possible to maintain aerodrome ground control visually. *Photo*: A. Kazda.

It is important to recognise that complex systems are required and are economically viable only at aerodromes where visibility conditions, complexity of aerodrome movement areas, traffic density and any combination of these factors present a problem for the ground movement of aircraft and vehicles.

17.4.   Flight Inspections and Calibrations[2]

Any air navigation system, from the time of its installation up to its decommissioning, is subject to extensive tests and inspections. While the word 'test' is used when

---

2. The authors acknowledge the kind help received from Ivan Ferencz from the Navigation Division of the Slovak CAA in the preparation of this section.

one or more parameters are measured with appropriate reference equipment, 'inspection' refers to a series of tests carried out by a State or by an organisation on behalf of the State. The output of a test is data and the output of an inspection is a statement as to whether the inspected system is fully operational or not. Depending on the testing methods there are two basic sets of test or inspections: ground and flight tests. Ground tests or ground inspections can be carried out only in the close proximity of the navigation facility and the system has to pass successfully a series of specified ground tests providing signals in space of the required quality.

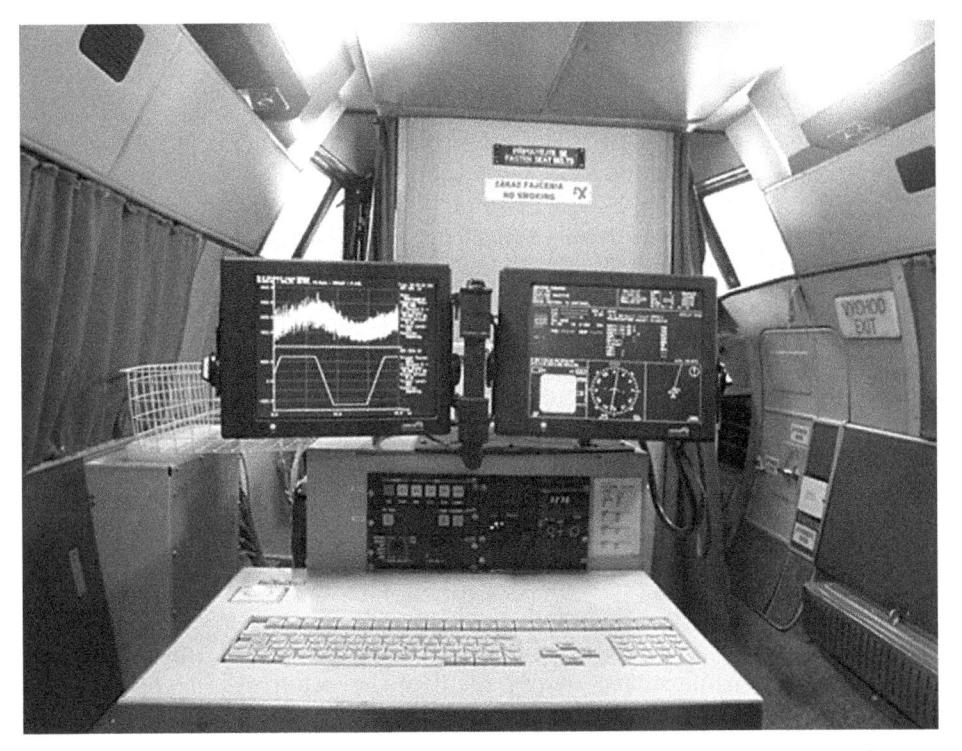

Figure 17.9: Flight inspection aircraft with measuring systems. *Photo*: J. Šulc.

Air navigation facilities are critical for the safety of civil aviation and it is not enough just to 'assume' that a system is safe. There is no room for compromises when testing reliability, accuracy or integrity of radio navigation equipment or systems. Therefore in most states the flight inspections of navigation facilities are carried out by an organisation under the State's responsibility. They are usually performed by a dedicated flight inspection unit designated to issue a facility Status Classification based on flight inspection results.

There are usually three elements of a flight inspection unit: aircraft, measuring systems and crew. Each of these elements has to comply with special requirements.

The flight inspection aircraft is usually a twin engined turboprop or jet with endurance of at least 3 hours, a wide range of speeds including the capability to fly at 105 knots or less, with suitable vertical climb speed and ample space on board for the measuring systems' installations. There are many different antennas installed on the top and bottom of the aircraft's fuselage.

Flight inspection systems are permanently or temporarily installed on board the aircraft and consist of sets of sensors, position reference systems, data processing units and human-machine interfaces. The range of sensors depends on the tasks performed by a particular aircraft. There are usually VOR/ILS, MKR and ADF receivers, SSR transponders, VHF transceivers or GNSS receivers complemented with spectrum analysers and oscilloscopes. Position reference provides autonomous information about the aircraft's position in space. For this purpose additional ground equipment is usually used. The required aircraft positional accuracy depends on the inspected system. Generally, it should be five times higher than the tolerances of the measured parameter. Typical position reference systems are GNSS with or without augmentation and with phase tracking capability, theodolite, laser or infrared tracker with telemetry uplink between the ground station and the aircraft (Figure 17.9).

The positioning systems can also be combined with an inertial reference system. This can further enhance the accuracy or continuity of positional information. Data from sensors and position reference systems are processed in a data processing unit which is running on a real-time computer. Where required, outputs of sensors are compared with reference data to specify ground equipment or a system's differences or errors. The results are presented in graphical or alphanumeric form to the flight inspection engineer who operates the system. A flight inspection crew consists of two pilots, one or two engineers and, when necessary, a ground reference systems operator. Because of special flight inspection procedures and a much higher workload on the test pilots, the crew must have special training and in some states also additional pilot's qualifications. Despite the flight inspection systems being highly sophisticated and most of the data being generated automatically, the flight inspection engineer must take into consideration specific limitations of sensors, performance of position reference systems, weather conditions during the flight and many other factors. The final decision whether the system is operational or not is the responsibility of the flight inspection engineer who declares the facility status.

18

AIRPORT WINTER OPERATION

Tony Kazda and Bob Caves

18.1. SNOW AND AIRCRAFT OPERATION

Snow means pleasure for children but it is also the source of problems for airport administrators and airlines. The obligation to maintain a clear surface of the runway and other movement surfaces of the airport has been imposed upon the airport operator in Annex 14, Volume I. Contamination of the movement areas, which include, for example snow and ice, means limiting or closing an airport's operation. Icing on the aircraft and its dispersal may decrease utilisation and disruption of the flight schedule. This results in loss of revenue for the airports and airlines and also in costs connected with clearing the snow and ice. Each airport has to:

✈ provide an effective snow plan and

✈ provide regularity of flight operation in the winter despite adverse meteorological conditions.

When clearing ice by chemical means, the negative impact of chemical substances on the environment must be minimised.

A layer of snow on the runway surface causes:

✈ resistance acting on the aircraft's wheels during the take-off run, the magnitude depending on the density and thickness of the snow layer, characteristics of the aircraft undercarriage and the speed and mass of the aircraft,

✈ increase of drag and decrease of lift of the aircraft during the take-off run due to snow being thrown up from the wheels, particularly the nose wheel and

✈ decrease of braking effect from the runway surface friction, particularly when the contamination is ice, increasing the possibility of exceeding the available distances for landing or rejected take-off.

Icing on the aircraft, particularly on the lifting surfaces, changes the aerodynamic characteristics and flight performance, and it can block or impair flight controls and even increase the weight of the aircraft so it may exceed the maximum allowable structural taxi mass. Iced sensors can cause the pilot to receive the wrong information about speed or engine condition. Therefore before take-off all the ice has to be removed.

442 Airport Design and Operation

The effect of contamination of the movement areas of the airport on the aircraft operations depends on several factors, in particular:

- air temperature
- runway temperature
- specific density of the snow.

The higher the air temperature the higher is also the specific weight of the snow. Three kinds of snow have been defined according to its specific weight, and slush which is not a snow but a mixture of water and ice crystals. Snow removal technology required depends on the kind of snow. Determination of the snow type must therefore be quick and simple:

Dry snow Snow which can be blown if loose or, if compacted by hand, will fall apart again upon release. Specific weight of dry snow is up to but not including 350 kg/m^3.

Wet snow Snow which, if compacted by hand, will stick together and tend to form a snowball. Specific weight from 350 kg/m^3 up to but not including 500 kg/m^3.

Compacted snow Snow which has been compressed into a solid mass resisting further compression and which will hold together or break up into lumps if picked up. Specific weight of compacted snow is above 500 kg/m^3.

Slush Water-saturated snow which, with a heel-and-toe slapdown motion against the ground, will be displaced with a splatter. Specific weight of slush is from 500 kg/m^3 up to 800 kg/m^3.

The easiest way to ascertain the specific weight is by taking a sample of a certain volume of snow and dissolving it. The higher the specific weight of snow or slush the higher is the resistance acting on the aircraft's wheels. A layer of slush on the runway of approximately 4 mm is enough to affect the performance of an aircraft, and if the slush layer is above 13 mm (half inch) the runway has to be closed and cleared. For comparison the runway must be closed if a layer of dry snow is 5 cm thick.

18.2. SNOW PLAN

The snow, slush or ice must be removed from the movement areas of the airport quickly and without residues in order to ensure safe aircraft operation. Economic factors must also be considered. The snow plan of an airport should specify the organisation, the provision of airport equipment and the necessary chemicals, and co-ordination of work with air traffic control. The possible scope of a winter plan is given in *Airport Services Manual, Part 2 Pavement Surface Conditions, Chapter 7, Snow Removal and Ice Control, 4th ed., 2002*, (ICAO Doc 9137-AN/898).

Preparation of the whole airport for the winter is very important. It includes not only full preparation of equipment, training of new workers and retraining of full-time personnel but also the maintenance of the movement areas. The basic requirement is that the airport and all equipment must be in perfect condition sufficiently in advance of the possible appearance of adverse weather conditions. The training of workers who will perform the winter maintenance should include the following:

- Radiotelephonic procedures. The workers must have necessary qualifications for the operation of the radio station; they must be familiar with the operation of the given type of transmitter and know the radio phraseology.
- Procedures for removing snow and ice. The procedures are different for each kind of snow and the runways in use.
- Knowledge of the characteristics of the chemicals used for runway and aircraft de/anti-icing.
- Knowledge of aircraft characteristics, de/anti-icing procedures and limitations.
- Knowledge of how to operate the equipment; each worker must control properly the assigned technical equipment in any weather by day or night without affecting safety of operation.
- Knowledge of the airport; the workers must know perfectly the layout of all parts of the movement areas of the airport even during night or low visibility conditions.

Many workers from various departments of the airport participate in the winter maintenance and it is necessary to co-ordinate their work. The snow co-ordination committee controlling individual activities of winter maintenance will include airport management, meteorological services, air traffic controllers and representatives of airlines.

When preparing the snow plan it is necessary to consider a number of factors such as topography, climatic conditions, airport location, types of aircraft that use the airport, frequency of operations and physical characteristics of airport movement areas. Different equipment and procedures are needed, for example at an regional airport located in subarctic conditions with 150–210 days of snow cover a year like Kiruna (Sweden) or a principal airport located in a warmer climate with extensive traffic but maximum 5 days of snow a year, such as Madrid Barajas.

Priorities for clearing snow and ice from the various parts of movement areas have been specified in Annex 14, Volume I, Chapter 10 and they must also be entered in the winter plan. They can be changed, but only after an agreement between the airport operator and air traffic control. Priorities for clearing movement areas according to *Airport Services Manual, Part 2 Pavement Surface Conditions, Chapter 7, Snow Removal and Ice Control* are as follows:

→ Runway(s) in use

→ Taxiways serving the runway(s) in use

→ Apron

→ Holding bays

→ Other areas.

However, when setting priorities the specifics of operation at each airport must be taken into account. For example, it would be useless to have a clean runway, if the aircraft would not be able to taxi to the de-icing pad.

In addition to the movement areas, attention has to be paid to clearing snow from the vicinity of antennas, radio navigation equipment, and particularly from the vicinity of the glide path ILS antenna, the signal of which is very sensitive and could be distorted by the layer of snow. Depending on weather conditions and operational possibilities, other movement areas and access roads will also gradually be cleared as the opportunity arises.

Because snow is hygroscopic, its weight can grow quickly when the temperature is around zero. When clearing of snow is being planned, it is necessary to take account of this property. Therefore with the fall of the first snow flakes, clearing of snow may already be started on the most important areas. Usually it is still possible to allow some aircraft movements before the runway will have to be closed and the centre of the runway be cleared. The procedure used for the first clearing of the runway depends on the equipment which is available, the kind of snow (dry, wet, compacted or slush) speed and direction of the wind and other factors. Which of the runways shall be cleared will usually be specified by the airport duty manager in co-operation with air traffic control on the basis of a meteorological forecast. If critical snowbanks are not formed (see Figure 18.1) it will be possible to open the airport for operations within a relatively short time. If snow continues to fall, further closure and clearing of the runway become necessary. It is no use clearing the runway during a snowstorm; the snow will be blown onto the runway as fast as it is being removed. After the snowstorm is over, and the airport has to be opened as quickly as possible, the personnel are tired and the equipment needs replenishing and inspection. If two runways are available and if this is made possible by the meteorological conditions, it is advantageous to open one runway and continue to clear snow from the other.

It is not economically realistic to maintain a sufficient quantity of equipment to cope with the worst predicted weather. Therefore, if those extreme conditions occur, the airport will have to be closed.

The aircraft operators are informed about the up-to-date condition of the airport or its closure by SNOWTAMs, which are distributed by means of the AFTN telex network (Figure 18.2).

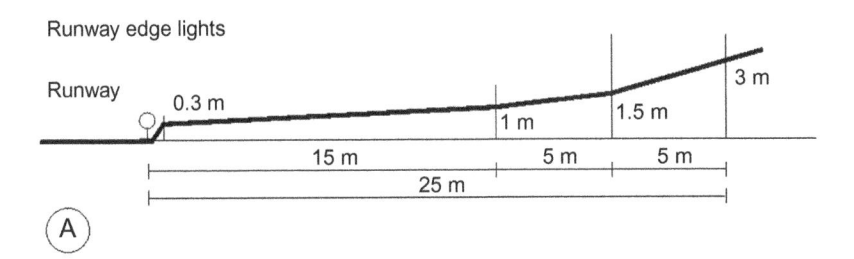

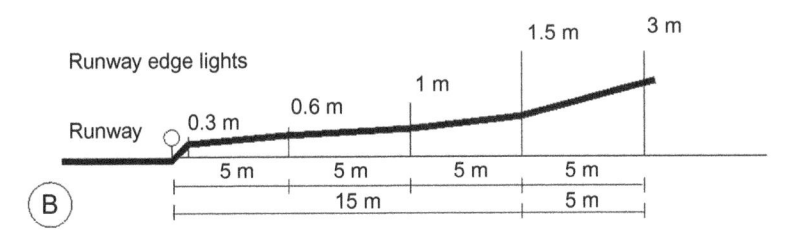

Figure 18.1: Maximum height of snowbanks. (A) Runways used by very large aircraft (such as B-747, DC-10, L-1011). (B) Runways used by other than very large aircraft. *Source*: *ICAO Airport Service Manual, Part 2 Pavement Surface Conditions*.

If the airport has only one runway, it is often necessary to use equipment which enables the snow to be cleared at high speed. For other areas with lower priority it is possible to use normal procedures of clearing snow and ice, or less used areas can be temporarily closed.

Although keeping the aerodrome movement areas open has the highest priority, it is necessary to ensure that the airside and landside roads and parking places are available in order to ensure transportation to and from the airport. It is sensible to contract-out the clearance of such areas, for example by agricultural companies or by construction companies, since their equipment is little used in the winter.

446   Airport Design and Operation

Removal of snow and ice from the movement areas may be by mechanical, chemical or thermal means.

(COM heading)	(PRIORITY INDICATOR)	(ADDRESSES)			⇐
	(DATE AND TIME OF FILING)	(ORIGINATOR'S INDICATOR)			⇐
(Abbreviated heading)	(SWAA* SERIAL NUMBER) S W * *	(LOCATION INDICATOR)	DATE/TIME OF OBSERVATION	(OPTIONAL GROUP)	⇐ (

SNOWTAM (Serial number) ⟶		
(AERODROME LOCATION INDICATOR)	A)	⟶
(DATE/TIME OF OBSERVATION *(Time of completion of measurement in UTC)*)	B)	⟶
(RUNWAY DESIGNATORS)	C)	⟶
(CLEARED RUNWAY LENGTH, IF LESS THAN PUBLISHED LENGTH *(m)*)	D)	⟶
(CLEARED RUNWAY WIDTH, IF LESS THAN PUBLISHED WIDTH *(m; if offset left or right of centre line add "L" or "R")*)	E)	⟶
(DEPOSITS OVER TOTAL RUNWAY LENGTH *(Observed on each third of the runway, starting from threshold having the lower runway designation number)* NIL — CLEAR AND DRY 1 — DAMP 2 — WET or water patches 3 — RIME OR FROST COVERED *(depth normally less than 1 mm)* 4 — DRY SNOW 5 — WET SNOW 6 — SLUSH 7 — ICE 8 — COMPACTED OR ROLLED SNOW 9 — FROZEN RUTS OR RIDGES)	F)	⟶
(MEAN DEPTH *(mm)* FOR EACH THIRD OF TOTAL RUNWAY LENGTH)	G)	⟶
(FRICTION MEASUREMENTS ON EACH THIRD OF RUNWAY AND FRICTION MEASURING DEVICE MEASURED OR CALCULATED COEFFICIENT *or* ESTIMATED SURFACE FRICTION 0.40 and above GOOD — 5 0.39 to 0.36 MEDIUM/GOOD — 4 0.35 to 0.30 MEDIUM — 3 0.29 to 0.26 MEDIUM/POOR — 2 0.25 and below POOR — 1 9 — unreliable UNRELIABLE — 9 *(When quoting a measured coefficient, use the observed two figures, followed by the abbreviation of the friction measuring device used. When quoting an estimate, use single digit))*	H)	⟶
(CRITICAL SNOWBANKS *(If present, insert height (cm)/distance from the edge of runway (m) followed by "L", "R" or "LR" if applicable)*)	J)	⟶
(RUNWAY LIGHTS *(If obscured, insert "YES" followed by "L", "R" or both "LR" if applicable)*)	K)	⟶
(FURTHER CLEARANCE *(If planned, insert length (m)/width (m) to be cleared or if to full dimensions, insert "TOTAL")*)	L)	⟶
(FURTHER CLEARANCE EXPECTED TO BE COMPLETED BY . . . *(UTC)*)	M)	⟶
(TAXIWAY *(If no appropriate taxiway is available, insert "NO")*)	N)	⟶
(TAXIWAY SNOWBANKS *(If more than 60 cm, insert "YES" followed by distance apart, m)*)	P)	⟶
(APRON *(If unusable insert "NO")*)	R)	⟶
(NEXT PLANNED OBSERVATION/MEASUREMENT IS FOR) *(month/day/hour in UTC)*	S)	⟶
(PLAIN-LANGUAGE REMARKS *(Including contaminant coverage and other operationally significant information, e.g. sanding, de-icing)*)	T)	)⇐

NOTES: 1. *Enter ICAO nationality letters as given in ICAO Doc 7910, Part 2.
              2. Information on other runways, repeat from C to P.
              3. Words in brackets ( ) not to be transmitted.

SIGNATURE OF ORIGINATOR *(not for transmission)*

Figure 18.2: SNOWTAM format. *Source*: ICAO Annex 15, Aeronautical information services.

18.3. Mechanical Equipment for Snow Removal and Ice Control

Mechanical equipment for snow and ice clearing from the runway surface should always, whenever possible, be preferred. Its primary advantages, in comparison to chemical and thermal means, are lower operational costs and negligible environmental impact. Mechanical equipment is used mostly for snow clearing. There is little that can be done mechanically to treat a layer of ice but, under specific conditions of Nordic winters, it is possible to improve the braking action by sanding the ice layer.

The speed and quality of snow clearing from movement areas of the airport depends to a great extent on the number and capacity of the equipment. When selecting the equipment the airport operator must consider a number of factors:

✈ Scope of operation

✈ Significance of maintaining airport operations

✈ Capital and operating cost of the equipment

✈ Dimensions of airport areas

✈ Availability of spare parts and possibility of repairs

✈ Climatic conditions.

At small airports with general aviation operations or with only a few scheduled traffic movements a day, it may be best to outsource snow clearing. Alternatively, if it is an airport where it snows only occasionally, it is possible to close the airport for a few hours or even days.

Airport manoeuvring areas have different physical characteristics from roads, so they require special equipment, which enables faster and higher quality clearing of the runway surfaces. The airport area is flat and the runways are wide. In the runway there are usually inset lights protruding above the surface of the runway. The terrain behind the runway edge lights, which is often not paved, must also be cleared. The snow layer on the airport runway is usually not deep, but the time for clearing the runway must be substantially shorter than for roads. All these are characteristics which have to be considered when selecting the mechanical snow removal equipment.

The number of pieces of equipment for winter service can be specified on the basis of the average height of snow cover, amount of snow falling within one snowfall, area of the movement surfaces which are to be cleared, and the type and intensity of the air operations. It is usually not worthwhile to equip airports having a total snowfall below 40 cm a year and up to 5 cm of snow in one snowfall with a great number of expensive and high performance pieces of winter service equipment and vehicles.

The snow accumulated at the edge of the runway must be removed. Snow blowers are normally used for this. The performance of the snow blower is a critical parameter in the fleet of winter service equipment and vehicles, and it is also the most expensive equipment. At some airports in temperate climates and only a small layer of snow cover a year, it is sometimes possible to use a special type of plough which

448 Airport Design and Operation

clears snow at high speed, rather than a snow blower (Figure 18.3). The use of such a plough has to be considered separately for each case, because several other factors have to be taken into account.

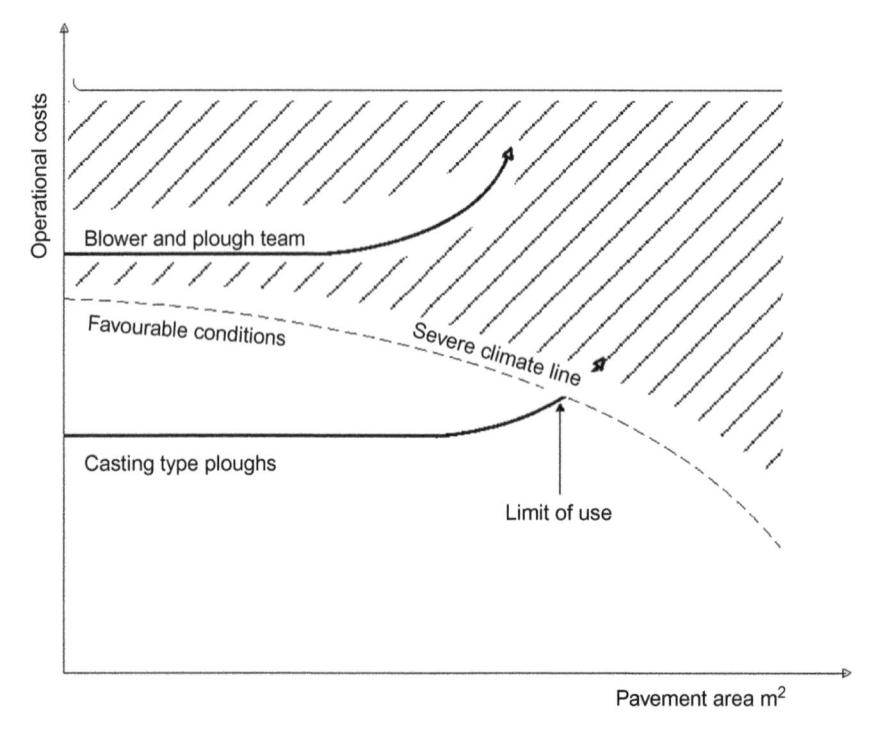

Figure 18.3: Casting type ploughs. *Source: ICAO Airport Services Manual, Part 2 Pavement Surface Conditions.*

Other winter service vehicles and equipment which are used on larger airports are air blower machines, ploughs, sand/aggregate trucks, chemical spreaders, tankers and loaders.

When removing small layers of snow it is best to use a rotary brush sweeper and air blower which clears the entire swept area by directing cold air onto the snow, since they will clean the surface of the runway properly. The height of the rotary brush above the surface of the runway can be electronically controlled so as to clean the runway thoroughly without the brush being excessively worn.

There are many types of special ploughs for airport use. The widest are wider than any of the road ploughs, at 4.5 m and more. The blade portion of some airport ploughs has been divided into several sections, each of them being separately spring or hydraulically loaded. If it contacts an obstacle, for example an inset light, the appropriate section will lift. Modern blades are produced of light plastic materials, which have four main advantages. They are substantially lighter than the common blades made of steel, their friction coefficient with the surface of the runway is low, they do not corrode and don´t damage the pavement surface or inset lights. The low weight of the blade and low friction coefficient contribute to decrease fuel consumption. However, special winter maintenance equipment, despite having higher productivity, is also more expensive and its purchase at smaller airports with moderate traffic must be carefully considered (Figure 18.4).

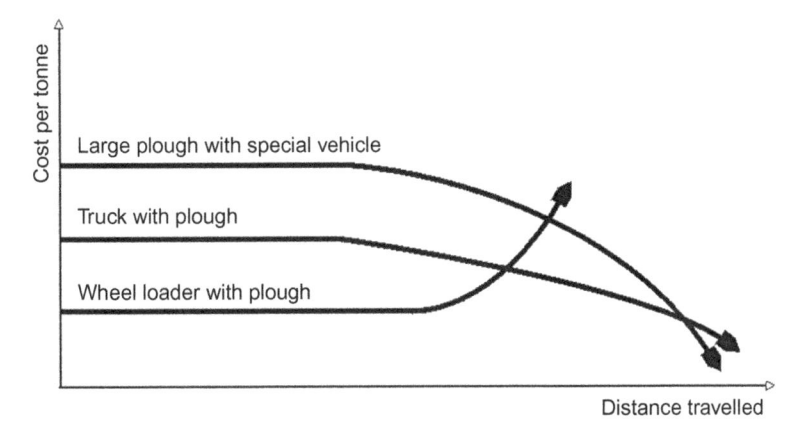

Figure 18.4: Ploughing costs/distance. *Source*: *ICAO Airport Services Manual, Part 2 Pavement Surface Conditions*.

An airport with regular air transportation should be equipped with one or more high performance snow blowers, which are capable of throwing snow with a specific weight of 400 kg/m^3 a distance of at least 30 m. On airports with regular commercial transport, there should be sufficient equipment to ensure removal of a 2.5 cm snowfall from the main runway and from two most used taxiways connecting the main runway with the apron. On the basis of operational experience, clearing the taxiways requires approximately 25% more time than clearing a runway of comparable area.

According to *Airport Services Manual, Part 2 Pavement Surface Conditions* a 2.5-cm thick layer of snow should be removed from the main runway and from the two taxiways within the following time limits with the stated number of air transport movements (atm):

+ 40,000 and more atm, the snow should be removed within 30 minutes
+ 10,000–40,000 atm, the snow should be removed within 1 hour
+ 6,000–10,000 atm, the snow should be removed within 2 hours
+ 6,000 and less atm, the snow should be removed within 2 hours, where possible.

450   Airport Design and Operation

Two ploughs should be included in the work group in front of each snow blower, the performance of which should correspond to the performance of the snow blower. Oslo Gardermoen airport uses 10 units to form snow removal 'trains' and then it takes 5–10 minutes to clear a runway for all snow.

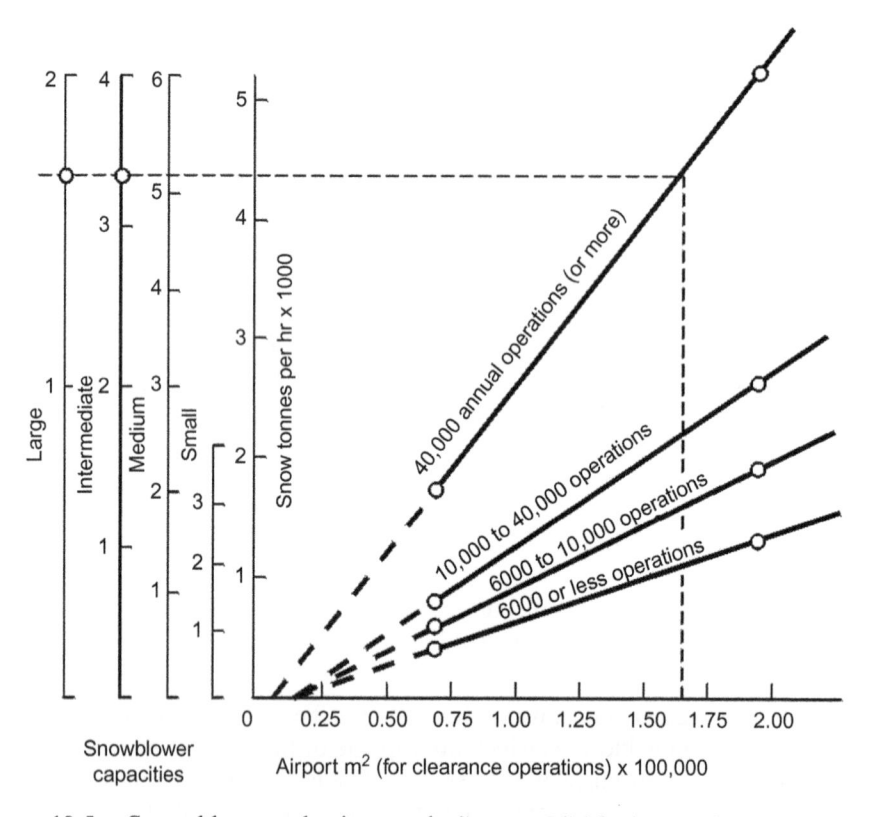

Figure 18.5:   Snow blower selection card. *Source*: *ICAO Airport Services Manual, Part 2 Pavement Surface Conditions*.

Similarly the equipment for a general aviation airport, which mostly serves aircraft up to 5,700 kg, vehicles and equipment for winter operations should ensure the removal of a 2.5-cm snow layer from one runway, a taxiway connecting the runway with the apron and from 20% of the apron. The airport should be equipped by one snow blower capable of throwing snow of a specific weight 400 kg/m^3 at least 15 m. The snow layer 2.5 cm thick must be removed as follows:

✈ 40,000 and more movements a year, the snow should be removed within 2 hours

✈ 6,000–40,000 movements a year the snow should be removed within 4 hours

✈ 6,000 and less movements a year the snow should be removed within 4 hours, where it is possible.

The airport should be equipped with at least one plough with the performance equal to that of the blower. A higher value than 2.5 cm must be used at airports with extreme weather conditions as the critical height of snow conditions, the value corresponding to the real climatic conditions.

The performance of winter service vehicles and equipment must match the total volume of snow which needs to be removed from the movement areas. Because each airport has a set time for snow removal according to its position in the appropriate group and the reference height of snow cover of 2.5 cm has also been fixed, the area of the movement areas from which the snow is to be removed is the determinant of the capacity of winter service vehicles and equipment. The relationship is linear, as shown in Figure 18.5. The required performance of other vehicles and equipment can be determined in a similar way to the performance of the snow blower (Figure 18.6).

Figure 18.6: Snow blower is usually the most expensive winter equipment. *Photo*: Courtesy Bratislava M.R. Štefánik Airport.

For example as given in Table 18.1 the performance of the snow blower must be at least 4,400 tonnes of snow per hour. The performance data of snow blowers given by the producers must be considered as approximate and in most cases they are optimistic. The performance must be verified by a practical test under the real winter conditions in the airport unless references can be obtained from other airport operators. In practice, the capacity of the snow blower can be 40%–50% lower than the given capacity.

The performance of other equipment must be higher that the performance of the snow blower. Their selection is affected by the same factors as the selection of the snow blower.

Table 18.1: Determination of snow blower capacity.

1. Pavement area to be		
Main runway	3,000 × 45 m	135,000 m²
Parallel taxiway with connectors	3,000 × 23 m	69,000 m²
Two connector taxiway	400 × 23 m	9,200 m²
Holding areas	2 × 40 × 60 m	4,800 m²
Apron	25% of 16,000 m²	4,000 m²
Total area to be cleared		222,000 m²
2. Other parameters		
Snow removal time (1st group of airports)		30 minutes
Temperature		−4 °C
Snow density		400 kg/m³
Depth of snow		2.5 cm
3. Determination of snow blower capacity		
Snow volume	222,000 × 0.025	5,550 m³
Snow mass	5,500 × 400	2,200,000 kg
Snow blower capacity per hour		4,400,000 kg
Snow blower capacity per hour in tones		4,400 tonnes

In some Nordic countries the use of sand is permitted to improve braking action on the runway in specific conditions. However, a small grain of sand can damage an engine blade if sucked into it, so sand should be used as the last resort when it is not possible to use other chemical or mechanical means to remove the ice. Therefore sand is used primarily at very low temperatures, when the use of chemicals is not effective or during heavy snowfalls. The kind of sand must be carefully checked. Fine sand which can fall through a sieve with 0.297 mm mesh does not guarantee the required improvement in braking, and it can be easily blown away by the wind from the surface of the runway. Coarse material, which does not get through a sieve with 4.76 mm mesh can damage the jet engine blades and it also damages the leading edges of propeller blades. Individual sand particles must be sufficiently hard to resist the shear forces upon braking, but soft enough to minimise the damage of the metal surface of the aircraft structure. The sand must not contain small rocks, impurities, salts and other corrosive substances. The most suitable material is crushed limestone with knife-edges, but the type of sand depends largely on local conditions.

In order to increase braking action significantly and at the same time to minimise the possibility of sucking the sand into the engine, individual grains of sand should be embedded into the surface of the runway. This is possible only if the temperatures are well below zero. The possible manners of embedding the sand are:

✈ heating the sand before distributing it,

✈ melting the ice sprinkled with sand by burners and

✈ spraying water on ice sprinkled with sand.

After the grains of sand have been embedded into the ice, for which several tens of minutes are necessary, free grains of sand must be swept away and vacuum cleaned. The runway has to be cleaned in the same way after a thaw.

18.4. CHEMICALS FOR RUNWAY DE-ICING

Chemicals used for removing ice or for the prevention of icing on the surface of the runway must meet a number of requirements, which are often conflicting. They must be cheap and effective. They must damage neither the aircraft structure nor the runway. They must not be toxic and their harmful effects on the environment must be minimal. After de-icing chemicals are applied, a thin layer of water is formed over the remaining sheet of ice. Such a surface is the most slippery possible and the braking action is practically zero. The time necessary for completely dissolving the ice depends not only on the kind and concentration of the chemical but also on the meteorological conditions, thermal condition of the runway and the thickness of the ice layer.

Chlorides cannot be used for de-icing of airport movement areas because of corrosion, although they are cheap and effective. Sodium chloride may be used mixed with crushed gravel for access roads in some countries. If it is applied in its solid state and particularly if applied liberally, sodium chloride damages the pavement surface.

One of the most common chemicals still used at airports is urea. The main advantage of urea is its price. In the words of the operational manager of one airport, 'we like to use it, but it is better not to talk about it'. Among major shortcomings is its ineffectiveness at low temperatures (practically up to −8 °C) and impacts on the environment. The technical term urea is used for the amid of carbonic acid — carbamid with chemical composition $CO(NH_2)_2$. Urea is a non-toxic substance, and its shelf life is unlimited.

The theoretical efficiency of urea is high; up to −11.5 °C, the eutectic temperature, at 24.5% concentration of the solution. Practically it is possible to use urea up to −5 °C. Its advantage is that it can be applied as a de-icing or as a preventive anti-icing chemical. Urea can be applied in the form of a solution, preferably warm, or by sprinkling granules. Sprinkling can be performed dry or the granules can be mixed with water and sprayed. Before application of urea as a de-icing fluid, it is necessary to clean the runway as much as possible by mechanical equipment. The spraying will take effect about 30–60 minutes after application. After the ice has softened it is necessary to complete the cleaning of the runway mechanically.

It is better to use urea as a preventive chemical in a 35% solution, which is safe against recrystallising up to −8 °C. When used in this way for anti-icing, the urea concentrations are lower, as shown in Table 18.2. This results not only in lower costs but also in smaller impact on the environment, as discussed later.

Table 18.2: Recommended concentrations of urea.

Urea application	Urea solution (l/m²)	Urea granules (g/m²)
Preventive	0.05–0.1	15–20
De-icing	0.15–0.35	30–70

Higher doses of urea for de-icing pose a greater threat to the surface of the runway. This is particularly so for concrete surfaces. After the ice layer has been sprinkled with urea granules, a system with two components — k ($CO(NH_2)_2$ and H_2O) in three phases — f (urea, ice and water solution) will be formed, as shown in Figure 18.7.

According to Gibb's phase rule, which determines the degree of freedom of system v, that is the number of dimensions determining the condition of the system (temperature, pressure and composition), this is a monovariant system, that is the degree of freedom is only one according to the equation:

$$v = k + 2 - f$$

where v is the number of degrees of freedom; k is the number of components and f is the number of phases.

The magnitude of the degree of freedom is the same as the degree of freedom of a eutectic mix. This means that under the given pressure and temperature this system is not in balance. Therefore the ice will start thawing and its temperature as well as the temperature of the runway surface will fall sharply as a result of the consumption of the latent heat of thawing. In the growing water volume, more urea will be thawed. This will continue to happen until the temperature of the mix decreases to the eutectic temperature. Of course, with a concentration of urea less than eutectic, the achieved temperature will be higher. This process leads to rapid cooling of the runway surface and to large differences in temperature inside the upper layer of the concrete slab. This results in high internal tensions, which cause peeling of the top 3–5 mm of the concrete runway. If urea is applied in a warm solution the necessary thermal energy will partially be taken from the solution and the surface of the runway is not so stressed as in granular application.

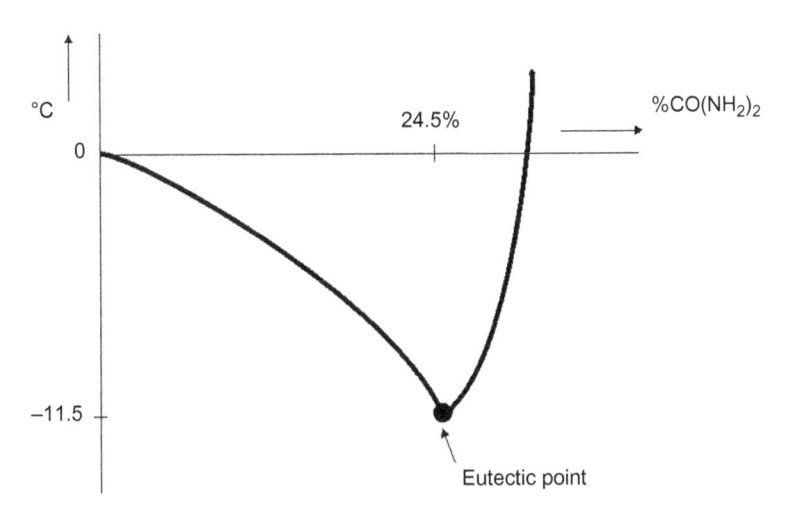

Figure 18.7: Ice — urea de-icing system.

This peeling occurs particularly with new rigid runways, the surface of which is porous and absorptive. At temperatures slightly below zero the water in capillary pores of the runway surface is not frozen. The rapid cool down as a result of application of the urea causes the water in the capillaries to freeze and this causes a further increase of stress in the surface layer of the runway. Therefore before winter, new runways have to be impregnated with a protective coating, which reduces the penetration of water into the surface.

Urea was originally used as a fertiliser. Nitrogen accounts for about 45% of the total weight of urea. Application of urea increases the concentration of nitrates in ground and surface waters. There is then significant growth of algae in the watercourses and this disrupts the ecosystem, particularly where it is not possible to achieve sufficient dilution when the waste waters are discharged. Therefore the use of urea is forbidden or limited in some airports. The concentration of nitrates is often monitored in several parts of the drainage system of the airport and in outlets into watercourses. Limiting values of contamination of waste waters of NH_4 max. 3.5 mg/l, NO_3 max. 15.0 mg/l and dissolved materials 300 mg/l should be met when checked 3–5 hours after application of de-icing chemicals on the airport.

In order not to exceed the appropriate limits of nitrate concentration in surface waters, retention pools have been built on some airports, in which contaminated waters are trapped and then gradually and slowly discharged to watercourse. Alternative solutions are the construction of a water treatment plant or the combination of both solutions. The volume of waste waters from two runways at Munich airport is so large that the considerable capacity of a water treatment plant would not be utilised in the periods when the ice is not melted. Therefore two retention pools have been built on the airport. The underground tank has a volume of 60,000 m^3 and the surface pool has a volume of 20,000 m^3. Water from retention pools is admitted into a water treatment plant in Eitting in precisely specified volumes depending on the chemicals' concentration, where it is treated together with municipal sewage.

The use of urea must therefore be always considered according to the local conditions. If its use is unsuitable, it is possible to use more expensive but environmentally friendly acetate or formate — based chemicals.

Airport operators are now rejecting traditional chemicals in favour of those based on acetates. These are more effective at lower temperatures, they act for longer, they leave a less slippery surface, their storage is easier, they are non-toxic and do not damage the environment. Potassium acetate-based de-icing fluids are theoretically effective up to the temperature of −60 °C. In fact they have just one disadvantage — a high price.

Acetate-based chemicals are recommended by the environmental authorities in the United States, and in several countries of Western Europe. In Scandinavia they are using potassium formates more than acetates because formates are even more environmental friendly. They are easily biodegradable with a small consumption of free oxygen in water. When they decompose, carbon dioxide and water are generated. The solution reaction is alkali which is preferred in most cases, since the soils mostly have an acid reaction. They are, however, more expensive than the more traditional de-icers. It is probable that for those airports that have already built expensive facilities for the waste water treatment from movement areas of the airport or retention pools as at Manchester, it will be more advantageous to carry on using urea or glycol, which are cheaper in comparison to the acetates. Because acetates have lower viscosity, they require different spraying devices than, for example glycol, which means another additional investment for many airports. Therefore each airport must consider the advantages and disadvantages of individual chemicals in its specific situation.

Potassium acetate-based chemicals are used in solution for anti-icing. For runway de-icing it is better to use a combination of granules of sodium acetate, or more commonly today, granules of solid sodium formate, and a solution of potassium acetate. The granules of sodium acetate are applied first. As the granules dissolve, heat is released and this creates holes in the ice. When the solution of potassium acetate is applied, it will penetrate through the holes to the surface of the runway, where it acts from below and releases the ice from the surface of the runway.

Formates or formic acid salts have similar properties to those of the acetates. Potassium formate is effective in low temperatures, has low toxicity and is easily biodegradable. However, specific forms of damage have been perceived on the asphalt pavements. Damage in pavements is not the only issue which has been encountered after the implementation of acetates and formates. The high density and low interfacial tension of formate makes it easier for de-icing chemicals to penetrate between bitumen and aggregates. In addition formates are highly hydroscopic, which means that a treated pavement stays wet always. This can cause faster aging of the pavement. Issues caused by chemicals have been perceived also in aircraft and airfield equipment. Glycol-based de-icing chemicals have two disadvantages. Ethylene glycol and diethylene glycol are toxic. Moreover, ethylene glycol is considered to be carcinogenic. This can be overcome by using mono-propylene glycol-based chemicals, which is considered non-toxic in many countries. Despite this, glycol is used on its own or in combination with urea because of its easy manipulation, storage and the price.

The disadvantage of all kinds of glycol is that upon decomposition in the water they consume a high volume of oxygen and this seriously endangers life in the watercourse. The environmental impact is mainly linked to the biochemical oxygen demand (BOD) and the time it takes for the chemical to break down. Therefore waste waters must be treated before being discharged into the watercourse.

Ice can also be removed from a runway with high pressure water. A few companies offer this technology. The water is sprayed under high pressure through nozzles against the runway. It penetrates through the layer of compacted snow or ice, disturbs it and separates it from the surface of the runway. Then the runway is mechanically cleaned. It is then treated by a small volume of glycol or by another de-icing chemical to prevent further icing.

18.5. THERMAL DE-ICING

Thermal procedures for removing of snow and ice are not so widely used as use of mechanical means and chemicals. The reasons are the growing price of energy and the problems with maintenance of some types of facilities. None-the-less, it is possible that in some airports the local conditions can be favourable, for example the utilisation of waste heat or geothermal energy for removing of snow and ice from apron stands or runways. Some airports are considering installing heated pavements at stands in order to increase safety during aircraft handling as slippery apron surface may cause injuries of ground personnel or aircraft or equipment damage.

The technical feasibility of using earth heat in combination with heat pipes for de-icing and removing snow from pavement surfaces has been demonstrated (in the Baltimore/Washington climate) in testing conducted at the Fairbank Highway Research Station (located at McLean, Virginia) during two winters. A report explains the analytical models constructed to analyse earth heated pavement systems, and the validation of these models by conducting tests.

458 Airport Design and Operation

In the case of asphalt pavements, a special conductive asphalt — graphite mixture could be used for an electrically heated surface layer (Figure 18.8). The system can prevent ice and/or snow surface build-up and, depending on the system, it is able to melt approximately 2.5 cm of falling snow per hour. Operating power usage is, depending on the weather conditions, between 200 and 500 Wm^2. Typical installations are small areas like heliports on constructions where the snow or ice clearing would be problematic. One of these systems was installed at Chicago, Illinois, the United States — O'Hare Airport taxiway by a company Superior Graphite under a trademark SNOWFREE.

Figure 18.8: Electrically heated pavement — emergency heliport on the roof of the regional hospital at Salzburg. *Photo*: A. Kazda.

Melting of snow is hardly used anymore because of high energy consumption and low efficiency. Also at most airports the use of jet blowers, jet engines fixed on the undercarriage of the truck, is forbidden. They are expensive due to the price of fuel, to noise and air pollution, and also because of the damage to the runway surface or runway lights. Thermal de-icing by jet blowers can be effective if not only will the ice be melted but also the surface of the runway must be completely dried. The use of a jet blower causes an increase of local runway temperature, which will cause high internal stress in the concrete slab and the later appearance of cracks. The combination of high temperature and kinetic effects of jet exhaust gases can cause the destruction of the covering layer of asphalt runways.

18.6. RUNWAY SURFACE MONITORING

According to the requirements of Annex 14, Volume I the surface of the runway must be maintained in a clean state to ensure good braking action. However, from the point of view of protecting the environment and for economy, it is necessary to minimise the volume of chemicals used for removing ice. The application procedure, the application time and the necessary volume of the chemicals are affected by a number of factors, the effect of which can be difficult to assess except on the basis of operational experience. For example, a sufficient amount of heat can be accumulated in the runway pavement after a sunny day so that the pavement does not freeze even during a wet and cold night. An optimum decision can be made on the basis of data from the monitoring system, which monitors meteorological conditions and the condition of the runway and enables a forecast of the condition of the runway surface to be made. Because the monitoring system allows a substantial decrease in the requirements for winter maintenance and in the volume of chemicals used by up to 70%, this investment usually has a high rate of return.

The monitoring system must be capable not only of indicating the appearance of ice on the runway surface but also to forecast conditions leading to icing. Modern systems use data from the sensors of automatic stations, which monitor the air temperature, dew point temperature, amount and ceiling of clouds, speed of wind, precipitation, temperature of the surface of the pavement, temperature under the surface of the runway and the residues of the de-icing fluid on the runway surface which could be detected indirectly by a so-called active sensor. The temperature on an active sensor can be decreased to a temperature such that the surface becomes iced up. This allows the chemical residues on the pavement to be detected indirectly and a decision made as to whether a new application of chemical is necessary. The system is capable of forecasting the runway condition for 24 hours ahead. Usually it is possible to enter the weather forecast into the system manually, and in this way to obtain various alternative forecasts.

On the basis of the forecast the following can be estimated:

- Whether the temperature of the runway surface will fall below 0 °C
- When the temperature of the runway surface will fall below 0 °C
- Whether the surface of the runway will be wet when the temperature is below 0 °C
- How long the temperature of the runway surface will remain below 0 °C
- Whether the residues of de-icing material from the last application on the surface of the runway will prevent the formation of ice or if preventive anti-icing of chemicals could be recommended.

On the basis of experience from different airports, a monitoring system that is too complicated is more of a burden than a useful aid. The number of sensors and their

460   Airport Design and Operation

locations could be optimised by mapping the runway surface with a thermovisual camera. On each surface there are critical 'cold' places, which freeze before other surfaces on the runway.

The runway surface is scanned by the thermovisual camera under three different conditions. During a cloudless and windless night it is possible to see the maximum contrast between the 'cold' and 'warm' places on the runway (Figure 18.9). Average conditions are 5/8 cloud cover and a 2.5–10 m/s wind speed. A wet night with more than 6/8 cloud cover with the cloud base of less than 600 m gives the smallest temperature differences between individual places of the runway. After evaluation of the mapping, the sensors are placed into 'cold' places and for comparison into places which, from the point of view of temperature, are the most stable. In this way it is possible to simplify the monitoring system, to increase its reliability and to decrease the construction costs.

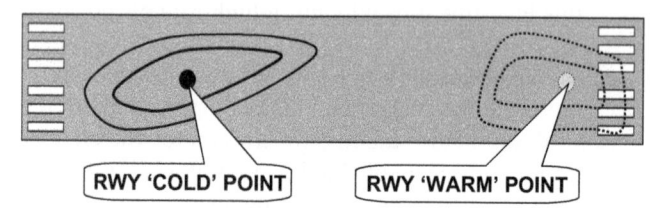

Figure 18.9:   Location of sensors in the 'cold' and the 'warm' point of the RWY.

18.7.   AIRCRAFT DE-ICING

It is dangerous for an aircraft to take-off while it is contaminated by snow or ice. Creating sufficient lift requires high values for the lift coefficient.

This is achieved by using flaps and slots and flying the aircraft at high angles of attack. The contamination reduces the available lift due to roughening of the airfoil surface, from the premature separation of the airflow from the wing surface and from the increase of aerodynamic resistance and aircraft mass. The situation is critical during and shortly after take-off at low speeds.

Research has shown that even a layer 0.5 mm thick covering the whole upper area of the wing can decrease the maximum lift coefficient by up to 33% and the stalling angle of attack from 13% to 7%. This means that even a small layer of ice is not 'insignificant'. Because of this in the majority of states it is strictly required that de-icing is carried out prior to take-off if there is frost, ice or snow adhering to any of critical surfaces of an aircraft. According to the Association of European Airlines recommendations there are two different procedures for the treatment of aircraft surfaces.

De-icing is a procedure by which frost, ice, slush or snow is removed from an aircraft in order to provide clean surfaces while anti-icing is a precautionary procedure which provides protection against the formation of frost or ice and accumulation of snow or slush on treated surfaces of the aircraft for a limited period of time (holdover time).

For the de-icing procedures, fluids based on propylene glycol are mostly used. However, if an aircraft is covered with thick layers of snow, slush etc. some airline companies are using an effective and cheap 'mechanical' method — snow is simply brushed away or, now more often, removed by forced air. Forced air could be used for dry — powder snow, which may occur if it snows at low temperatures and there is no wind. Powder snow does not stick to the wing and can be easily blown off. This leads to considerable reduction of de-icing/anti-icing fluids usage. It should be emphasised that aircraft de-icing with de-icing fluid and water is primarily a thermal process. Definitely there is a certain effect of the mechanical energy exerted by the fluid stream, but the maximum allowable skin pressure, which is defined in the aircraft maintenance manual, limits the mechanical influence.

De-icing and anti-icing fluids must meet a number of requirements. They must not be corrosive, they must not damage materials and aircraft coatings and must be non-flammable. Fluids specifications are defined by the Society of Automotive Engineers standards (SAE AMS 1428 and AMS 1424). Four types of aviation de-icing/anti-icing fluids are used.

Type I — de-icing fluids are used for removing ice from the aircraft surface. However, they do not provide long term protection against re-icing of the surface because they quickly flow off surfaces after use. They are almost exclusively used for de-icing in the United States but also extensively used in Europe (and CIS and Canada). They contain sometimes up to 88% of glycol, but typically close to (but above) 80%. The remainder consists of water, corrosion inhibitors, wetting agents and dye. Inhibitors ensure protection of the airframe, and wetting agents ensure the even coverage of the whole surface. However, the biggest contribution to removal of contamination is the heat energy. In practice, the maximum achievable temperature is about 80–85 °C of Type I de-icing fluid at the nozzle. It is then a matter of reducing the spray distance between nozzle and aircraft, as a fluid stream through cold air is bound to cool off with distance (Figure 18.10). So the distance between the nozzle and aircraft surface should be minimised, but on the other hand the nozzle must be far enough away to guarantee safe operations. The width of the liquid stream is also important. The wider the spray pattern, the more quickly is the temperature of the liquid reduced (Figure 18.11).

Before using the de-icing fluid, it should be diluted with water according to outside temperature and meteorological conditions (e.g. snow, freezing rain) in order to ensure the thermal/operational but also cost efficiency of a de-icing fluid mix and its maximum efficiency corresponding to the concentration and temperature of the eutectic point (see Figure 18.12). There are different Type I fluids on the market. Depending on the kind of glycol and the manufacturer, some are concentrates with the intention to dilute them. These fluids must not be applied undiluted on the wing and must be diluted to a concentration according to the manufacturer. On some types of de-icing trucks the fluid concentration could be pre-selected. There are also the 'ready to use' fluids already diluted to the desired concentration, for example 55%. Type I de-icing fluids must be orange.

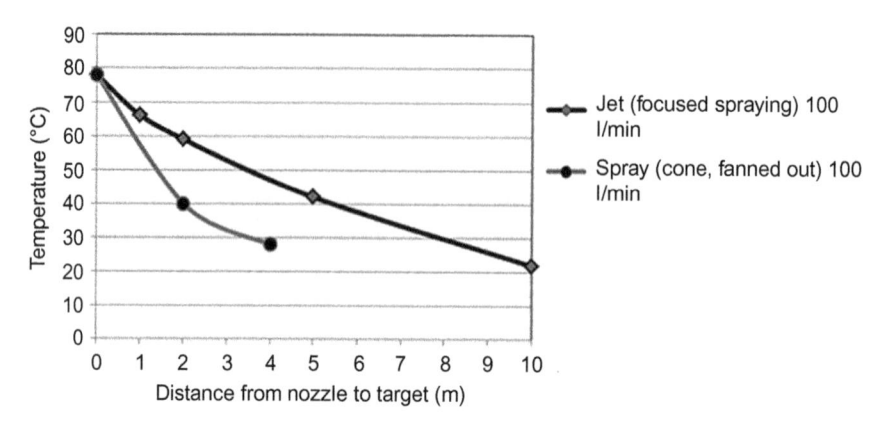

Remarks: Temperature at nozzle 78 °C
Outside air temperature 2 °C

Figure 18.10: Heat loss versus distance of nozzle from the aircraft surface. *Source*: Vestergaard Company A/S tests data; *Courtesy*: Vestergaard Company A/S. *Notes*: Temperature at nozzle 78 °C; outside air temperature 2 °C.

Type II — protective (anti-icing) ***fluids*** can also be used for removing ice from the aircraft surface but mainly to protect the aircraft from re-freezing again during taxiing and waiting before take-off. In the past thickened fluids were more expensive (being technologically more complex) but Type I's glycol content mainly influence the price of the main raw material for glycol, that is crude oil. This offsets the higher 'technological complexity' and lower production volumes of Type II/IV liquids. Presently, the manufacturers' selling price is approximately the same for thickened as for un-thickened de-icing fluids, often actually with Type I being higher-priced than Type II. Because of the lower price but also time savings thickened anti-icing fluids are sometimes used also for de-icing. However, this is an exclusively European practice and seems to be on the decline due to identified flight safety hazards caused by polymeric residues left in certain parts of some aircraft types (see also below).

Protective fluids are normally applied to the cleaned surface of the aircraft. As with the Type I fluids, Type II fluids contain glycol, water, inhibitors wetting agents and dye. In addition to this they contain a polymer-based thickener. The thickener increases the viscosity of the protective fluid so that it adheres to the aircraft surface and does not flow away. The thickener is dissolved in water. The composition of protective fluids corresponds to minimum freezing point. Most Type II fluids are not diluted any more. However, in general a thickened fluid can be diluted to 75/25 and 50/50. These dilutions can be found in the Holdover time table with corresponding holdover time for each dilution and weather condition Type II fluids are dyed light yellow.

Figure 18.11: Aircraft de-icing at Vienna airport, Type I fluid is used. *Photo*: A. Kazda.

Type III — protective (anti-icing) ***fluids*** have characteristics that lie between Types I and II/IV fluids. They can provide longer holdover time than a Type I fluid but less than Type II/IV fluids. They are not frequently used today and are available at very few airports only. They are also offered by a few de-icing fluid manufacturers. In fact, as per 2014, only one was certified according to SAE standards. They are designed especially for turboprop aircraft performance with lower rotation speeds with V_R less than 100 knots. They must be coloured bright yellow.

Type IV — protective (anti-icing) ***fluids*** are very similar to the Type II fluids but they have significantly longer holdover time (see Tables 18.3 and 18.4). Type IV fluids are dyed emerald green. The holdover time (HOT) is defined as the period of time during which a thin layer of anti-icing fluid stays on the aircraft to protect the surface against precipitation. It starts immediately the application of fluid is begun (one step) or with the beginning of the second step in a two-step procedure.

464   Airport Design and Operation

Protective fluids are classed as non-Newtonian, that is viscous fluids. The viscosity of a non-Newtonian fluid changes depending on the shear on the fluid surface, as shown in Figure 18.13. The fluid has high viscosity at small values of shear and small viscosity if the shear on the fluid surface grows. This property is very important for protective fluids. When the aircraft is stationary or is taxiing, the fluid viscosity is high and the fluid adheres to the surface of the aircraft. During the take-off run, the fluid experiences higher and higher shear, its viscosity is suddenly reduced and it flows from the surface of the aircraft. Ideally the fluid viscosity should be high up to a speed of about 110 km/h and then it should be reduced rapidly so that at the nose wheel lifting speed, the aircraft surface is clean.

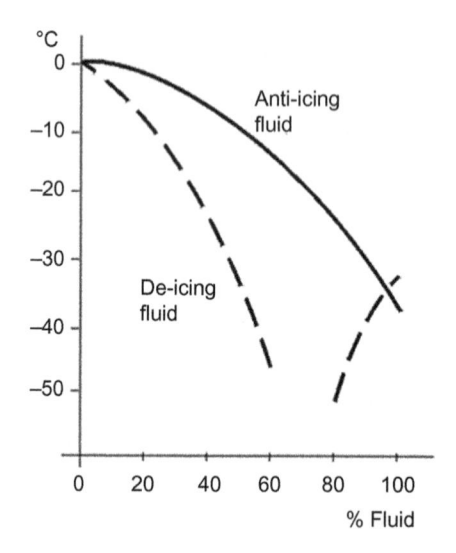

Figure 18.12:   Comparison of the freezing point diagrams for aqueous solutions of Type I — de-icing fluids against Type II — protective fluids.

The problem, however, is that not all of the fluid is always shorn-off completely and a small amount of the fluid may remain on the aircraft, particularly in the areas where the speed of the airflow is not sufficient to blow it off. After the glycol evaporates and only the thickening agent is left, residues can be formed. The thickening substance has practically no anti-freeze properties and is very hygroscopic. It can form gel-like deposits which could result in blockages of flight controls on certain types of aircraft. The thicker the fluid is (i.e. Type IV liquids), the more deposits are formed. However, there are specific pass/fail criteria within the SAE AMS 1428 standard that checks the gel residues. Each fluid on the market has to pass specified limits which reduce the gel formations to a minimum. Furthermore, the use of hot Type I liquids washes the gel residues and helps to decrease the possibility of gel formations. There is a trend of airports moving toward being able to offer a Type I fluid based de-icing, because of the inherent problems with the polymers forming residues — and potentially blockages — in aircraft's aerodynamically 'leeward' — hidden areas.

Table 18.3: FAA holdover time guidelines for SAE Type I fluid on critical aircraft surfaces composed predominantly of aluminium.

Outside air temperature[a,b]		Wing surface	Approximate holdover times under various weather conditions (hours: minutes)							
Degrees Celsius	Degrees Fahrenheit		Freezing fog or ice crystal	Snow, snow grains or snow pellets[c]			Freezing drizzle[e]	Light freezing rain	Rain on cold soaked wing[f]	Other[g]
				Very light[d]	Light[d]	Moderate				
−3 and above	27 and above	Aluminium	0:11–0:17	0:18–0:22	0:11–0:18	0:06–0:11	0:09–0:13	0:02–0:05	0:02–0:05	
Below −3 to −6	Below 27–21	Aluminium	0:08–0:13	0:14–0:17	0:08–0:14	0:05–0:08	0:05–0:09	0:02–0:05	CAUTION: No holdover time guidelines exist	
Below −6 to −10	Below 21–14	Aluminium	0:06–0:10	0:11–0:13	0:06–0:11	0:04–0:06	0:04–0:07	0:02–0:05		
Below −10	Below 14	Aluminium	0:05–0:09	0:07–0:08	0:04–0:07	0:02–0:04				

THE RESPONSIBILITY FOR THE APPLICATION OF THESE DATA REMAINS WITH THE USER.
CAUTIONS:
↦ THE TIME OF PROTECTION WILL BE SHORTENED IN HEAVY WEATHER CONDITIONS. HEAVY PRECIPITATION RATES OR HIGH MOISTURE CONTENT, HIGH WIND VELOCITY OR JET BLAST MAY REDUCE HOLDOVER TIME BELOW THE LOWEST TIME STATED IN THE RANGE. HOLDOVER TIME MAY BE REDUCED WHEN AIRCRAFT SKIN TEMPERATURE IS LOWER THAN OAT.
↦ SAE TYPE I FLUID USED DURING GROUND DE/ANTI-ICING IS NOT INTENDED FOR AND DOES NOT PROVIDE PROTECTION DURING FLIGHT.
↦ THIS TABLE IS FOR DEPARTURE PLANNING ONLY AND SHOULD BE USED IN CONJUNCTION WITH PRETAKEOFF CHECK PROCEDURES.

[a]Type I fluid/water mixture must be selected so that the freezing point of the mixture is at least 10 °C (18 °F) below outside air temperature.
[b]Ensure that the lowest operational use temperature (LOUT) of the fluid is respected.
[c]To determine snowfall intensity, the SNOWFALL INTENSITIES AS A FUNCTION OF PREVAILING VISIBILITY table (Table 1C) is required.
[d]Use light freezing rain holdover times in conditions of very light or light snow mixed with light rain.
[e]Use light freezing rain holdover times if positive identification of freezing drizzle is not possible.
[f]No holdover time guidelines exist for this condition for 0 °C (32 °F) and below.
[g]Heavy snow, ice pellets, moderate and heavy freezing rain, small hail and hail.

Table 18.4: FAA holdover time guidelines for SAE Type IV fluids.

Outside air temperature[a]		Type IV fluid concentration neat-fluid/water (volume %/volume %)	Approximate holdover times under various weather conditions (hours: minutes)					
Degrees Celsius	Degrees Fahrenheit		Freezing fog or ice crystals	Snow, snow grains or snow pellets[b, c]	Freezing drizzle[d]	Light freezing rain	Rain on cold soaked wing[e]	Other[f]
−3 and above	27 and above	100/0	1:50–2:55	0:35–1:10	0:50–1:30	0:35–0:55	0:10–1:15	
		75/25	1:05–1:45	0:30–0:55	0:45–1:10	0:30–0:45	0:09–0:50	
		50/50	0:20–0:35	0:07–0:15	0:15–0:20	0:08–0:10	CAUTION: No holdover time guidelines exist	
Below −3 to −14	Below 27–7	100/0	0:20–1:20	0:25–0:50	0:20–1:00[g]	0:10–0:25[g]		
		75/25[h]	0:25–0:50	0:20–0:40	0:15–1:05[g]	0:10–0:25[g]		
Below −14 to −25 or LOUT	Below 7 to −13 or LOUT	100/0[i]	0:15–0:40	0:15–0:30				

THE RESPONSIBILITY FOR THE APPLICATION OF THESE DATA REMAINS WITH THE USER.
[a] Ensure that the lowest operational use temperature (LOUT) of the fluid is respected. Consider use of Type I fluid when Type IV fluid cannot be used.
[b] To determine snowfall intensity, the SNOWFALL INTENSITIES AS A FUNCTION OF PREVAILING VISIBILITY table (Table 1C) is required.
[c] Use light freezing rain holdover times in conditions of very light or light snow mixed with light rain.
[d] Use light freezing rain holdover times if positive identification of freezing drizzle is not possible.
[e] No holdover time guidelines exist for this condition for 0 °C (32 °F) and below.
[f] Heavy snow, ice pellets, moderate and heavy freezing rain, small hail and hail (Table 10 provides allowance times for ice pellets and small hail).
[g] No holdover time guidelines exist for this condition below −10 °C (14 °F).
[h] For Cryotech Polar Guard temperature is limited to −5.5 °C (22 °F).
[i] For Cryotech Polar Guard and Clariant Max Flight 04, temperature is limited to −23.5 °C (−10.3 °F). If the fluid specific brand is unknown, all of the temperature limitations in this and the preceding note apply.

Depending on the meteorological conditions, and on airport and state practices, one-step or two-step procedures are used for aircraft de/anti-icing.

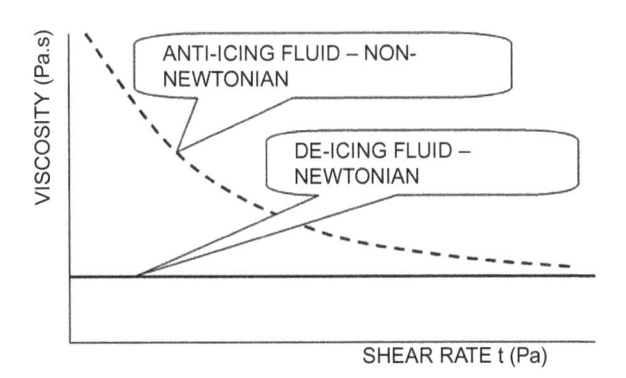

Figure 18.13: Viscosity against shear rate profiles for Newtonian and non-Newtonian fluids.

During a *one-step procedure* and depending on actual meteorological conditions Type I only or thickened Type II/III/IV fluid is used for snow/ice removal from the aircraft's surfaces and the same time to provide the protection to those surfaces. In fact at a number of airports in Central Europe with humid oceanic or humid continental climate, up to 70% of all de-icings are with Type I only, in a one-step procedure. This due to a holdover time of typically 45 minutes for Type I during clear skies, no precipitation and just sub-zero temperatures.

However, during one-step procedures it is also permissible to apply heated thickened fluids. Warm Type II applications are typically for South Europe, where the Type II fluid's higher freezing point is not a problem. In these cases heated Type II is typically used for *anti-icing* and not *de-icing*. This is possible with certain types of de-icing equipment and their tank configurations. During the heated Type II de-icing the fluid temperature should be significantly lower compared to a Type I application. Operators must pay special attention to not over-heating Type II (both in terms of temperature level, as well as heating duration). In the above case, the water is typically kept heated at 80–85 °C, where the thermostat for the Type II should not be above, say, 45–50 °C.

In a **two-step procedure** the aircraft is cleaned by the Type I fluid first and then, within 3 minutes, the surface should be treated by the Type II or IV fluid. To make the cleaning of the aircraft easier, the de-icing fluid is applied hot and under high pressure. At a temperature of up to −7 °C some airports use hot pressurised water for cleaning of the aircraft. Protective fluids are less stable and must not therefore be exposed to higher temperatures. They are applied cold.

In order to avoid disrupting the function of the polymerising fluid and reduction of the fluid viscosity, the protective fluid must not be pumped by common types of rotary or piston pumps. Instead, membrane or screw pumps must be used.

In the United States and Canada, the two-step procedure is usually used but, in the United Kingdom the one-step method is prevailing instead. One of the reasons is that most of the airports in North America use a centralised system of de-icing on special pads located close to the runway holding points. Use of Type I fluid in the United Kingdom is limited, although British Airways at Heathrow is now also switching towards Type I. In the slightly sub-zero conditions it is possible to carry out the first step using a very weak Type I mixture for the de-icing and thereafter a Type II or IV for anti-icing treatment. This results in a reduction in the amount of de-icing fluid as Type I's inherent 80% glycol content allows for much more 'cheap, added water' when mixing the fluid on board the de-icer, compared to only 50% in Type II/IV fluids. Operators from Scandinavia and Baltic countries seem to be using Type I on a much larger scale.

According to FAA and AEA (Association of European Airlines) recommendation about 1 l of fluid should be used for one square metre for aircraft treatment (i.e. 1 mm/1 m^2). The total area of wings and tail surfaces of B 737 800/900 is approximately 158 m^2. Depending on meteorological conditions about 150–250 l of fluid is used for de/anti-icing on average for a B 737 or A 320. When anti-icing fluid is applied a 'natural thickness' of 1 mm fluid thickness coverage should be applied so it takes 1 l of fluid to cover 1 m^2 to a depth of 1 mm. Since application is never perfect, it may take more than 1 l/m^2 to achieve this 1 mm fluid thickness. Usually the rest of the fluid starts to drip off the wing. Some types of wing contamination are very difficult to detect. For example clear ice is nearly invisible and a hands-on check is required to feel the build-up. The tactile check is the only known method to verify if the treated surface is clean of ice.

De-icing of an aircraft can be performed in either centralised or decentralised ways. The latter is by mobile vehicles on the apron and the former way is to use one or several specially constructed stands located as close as possible to the take-off runway threshold. Both solutions have advantages and disadvantages depending on the particular airport layout, traffic volume, type of operation and climate. However, environmental protection and fluid collection must be guaranteed both on stand or on special pads de-icing operations.

In the case of decentralised de-icing (at the apron) both Type I and II/IV fluids must be used so that, during taxiing to take-off, the aircraft does not freeze again. The residues of de-icing fluids remain on the apron, and they penetrate into the substratum, where they contaminate soil and ground water. The residuals of the Type II or IV fluids together with the remaining snow or ice are very slippery, which reduces safety on the apron. This cannot be avoided completely even on airports where de-icing is performed centrally, because propellers and the inlets of jet engines must be de-iced on the apron stands. To guarantee safe and non- contaminated working conditions on stands during turn around handling, some airports clean the stand after each turn round by big vacuum cleaners (e.g. Stockholm Arlanda).

To ensure the protection of the environment, and to increase the safety of air operation by shortening the time between de-icing of the aircraft and its take-off, areas for central de-icing of aircraft are being built at larger airports. The design of the surfaces prevents the escape of the de-icing substance. Glycol is trapped and recycled. Recycling is in most cases performed in the production plant, because the recycled fluid must have the same properties as a new fluid, but for example in the case of the Munich airport recycling is performed directly on the airport which proved to be an effective solution. On average about 51%–55% of the glycol used is collected. The rest stays with the aircraft, is blown away by the engine exhaust or sent to the sewage treatment plant when the content of glycol collected is lower than 5%. Also for central de-icing pads mobile de-icing facilities are used. Portal type de-icing facilities which were in use on Luleå (Sweden) and Munich airports were dismantled. On the one hand they were able to de-ice an aircraft in a short time, but lacked the necessary operational flexibility.

Depending on the type of operation and airport size, there are advantages of both stand de-icing and special pads de-icing. However, the basic requirement is to guarantee environmental protection so perfect shielding of the subgrade created by plastic membrane and collection of contaminated waters must be safeguarded. One advantage of centralised de-icing is that de-icing can be performed immediately before take-off so that in most cases it is not necessary to then protect the plane by Type II/IV fluid. This requires close co-ordination between air traffic control and the airport administration or the handling agent that provides the de-icing. Otherwise, the aircraft which has been de-iced might have to wait too long before take-off. Co-ordination with air traffic control is just as important in de-icing of an aircraft on the apron in order to avoid long delays to the aircraft after de-icing (see Section 9.3).

De-icing of an aircraft is a professional and specialised activity which has a direct effect on the safety of air transport. The workers who perform de-icing must complete specialized training, be perfectly familiar with the de-icing equipment and with the properties of the fluids used. Also annual recurrent training is required for procedures practicing and familiarisation with novelties. For example aircraft with high composite contents have significantly different de-icing characteristics than the aluminium ones because of different heat-transfer properties. Simulators are used more and more for training of skills to operate de-icing equipment but also for recurrent training. Increased attention has to be paid to the protection of these workers, both because of adverse weather conditions, during which the de-icing of aircraft is most frequent and necessary, and because glycol is harmful to the health.

470 Airport Design and Operation

The toxicity of de-icing fluids is an environmental concern, and research is underway to find less toxic (i.e. non-glycol-based) alternatives, for example glycerol as an anti-freeze agent is tested. At the moment, only very few such products exist (in Europe, e.g. 'Defrosol'), but still mostly in the development phase. The main advantage is that glycerol is not toxic, though the other elements in the de-icing fluid, such as surfactants and corrosion inhibitors can be toxic.

19

AIRPORT EMERGENCY SERVICES

Tony Kazda and Bob Caves

Motto: Know safety — no pain; No safety — know pain.

19.1. ROLES OF THE RESCUE AND FIRE FIGHTING SERVICE

Airport administrations spend millions of dollars providing and operating fire fighting equipment, hoping that they will never have to use it. A Rescue and Fire Fighting Service (RFFS) produces costs without explicit benefits for airport administrations. As with other departments that generate no revenue, when there is a squeeze on finances at an airport, management tends to minimise spending on the RFFS. From a long-term viewpoint, the airport image may suffer from such an attitude. The safety standard of an airport is high on a list of priorities of airlines. As a rule of thumb, in the event of a serious aircraft fire, each fire fighter might expect to save up to five passengers. If the RFFS is under-funded, the airlines may decide to fly elsewhere. From that viewpoint, the aerodrome RFFS is one of the inseparable parts that go to create the complete 'airport product'. If the services are of high quality, they may be an important factor in the marketing campaign of the airport.

Air transport is one of the safest means of transport. Nonetheless, there is a finite probability that an accident will occur sooner or later. The accident statistics show that the majority of aircraft accidents happen during take-off and landing, that is to say, in the vicinity of airports. The airport operator must be prepared for such an eventuality. During take-off, there is a maximum quantity of highly inflammable fuel on board. A severe accident usually causes a break in the integrity of the aircraft's fuel tanks, leading to the escape and atomising of the fuel and a subsequent fire. Those are classified as 'Class B' fires (whose fuel is flammable or combustible liquid or gas). An extensive fuel fire brings the greatest probability that the passengers will not survive the accident. The principal task of the RFFS is to save human lives. The necessary level of provision of RFFS does not depend on the aerodrome size, for example with regards to the runway length to be safeguarded in order to reach an accident site, but principally on the size of the aircraft. The demands upon the RFFS at international airports have been gradually increasing because ever larger aircraft with greater fuel volumes and greater seat capacity have been put into operation. On the other hand, new more effective

472 Airport Design and Operation

extinguishing agents were developed which reduce the required water quantities for foam production.

Fortunately, serious accidents with fire, and hence a heavy death toll, occur only rarely (see Table 4.3). More often, there are smaller incidents like spilled fuel and incipient emergency situations like undercarriage failures that the RFFS must attend. Most events are not connected with an aircraft at all, but concern fire alerts in the terminals, road accidents and other emergencies that would be expected in any city.

Changes in the rescue and fire fighting in recent years have mainly been driven by the introduction of large aircraft and the introduction of a high proportion of composites in aircraft constructions. Composites have been used on aircraft for decades, for example wing-to-body fairings, rudders and, on the Airbus A 300-600 and A 310, the entire tail sections are composites. Interior components also contain composites. What is different is the scope of usage. For example in Boeing 787 composite content is 50% of its weight and the Airbus A350 which was introduced to regular service in 2014 has a composite fuselage with 53% of the aircraft weight being composite.

According to the US Air Force which studied B 2 composite fires it is necessary: 'to identify new tools and techniques that will allow firefighters to more efficiently cut, penetrate and extinguish burning aircraft composite materials'. Composite fires are challenging because:

- hidden interior fires are difficult to extinguish
- composites smoulder and reignite
- fuselage penetration is virtually impossible with an axe and difficult with a fire fighting saw.

However, the composites used for civilian aircraft have considerably different properties from the military ones and according to FAA tests the fire behaviour does not significantly differ from the traditional construction.

During the B 787 certification it was demonstrated that the level of fire safety in the B 787 was equivalent to a conventional (aluminium) aircraft. FAA also conducted research and testing to characterize and understand the fire behaviour of this type of composite structure and to support the certification process.

The composite burns primarily from the vaporization of its resin. When it burns, the resin vapour is forced out of the fibre pores, and pressure causes the material to swell to over twice its volume. However, in almost all cases studied, the composite maintained its rigidity, but its structural strength was not examined after degradation. Composite fuselage structures exhibit longer burn-through times than aluminium structure.

However, the key hazards are the dust and fibres in the air from structures shattering during impact, explosion of fire leading to potentially serious skin and eye irritation. Rescuers must therefore wear personal protective equipment and full face dust and mist respirators or self-contained breathing apparatus.

19.2. Level of Protection Required

19.2.1. Response Times

The required level of provision of the RFFS is established in Annex 14 and in Airport Services Manual: Part 1 — Rescue and Fire Fighting, 4th ed., 2014.

The most important requirements are the speed and the effectiveness of the response to the emergency, which are defined by maximum times for the response to the emergency and to control the fire.

The maximum response time criterion is that the RFFS should reach any location in the operational runway system and any other part of the movement area where the accident may have occurred and begin the rescue operation preferably within 2 minutes and not exceeding 3 minutes in optimum visibility and surface conditions, that is daytime, good visibility, no precipitation with normal response route free of surface contamination, for example water, ice or snow. For example in the case of Air France Flight 358 Airbus A340-313X emergency at Toronto Pearson International Airport on 2 August 2005 the fire fighting teams responded to the incident 300 m beyond the end of runway 24L and were on site within 52 seconds of the crash occurring.

The response time is the time between the initial call to the RFFS and the time when the first emergency vehicle is in position to discharge the extinguishing agent with at least 50% of its maximum rate. The other vehicles must arrive on the spot not later than 1 minute after the first fire engine to provide uninterrupted fire extinguishing.

The control time criterion is that 90% of the fire's extent should be extinguished within 1 minute from beginning to fight the fire.

Extinguishment time is the time required from the application of the agent of the first fire fighting vehicle to the time the fire is extinguished.

Although it may seem that the response time criteria are strict, the times have been determined on the basis of real conditions, statistics of accidents and the subsequent fires. They also closely match the certification criteria for the aircraft, which require a demonstration that all the passengers be able to exit the aircraft within 90 seconds using only 50% of the exits. The flames themselves are not the only threat to life. The high temperature raises the concentration of asphyxiant and irritant gases, and lowers the level of oxygen in the cabin. The evacuation of the passengers is aggravated by a rapidly declining visibility as a consequence of thick smoke. In some cases the visibility declines from the full cabin length to only the cabin width within 15 seconds.

19.2.2. Aerodrome Category for Rescue and Fire Fighting

The required level of fire protection to be provided at an aerodrome should be determined by the aircraft dimensions and number of movements of the aircraft using the aerodrome. The aerodrome shall be ranked into one of ten categories based on the fuselage length and width of the longest (critical) aircraft (see Table 19.1) except that, where the number of movements of the aircraft in the highest category normally using the aerodrome is less than 700 in the busiest consecutive three months, the level of protection provided shall be not less than one category below the determined category. If the width of the critical aircraft fuselage is greater than the maximum allowed in Table 19.1, after selecting the category appropriate to the longest aircraft, then the actual category shall be one number higher.

Table 19.1: Aerodrome category for rescue and fire fighting.

Aerodrome category	Aircraft overall length	Maximum fuselage width (m)
1	0 m up to but not including 9 m	2
2	9 m up to but not including 12 m	2
3	12 m up to but not including 18 m	3
4	18 m up to but not including 24 m	4
5	24 m up to but not including 28 m	4
6	28 m up to but not including 39 m	5
7	39 m up to but not including 49 m	5
8	49 m up to but not including 61 m	7
9	61 m up to but not including 76 m	7
10	76 m up to but not including 90 m	8

Source: Annex 14, Volume I, Aerodromes, Chapter 9; 6th ed; July 2013.

During low traffic periods, the level of rescue and fire fighting may be reduced so that it corresponds to those types of aircraft which are intended to use the aerodrome at that time. The level of fire protection at all-cargo aerodrome may be reduced. It is based on the protection of the area around only the cockpit of an all-cargo aircraft. Details on the respective category reduction and method of critical area calculation for the water quantity can be found in the Doc 9137, Airport Services Manual: Part 1. The amounts of extinguishing agents or the amounts of water necessary for their preparation (see Table 19.4) and the numbers of fire fighting vehicles (Table 19.5) at the aerodrome shall be determined by the aerodrome's rescue and fire fighting category.

19.2.3. Principal Extinguishing Agents

Both principal and complementary extinguishing agents should be provided at an aerodrome. The exceptions are aerodromes ranked in the 1st and 2nd category, where the entire amount of water for production of principal extinguishing agents may be replaced by complementary agents. In this case 1 kg of complementary agent is equivalent to 1 l of water for production of a foam meeting performance level A.

The principal extinguishing agents facilitate the extinguishing and long-lasting control of fire. The principal extinguishing agents are foams with performance levels A, B or C. It is recommended that for aerodromes in categories 1–3, 'B' or 'C' performance level foams should be used.

The traditional principal extinguishing agent was a protein foam. It creates a resistant and stable foam cover consisting of an aggregation of bubbles of a lower specific weight than the fuel or water, suitable for covering, extinguishing and cooling the hot aircraft fuselage. The foam should have strong cohesive qualities and be capable of covering and clinging to vertical and horizontal surfaces. Aqueous foam cools hot surfaces by its high water retention ability and must flow freely over a burning liquid surface to form a tough, air-excluding blanket that seals off volatile flammable vapours from access to air or oxygen. Good-quality foam should be dense and long lasting, capable of resisting disruption by wind or draught, stable to intense thermal radiation, and capable of re-sealing in the event of mechanical rupture of an established blanket. The basic component of the foam concentrate is a hydrolysed protein complemented with stabilising additives suppressing decomposition of the foam at high temperatures, inhibitors of corrosion, antibacterial additives and antifreeze additives. It shall be mixed with water in a concentration of 3%, 5% or 6% according to the type of extinguishing device and the kind of extinguishing agent and its performance criteria. A problem may arise with incompatibility of the protein foam with some types of extinguishing powders which may decompose it. The protein foam usually complies with the criteria for the 'A' level principal extinguishing agents (Table 19.2).

The Aqueous Film Forming Foam (AFFF) is very effective for spilled burning fuel. It is a surface-active agent consisting of perfluorinated surfactants with a foam stabiliser and viscosity control agents. The AFFF does not have the appearance of a protein or fluoroprotein foam. At high temperatures the layer of this agent decomposes. It is unstable. The AFFF is not suitable for fires containing large quantities of hot metal. However, it acts both as a barrier to exclude air or oxygen and, in addition, it produces an aqueous film on the fuel surface capable of suppressing the evolution of fuel vapours. Ideally, the foam blanket produced by the AFFF should be of sufficient thickness so as to be visible before firefighters rely on its effectiveness as a vapour suppressant. The concentrate shall be mixed with water in concentration of 1–6%. The AFFF usually complies with the criteria for the 'B' and 'C' level principal extinguishing agents.

Table 19.2: Foam specifications.

Fire tests	Performance level A	Performance level B	Performance level C
1. Nozzle (air aspirated) (a) Branch pipe (b) Nozzle pressure [kPa] (c) Application rate [l/min/m^2] (d) Discharge rate [l/min]	UNI 86 — foam nozzle 700 4.1 11.4	UNI 86 — foam nozzle 700 2.5 11.4	UNI 86 — foam nozzle 700 1.56 11.4
2. Fire size	2.8 m^2 (circular)	4.5 m^2 (circular)	7.32 m^2 (circular)
3. Fuel (on water substrate)	Kerosene 60 l	Kerosene 100 l	Kerosene 157 l
4. Preburn time (seconds)	60	60	60
5. Fire performance (a) Extinguishing time (seconds) (b) Total application time (seconds) (c) 25% re-ignition time (minutes)	≤ 60 120 ≥ 5	≤ 60 120 ≥ 5	≤ 60 120 ≥ 5

Source: Doc 9137, Airport Services Manual: Part 1 — Rescue and Fire Fighting and ICAO Fire Fighting Foam Testing.

The fluoroprotein foam is a combination of the protein foam with fluorocarbonate. This increases the fluidity of the foam and improves re-ignition characteristics. It is very effective at extinguishing deeper layers of fuel, for example for subsurface application in fuel tanks, because it inhibits penetration of the burning fuel to the surface through individual foam bubbles. In comparison with the protein foams, it demonstrates better compatibility with extinguishing powders. Although the fluoroprotein foams are generally more expensive than the protein ones, their advantage is in higher efficiency and lower consumption of water. Depending on the type of extinguishing device, it shall be mixed with water in a concentration of 3–6%. It is an agent complying with the criteria for the 'B' and 'C' level principal extinguishing agents.

The Film Forming Fluoroprotein Foam (FFFP) is a modern extinguishing agent, highly effective for spilled fuel. The surface tension in the expanded foam is reduced so it can spread across the surface of a fuel. It is a combination of protein foam with a surface-active agent on a base of fluorocarbonate. The extinguishing agent forms a water solution with properties of an oily material on the surface of spilled fuel. Therefore the extinguishing agent rapidly spreads on the burning fuel surface, inhibiting the access of air and suppressing evaporation of the fuel. As with other agents, it is used in a concentration of 3–6% and can be applied either with fresh or sea water. It is compatible with the majority of extinguishing agents. It complies with the criteria for the 'B' level principal extinguishing agents.

Synthetic foams are based primarily on petroleum products — alkylsulphates, alkylsulphanates, alkylarylsulphanates, etc. As with the other foams they also include stabilisers, anti-corrosives, and components to control viscosity, freezing temperature and bacteriological decomposition. Compatibility with complementary agents — that is powders must be checked before their intended usage. Fluorine and organohalogen free foams complying with levels A, B and C requirements may be available under the heading of synthetic fire fighting foams, too. These products cause less long-term damage to the environment and are stable against polar and non-polar hydrocarbons. They are pseudo plastic materials containing surface-active agents, polymer film formers, co-surfactants, stabilisers and antifreeze compounds. Polymer film formers prevent the destruction of produced foam. These products can be used in concentrations up to 6%, with foam compatible dry powders and with aspirating and non-aspirating equipment.

All principal agents are so called 'mechanical foams' produced similarly to whisked egg white.

The synthetic foams are produced from petroleum products. They are convenient for extinguishing fires in closed rooms and to some extent for spilled fuel.

The specifications for 'A', 'B' and 'C' principal extinguishing agents include physical properties and the performance of the foams under fire test conditions. These include the data on the discharge monitors, quantity and velocity of delivery of the extinguishing agent, the extent of the fire, the speed of extinguishing the fire, and on the resistance of the extinguishing agent against a new flare-up of the fire. The amounts of water for foam production are specified for an application rate of 8.2 $l/min/m^2$ for a foam meeting performance level A, 5.5 $l/min/m^2$ for a foam meeting performance level B and 3.75 $l/min/m^2$ for a foam meeting performance level C.

If it is necessary to raise the aerodrome category for rescue and fire fighting, and an extinguishing agent complying with the 'A' or 'B' level criteria is used at the aerodrome, the agent may be replaced by an extinguishing agent complying with the 'C' level criteria which requires smaller quantities of water.

19.2.4. Complementary Extinguishing Agents

According to the Annex 14, Volume I, Chapter 9 requirements, the complementary extinguishing agents should be dry chemical powders. The halon-based agents have strong ozone layer depletion potential. The Montreal Protocol required their earliest production and import phase-out in the United States by 1994. However, there are complementary extinguishing agent substitutes available fulfilling also FAA Minimum Performance Standards and US Environmental Protection Agency criteria as part of its Significant New Alternatives Policy (SNAP). *As replacement fire suppression agents for gaseous or flammable liquid fires, chemicals based on halocarbons such as Heptafluoropropane are used for example in telecommunication or computer facilities.* Halons are therefore, no longer discussed in this book but may still be found in some aircraft fixed installations. Carbon dioxide is still traditionally used in aircraft RFF operations as a means of rapid 'knockdown' (i.e. extinguishing a fire). It is used to extinguish or control specific small fires or as a flooding agent in reaching concealed fires in areas inaccessible to foam for example wheel wells. It is not intended for use on an out-of-control fire. It should not be used on fires involving flammable metals (e.g. magnesium fires). The most effective application of CO_2 is at high rates and low pressure.

The complementary extinguishing agents, in particular powders, are very effective and facilitate quick suppressing of the fire. However, they are effective only while being applied, and do not protect against a new flare-up of the fire. Therefore they are used for the first strike, before the arrival and intervention of the major fire engines. They are also effective within enclosed rooms like baggage lockers, spaces under the aircraft wings and undercarriage bays, where the foams are not effective. Their further use is during specific local fires, such as to tyres, wheel discs, engine that is Class D fires of combustible metals, etc.

Dry chemical powders are among the most effective complementary extinguishing agents. They are the most convenient agents for extinguishing fuel spills and three dimensional fires. A disadvantage of the powders is that they do not cool the extinguished object when they are applied. If a greater amount of powder is used, it may form a cloud which reduces visibility and causes breathing difficulties. Powders are based on potassium hydrocarbonate, potassium chloride and ammonium phosphate. Carbamide powder, a product of the reaction of potassium hydrocarbonate and urea and offered in the market under the name Monnex (potassium allophanate). Monnex is about 4–5 times more effective against flammable liquid fires than carbon dioxide, and more than twice that of sodium bicarbonate. Dry powders should comply with the ISO 7201 standard.

19.2.5. The Amounts of Extinguishing Agents

The amount of water for particular aerodrome categories shall be derived from a theoretical critical area which should be extinguished. The theoretical critical area is an area around the aircraft in which the fire should be extinguished, and which allows survival and escape of the passengers from the aircraft. However, it is not intended to represent the average, maximum, or minimum spill fire size associated with a particular aircraft. It expresses the magnitude of fire hazard potential.

The dimensions of the critical area have been determined experimentally. The theoretical critical area has a rectangular shape. One of its sides is equal to the length of the aircraft fuselage. For aircraft whose fuselage length is equal to or greater than 20 m, the other side of the critical area is equal to the greatest width of the aircraft augmented by 24 m on one side and 6 m on the other side of the fuselage in a cross wind of 16–19 km/h. For aircraft with a fuselage length shorter than 12 m, the critical area shall be widened by 6 m on both sides from the greatest width of the aircraft fuselage. A transition is used when the fuselage length is between 12 and 24 m (see Table 19.3).

Table 19.3: Theoretical critical area specifications.

Overall length [m]	Theoretical critical area [m^2]
$L < 12$ m	$L \times (12 \text{ m} + W)$
$12 \text{ m} \leq L < 18$ m	$L \times (14 \text{ m} + W)$
$18 \text{ m} \leq L < 24$ m	$L \times (17 \text{ m} + W)$
$L \geq 24$ m	$L \times (30 \text{ m} + W)$

Source: ICAO (Doc 9137-AN/898) Airport Services Manual, Part 1 – Rescue and Fire Fighting 4th ed; 2014).
Note: L is the overall length of the aircraft; W is the maximum width of the aircraft fuselage.

In the majority of incidents, only two thirds of the theoretical area — the practical critical area — are caught by the fire. The total amount of water shall be determined from the practical critical area, the amount of water required per square metre per minute and the extinguishing time, with 60 s as the control time. The water quantities specified for foam production are calculated with an application rate of 8.2 l/min/m^2 for a foam meeting performance level A, 5.5 l/min/m^2 for a foam meeting performance level B and 3.75 l/min/m^2 for a foam meeting performance level C. These application rates are considered to be the minimum rates to reduce the initial intensity of the fire by 90%. Then the amount is augmented by the water required for extinguishing the remains of the fire. This additional amount of water cannot be calculated exactly and it depends on a number of factors such as maximum aircraft mass, maximum passenger capacity, fuel load and previous operational experience. The minimum amounts of water for production of extinguishing agent for particular aerodrome categories are given in Table 19.4.

The quantities of water are determined by the average overall length of aircraft in a given category. If an aircraft with larger dimensions operates from an airport where the airport category for rescue and fire fighting is reduced, the minimum quantities of water must be recalculated. For example if the Airbus A380 (category 10) is operating infrequently at a B747 airport (category 9) and the number of A380 movements is less than 700 in the busiest consecutive 3 months, the aerodrome is allowed to provide a category 9 level of protection. However, the recalculated water quantity will be more than for category 9 but less than for category 10.

It is recommended that the quantity of foam concentrates on a vehicle should be sufficient to produce at least two charges of foam solution. There must be a reserve supply of foam concentrate and complementary agents at the airport at least equivalent to 200% of required amounts to be provided in the fire engines for vehicle replenishment purposes.

In many cases, considerably larger amounts of water are required for extinguishing an aircraft fire than those recommended in Table 19.4. In the majority of events the amount may even double. This, but also the advent of new larger aircraft in commercial operations carrying more fuel (see Chapter 10) and passengers, speeded up the development of a new generation of fire fighting foams that are more effective and allow cheaper rescue and fire fighting operations. Amendment 11 to Annex 14, Volume I Aerodrome design and operations dated 14 November 2013 includes provisions for a new type of principal extinguishing agent that is the performance level C foam. For example A 380 belongs to the RFF category 10, a performance level A foam requires 48,200 l on the other hand performance level C foam requires 22,800 l of water.

It must be emphasised that the determined minimum amounts of extinguishing agents should be constantly available on rescue and fire fighting vehicles. In the event that all the rescue and fire fighting vehicles participate in the intervention, the aerodrome should be closed. If only a certain number of the rescue and fire fighting vehicles participate in the intervention, the aerodrome category for rescue and fire fighting should be temporarily downgraded. The administrations of large airports make efforts to provide a continuous operation, and if possible, without any limitations. In that case they must provide a greater quantity of extinguishing agents and numbers of rescue and fire fighting vehicles than required for a single emergency.

Table 19.4: Minimum usable amounts of extinguishing agents.

| Aerodrome category | Foam meeting performance ||||||| Complementary agents ||
| | Level A || Level B || Level C || | |
	Water [l]	Discharge rate foam solution/minute [l]	Water [l]	Discharge rate foam solution/minute [l]	Water [l]	Discharge rate foam solution/minute [l]	Dry chemical powders [kg]	Discharge rate [kg/s]
1	350	350	230	230	160	160	45	2.25
2	1,000	800	670	550	460	360	90	2.25
3	1,800	1,300	1,200	900	820	630	135	2.25
4	3,600	2,600	2,400	1,800	1,700	1,100	135	2.25
5	8,100	4,500	5,400	3,000	3,900	2,200	180	2.25
6	11,800	600	7,900	4,000	5,800	2,900	225	2.25
7	18,200	7,900	12,100	5,300	8,800	3,800	225	2.25
8	27,300	10,800	18,200	7,200	12,800	5,100	450	4.5
9	36,400	13,500	24,300	9,000	17,100	6,300	450	4.5
10	48,200	16,600	32,300	11,200	22,800	7,900	450	4.5

Source: Annex 14, Volume I, Aerodromes, Chapter 9; 6 th ed; July 2013)
Note: The quantities of water are based on the average overall length of aircraft in a given category.

19.3. RESCUE AND FIRE FIGHTING VEHICLES

The task of the rescue and fire fighting vehicles is to reach the accident location as soon as possible, to provide escape routes, to put the fire out, and to begin rescue actions.

The minimum determined numbers of rescue and fire fighting vehicles must facilitate the transport of the required amounts of extinguishing agents to the fire's location and their effective application. The number of rescue and fire fighting vehicles depend on the aerodrome category, as shown in Table 19.5. In determining the number of rescue and fire fighting vehicles, not only the amount of extinguishing agents carried by them is important, but also the possibility of applying them so that the entire aircraft fuselage can be covered. If, for example, 3,600 l of extinguishing agent should be provided at the aerodrome, is it more appropriate to have available two vehicles, each of them carrying 1,800 l, with which the attack may be conducted from two points, than only one with the entire amount. At the same time, it is necessary to assess the capital cost of the rescue and fire fighting vehicle, and their replacement approximately every 10–15 years, and the operating costs, which will be higher with two vehicles and also higher operational costs because more crew is usually needed for two fire engines. The capacity of the vehicle tank for foam concentrate should correspond to at least two full loads of such quantity of water as is necessary for foam production at the determined concentration.

Table 19.5: Number of rescue and fire fighting vehicles.

Aerodrome category	Rescue and fire fighting vehicles
1–5	1
6–7	2
8–10	3

Source: Annex 14, Volume I, Aerodromes, Chapter 9; 6 th ed; July 2013.

The original ICAO concept determined separately the numbers of the so called rapid intervention vehicles (Figure 19.1) equipped in the majority of cases with complementary agents, powders in particular, and major vehicles which carried the load of water for the preparation of principal extinguishing agents. The rapid intervention vehicles reached the accident location first and began the fire fighting intervention.

The major vehicles had to reach the incident location before the rapid intervention vehicle consumed all its extinguishing agents. That concept was invalidated by the technical limitations of the older types of major vehicles with which it was not possible to fulfil the required response time. At present parameters are determined separately for vehicles with a volume of water up to 4,500 l and above 4,500 l (see Table 19.6). The fire fighting intervention should be started within the limit of the response time with at least 50% of the performance given in Table 19.4. All the remaining vehicles must reach the accident location within 1 minute after the first vehicle.

Figure 19.1: Rapid intervention vehicles must reach the accident location first.
Photo: A. Kazda.

The majority of vehicles are custom-made in close co-operation with the customer. Vehicle specifications are usually defined in a few steps. Besides the basic equipment, the customer may order other special equipment which exceeds the current standard. In an attempt to reduce the costs of RFFS, personnel costs in particular, efforts have been made to use more and more sophisticated rescue and fire fighting vehicles, so reducing the number of employees. Most current vehicles allow operation at full capacity with one operator although some airports prefer two-member crews. This consists of a driver and a monitor operator, which provides a more effective task distribution and reduces stress and workload. Vehicle equipment should also reflect the type and extent of airport operation. For example in frequent low visibility operations at large airports the vehicles could be equipped with low visibility enhanced vision using a forward looking infrared (FLIR) device or similar system to improve visual awareness in smoky, foggy or dark environments combined with GPS navigation or radar system. Such vehicles substantially ease the intervention and make it more effective, or minimise human factor errors. On the other hand, it increases costs of acquisition and maintenance. The failure of any of the sophisticated components might cause the vehicle to be unavailable. If a failure occurs of the main systems of the rescue and fire fighting vehicle during a rescue and fire fighting intervention, for example the monitor or valves which are equipped with a remote control with boosters, it should be possible to override the remote control and to control the system manually. Therefore, when specifying the requirements for a new vehicle, a compromise must be made between the vehicle's simplicity, the extent of the additional equipment, price, operating costs and its availability.

Table 19.6: Minimum characteristics recommended for rescue and fire fighting (RFF) vehicles.

Parameter/equipment	RFF vehicles up to 4,500 l	RFF vehicles over 4,500 l
Monitor	Optional for categories 1 and 2 Required for categories 3–10	Required
Design feature	High discharge capacity	High and low discharge capacity
Range of monitor	Appropriate to longest aircraft	Appropriate to longest aircraft
Handiness	Required	Required
Under truck nozzles	Optional	Required
Bumper turret	Optional	Optional
Acceleration	80 km/h within 25 seconds	80 km/h within 40 seconds
Top speed	At least 105 km/h	At least 100 km/h
All-wheel drive	Required	Required
Automatic or semi-automatic transmission	Required	Required
Single rear wheel configuration	Preferable for categories 1 + 2 Required for categories 3–10	Required
Minimum angle of approach and departure	30°	30°
Minimum angle of tilt (static)	30°	28°

Source: Airport Services Manual: Part 1 — Rescue and Fire Fighting, 4th ed., 2014.

The assessed optional characteristics and equipment of the vehicle may be classified into two basic groups:

✈ fire fighting and rescue equipment
✈ performance and driveability of the vehicle.

One of the basic requirements is a perfect view from the vehicle facilitating an effective performance of the rescue and fire fighting intervention even in bad visibility. Nozzles that spray the extinguishing agent under the vehicle may permit the vehicle to pass through the fire, though some would argue that the fire should have been knocked down before the vehicle penetrates the area. The monitor on the front bumper facilitates fighting a fire in such spots that are beyond the reach of the main monitor installed on the roof of the vehicle, for example under the aircraft wing. If the aerodrome is used by aircraft with rear-mounted engines, or with the third engine installed in the fin of the aircraft, it is necessary to be able to apply extinguishing agents up to a height of 10.5 m.

It should be emphasised that during each rescue and fire fighting intervention the firefighters act under a great psychological pressure, sometimes they are overcome with stress to the extent that they need professional support after the emergency. Although the modern vehicles permit a single person to perform the rescue and fire fighting intervention, it is more convenient if one person is dedicated to driving and another to the control of the monitor. The cabin should protect the crew completely. A low level of noise allows communication between the crew members and by the means of a radio with the air traffic control, or other staff participating in the rescue and fire fighting intervention. The rugged front of the vehicle should facilitate its passage through obstacles, such as gates, fences and smaller trees. Ventilation or air-conditioning in the cabin is important as protection of the crew against heat. It should be possible to hold the control handles and levers with hands in gloves, and the symbols must be readable even in bad light conditions and under vibrations.

The majority of manufacturers of the rescue and fire fighting vehicles mount the special fire fighting equipment on different types of undercarriage. Thus the buyer has the possibility of unifying the vehicle stock and of reducing the costs of spare parts.

In addition, when contemplating the purchase of a new rescue and fire fighting vehicle, it is necessary to consider its size and mass relative to the parameters of the communications and equipment in the area of the aerodrome and in its close vicinity. In particular, the dimensions of underpasses, tunnels, the load bearing capacities of bridges and pavements which may be used by the vehicle during a rescue and fire fighting intervention should be noted. In the majority of events, the vehicle needs to be capable of coping with poor terrain and thus have an all-wheel drive.

If in the aerodrome vicinity there is difficult terrain (Figure 19.2), special vehicles and equipment should be available. The terrain may include:

+ large water surfaces, in the area of approach and take-off in particular
+ swamps and similar areas, including river estuaries
+ mountain areas

486   Airport Design and Operation

✈ deserts

✈ areas with large quantities of snow in winter.

Figure 19.2:   Special equipment is required for rescue operations in difficult environments. *Photo*: A. Kazda.

Helicopters, hovercraft, boats, amphibians and land rovers may be utilised as special means and vehicles. However, the airport can co-operate with special rescue services to be able to respond to these types of emergencies.

The RFFS requires a whole range of ancillary equipment if it is to function successfully in all circumstances. These include breathing apparatus and qualified staff to use it, extending ladders and floodlighting masts.

To guarantee readiness and efficiency of fire fighting but also safety of personnel to maximise deployment of resources, a complex maintenance programme must be introduced at each airport. This should include preventive maintenance of fire engines but also of every piece of fire fighting equipment including personal equipment and protective clothing. The programme should also include rechecking after maintenance and regular testing of equipment and maintenance record keeping.

19.4. AIRPORT FIRE STATIONS

The location of a fire station at an airport is crucial for fulfilling the response time requirements (Figure 19.3). All the remaining criteria which may influence the location of the fire station among the other facilities in the airport are subordinate to this. In order to fulfil the response time limit, two or more fire stations should be provided in airports with an extensive runway system. Each of them should be located near the ends of the instrument runways, where accidents occur most. If more airport fire stations are provided, one of them is usually established as the main fire station, and the remaining are designated as satellite sub-fire stations. One or more rescue and fire fighting vehicles are located in each airport fire station. The total number of extinguishing agents is divided so as to be sufficient for the first intervention. Vehicles from the second airport fire station should reach the place of accident within 1 minute after the first rescue and fire fighting vehicle. For example to provide fire protection and maintain optimum risk management of the two main runways Singapore Changi Airport operates four stations in and around the airport. Both Fire Station 1 and Fire Station 2 provide fire protection up to Category 10 requirements. Fire Sub-Station 3 is centrally located within Changi Airport and provides fire service cover at three passenger terminals and other buildings in the airport area. There is also the Sea Rescue Base to provide search and rescue in the event of an aircraft ditching incident at sea, particularly off the coast of Changi Airport.

When the location of a new fire station is considered, an assessment should also be made of the further development of the airport, and of the runway system in particular. If possible, the access from the fire station into the runway should be direct, without unnecessary turns.

In the past, airport fire stations were designed very functionally with minimum facilities for the crews. On the basis of investigations and operational experience, it has been found that the precondition for a successful rescue and fire fighting intervention is not only perfect technical equipment but also the psychological and physical well-being of the rescue crew. The extent and size of particular rooms should differ according to whether it is the main airport fire station or a satellite one, but in the majority of cases there will be the following facilities:

↦ garages for the rescue and fire fighting vehicles allowing their regular daily maintenance

↦ work and rest facilities for the staff

↦ communication and alarm systems

↦ storage rooms for technical support of the rescue and fire fighting vehicles, provision of extinguishing agents and fire fighting outfit and equipment

↦ training facilities and lecture rooms.

The garages should be designed in such a way that there is enough space around the vehicles, at least 1.2 m on each side, even if bigger vehicles are acquired in the future. The garages should be heated to a temperature of 13 °C as a minimum. The rescue and fire fighting vehicles require connection with a supply of direct current for accumulator charging and permanent heating of oil and compressed air. When the vehicle starts, the connectors must be automatically disconnected. In order that it should not be necessary to leave the garage with the vehicles to make the regular engine checks, ventilation should be provided for the exhaust gases. It must be possible to open the door quickly, preferably by a remote control from the watchtower. In the event of a failure of its automatic system, it should be possible to open it manually. A broken-down vehicle must not block the exit for the others.

Figure 19.3: The main fire station must allow fire vehicles garaging and daily maintenance. *Photo*: A. Kazda.

The work and rest facilities for the staff should include locker rooms and sanitary installations, a multifunctional room for training and resting, with a kitchen for the preparation of meals. The size of the offices will depend on the extent of tasks to be provided by the respective airport fire station.

A view should be provided over the largest portion possible of the movement areas from the watchtower. The alarm is usually given from the watchtower, and activity with other units co-ordinated from there during a rescue and fire fighting intervention. The communication systems of the watchtower should provide:

- direct telephone connection with air traffic control
- radio connection at the frequencies of the air traffic control for monitoring the radio correspondence, or for a direct communication with the crew of the aircraft
- radio connection with the rescue and fire fighting vehicles, or a connection between the main and satellite airport fire stations (if constructed)
- dedicated telephone connection with fire fighting units in the airport vicinity and other services participating in a rescue and fire fighting intervention (police, hospital)
- a siren or other alarm system and a public address radio system.

The majority of airports are not prepared to receive the great number of telephone calls which follow a flight accident. For example, after the destruction of the B 747 Pan American airliner, Flight 103, over Lockerbie in December 1989, approximately 15,000 calls were registered every 15 minutes at Heathrow Airport. In the event of an accident to an aircraft with 140 seats, on the first day, approximately 10,000 to 12,000 telephone calls may be expected at the airport. On the second day, that number drops by one half and on the third day by one half again. Thus the callers block the telephone exchange which is needed for the connection with other units according to the alarm schedule. There are computer systems available which allow the telephone lines to be used only for outward calls, and according to a programme, they dial sequentially the particular units that are supposed to participate in the rescue and fire fighting intervention. Provision should be made for the reception and briefing of the meeters and greeters who will need information on their relatives and friends.

Storage areas and other areas of technical support should allow the storing of extinguishing agents, their loading into vehicles and the storing and maintenance of rescue and fire fighting suits and equipment. A facility should be constructed for the drying of hoses and other equipment. The airport fire station should have its own standby source of electrical power or a connection to another standby source. Emergency water tanks should be located near the runway thresholds to replenish the tenders.

A training area is required, with either an old wreck or a special rig that can be adapted to represent a variety of different aircraft types (e.g. under-wing or fuselage-mounted engines). It should be sited where fire fighting activities will not interfere with the normal operation of the airport. It should be designed so that none of the contaminants generated by the exercise can enter the ground water.

19.5. EMERGENCY TRAINING AND ACTIVITY OF RESCUE AND FIRE FIGHTING UNIT

19.5.1. Personnel Requirements and Training

The number and the level of training of fire fighting personnel at the airport should correspond to the types of rescue and fire fighting vehicles used at the airport and to the maximum supply of the extinguishing agents for which the vehicles are designed. In determining the number of professional firefighters, consideration should be given to the types and number of movements of the aircraft using the aerodrome but if required also other factors like airport road accident and incident interventions. If there are only a few movements of aircraft during the day, the jobs may be done one after another by a single person, for example fire fighting and baggage handling. All the personnel intended to participate in rescue and fire fighting actions should have specialised training and periodical tests in order to be able to perform their duties. In addition, the personnel who are intended to operate the rescue and fire fighting vehicles should be familiar with the driveability of the vehicle on different types of surfaces, including soft terrain. The minimum number of qualified and competent personnel to deliver effective RFF services should be based on Task and Resource Analysis (TRA). The analysis should be based on a realistic, worst case, aircraft accident scenario.

Personnel should be trained so that, in the event of a rescue and fire fighting intervention, they are able to perform the activity entirely independently, without orders being given by the rescue commander. In critical situations, they should be capable of making decisions and of assuming responsibility for their decisions. They must be able to act independently without officer instructions but also as a member of team.

The training of the rescue and fire fighting unit personnel should include initial and recurrent instructions. The initial one may be divided into two categories:

- basic training in skills and use of the rescue and fire fighting equipment, familiarisation with the airport vicinity and aircraft equipment
- tactical training in extinguishing different types of fires and equipment, fire fighting and rescue actions (Figure 19.4).

The entire training programme should ensure that all members of the rescue and fire fighting unit are convinced that they are able to control fully the respective equipment, that the equipment functions without failures and that they are able to manage the assigned task. The acquired skills and abilities quickly degrade if not continuously practised.

The basic skill training comprises knowledge of different kinds of fire, types of extinguishing agents and their utilisation, technical parameters and control of all equipment to such a level that the usage and control of the equipment becomes fully automatic. An aspect of becoming familiar with the airport vicinity is the task of recognising both the most rapid and alternative routes to different areas under any type of weather conditions. All members of the rescue and fire fighting unit should, most importantly, know all the types of aircraft using the airport, in particular:

✈ location of the exits and emergency exits of the aircraft, and how to open them

✈ configuration of seats

✈ kind and amount of fuel and location of tanks

✈ location of accumulators

✈ position of the points designed for a forcible entry into the aircraft fuselage.

Figure 19.4: Tactical training in extinguishing different types of fires is also required. *Photo*: Courtesy of Bratislava M.R. Štefánik Airport.

The purpose of the operational training is to perfect rescue and fire fighting behaviour using models simulating real conditions of an aircraft and its parts, such as engines, undercarriages, aircraft fuselage, cabin, and the like, on fire under different weather and visibility conditions. The training should be repeated in intervals shorter than one month.

Depending on the airport size and the national regulations, the airport operator is obliged to perform a full fire emergency exercise once a year, simulating the response to a high extensity aircraft accident. The purpose of such training is to verify the competency of the rescue and fire fighting unit and the co-ordination of activity with the units of fire services in the airport vicinity, police and health and emergency service. These exercises should be complemented by much more frequent 'table-top' exercises.

19.5.2. Preparation for an Emergency Situation and Rescue and Fire Fighting Intervention Control

At every airport, an emergency plan must be prepared for different types of emergency situations. The main purpose of the emergency plan is to minimise the effects of an unexpected event while also minimising the effect on flight operations. The airport emergency plan comprises the procedures which should be followed during the given type of emergency situation. The procedures include the sequence and range of responsibilities of each of the units participating in the rescue and fire fighting intervention. A serious accident always requires co-ordination and close co-operation with the fire units and hospitals in the airport vicinity. The units participating in the rescue and fire fighting intervention may include air traffic control, fire units of the airport, local fire brigades, police, emergency service, hospitals and health centres, telecommunications, military, civil aviation authority and representatives of the airline company. The main communications are between the airport fire unit and the air traffic control unit.

The contents of the airport emergency plan may be divided into:

1. Preparation of the airport for emergency situations, including the issues concerning organisational structure, responsibilities, testing the functional status of the equipment and clothing, and training of the personnel.

2. The activity during an emergency situation. This section describes procedures and tasks of the particular units participating in the rescue and fire fighting intervention.

3. The activity after termination of the emergency situation. These are activities which must be completed, but do not require an immediate solution, and the return of the airport to its regular operation (Figure 19.5).

The airport emergency plan should cover the issues concerning accidents both on the aerodrome and also outside the aerodrome boundary. The control of the rescue and fire fighting intervention at an accident on the aerodrome is usually within the competencies of the airport administration. The control of the rescue and fire fighting intervention at an accident out of the aerodrome depends on the laws and regulations in the given country and on the agreements between the airport administration and rescue and fire fighting units in the airport vicinity. In line with the Airport Emergency Planning Manual recommendations an area under approach and departure routes within 1,000 m of the runway threshold should be subject to emergency procedures assessment.

Figure 19.5: The aircraft recovery after runway end overrun. *Photo*: J. Mach.

In some states the activity of the Rescue and Fire Fighting Service is limited to the area demarcated by the airport perimeter. In other countries, the area of responsibility is greater. When there is an accident of an aircraft off the aerodrome, the necessary extent and the speed of the rescue and fire fighting intervention is, in the majority of cases, beyond the capabilities of local fire units. In addition, it is a specialised activity. Therefore the airport fire unit often responds to an accident a considerable distance from the airport. For example, at an emergency landing of the SAS MD-81 airliner on 27 December 1991, 20 km north-west from Arlanda, three fire fighting vehicles were sent there despite the aerodrome category for rescue and fire fighting of the Stockholm Arlanda airport having to be temporarily lowered. In the event of an accident more than 4 km beyond the perimeter of London Heathrow Airport, the airport usually sends 50% of its rescue and fire fighting units to the place of the accident. Further than that, a special vehicle with cutting and extrication devices will be probably sent at the request of the intervention commander.

The emergency plan should comprise the maps of the airport and its surroundings with grids in two different scales. The more detailed grid maps shall include access roads to the airport, location of water sources, emergency assembly points and parking bays. In addition to those data, the other map shall include information about hospital facilities to a distance of approximately 8 km from the airport reference point. By the means of a grid map, it is possible to identify with sufficient exactness the place of accident or of any other point in the area in question. In conditions of low visibility, the rescue and fire fighting unit may be guided to the area of the accident by means of radar in co-operation with air traffic control, but other means for RFFS guidance like GPS or A-SMGCS systems are used also. The officer in charge shall establish three zones of activity. Access to the crash zone should be confined to the RFFS. The rescue zone should be close by and provide an immediate reception zone for survivors. The command post will normally be there. This will also be the rendezvous point for vehicles so that they can receive direct instructions if there is any sort of communications breakdown. The medical zone should be near hard standing, so that ambulances can drive up easily.

The activity of the rescue and fire fighting unit shall differ according to the category of emergency. The particular types are indicated as follows:

3rd degree of alert — *'aircraft accident'* — is declared if an aircraft accident happens in the airport or in its vicinity. It shall be declared by the air traffic control or by the watchtower of the fire fighting unit.

2nd degree of alert — *'full emergency'* — is declared if it is known, or if there is a suspicion that an aircraft on approach has problems which may lead to an accident. It is declared by the air traffic control.

1st degree of alert — *'local standby'* — is declared if it is known, or if there is a suspicion that an aircraft in approach has a failure, however not of such a nature that it could seriously affect the safety during landing. It could also include a bomb threat and other incidents. The local standby is declared by the air traffic control.

19.6. RUNWAY FOAMING

Passenger aircraft with the mass over 5,700 kg with damaged or retracted undercarriages cannot land on a grass strip for emergency landing. On a strip with a low load bearing strength, the aircraft would sink in and decelerate abruptly, with extensive damage to the structure and possible casualties. Therefore heavy passenger aircraft must use a paved runway for a gear up landing. In these cases, the runway or a part thereof is sometimes covered with a protein foam blanket. Other newer types of extinguishing foams are not suitable for covering of the runway because they quickly decay and this procedure currently is not widely used.

In the latest issue of the International Civil Aviation Organisation (ICAO) Airport Services Manual, Part 1 — Rescue and Fire Fighting (Doc 9137-AN/898) 4th ed.,

2014 the part on Foaming of Runways for Emergency Landings (former Chapter 15) which also specified runway foaming procedures and dimensions of the foam blanket was withdrawn. The FAA, in 1987, withdrew Advisory Circular 150/500-4 dated 12/21/66 entitled Foaming of Runways and has no plans to recommend this procedure in the future. Also neither Boeing nor Airbus has no recommended procedures concerning the application of foam to the surface of a runway when an aircraft may land with some, or all, landing gears either retracted or indicating not down and locked. The advantages of a foam blanket have never been entirely vindicated. There are some theoretical advantages of a foam blanket:

- reduction of deceleration forces
- reduction of damage to the aircraft
- reduction of the risk of creating sparks when the aircraft slides across the runway surface
- reduction of the risk of spilled fuel catching fire.

Analysis of emergency landings of aircraft with retracted undercarriages has not fully confirmed the theoretical advantages of the foam blanket. Neither is there any evidence to prove that laying the blanket has any psychological advantages for the pilot.

When an assessment is made of the possibility of laying a foam blanket, besides the theoretical advantages, consideration should be given to a whole range of operational issues, affecting either the aircraft engineering or the airport and its operational problems. As far as the aircraft is concerned, the extent of damage to, or failure of, its undercarriage or other damage is important. Consideration should also be given to the pilot's experience, weather conditions, particularly the visibility, and the runway equipment. It is not always possible for the pilot to use a preferred runway, as shown by the accident at Amsterdam with the El Al Boeing 747 Freighter.

When an evaluation is made of the possibilities of the airport using the foam blanket, the factors include the number of runways, runway length on which the blanket is intended to be laid, traffic density, and consequences of closing the runway for several hours or days, or cleaning and repair of the runway after the disabled aircraft removal. Special equipment should be available for runway foaming. Taking into account its price and the rarity of such an event, not many airports carry such a supply of foam. It is usually convenient to equip one airport in a given region. When a selection is made of which airport, besides the operational factors, consideration should be given also to the characteristics of the airport surroundings as far as the hospital capacities, approach communications, etc. are concerned. Extrication devices must be available, together with cranes and facilities to repair aircraft. Airports that do not have special equipment available should not try to provide runway foaming. The rescue and fire fighting vehicles should be constantly prepared for an intervention and have available specified quantities of extinguishing agents. The captain of the aircraft with the emergency may be forced by circumstances to execute the landing before runway foaming is finished, and the rescue and fire fighting vehicles should be prepared for this.

It is also necessary to assess the time which is available for foaming. Even with good organisation, the preparation and laying of the foam blanket will take 1 hour or more. If the aircraft has enough fuel, it may fly off, reduce the mass and thus also the risk of a subsequent fire.

Weather conditions also have an effect upon the quality of the foam blanket. At high temperatures and overheated runways, the foam may decay more quickly. Conversely, at temperatures below zero, the water from the decaying foam freezes and considerably reduces the braking effect. It is not possible to lay the foam blanket in heavy rain or snow.

Foaming may be requested by the aircraft captain or by a representative of the air carrier. However, they should have sufficient flight and operating experience to consider in advance all these points, and also other factors, which have an effect on that decision.

The touch-down point of an aircraft with retracted undercarriage on the runway is, in the majority of cases, 'longer' by 150–600 m than normally, as a consequence of a more significant 'ground effect'. The width, length and manner of runway foaming should correspond to the type of aircraft, to the kind of undercarriage failure and to the requirements of the pilot in command. In restricted visibility, when it is difficult to distinguish where the blanket begins, it should be marked in an agreed way.

After foaming, it is necessary to leave the foam 'resting' for 10–15 minutes, so that the runway surface may dampen. However, the foam must not lie on the runway for a very long time, for example more than 2 1/2 hours, particularly in summer, since water from the blanket runs out and the foam dries up. The foam blanket thickness should be approximately 5 cm.

19.7. POST EMERGENCY OPERATIONS

After the airport emergency is finished, the operations of the rescue and fire fighting units should be organised in such a way that evidence is not destroyed which may be pertinent to the subsequent investigation. Ambulances and rescue and fire fighting vehicles should not cross the track that the aircraft left during the emergency landing, but they should, if possible, move along and to the side of it.

The state aviation authority should be immediately advised about the aircraft accident, and a committee should be appointed to investigate it. Investigation of the accident usually requires several hours or even days during which a plan for the removal of the aircraft may be established and the required equipment may be provided. Until the committee terminates the investigation and gives approval, the wreck of the aircraft must not be moved. The only exceptions are when the wreck creates an obstacle and is a serious danger to air traffic. If the wreck is blocking the runway end, consideration should be given to the possibility of operation on a shortened runway.

The disabled aircraft usually disrupts the regular operation of the aerodrome, and therefore the airport administration requires its removal as soon as possible. The damaged aircraft is the property of the flight operator and the insurance company. The airport administration may in no way interfere with the wreckage, otherwise it would expose itself to the possibility that it will be accused of further damaging the aircraft as a consequence of unauthorised and unprofessional manipulation of the aircraft. Therefore the management of the disabled aircraft removal shall be in the exclusive competence of the flight operator and the insurance company, with whom the airport administration closely co-operates. The removal of small aircraft is a relatively simple matter. Upon agreement with the proprietor, the airport administration may undertake to do it.

The recovery of large passenger aircraft is technically and financially demanding (Figure 19.6). It is often dangerous and requires special equipment and knowledge in order that it should not be damaged. The airport administration usually has not the necessary equipment and mechanisms for such an activity. A specialist team shall be established for the removal of the disabled aircraft, to which representatives of the operator, insurance company and manufacturer shall be appointed. The specialist group shall elaborate a plan for the removal of the aircraft, which comprises in particular the following:

- a list of the required equipment, personnel and schedule of the activities
- preparation of the access roads for heavy-duty equipment
- surveying the accident location
- safeguarding the accident location
- the manner of leasing the special equipment and mechanisms
- data and recommendations of the manufacturer
- defuelling, and draining of the oils and other liquids from the aircraft.

In order that the disabled aircraft can be removed, it is usually necessary to lift it, insert an emergency undercarriage under it, and then remove it on a temporary strengthened surface. Special air bags are most often used for lifting aircraft

498   Airport Design and Operation

(Figure 19.7). The air bags are inserted under the wings, and the front and rear parts of the fuselage so that the aircraft stability can constantly be ensured. Lifting is controlled by gradual inflation of the individual bags from a compressed air generator.

Figure 19.6: Post emergency activity requires close co-operation with local companies who own heavy-duty mechanisms. *Photo*: J. Mach.

It is usually possible to lease the heavy-duty mechanisms required for removal of the aircraft, such as cranes, lorries and machines for treatment of the access roads from local companies. The airport administration should have available the information about readily obtainable particular kinds of heavy-duty plant.

The other group shall be formed of special equipment usable for all types of aircraft. These are hydraulic lifts, air bags, compressors, lighting equipment, etc. Since it would be uneconomic to have such equipment at every airport, the IATA International Airlines Technical Pool programme was established, by means of which the special equipment is made available. The equipment may be used by the member airlines, and by the non-member airlines for a charge. The equipment is kept at 11 airports (Bombay, Chicago, Honolulu, Johannesburg, New York, London, Los Angeles, Paris, Rio de Janeiro, Sydney and Tokyo). The equipment is laid on pallets and may be supplied to any airport within 10 hours, and in the majority of events, the time does not exceed 5 hours.

Figure 19.7: Lifting of aircraft by special air bags. *Source*: Vepro company.

19.8. EMERGENCY SERVICES AND ENVIRONMENT PROTECTION

At the majority of airports, the fire fighting unit is charged even with interventions at ecological accidents or incidents, and sometimes with a preventive activity in connection with environmental protection. An aircraft accident and an intervention by fire fighting units may result in an ecological disaster. Therefore the emergency plan of the airport must address the issues of environmental protection during the training of the rescue and fire fighting unit and during the intervention.

A long-term goal is to substitute harmless or biologically degradable material for extinguishing agents that adversely impact the environment. Some types of halons which deplete the ozone layer can be substituted by some complementary extinguishing agents. Many airlines participate in the HAL project, which is resolving the issues of a 100% recycling of halon gases from fire-extinguishers. There are ongoing projects aiming at development of modern environmental friendly and high performance extinguishing agents for substitution of the foam forming surface-active agents with a base of fluorine. Similarly, during the training of fire fighting units, attention should be paid to possible pollution of the Earth's atmosphere and underground water. During free combustion of kerosene, thick black smoke is created which may deplete the ozone layer. At specialised training centres for rescue and fire fighting units kerosene fuel is being substituted by natural gas which is combusted in a special ring. The training polygon in the Birmingham UK airport complies with the stringent criteria of environment protection.

500   Airport Design and Operation

All the facilities designed for training (an aircraft fuselage, undercarriage and engine, steel wall) should be situated in an impermeable basin, covered by a layer of gravel. Extinguishing agents and the remains of kerosene are drained off from the basin and treated. If kerosene is used for in the fire fighting training facilities it should be recycled.

19.9.   FINAL THOUGHTS

Besides the training and equipment of the airport rescue and fire fighting units, the quality of an airport fire emergency service may be further enhanced by improved cabin crew emergency procedures training and an increase in the fire resistance of the aircraft. The issues of aircraft construction, although very important with regards to the fire emergency service, are not a subject of this book.

Figure 19.8: Passenger safety can be enhanced by cabin crew emergency procedures training. *Photo*: M. Kvapil, CSA Training Centre.

Every flight operator is obliged to provide the passengers with a briefing on safety. The majority of them consist of a presentation given by the cabin staff, instructional cards which the passengers have available, or with a video film. It is questionable whether under the conditions of a real accident, a passenger will be able to open the emergency exit on the basis of such a briefing. Many passengers would be uncertain of how to open the door, asking themselves questions like: 'Which direction is it to be opened, inwards or outwards?' 'What strength is to be exerted for turning the handle?', 'How heavy is it?' and so on. The crew have to make themselves perfectly familiar with the emergency equipment of an aircraft during their training in emergency procedures (Figure 19.8). The passengers ought to have a similar opportunity in 'safety centres' established at airports. The emergency equipment of an aircraft most frequently operated at the airport may be installed in those safety centres. The passenger designated to sit next to the emergency exit could 'touch' and try to open it. The question is what attitude the airlines may adopt with regards to the idea of safety centres. Some may think that they could emphasise the fact that accidents happen from time to time, and discourage a percentage of the passengers from using air transport.

20

ENVIRONMENTAL CONTROL

Tony Kazda, Bob Caves and Milan Kamenický

Inscription on the gate of an American airbase: 'Pardon Our Noise it is the Sound of Freedom'.

20.1. BACKGROUND

In the relationship of air transport with the environment, a number of conflicts have been identified. Air transport clearly has a negative effect on the environment. In comparison to other means of transport, however, the local environmental impact is taken only on the surroundings of the airport whereas with road or rail transport the area along the whole route is affected by noise and gaseous emissions. In compensation for the environmental disbenefits, air transport supports economic growth, as indeed do the other modes. A medium size airport provides directly several thousand jobs and indirectly tens of thousands. It is 'an economic catalyser' for its catchment area.

One of the basic human rights is the right of an individual to a healthy environment. From this point of view air transport has been the subject of negative criticism. In the last 15 years the standard of living of people in the developed countries has increased rapidly. The real income has increased and this opened up more free time and the wealth to use it actively. People use air transport more and more for holidays. Air transport has at the same time contributed considerably to further economic growth. As wealth increased, sensitivity to the quality of the environment grew even faster. In other words, after people had achieved their fundamental goals, they became more sensitive to their environment. The class of people with a high standard of living use air transport more and more, while at the same time more and more criticising its negative consequences.

The ministers of transport of the ECAC states have concluded that the environment is one of the decisive factors in setting the future limits of airport expansion. The main target of ECAC strategy for airports is the improvement of the potential capacity of European airports and airspace while preserving the standard of safety and respecting the requirements of the protection of the environment, taking into account the forecast growth of air transport in the ECAC states.

The basic strategy for dealing with the environmental issue in accordance with the ECAC conclusions can be characterised for the airports as follows:

To specify the maximum capacity in the remote future in terms of the environmental carrying capacity as well as the physical capacity of the airport site.

In line with ACARE[1] 2020 Strategic Research Agenda — High Level Target Concepts, air transport should be ultra-green in the future. Goals set by ACARE, with a 2020 deadline, include a 50% reduction of carbon dioxide CO_2 emissions through a drastic reduction of fuel consumption, an 80% reduction in emissions of oxides of nitrogen NO_x and a 50% cut in external noise from the 2006 level. It also aims for a 'green' product life cycle design covering manufacturing, maintenance and disposal. The impact of an airport on the environment must be based on an Environmental Impact Assessment (EIA) study. After an EIA has been performed by a specialised agency, and considered by the responsible state administrators after deliberation with other affected state administration bodies, a process of public hearings takes place with the affected communities. The length and scope of this process depends on the, extent of the project and character of the surrounding area, for example it will differ for large urban communities, compared with for nature conservation areas or low quality uninhabited areas, the latter being hard to find in developed countries anywhere near a centre of population. The EIA must, above all, be based on a deep knowledge of the affected area, the character of other adjacent areas, environmental carrying capacity and restrictions, and on the knowledge of land use and compatibility planning of the site and in the communities in the surroundings of the airport.

A list of factors related to air transport that have a negative effect on the environment is given below, divided into seven groups. The impact of the airport on the environment can be compared to the impact of a medium size or large industrial enterprise, with an additional special impact from the air transport operation.

Noise of aircraft

- noise of aircraft in the vicinity of the airport
- tests of aircraft engines

Air pollution in the vicinity of airports

- emissions of aircraft engines
- emissions from the operation of vehicles on the airport
- emissions from ground transportation to and from the airport

---

1. ACARE — Advisory Council for Aeronautics Research in Europe.

- emissions from other sources on the airport (e.g. heating plant, unburned hydrocarbon emissions from fuel storage)

Factors with a global impact
- long distance transfer of air pollution (e.g. acid rains)
- greenhouse effect — contribution to global warming due to deposition of carbon dioxide in or near the tropopause
- depletion of ozone layer
- contrails global 'darkening'
- supersonic boom
- noise of aircraft en-route

Airport construction
- excavation of soil
- soil erosion
- interference with ground water channels and rivers
- impact on flora and fauna
- visual intrusion

Contamination of waters and soil in the vicinity of airports
- imperfect treatment of waste waters
- leaks of oil products
- de-icing of airport pavements and aircraft

Waste management
- storing and disposal of dangerous substances used in maintenance and repairs of aircraft
- waste from the airport operations and from arriving aircraft

Aircraft operation including accidents or incidents
- wake vortices and their impact on the neighbouring dwellings
- blue-ice phenomenon (it is created when toilet tanks or drain tubes on an aircraft leak)
- accidents or incidents of aircraft with hazardous cargo
- emergency procedures connected with aircraft de-fuelling

✈ other negative effects to the environment connected with an aircraft accident (e.g. fuel leakage, leakage of extinguishing substances, aircraft safety of the communities around airports).

Some airports voluntarily collect and report data in line with the Global Reporting Initiative (GRI). The reports use a Sustainability Reporting Guidelines template and cover not only air transport environmental issues but also other areas to allow an airport complex sustainable management overview. Guidelines set out the reporting principles, disclosures on management approach and performance indicators for economic, environmental and social issues. The guidelines permit reporting on different levels depending on the airport size and local conditions. The report can also cover key sector-specific issues, including the impact on local communities through employment and business development; measurement and monitoring of noise; impacts on biodiversity; management of water and effluents; implications for labour practices related to pandemic diseases; exposure to pollutants and aircraft noise; human trafficking; resettlement of local communities; assessing business continuity and emergency preparedness.

Depending on the situation, any of the impacts could be the deciding factor in determining the outcome of an EIA, but noise is the factor that normally gets the most attention. Therefore the first part of this chapter deals with noise. It then deals with the issues of exhaust gases and ground water. The issues of ground water and soil protection are also dealt with in Chapters 10, 18 and 19.

20.2. Noise

Noise has long been a problem around airports, Newark in New York being closed in 1952 after demonstrations about it. Reaching a compromise between the industry and the local communities has required progress on three fronts: planners have needed to understand the attitude of the communities for whom they are responsible; the industry has attempted to ensure that the best feasible levels of technology are being used; and airports have had the uncomfortable task of trying to reconcile the resulting situations by mitigating the impacts.

20.2.1. Characteristics

Sound is a mechanical wave motion spread from a source. It can move through gases, liquids and elastic solids and it is connected with a transfer of energy. The speed at which a sound moves depends on the medium through which it moves.

A physical definition describes sound as: 'A change of the pressure that is received by the human ears and interpreted as a sensation by the brain'. The frequency could be at the minimum 20 and at the maximum 20,000 cycles per second, or Hertz (Hz). If a sound wave's frequency is higher than 20,000 Hz, it is called ultrasound. If it is lower than 20 Hz, it is called infrasound.

What is the difference between sound and noise? Clearly it cannot be the loudness only, although an aircraft can be so noisy as to injure the ear if one is too close without hearing protection. A mosquito at night-time can be more than enough to spoil a good night's sleep. Instead of considering only the physical characteristics of noise, it is possible to think of it as *'unwanted sound'*. This has certain implications. Because noise is unwanted sound, it is closely connected with the feeling of annoyance. Noisiness is related to the loudness of a sound, which in turn depends on when and where it occurs. Mosquitoes cause less annoyance during daytime, at least from the point of view of noise. If noise becomes loud enough, the primary concern will be the risk of hearing impairment, not the annoyance.

Noise is a specific form of pollution. It is more a sociological element than an economic one. Too loud a sound or one that is generated too often, at an improper time, at an improper place or in an improper situation is designated as noise. Aircraft noise has a disturbing effect in particular on the inhabitants in the vicinity of aerodromes and under arrival and departure tracks.

Social research has shown that people differ enormously in their attitude to noise, but that their annoyance is related to both the noise intensity of a flight and the number of flights they hear. This led to the development of noise indices to represent the impact on a community, each country developing its own indices, but most containing terms reflecting both loudness and frequency.

Modern aircraft technology has considerably reduced the noise from the engine sources. The area on the ground affected by a given level of noise from modern jets is approximately five times smaller than that affected by the jets in the 1960s. This positive trend will be in future partially diminished by overall growth of traffic and aircraft movements at major airports. However, it is important that, with the withdrawal of the noisiest aircraft from operation, the maximum noise levels for individual movements will be reduced. These peak individual noise events are one of the main reasons for complaints among the communities around airports.

Noise around airports can be controlled in four ways. The ***first way*** is to reduce the noise at source through improved technology, and aircraft noise certification to ensure that the available technology is employed. This is not dealt with further in this book. The ***second way*** is to control the aircraft operational procedure. This includes optimisation of flight procedures and distribution of movements between runways. The ***third way*** is by limiting operations by type and time of day. The process of noise monitoring in selected points of the airport vicinity is an important part of the optimisation of operations. The ***fourth way*** is land use and compatibility planning around the airports, particularly with regard to urbanisation zones.

These measures comprise the so-called 'balanced approach' promoted by ICAO. An approach to noise control on a system basis in all of the four areas should ensure an acceptable noise load on the inhabitants in the vicinity of aerodromes and under the arrival and departure routes.

20.2.2. Descriptors Used for Aircraft Noise Rating

Sound pressure is the basic metric for assessment of noise. Sound pressure represents time fluctuations of pressure around its static value. In a common air environment, static pressure represents barometric pressure. For common sounds, the range of acoustic pressure values is approximately from 2×10^{-5} Pa (pascals) up to 20 Pa. A value of 2×10^{-5} Pa is the weakest sound to be noted by the human ear — so-called **threshold of audibility**. That value is 5×10^9 times smaller than the normal value of barometric pressure and would be created by a deviation of the tympanum over a distance smaller than the diameter of one atom. The values of acoustic pressure of common sounds are relatively small but their range is very big ($1:10^6$). Therefore, since sound is perceived by a human being, the pascal is not very convenient for measuring it. Further, this is also not convenient because human perception of sound is on the basis of Webber–Fechner's physical law:

When physical intensity of the tone '*i*' of the given frequency grows in a geometric series, its subjective effect '*h*' on the human ear (tone noisiness level) increases approximately only in an arithmetic series. An approximately correct mathematical expression of tone intensity dependence on its noisiness level has the form:

$$i = k \cdot a^h$$

where '*k*' and '*a*' are constants.

A logarithmic value, the so-called level, is used for expressing the sound pressure and other acoustic quantities. The sound pressure level is expressed with the unit of dB, called decibel

$$L_p = 10 \log(p^2/p_0^2)$$

where the quantities with a subscript zero are reference quantities related to the threshold of audibility. The reference for sound pressure in air is 2×10^{-5} Pa, which is equivalent in the decibel scale to 0 dB. By converting sound pressure in pascal to decibels, a scale spanning $1:10^6$ is conveniently reduced to 1:120. Figure 20.1 indicates typical sound pressure levels for 'everyday' incidents.

Physically doubling the sound pressure means to increase the level by 6 dB, while a 10-fold increase means to increase it by 20 dB. On the other hand, if the sound pressure is reduced by 50%, the level has been reduced by 6 dB, and a reduction to 1/10 corresponds to a decrease of 20 dB.

It was mentioned above that the range of normal human hearing is approximately 20 Hz to 20 kHz. However, human hearing is not linear with respect to the sound frequency. If there are two signals with a different frequency but with the same level they are perceived differently in terms of noise. In other words, sound is frequency dependent. Further, this dependence also depends on the level. This complicated phenomenon can be clarified by curves of equal loudness level, called the Fletcher–Munson's curves. The curves have been standardized according to ISO 226 since 1961. Based on extensive research, this standard has been revised in 2003 and new curves of the same loudness level were developed (Figure 20.2).

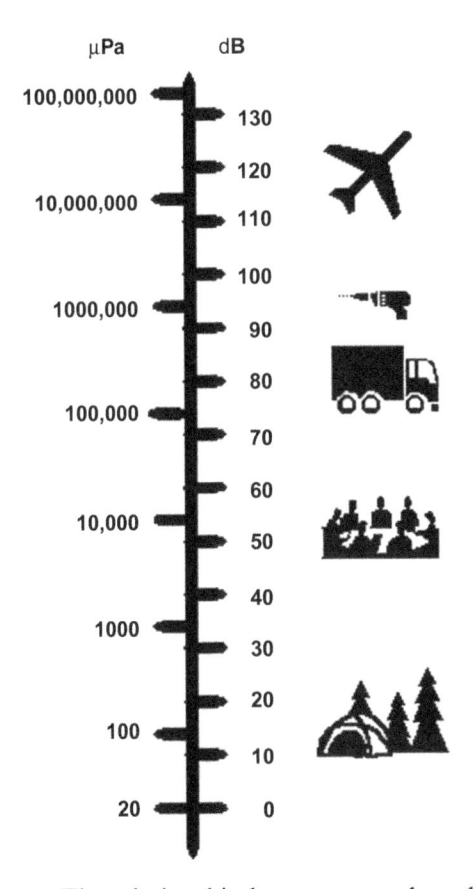

Figure 20.1: The relationship between pascals and decibels.

By reversing these curves we get the frequency dependence of sound sensitivity. On the basis of these curves, weighting curves were created and the corresponding weighting filter used for a sound level metre. The sound level metre evaluates sound in all frequencies that are perceived by human hearing. Three basic weighted curves

A, B and C, the most used being the A-weighted curve and filter, were created in a way that all three chosen curves represented the same loudness of 40 Phon.[2]

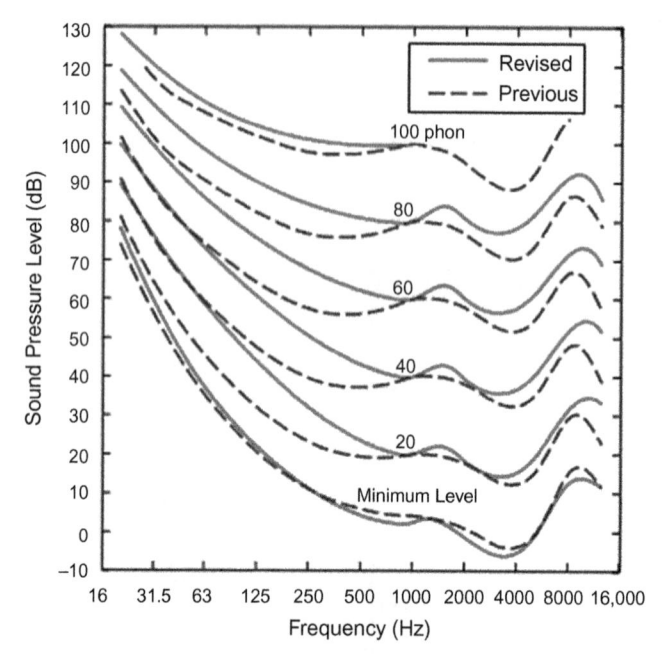

Figure 20.2: The equal loudness curves (original Fletcher–Munson curves in dashed lines).

Other weighted curves were also designed; of which the most used for evaluating airport noise is D. This curve was designed for evaluating sound events that occur when the aircraft overflies the measuring point. This is to capture the more disturbing part of this noise as perceived by a human being and is the Perceived Noise Level (PNdB).

The energy content of the sound is a fundamental aspect of risk assessments of hearing impairment. The Root Mean Square (RMS) is a special kind of mathematical

---

2. Phon is a unit of loudness level for pure tones. This unit considers the frequency and amplitude dependency of sound perception by humans. The reference frequency for considering different sound perception is 1 kHz. Thus a 1 kHz tone with sound pressure level of 40 dB also has a loudness level of 40 Phon.

average value which is directly related to the energy content of the sound. It is one of the most important and most used measures, the sound pressure level being defined from an RMS value of sound pressure.

Loudness is not only a function of the frequency and the level of the sound, as has already been mentioned, but also of the sound duration. The risk of hearing impairment depends not only on the level of the sound, but also on the amount of sound energy entering the ear. For a given sound level the amount of energy entering the ear is directly proportional to the duration of exposure. Sounds of short duration are perceived to be of a lower level than steady continuous sound of the same level. To assess the potential for damage to hearing from a given noise environment, both the level and the duration must be taken into account. However, if the level is very high the duration is irrelevant. The hearing impairment will occur almost instantly with a sound pressure level over 120 dB(A). The influence of time as a factor of the noisiness effect is expressed by the quantity Equivalent Continuous Sound Level, called L_{eq}.

The equivalent continuous A-weighted sound level, L_{Aeq}, equals the constant A-weighted sound level whose acoustic energy is equivalent to the acoustic energy of a fluctuating A-weighted sound over some time interval.

The mathematical formulation is:

$$L_{Aeq} = 10 \times \log \left(\frac{1}{T} \int_0^T \left(\frac{P_{A(t)}^2}{P_0^2} \right) dt \right)$$

where T is the duration, $P_{A(t)}$ the actual A-weighted sound pressure, P_0 is the reference pressure.

This is similar to the RMS value. Although they both express an equivalent constant signal containing the same amount of energy as the actual time-varying signal itself, they cannot be substituted for one another. The L_{eq} expresses the linear energy average, while the RMS value expresses a weighted average where more recent events have more weight than older ones.

The equivalent continuous level is not the only parameter that could be used for energy content assessment. An alternative parameter is the Sound Exposure Level (L_{AE}), also called the Single Event (Sound) Exposure Level, often known for short as the SEL (Figure 20.3). The L_{AE} is defined as the constant level acting for one second which has the same amount of acoustic energy as the original sound. The L_{AE} is the L_{eq} normalised to a one second interval. L_{AE} measurements are often used to describe the noise energy of a single event such as an aircraft flyover. L_{eq} is normally used to integrate the succession of single events over a longer time period.

512 Airport Design and Operation

The time course of A-weighted sound pressure level ($L_{P,A}$) during aircraft passing is shown in Figure 20.4 (where $L_{P,AP}$ is residual noise level, $L_{P,ASmax}$ is maximal AS-weighted sound pressure level, T_E is duration of the sound event caused by aircraft flyover, T'_E is 10 dB down time).

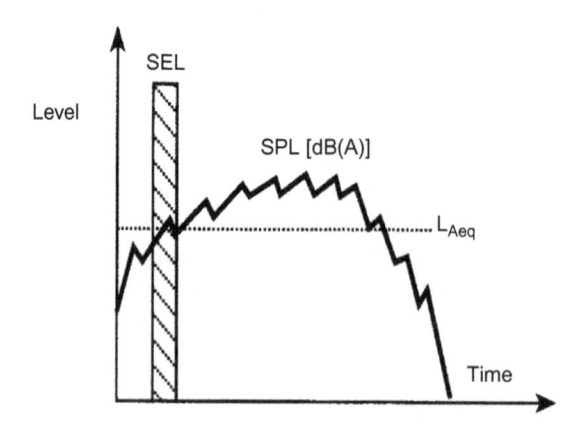

Figure 20.3: The relationship between sound pressure level (SPL), L_{Aeq} and SEL.

In the majority of countries, descriptors for aircraft noise issue from mathematical quantities that are defined in the standard reference:

ISO 20906:2009 Acoustics — Unattended monitoring of aircraft sound in the vicinity of airports with Amd1:2013, which replaced the old ISO 3891 'Acoustics, Procedure for describing aircraft noise heard on the ground'. In Great Britain, the equivalent norm is the BS ISO 20906:2009 standard. In this standard, the basic quantity for evaluation of the aircraft noise is A-weighted Equivalent Sound Level L_{Aeq} for a specified period. The standard gives an exact description of the transformation of the measured quantities to those used for evaluation. A-weighted Equivalent Sound Level L_{Aeq} for a specified time period is determined by Sound Exposure Level (L_{AE}). Relationship between L_{Aeq} and L_{AE} is:

$$L_{Aeq} = L_{AE} + 10 \times \log(T_E)$$

It is also recommended that $L_{p,ASmax}$ be determined for each sound event caused by air traffic, in addition to L_{AE} (this parameter is used in many countries as complementary to the assessment of air traffic noise). The latest amendment of the ISO 20906 standard is also dealing with:

- requirements for installation and operation of air transport noise continuous measuring equipment
- recommendations for noise data recording from aircraft operations and individual aircraft movements

✈ air transport noise descriptors requirements
✈ noise monitoring protocols requirements
✈ determination of uncertainties during noise monitoring.

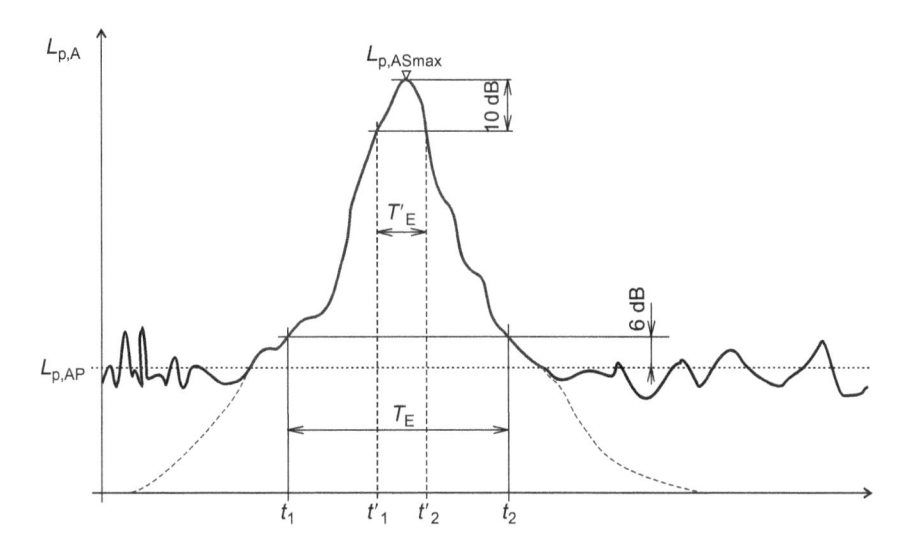

Figure 20.4: The time course of A-weighted sound pressure level ($L_{p,A}$) during aircraft overflight.

The latest amendment of the ISO 20906 standard corresponds to the new standard IEC 61672-1 which describes technical characteristics and requirements for air transport noise measuring equipment.

Certification of aircraft in terms of noise has been legislated internationally for more than 40 years. It is now provided by the ICAO international regulation, the Annex No. 16 to the Convention on International Civil Aviation.

Following Annex 16, the aircraft are divided into different categories — according to the first introduction to service (certification), Maximum Take-off Mass, type of propulsion unit and whether a fixed wing aircraft or a helicopter. Depending on the aircraft category, different procedures and different parameters are used for certification (Effective Perceived Noise Level, maximal AS-weighted sound pressure level, or the A-weighted sound exposure level). The measurement locations and operations through which the outside noise of aircraft is assessed for certification are different for each category.

For example, the current civil subsonic transport aircraft and propeller aircraft with a maximum take-off mass exceeding 8,618 kg certified after 1 January 2006 are evaluated using the Effective Perceived Noise Level (in EPNLdB). Measurements are made at three locations (see Figure 20.5).

514   Airport Design and Operation

However, all these metrics are only poor indicators of people's actual disturbance from noise events, this being indicated by the poor correlation of the number of complaints with the noise contours generated by the metrics. They do not adequately treat problems of noisy peak operations or noise at night, particularly in the period before waking, nor for the whining tone of the latest generation of engines, despite being corrected for the ear's sensitivity to different frequencies. There is at least a suspicion that noise complaints are actually a surrogate for fear of accidents, instanced by complaints being made even though the aircraft has been seen but not heard.

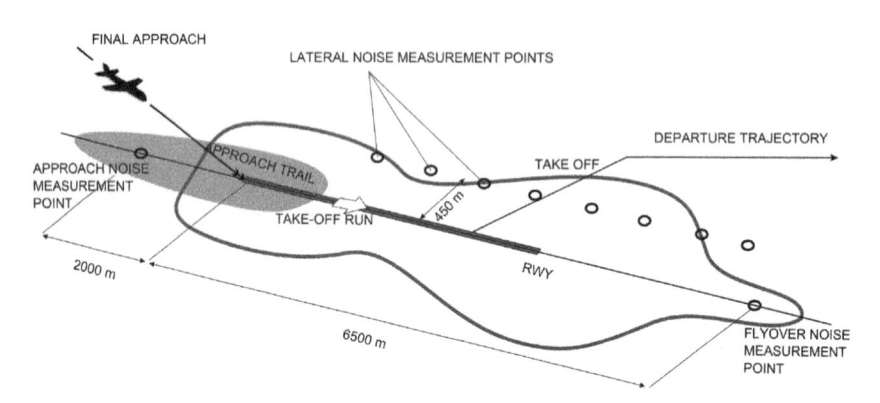

Figure 20.5: Measurements for the certification of transport propeller and jet aircraft by ICAO.

20.2.3. Evaluation of Noise in the Vicinity of Airports

Not all countries use the same methods to evaluate the effects of noise. In the majority of countries, however, only sound events that are connected with aircraft movement or other aircraft activity such as engine tests and taxiing are evaluated. Such a sound event will increase the respective acoustic indicator that describes aircraft noise against an acoustic background.

Since 1992, Great Britain has used the equivalent A sound level L_{Aeq} from sound events between 0700 and 2300 hours as the main noise evaluation tool, supplemented by a night-time L_{Aeq}, usually over the busiest three month period.

In Germany the procedures for assessing the noise load rating in the airport vicinity are elaborated in detail. They reflect experience from long-time monitoring of aircraft noise. The assessment is carried out on the basis of the Act on Protection against Aircraft Noise and the new DIN 45643:2011. Equivalent continuous sound level determined by sound events from air transport in a time period of 6 months is the evaluation descriptor used. A new amendment of the Aircraft Noise DIN 45643 standard was adopted where the equivalent A sound level L_{Aeq} is used for the noise evaluation, except during the night period when the maximum A sound level L_{Amax} must be used.

Besides using the equivalent continuous level, the DIN standard recommends, on the basis of an analogy with ISO 3891 standard, also measuring and evaluating equivalent A-weighted sound level. The long-run equivalent continuous level is computed from measured A-weighted sound levels corrected to equivalent pressure levels over one second, the result corresponding to sound exposure level. The equivalent sound level is derived for the 6-month period for which the traffic level is highest for the airport in question. This rating level is used for comparison with noise levels from other kinds of transport and industry. In addition, the DIN standard recommends the use of the maximum A-weighted sound level measured with a time constant as an auxiliary descriptor.

In Norway two quantities are used for aircraft noise evaluation at airports. Equivalent continuous A-weighted sound level averaged over 24 hours with additional weighting for night and evening flights, similar to CNEL and L_{AeqDN} in the United States as described below, and maximum A-weighted sound level in one week. A third A-weighted sound level, maximum level, is also monitored for a period of time. Furthermore, equivalent level is averaged for the whole week with variable additional weighting of flights during Saturday and Sunday.

In France a quantity 'Indice psophique' R is used for airport noise evaluations. It is computed as a sum of noise peaks of all aircraft movements, while perceived noise level (PNL) is measured. During computation and evaluation, night flights between 2200 and 0600 hours are penalised with a value of 10 dB. For computation the following relation is used:

$$R(IP) = 10 \times \log\left(\sum 10^{L_{PN,d}} + 10 \times \sum 10^{L_{PN,n}}\right) - 32$$

In the Netherlands a 'Rating Index' B is used. Maximum A sound levels of each sound event caused by air transport are used. The following equation determines the evaluative index B:

$$B = 20 \times \log\left(\sum g_i \times 10^{L_{A\,max}/15}\right) - 15$$

$g_i = 1$ for interval of 8–18 hours; 2 for 18–19 hours; 3 for 19–20 hours; 4 for 7–8 and 20–21 hours; 6 for 21–22 hours; 8 for 6–7 and 22–23 hours, 10 for 23–6 hours.

In Switzerland the Noise and Number Index NNI(A) is used for aircraft noise rating. It is an analogy to NNI that used to be used in Great Britain, but it is derived from maximum A-weighted sound level measurement (with a SLOW time constant) of sound events caused by air transport. For computation only those sound events whose maximum levels exceed 68 dB(A) between 0600 and 2200 hours are taken into consideration. Besides the averaged maximum sound level of all considered sound events, the average daily number of movements of aircraft is also taken into account in computation. NNI(A) is determined by means of the following relation:

$$NNI(A) = (L_{Amax})_{en} + 15 \log(N) - 68$$

where N is $n/365$, n the number of operations 0600–2200 hours at which L_{Amax} exceeds 68 dB(A).

In the former USSR air transport noise is rated by maximum A-weighted sound level and equivalent A sound level.

In Slovakia the equivalent A sound level L_{Aeq} and the maximum A sound level L_{Amax} with a SLOW time constant are used for noise evaluation. In the Czech Republic air transport noise is evaluated by equivalent A sound level.

The existing European regulations do not solve the long-run issues of airport noise evaluation. Regulations ES 80/51 and ES 83/206 determine only maximum limits for subsonic aircraft, similar to those from aircraft certification. There are discussions about implementation of a unified evaluative quantity Weighted Equivalent Continuous Perceived Noise Level (WECPNL). In substance, it is the Equivalent Perceived Noise Level — L_{PNeq} recommended in ISO 3891, but with penalties for aircraft movements in the evening from 1900 to 2200 of 3 dB and at night from 2200 to 0600 of 10 dB. WECPNL is determined for a time range of 24 hours. That descriptor is at present recommended also by ICAO.

In the United States the issues of noise have been dealt with in detail since the 1950s. Several methods have been proposed and several rating quantities have been recommended during the development. In the majority of cases they issued from the US Environmental Protection Agency (EPA) recommendations. Indicators for general evaluation of noisiness in the environment were adapted to aircraft noise evaluation. At present several indicators are in use.

The Noise Exposure Forecast (NEF) is an improved older quantity, the Composite Noise Rating (CNR), which is analogous to NNI. NEF is derived from the effective perceived noise level L_{EPN} for individual noise events caused by aircraft movement. Noise events caused by a movement at night are multiplied by a coefficient of 16.67. For NEF = 0, a concept of 'threshold of perceived aircraft noise' was introduced. The NEF is now used to determine noise load from measurements, but was originally calculated from predictive methods.

Equivalent A-weighted sound level for Day-Night — L_{AeqDN} is another quantity used in the United States. It is used for general evaluation of noise load in the environment. It is derived from a 24-hour evaluation of equivalent level. Levels for day are multiplied by a coefficient of 0.625 over 15 hours and those for night by a coefficient 0.375 over 9 hours. In addition, levels for night are penalised by 10 dB. When an evaluation is made of aircraft noise, this quantity comes from particular sound events, which are rated by A-weighted sound exposure level (L_{AE}). Simplified computation of L_{AeqDN} may be done by means of the following relation:

$$L_{AeqDN} = (L_{AeqD} - 1.8 \text{ dB}) + (L_{AeqN} + 5.2 \text{ dB})$$

Community Noise Equivalent Level (CNEL) is an improved version of L_{AeqDN} used predominantly in the state of California. It is also derived from 24-hour averaging of A sound levels. For aircraft noise however it is modified relative to L_{AeqDN} so that the sound events between 1900 and 2200 are in addition penalised by 3 dB. When CNEL is determined, it is derived, as with L_{AeqDN}, from A-weighted sound exposure levels (L_{AE}).

In special cases, daytime equivalent A-weighted sound level evaluated from 0700 to 2200 hours and a night-time one evaluated in a period of time from 2200 to 0700, are also used.

It may be stated that from the viewpoint of evaluation of the noise caused by aircraft operation, there is no consistent international regulation that would be implemented into the legislation of particular states. In every state the evaluation is made on the basis of national regulations, recommendations or laws, though some national regulations do issue from the standard ISO 3891. However, some states have calibrated their own descriptors satisfactorily against surveys of community annoyance.

20.2.4. Land Use and Compatibility Planning

In the majority of states there is an effort to provide a healthy and high quality environment, particularly in the residential areas and for noise sensitive locations such as schools, hospitals and recreational zones. Planning of built-up areas in the airport vicinity is an efficient instrument for preventing complaints of inhabitants in the future, as long as the plans can be reinforced. States are then able to prevent additional settlement of new inhabitants in the noise-affected zones in the airport vicinity. However, in the majority of cases, there is a question over the limitations of the owners' rights to their property and therefore it is an especially sensitive area of legislation. For that reason there is an effort to solve disputes by an agreement in terms of various forms of compensation or repurchase of the estate rather than by expropriation of the estate even if the right of expropriation exists in the legal system.

In **Germany** the Act for Protection against Aircraft Noise determines also the limits that restrict two noise zones in the airport vicinity:

Zone 1 — equivalent noise level is higher than 75 dB(A)

Zone 2 — equivalent noise level is lower than 75 dB(A), but higher than 67 dB(A).

The act specifies which activities may be allowed in the respective zone, for example there is a strict prohibition on building schools, hospitals and new residential buildings in both of the zones. Noise limits according to the new amended standard are also defined for two zones but, for airports, four different cases are specified with respect to the type of an airport and its development:

1. *New or, by a re-construction, considerably expanded civil aviation airports*

    Daytime limits:
    Zone 1 L_{Aeq} = 60 dB(A), Zone 2 = 55 dB(A)

    Night operation limits:
    (a) Till 31 December 2010, L_{Aeq} = 53 dB(A), the L_{Amax} = 57 dB(A) could be exceeded six times during night
    (b) From 1 January 2011, L_{Aeq} = 50 dB(A), the L_{Amax} = 53 dB(A) could be exceeded six times during night

2. *Existing civil aviation airports*

    Daytime limits:
    Zone 1 L_{Aeq} = 65 dB(A), Zone 2 = 60 dB(A)

    Night operation limits:
    L_{Aeq} = 55 dB(A), the L_{Amax} = 57 dB(A) could be exceeded 6 times during night

3. *New or, by a re-construction, considerably expanded military airports*

    Daytime limits:
    Zone 1 L_{Aeq} = 63 dB(A), Zone 2 = 58 dB(A)

    Night operation limits:
    (a) Till 31 December 2010, L_{Aeq} = 53 dB(A), the L_{Amax} = 57 dB(A) could be exceeded six times during night
    (b) From 1 January 2011, L_{Aeq} = 50 dB(A), the L_{Amax} = 53 dB(A) could be exceeded six times during night

4. *Existing military airports*

    Daytime limits:
    Zone 1 L_{Aeq} = 68 dB(A), Zone 2 = 63 dB(A)

    Night operation limits:
    L_{Aeq} = 55 dB(A), the L_{Amax} = 57 dB(A) could be exceeded 6 times during night.

Note: Equivalent noise level daytime limits are specified for a period 06h00 and 22h00.

Equivalent noise level night limits are specified for a period 22h00 and 06h00.

Norwegian legislation determines four protective zones for equivalent and maximum levels:

1. zone — $L_{Aeq} = 55-60$ dB(A), $L_{Amax, DAY} = 85-95$ dB(A)
2. zone — $L_{Aeq} = 60-65$ dB(A), $L_{Amax, DAY} = 95-100$ dB(A)
3. zone — $L_{Aeq} = 65-70$ dB(A), $L_{Amax, DAY} = 100-105$ dB(A)
4. zone — $L_{Aeq} > 70$ dB(A), $L_{Amax, DAY} > 105$ dB(A).

In particular zones it is exactly determined under which conditions what structure might be built.

Austria deals with these issues by legislation as in Germany. Three zones are determined in the airport vicinity restricted by equivalent sound levels A — 70, 65 and 60 dB(A) for daytime and 60, 55 and 50 dB(A) for night operation. A draft of the governmental order was published in 2009 — Protection Against Immissions from Air Traffic (from the Federal Ministry of Transport, Innovation and Technology). The standard proposal was discussed for a long time. However, the order has not been accepted yet. Currently in Austria, there are no noise zones legally defined for air traffic noise. Calculations are carried out in accordance with instructions in OAL Richtlinie 24, Blatt 1-5.

In the **United States**, legislation determines limiting values for L_{AeqDN} as follows:

1. L_{AeqDN} lower than 65 dB(A) — normally acceptable, permitted development without restriction.
2. L_{AeqDN} higher than 65 dB(A) but lower than 75 dB(A) — normally unacceptable for all but specially designed houses.
3. L_{AeqDN} higher than 75 dB(A) — absolutely unacceptable zone, with prohibition to build any new structures other than for airport-related purposes.

Similar noise criteria are defined in the United Kingdom but using L_{Aeq}. Permission for new dwellings will normally be granted below 57 dB, noise should be taken into account between 57 and 66 dB, permission would not normally be given between 66 and 72 dB, and it would normally be refused in zones with noise above 72 dB. Similar guidelines apply to night noise, but the limits are 8 or 9 dB lower. In its White Paper on Airport Policy in December 2003, the UK government stated that it wished to see airports offer to assist households to relocate if they are subject to 69 dB or more due to a development, and to offer acoustic insulation to houses, schools, hospitals, etc. if they are subject to 63 dB or more. The UK CAA is now promoting the concept of a Noise Envelope (CAP 119, Noise Envelopes, December 2013), such that an airport would guarantee not to let operations exceed the envelope in the future. It also favours increasing the funding of noise mitigation to the levels recently seen at Amsterdam for its sixth runway and Frankfurt for its fourth runway (CAP 1165, Managing Aviation noise, 2014).

In *Slovakia* legislation determines limits for four different categories of land use for day (06h00—18h00), evening (18h00–22h00) and night (22h00–06h00) air transport operations (Table 20.1).

Table 20.1: Land use category noise limits.

Land use category	Noise limits [dB]					
	Day		Evening		Night	
	L_{Aeq}	L_{ASmax}	L_{Aeq}	L_{ASmax}	L_{Aeq}	L_{ASmax}
Zones with special noise protection, hospitals, spa, etc.	50	–	50	–	40	60
Urban areas, schools, etc.	55	–	55	–	45	65
Areas in the vicinity of airports and air routes	60	–	60	–	50	75
Industrial areas	70	–	70	–	70	95

If an airport operates a continuous noise monitoring system the equivalent A noise level can be increased by 5 dB but the weekly average level cannot be exceeded. Also L_{ASmax} limit could be exceeded twice times during the night.

In the *Czech Republic* the legislation for air transport noise specifies limits for the equivalent A noise level for a daytime (06h00–22h00) and night (22h00–06h00). The highest equivalent noise level is determined by summation of the basic noise level $L_{Aeq,T} = 60$ dB for daytime and 50 dB for night. The values are determined for a typical flying day for the period from 1 May to 31 October.

The land use and compatibility planning is, in the long run, the most effective way to control noise around the airports. The Airports Council International (ACI) adopted a resolution in January 2000 which urges the co-ordination and unification of EU policy in this field.

20.2.5. Aircraft Noise Measurement

The measurement of air transport noise should reflect the choice of indicator used to evaluate the noise load caused by air transport. This will determine the type of instrument and the time for which the measurements are taken. The method and equipment used for long period monitoring will be different from those for short-term tasks such as week-long surveys. However, the results from both sets of measurements should be compatible.

In choosing the method of measurement, it is important to take into account the purpose for which the evaluation will be performed: control of the legislated noise limits, control of adherence to the chosen flight procedures, verification of the predicted data. The choice of the location of the noise metres also depends upon the purpose of measurement. For verification of prediction and control of the protective noise zones, it is necessary to choose measuring points with the lowest acoustic background possible, away from populated areas. When the purpose is to control adherence to health or annoyance limits, the measuring points are located in the inhabited zone and in the areas where noise sensitive activities are situated. In such cases it is often very difficult to identify noise that is caused solely by air transport and special measuring apparatuses and special rating techniques are required.

20.2.5.1. Short-Term Measurement

Short-term measurements should serve to control adherence to the set limits for a given category of activity, to control and verify the predicted noise load in the vicinity of airports and flight corridors, and to make special measurements such as comparing noise from an aircraft with its certified values. In a short-term measurement it is necessary to observe whether individual sound events are caused by air transport, or whether they are caused by other sound sources. The measurements should therefore always be performed with staff in attendance who identify the type of sound event as it is being measured, unless it is possible to correlate the timing of aircraft movements very accurately from radar traces. A measuring system may also be used that only provides a record of each noise event that exceeds the background noise level. The record should be qualitatively and quantitatively in such a form that will allow to the noise event to be positively linked to the cause.

The level of sound at which the recording should start may be selected. This will depend on the background level, the time for which it is exceeded, and other characteristics of the noise trace.

Short-term measurements should be performed with sound level metres of class 1 accuracy, in compliance with IEC 61672 standards. Measuring apparatus should be able to measure simultaneously maximum A-weighted sound level and equivalent A-weighted sound level in a certain time interval. Simultaneous measurement of A-weighted sound exposure or time level should also be possible.

If a sound level metre that can measure these quantities simultaneously is not available, the sound signal may be recorded, in such a way that the time, frequency and dynamics are unchanged, in a suitable recording medium, for example by means of an analogue or digital tape recorder. Alternatively, the trace can be transformed into a digital form and saved in computer memory. Then the chosen indicators may be evaluated later with a suitable sound measuring apparatus. The recording equipment for saving the original signal should fully comply with the requirements specified by the IEC 561 standard.

If descriptors derived from perceived noise level, for example in aircraft certification, are used for noise evaluation, it is essential to use a multi-spectral real time analyser with spectrum sampling of 500 ms as a minimum if detailed measurement is required.

Microphones for free field should be used in compliance with IEC 61672, with a cover against wind or rain. The microphone should be located at least 1.5 m above the terrain level, or at least in a minimum of 1.2 m from the outside wall of a building if the measurement is performed in the building facade.

Before and after each series of measurements, an acoustic calibration must be performed with a microphone calibrator. When the measurement is performed during several days, the calibration must be carried out at least once a day. In the course of the measurement, temperature, air humidity and wind speed must be monitored. After measurement, the results are processed in a suitable manner into the prescribed form. They may also be used in verification of predictive methods and development of mathematical models for aircraft noise simulation.

20.2.5.2. Long-Term Noise Monitoring

Long-term noise monitoring is performed for the purpose of a long-term control of adherence to the defined limits of noise load for a certain location, for land use and compatibility planning and for monitoring actual air traffic activity which from time to time exceeds the noise limits (Figure 20.6). In such measurement it is also necessary to monitor aircraft movement through the same airspace where the noise is being generated. Aircraft movement monitoring may be provided automatically by radar information processing. The data on the aircraft movements are matched with the noise events detected by the noise monitoring. The noise load originating only from air transport may then be separated from other noise sources.

Unification of the results from aircraft noise Monitoring and Track Keeping (NTK) radar data on aircraft movement may be performed in real time or additionally by appropriate data processing by computer software.

Figure 20.6: Terminal for permanent aircraft noise monitoring. *Source*: Sound Analyser NOR 140 with a multi-spectral analyse and audio recording from a sound events of flyover, NORSONIC AS Norway, control unit and post-processing software EUROAKUSTIK Ltd., Slovakia; *Photo*: M. Kamenický.

In accordance with the above-mentioned, long-run monitoring of aircraft noise may be performed by means of two types of equipment:

1. Measurement is provided by means of instruments that simultaneously measure the required rating quantity (e.g. maximum and equivalent A sound levels) in the specified time interval.

2. Measurement is provided by means of instruments that simultaneously measure and record the continuous course of evaluation descriptors averaged and evaluated from a short-term interval.

Both of the systems may transmit the data from the noise monitoring units to the central control location for processing in two ways:

1. Data transmission is provided in real time, that is each instrument is constantly directly connected with the central measuring point and the data are continually transmitted to that point. Connection may be provided by a local network or by means of telecommunication lines. Output may be shown also on monitors in selected neighbourhood centres or airport terminals. The UK's East Midlands Airport (EMA) was the first in Europe which allowed the public to access the system output on the internet with a 24-hour delay, using a package called WebTrak. The viewer can see the track being flown, the flight number and the altitude of all operations, allowing them to identify aircraft they believe to have been off track or too low (http://www.eastmidlandsairport.com/cms/173/webtrak.html). EMA, with nearly 4 mppa, finds it necessary to employ four people to manage the environmental control system. This includes a complaint procedure; until recently, half of the complaints were generated by one citizen. Now this same information on aircraft movements can be obtained for any airport on a smart phone with an app (Flightracker 24).

2. Data transmission is performed in discrete batches after certain time intervals. Data from noise measurement are saved in the local memory of the noise measuring equipment and, at the specified time, as a rule once in 24 hours, they are transmitted to the central processing point. Again, a local network or the public telecommunication network may be used for that purpose.

In the majority of cases, a combined system is used in big airports. Some measuring places are continuously connected with the central control point (Figure 20.8), while the data transmission from others is intermittent. The majority of systems for aircraft noise monitoring are augmented with one or more mobile measuring stands, which are most often located in a dedicated vehicle.

Simultaneously with the sound signal measurement, it is necessary to measure also barometric pressure, air temperature and humidity, wind direction and speed. The airport operator usually performs these meteorological measurements as well as the noise monitoring (Figures 20.7 and 20.8).

20.2.6. Prediction of Air Transport Noise

Mathematical models are used as the basis for computer programmes that predict the noise load in the vicinity of airports. The load may be predicted in the form of noise contours of equal levels of the chosen descriptors, and also of maximum noise levels at specific locations. Just as there is no consistent methodology for selection of the descriptors and the method of evaluating noise around airports, so there is no consistent international recommendation or directive for the prediction of aircraft noise load in the vicinity of airports.

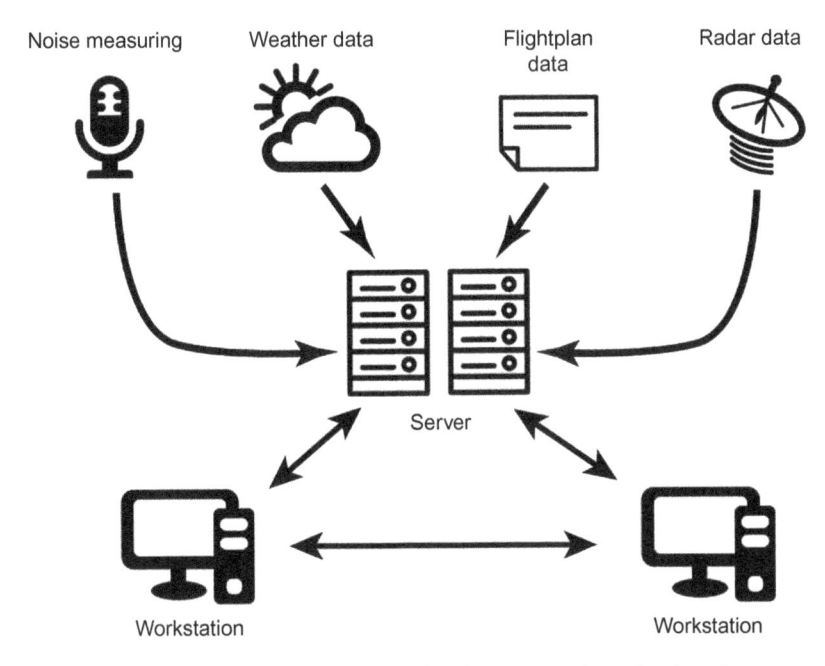

Figure 20.7: Aircraft noise monitoring network — basic scheme.

In Germany the method of evaluating noise by prediction is determined by the Act on Protection against Aircraft Noise (new ICAN/AzB from Year 2008). Long-run equivalent A-weighted sound level is the descriptor used for day and night. For night-time, the number of exceedances of the limit value of the maximal AS-weighted sound pressure level ($L_{p,ASmax}$). The method is based on the categorisation of aircraft into noise classes. Acoustic pressure levels in octave bands, which are determined on the basis of measurement for the given aircraft categories, are used as inputs. They are integrated through the six months during which the airport trafic is the highest. Protective airport noise zones are determined by computation, as described above. The computation and projection of the contours are realized by means of a polynomial formula. Noise load evaluations by prediction in the vicinity of all the larger airports, military ones included, are performed under the legislation. Each airport then has a legally valid map marked with both of the protective zones. The polynomial itself is also reported. A control of adherence to the determined noise levels in the protective zones is also defined. If an increase of more than 4 dB(A) is detected by means of long-run measurement in the outer edge of one of the zones, a new noise load evaluation by prediction should be performed in the airport.

Additionally to the aforementioned method, the procedure set out in DIN 45684th is also used in Germany. The DLR developed procedures for calculations named SIMUL and PANAM, which employ different principles which are both used in ICAN. The ICAN is used to calculate the trajectory segmentation of track flights (see Figure 20.10). SIMUL uses a semi-empirical description of the sources of aircraft noise at each of the points of the flight trajectory with directional characteristics of sound radiation. PANAM uses parameterisation of flight on the flight trajectory. Both DLR procedures are mainly used for tests and research. Currently DLR is preparing a draft of standard DIN 45689, which will describe the new method for the calculation of air traffic noise in the vicinity of airports, which should be useful also in everyday practice.

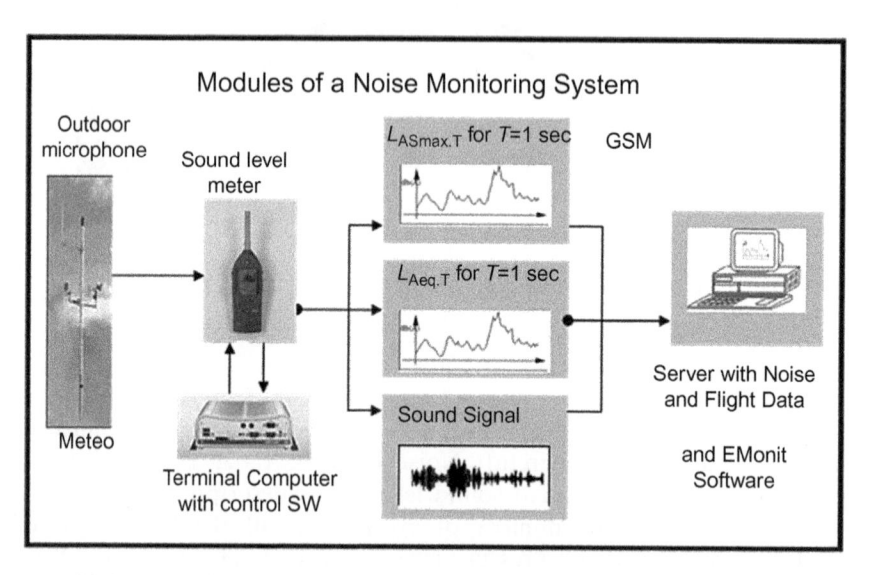

Figure 20.8: Principle of aircraft noise monitoring system. *Source*: EUROAKUSTIK, Ltd., Slovakia.

In the United States several methods for computing aircraft noise were elaborated in the past. The main descriptors used are L_{AeqDN}, NEF and CNEL. The US Government has financed though the Federal Aviation Administration the Integrated Noise Model (INM), which allows the noise load to be determined consistently in the vicinity of any airport. The programme allows the majority of the noise load descriptors to be calculated. A similar programme called NOISE MAP was developed also by the USAF for their own requirements.

In all these models, the path of an aircraft is approximated by means of a series of straight flight segments. In each segment, time-integral computation of target descriptors, which are computed from sound exposure level or perceived noise level of individual sound event, occurs. The maximum A-weighted sound level contours and maximum perceived tone-corrected noise level (L_{PNTM}), are computed directly from these descriptors.

The computation at individual points of the grid map is performed separately for each independent aircraft that is considered in the model. Noise characteristics of individual types of aircraft acquired from aircraft certification or from special measurement may be used as input data for computation. The data collected during certification are incorporated into the INM. However, these data are not part of the public dissemination of certification data, the latter only covering the noise results for the three measuring points specified in Annex 16. In the event that data on maximum A-weighted sound levels or maximum perceived tone-corrected noise level are not available for some types; these will be computed by means of an empirical formula based on linear regression. In the computation, the effects of flight characteristics, procedures and aircraft configurations during take-off and approach are taken into consideration.

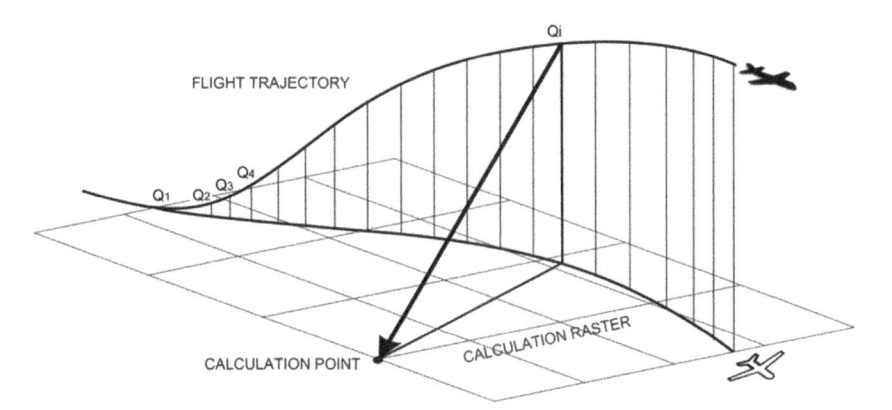

Figure 20.9: Simulation of the aircraft movement and calculation of the noise descriptors on the ground.

In Switzerland the programme FLULA is used for calculation of noise exposure in the vicinity of an airport. In this case a simple mathematical model is employed. The aircraft trajectory is replaced by a series of points in space which match one second intervals of aircraft movement. A discrete noise source with directional characteristics is placed at each point. These sources substitute moving aircraft noise emissions (Figure 20.9). Taking into account noise attenuation in the ground layer of the air at every point, the value of A-weighted sound exposure level and maximum AS-weighted sound levels during the aircraft movement are calculated. The value of equivalent A-weightedsound level derived from the value of the A-weighted sound exposure level for each aircraft type and the number of movements of each aircraft type considered on any projected trajectory is determined at each grid point for a given time interval. Similarly, for each raster point and time interval maximum AS-weighted sound level distribution is specified. Calculated data is used for determining noise contours for various descriptors such as the NNI(A).

528 Airport Design and Operation

The whole mathematical model is based on the directional noise characteristics of a particular aircraft type defined by multi-spectral noise measurements in a third-octave analysis of the noise. The analysis is worked out for different flight situations and the aircraft trajectory is continuously monitored by precision radar system and recorded. Based on these data, special directional noise characteristics for each aircraft are created and thereafter used in the noise modelling.

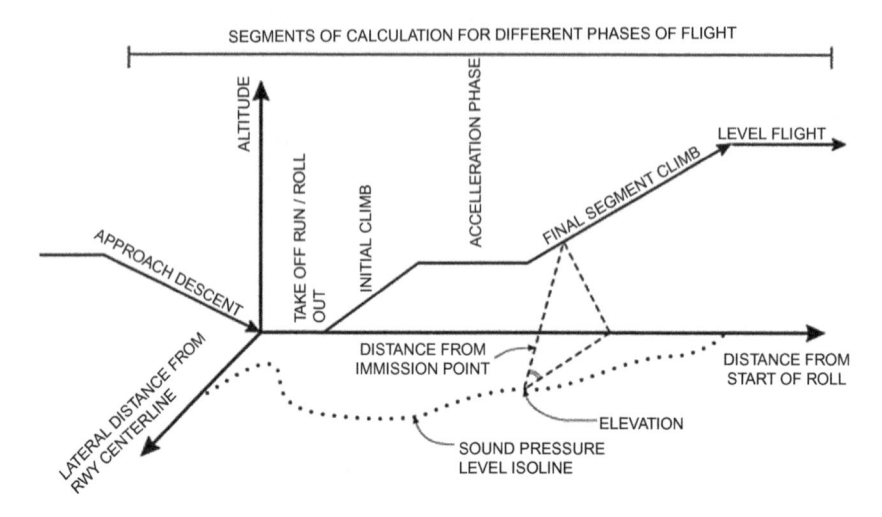

Figure 20.10:   Principle of segmentation-calculation in ICAN/INM/ECAC.

There is a continuous effort to unify air transport noise prediction methods. However, noise expert groups meetings (NATO CCMS-GROUP, Aircraft Noise in a Modern Society, Symposium on Aircraft Noise Abatement Receiver Technology) did not achieve that aim. Every country has different legislation based on its historical background and different noise descriptors for the air transport noise assessment. There is no common legislation even on the EU level. The report from the 21st meeting of ECAC in 1997 — Doc. 29 — Report on Standard Method of Computing Noise Contours Around Civil Airports described in detail a method of noise exposure from air transport in the vicinity of civil airports. This method uses two descriptors — maximum sound level A and sound exposure level A. The model is based on a noise database for different civil aircraft categories. The inputs for particular aircraft types are based on the mentioned descriptors for third-octaves with average band frequency between 50 Hz and 10 kHz for different aircraft speeds, power plant settings and minimum distance from the aircraft trajectory. The method is based on the descriptors calculations for the aircraft movement points defined by the aircraft performance characteristics. The method is based on the calculation of individual descriptors for every point of the grid for each of the aircraft movements. Each of these aircraft movements is defined by the aircraft performance characteristics. The final noise level calculation is done by simulation of individual flight segments. The noise contours are interpolated between calculated points. However, this document was not adopted by the European states as a standard. It is only a recommendation of The European Parliament and the Council — Directive 2002/49/EC from 25 July 2002 relating to the assessment and management of environmental noise. The document describes methods for noise prediction in the vicinity of civil airports and creation of noise strategic maps (Figure 20.11).

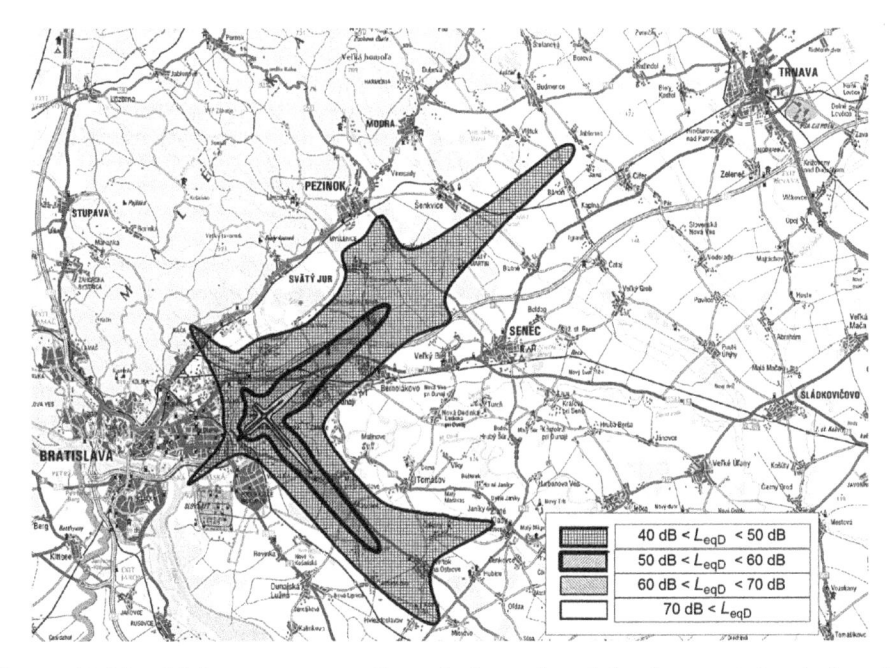

Figure 20.11: Noise contours of equivalent A-weighted sound level for day L_{AeqD} — prediction of noise rating around M.R. Štefánik — Bratislava airport for the New Terminal Project. *Source*: Prediction M. Kamenický, with CadnaA Software, ECAC. CEAC Doc. 29 and ISO 9613.

Within the projects IMAGINE, HARMONOISE and CNOSSOS financed by the European Union a new method of air transport noise calculation is being prepared. The mathematical model will be based on the aircraft trajectory segmentation principle (similarly to ICAN and INM) and should take into account the terrain morphology, attenuation in the ground layer and meteorological effects. The recommended calculation procedure is the calculation according to ECAC Doc. 29 3rd Edition with database of ANP (Aircraft Noise-Power). The ANP database is modified from the database NPD (Noise-Power-Distance) from the INM model. A significant change is in the extension of data about sound pressure level in one-third octave frequency bands and directional characteristics of radiation of the sound source. Aircraft types will be grouped according to power plant types, similar to

INM. The principle of the calculation of the noise from the air traffic around airports using segmentation (INM, ECAC Doc. 29, ICAN) is shown in Figure 20.10.

The Environmental Research and Consultancy Department of the Civil Aviation Authority (ERCD of CAA in Great Britain) developed and used the model ANCON2. The computing procedure in this model fully corresponds to the procedure ECAC Doc. 29 3rd Edition. This procedure also uses the database of NDP.

Recently there has been a considerable effort to unify the predictive methods used for determining aircraft noise at airports. Unfortunately no useful conclusion was reached. ICAO recommends the method for computing noise given in Annex 16. The use of this method allows comparison and evaluation of airports on an international basis. But this method of calculation is not commonly used.

20.2.7. Airport Noise Mitigation and Noise Abatement Procedures

Airports need to adopt noise mitigation measures in order to placate their local communities or simply to behave in an environmentally responsible way. Noise abatement operational procedures are being employed at many airports today to allow noise relief to urban developments around airports. Procedures are defined in the ICAO Doc. 8168 Procedures for Air Navigation Services — Aircraft Operations; Volume I — Flight Procedures for both departing and arriving aircraft. Guidance describes two noise abatement departure procedures (NADP) to mitigate noise close in the airport vicinity — NADP 1 or further from the airport depending on the local conditions. According to ICAO Review of Noise Abatement Procedure Research noise abatement procedures relate also to other phases of flight and aircraft operations which could break down as follows:

Noise abatement flight procedures

- Continuous Descent Approach (CDA)
- Noise Abatement Departure Procedures (NADP)
- Modified approach angles, staggered, or displaced landing thresholds
- Low power/low drag approach profiles
- Minimum use of reverse thrust after landing

Spatial management

- Noise-preferential arrival and departure routes
- Flight track dispersion or concentration
- Noise-preferential runways

Ground management

+ Sound-insulation of houses and engine run up management
+ APU management/usage ban
+ Taxi and queue management
+ Towing/push-back
+ Taxi power control (taxi with less than all engines operating).

While noise abatement procedures could have measurable environmental benefits, their operational implementation could be challenging. Every single procedure must be designed, developed, tested, evaluated and implemented by airports or airlines. Some of those issues are discussed further.

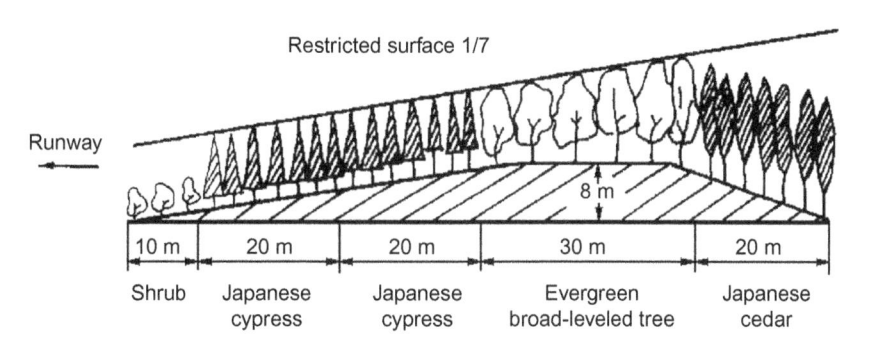

Figure 20.12: The noise protection wall. *Source*: Airport Planning Manual, Part 2, Land Use and Environmental Control.

The airport operators use measures in addition to noise abatement approach and departure procedures such as noise insulation of houses, different compensation schemes of residents or noise protection walls (Figure 20.12). They include displaced thresholds, preferential use of runways which demand more crosswind and tailwind movements than some airlines are happy with, restrictions on the use of reverse thrust, limits on daily movements, curfews and night quotas.

The take-off noise should mainly be a problem for the heavy four-engined aircraft, since the performance of the twin-engined aircraft could allow them to be at a comfortable altitude before crossing the airport boundary.

This is compromised by the use of reduced thrust and by intersection take-offs. These are two examples of the way today's operating procedures differ from those used in certification. A further example is the height at which the thrust is cut back, which is more important for the noise-critical large four-engined aircraft. In certification, this is done to minimise noise at the certification point 6.5 km from start of the take-off role, whereas it is normally controlled by specific noise abatement procedures, instructions in the operating manuals or operators' preferences. If this class of operations is to operate satisfactorily, it is essential that the procedures are chosen to minimise noise at the most sensitive locations.

The increasing use of high bypass twin-engined aircraft has transferred attention from the take-off noise to the final phases of approach, where the dominant noise affecting L_{eq} contours comes from either the front fan of an engine or aerodynamic noise from the airframe, and in particular the landing gear and flaps. This phase is difficult to improve, because the aircraft needs to be stabilised on a standard three degree glideslope. There is, however, also increasing concern about noise further out under the approach path, many complaints coming from these areas in the early morning.

The noise under the early stages of approach could be mitigated by changes to the flight procedures. The two segment approach produces the expected large benefits between 20 and 10 NM, a slight disbenefit at 7 NM being due to the relatively early selection of landing flaps. Both the Continuous Descent Approach and the Decelerating Approaches (DA) offer a good proportion of the two segment approach benefits, the latter being the more preferable, the benefits all coming before 8 NM from touchdown. These types of approach require much cooperation between the pilots and air traffic controllers, but initiatives at Heathrow to increase CDA have been successful.

Closer to the airport, where noise levels would be expected to cause more annoyance, one of the main measures used by national regulators and airport companies in attempting to limit arrival noise is the stipulation of minimum altitudes for joining the Instrument Landing System (ILS). At Heathrow this altitude is 2,500 ft during the daytime and 3,000 ft between 2,300 and 0700 hours. At Birmingham, Gatwick and Stansted the stipulated minimum is 2,000 ft.

If the approach can be flown at a higher angle all the way to the flare before touchdown, the noise benefit seen in the two segment approach can be extended through the areas of greatest noise. The potential benefits would come from the greater distance between the aircraft and the ground, and also the lower thrust levels required for greater descent angles in the landing configuration. Airbus has estimated that the A 310's 80 dB(A) contour area decreases from 1.45 to 0.51 km^2 when the approach increases from 3° to 5°. Approaches are regularly flown into Innsbruck Airport at 4° and into London City Airport at 5.5° with 100 seat jets, the latter airport having used 7.5° while only operating with turboprop aircraft. Although the glideslope has to be intercepted by 3,000 ft and careful attention has to be given to speed control, the approaches with BAe 146 aircraft are flown at the aircraft's normal approach speed, they can be performed with the speed brakes inoperative and the only modification to the aircraft is a button to turn off the rate of descent element of the Ground Proximity Warning System (GPWS). Despite the need for extra care, there is no evidence of a greater missed approach rate than with normal operations on the same aircraft.

Up to 4°, a conventional jet aircraft can be used at the normal approach speeds but lower thrust settings. Conventional aircraft could be used at still higher angles, particularly if the landing mass could be decreased, but this is unlikely on a regular basis. For angles greater than 4°, it would therefore be necessary to redesign the aircraft to reduce landing speed for a given mass. A problem with steeper approaches at the moment is therefore that not all aircraft could fly them, so it is necessary to segregate the traffic in some way. This may be by displaced thresholds or approaches to new short parallel runways, which would give fewer ATC problems. The UK CAA (CAP 1165) has also raised the issue of the limitations of the certification of precision approaches: Cat III constraints mean that 3.2° is the maximum that could be achieved, but even that is worth almost a decibel.

20.3. CONTROL OF GASEOUS EMISSIONS AND ENERGY CONSERVATION

Air transport consumes only 5% of the world consumption of oil products. Every year there is an improvement in the efficiency of energy utilisation of 3–4%. The turbo-jet engines built after 1982 produce 85% less unburned hydrocarbons (HC) than the engines built in the seventies. Carbon Monoxide (CO) exhaust gases decreased by 70% in the same period. Jet aircraft account for approximately 2–3% of the world-wide production of Nitrous Oxides (NO_x) and Carbon Dioxide (CO_2). Surface transport, by contrast, produces 22% of carbon emissions and Europe's merchant ships use around a third more carbon than aircraft do. Air transport contributes approximately 1% to global warming as a consequence of CO_2 emissions.

The procedures of emission certification of aircraft engines, limits of harmful substances and specifications of metering devices are included in Annex 16, Part II. Aircraft are certificated for engine emissions as calculated over the ICAO Take-off and Landing (LTO) Cycle. This assumes operational characteristics for approach from 1,500 feet, taxi in, taxi out and take-off and climb to 1,500 feet that are not necessarily the same as today's operations. In particular, there are differences in taxi time, in engine setting for taxi and take-off, and no account is taken of the start-up cycle. However, it is not the objective of this book to deal with the reduction of harmful substances at their source. There are operational measures that decrease the volumes of products of combustion at airports, including optimising the movement of aircraft on the apron and taxiways or taxiing after landing with the minimum necessary number of engines running. Depending on the size of airport and traffic split most of the greenhouse gases generated by airport operation are produced by passenger access and egress transportation and aircraft operation (Figure 20.13).

Reduction of emissions in airport operation can often be achieved by the same procedures which are used for decreasing noise pollution and fuel consumption. During the take-off and climb it is possible to achieve a reduction of emissions by selecting the optimum aircraft configuration and engine setting. Flight paths can be shortened during approach and landing. Descent with higher approach speed in 'clean configuration' and later with lower flaps settings reduces aerodynamic drag, so requiring less thrust and decreasing noise. It may also result in a reduction of exhaust products, depending on the product and how it varies with engine power. The amount of CO and HC produced per kilogram of fuel burnt reduce with increase of power, while NO_x increases.

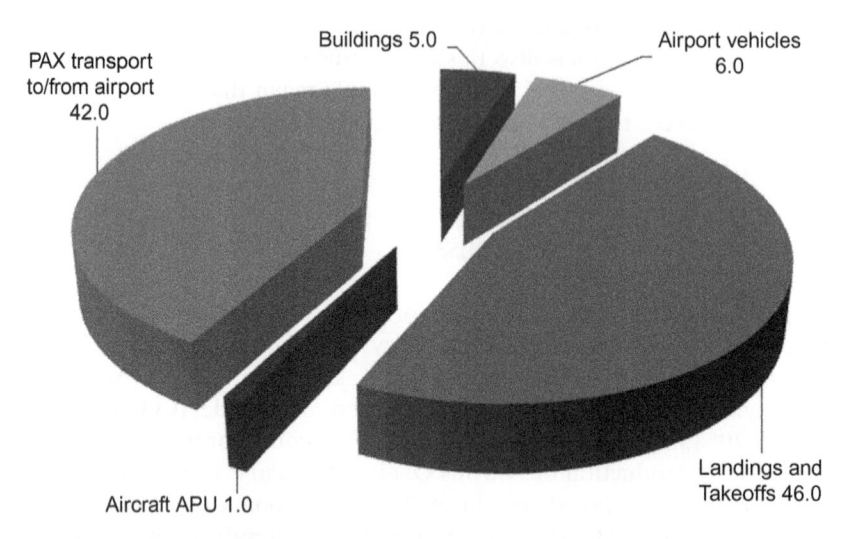

Figure 20.13: Air quality and greenhouse gases. *Source*: Newark Liberty International Airport; 2013 Sustainability Report.

Aircraft generally contribute much less to air pollution in the vicinity of an airport than the power generation at the airport and particularly surface vehicles on the airside and access transport. It is necessary to pay the utmost attention to the design of the system of surface transport to reduce the pollution as much as possible, as discussed in Chapter 14. The FAA's Emissions and Dispersion Modelling System (EDMS) can be used to assess and model the quantity of gaseous emissions produced in the vicinity of airports. Future levels can also be estimated if the characteristics of combustion and energy use can be predicted.

It can be assumed that the worst contamination of ambient air on the apron occurs in the operational peaks in the summer period. In the future it will be necessary to introduce a higher proportion of electrically powered ramp handling equipment. At the same time it will be necessary to decrease the use of aircraft auxiliary power units (APU) by using the central supply of electrical power, air conditioning and other media directly to the aircraft by lines located on passenger loading bridges or built into the apron.

The experience of most developed countries shows that the quality of ambient air in most cases is higher on airports than in the cities, the exception being by the drop-off and pick-up kerbs in front of the terminal, and sometimes in apron cul-de-sacs.

On the other hand the air transport industry's energy consumption has been growing faster than that of other polluting industries. Air transport will soon be central, not marginal, to the emissions issue. In principle there are two possible solutions for how to limit emissions: to tax them or to let polluters trade the right to emit.

Under the Kyoto climate change treaty, the EU committed itself to reducing carbon emissions by 8% of 1990s level by 2012, a commitment that is currently off track. However, international aviation is neither subject to Kyoto or other commitments nor to fuel tax or VAT. In line with the EU long-term strategy, Europe should take a leadership position in global aviation in order to reduce the climate change impact of aviation.

In November 2008 the European Parliament and The Council published Directive 2008/101/EC amending Directive 2003/87/EC so as to include aviation activities in the scheme for greenhouse gas emission allowance trading within the Community. This among other things endorsed a 30% reduction in the EU's greenhouse gas emissions below 1990 levels by 2020 as its contribution to a global and comprehensive agreement for the period beyond 2012. The Directive set the total quantity of allowances to be allocated to aircraft operators and defined method of allocation of allowances for aviation through auctioning from 1 January 2013. The principles are the same as other industrial sectors. To increase airport energy efficiency and a carbon footprint reduction, as a first step it is necessary to perform an energy audit to understand how much energy is used and how the usage compares with that of similar buildings and other facilities. An energy audit consists of a comprehensive assessment of electricity, water, gas and any other utilities in a building or facility. The most frequent measures airports introduce for energy conservation and the reduction of greenhouse gases (GHGs) are building insulation, installation of solar photovoltaic panels on roof space of airport buildings and lighting renovation with LED lights. LED technology is becoming also popular for runway and taxiway lighting (see also Chapter 15).

20.4. PROTECTION OF WATER SOURCES

Construction of the airport and in particular construction of the runway can disturb the ground water system not only in the area of the airport but also in the wider surroundings of the airport. Rainwater drains quickly from the paved areas, which can cause flooding in the water courses that are fed by the rain drains. Flood waters are therefore usually fed into retention tanks.

In the Terminal 2 of Frankfurt Airport the rainwater is being used for services. Rainwater is used for cooling of air conditioning, in fountains and in 800 lavatories, which consume approximately 8,000 m^3 of rainwater a year. Before use the water is chemically treated and cleaned in sand filters.

Most countries have standards that set the waste water quality. However, the limits set by airports are usually more stringent. For example in 1995 Manchester Airport imposed the following maximum concentrations limits: 10 mg/l BOD,[3] 50 mg/l SS,[4] 2 mg/l ammoniacal nitrogen, no visible oils.

Before entering the sewer the waste waters must be properly treated. This treatment of the waste run-off from the operations of washing of vehicles and aircraft, from the accumulator room, the boiler house, the degreasing workshop, the kitchen and dining hall must meet other health and safety requirements regarding the contents of heavy metals, chlorinated hydrocarbons and sedimentation substances.

Sewage water from the airport is cleaned in normal water treatment plants. Drainage from the movement areas of the airport require a special treatment due to the oil products or de-icing substances used in the winter. Rain water from the paved areas, in particular from the apron, can be cleaned in a special treatment plant on the airport, with separation of oil products usually by sedimentation, or the sewage collector can be connected to the local municipal treatment plant. Special procedures have to be prepared and put into operation in the case of a large oil spillage. In addition to the measures to cope with an oil spillage, the contingency plan should contain the following:

- procedure for removing spilt fuel
- procedure for closing or sealing the sewage system
- procedure for reporting the accident
- prohibition of routine and planned maintenance on the areas not equipped with a trap for oil products
- minimising the washing of aircraft on the apron.

Particularly dangerous areas, for example fuel stores, hangars and maintenance workshops, have to be equipped with traps for oil products and inspected regularly.

---

3. BOD — Biochemical Oxygen Demand.
4. SS — Settleable solids (mg/l) — suspended solid after one hour quiescent settlement.

At the same time it is necessary to pay attention to possible water or soil contamination during fire-fighting training. These training activities should be permitted only within an especially dedicated area.

One of the most important environmental issues facing many airports is the run-off and consequential contamination of waterways from glycol pollution (see also Chapter 18). According to the US Environment Protection Agency many airports in United States have updated equipment and glycol collection systems. The recycling plant for aircraft de-icing at Munich Airport in 1993 was the world's first plant of this size that has gone into operation. According to Munich 2013/2014 De-icing Season Report the total consumption of aircraft de-icing fluid (ADF) Type I Mix was 1,939 and 369 m^3 of Type IV were needed. A total of 58.7% of the Type I consumption was supplied by recycled fluids and 48.4% of total glycol consumption was returned to the recycling process. Collection of Type II/IV liquids is problematic as most of the fluid stays on the aircraft surface during application and taxiing and is blown off during the take-off. More pressure could be expected on airports in the coming years for the collection of de-icing fluids.

20.5. Landscaping

Air transport needs less than 8% of the land needed by rail transport and less than 1% of that needed by road transport. In relation to the number of transported passengers, the land is used five times more efficiently by air transport than by railways and six times more efficiently than by roads. None-the-less, airports can cause considerable visual intrusion. They are best placed on elevated ground in the interests of clear flight paths, so are likely to alter the skyline with levelled ground and tall buildings, particularly the control tower. At night the approach lights, approach flash system and the apron floodlights are a distraction, the more so because airports are normally sited in open countryside.

When planning the development of an airport, one has to consider not only the airport site but also its surroundings. The planning must be integrated into the urban planning of the region. In addition to other impacts, the impact of the airport on the appearance of the landscape must be assessed. It is particularly difficult to situate or hide large airport facilities like hangers and terminals so that they do not disturb the appearance of the landscape. Changing the landscape in the vicinity of the airport, such as tree planting, can be used to minimise the negative impacts of the visual intrusion. Landscaping by the creation of planted buffer zones can also help to control air, in particular dust pollution, but it is usually not able to limit the propagation of noise. However, the types and composition of the trees must comply with recommendations for discouraging birds. In this context, it is necessary to pay attention to the treatment of waste from the airport. Earthworks (bunds) can be built as acoustic barriers but can reduce the view and create their own visual intrusion.

Landscaping is also important in managing the habitats and hence the species of flora and fauna around the airport. Careful planning can avoid habitats being isolated by leaving movement corridors, provide habitats for endangered and protected species, and discourage larger birds. Where necessary, replacement sites need to be created for habitats, including wetlands, and historic buildings moved. Some projects have had to be abandoned when this has not been possible, or when too much built heritage would have had to be destroyed.

Together with improvement of the environment on the airport, airports are increasingly utilising their land as recreation zones not only for the visitors to the airport but also for their employees and inhabitants of adjacent cities. Sports grounds, golf courses, areas for water sports etc. have been located at or near airports. These activities, together with other airport services, could be a significant source of airport revenues, but it is necessary to avoid large congregations of people under the flight paths in Public Safety Zones.

20.6. WASTE MANAGEMENT

An airport with a capacity of approximately 5 million passengers a year can generate as much waste as a small town. Airports generate between 0.5 and 1.0 tonnes per annum per 1,000 passengers. To minimise the negative impact on the environment in connection with waste dumping, it is necessary to look for ways to decrease the quantity of waste particularly by recycling, composting, reuse and introduction of wasteless technologies. In developed countries the concepts of separation and recycling are supported by legislative measures. These measures include tax relief and direct state subsidies. This will further be accelerated by the fact that in all countries there will be a gradual increase of prices for the dumping of waste.

Airport waste contains a high proportion of substances and raw materials which really need to be sorted out and recycled. The sorting of waste must be considered not only as a measure for protecting the environment but also as a financial asset.

To improve waste management achievements in developing countries, it would be necessary to improve environmental education. However, it is a very long-term task. Economic growth is felt to be more important than the problems of waste management in these countries. It is therefore important to pay increased attention to the education of the airport employees in waste management, in particular in the key workplaces which are the main generators of waste.

The foundation of good waste management is ascertaining the composition of the waste. The information can be obtained from the waste disposal company. However, detailed data could be obtained only by comprehensive waste audit to identify opportunities for increased recovery of recyclable materials. A waste audit not only helps to unveil inefficiencies related to waste removal but usually also allows the comparison of the waste contractor's bill data with actual data.

Depending on the airport size and the location and type of the individual workplaces, the sorting of waste can be organised as centralised or decentralised or as a combination of both procedures. The centralised collection and sorting of waste requires high investment and personnel costs, but is more tidy and provides better separation of individual fractions and therefore also better opportunities for negotiations with consumers of classified waste. Decentralised collection can allow certain kinds of waste, such as paper and carton, to be separated at each location. Biodegradable organic materials, which can be composted, should be separated on the airport and this can decrease the total quantity of waste. For example a waste audit at Newark Liberty International Airport in 2013 revealed that of all the products, compostable waste comprised most of the waste stream, averaging 40%. A difficulty is that organic waste from aircraft must be considered as infectious material. Therefore in some airports it is sterilised before further processing.

The airport company can offer the waste management system to other firms in the airport, because sorting out and recycling of waste are ineffective for small enterprises. For example, Amsterdam Schiphol's administration offers the possibility of separation and processing of waste to individual organisations at the airport. The total costs for such collection are about 25% lower compared with the classical collection process, where most of the waste is burnt. Waste can be solid or liquid, organic or inorganic, non-hazardous or hazardous. As many as 10 kinds of materials may be separated, but most organisations use three to four categories of separated materials. The main separation parameters are mass and toxicity. In the future, the containers will be equipped with electrical chips, which will not only generate automate invoicing but in the case of contamination of separated material by other waste, for example paper by metal, plastic, etc., they will at the same time be able to trace the transgressor. However, the best waste management philosophy is not to let the waste be generated in the first place. It may be that the purchasing strategy should be reviewed to avoid unnecessary packaging or to emphasise materials with a low toxicity.

21

WILDLIFE CONTROL

Tony Kazda and Bob Caves

21.1. INTRODUCTION

The number of aircraft increases continuously. At the same time their speed increases and a great number of flights are performed at low altitudes by general aviation, military aviation and by commercial aircraft on landing and take-off. It is therefore logical that this increases the number of potential conflicts between aircraft and wildlife. From Table 21.1 it is apparent that most collisions are caused by bird strikes. Compared with bird strikes collisions with mammals make up only 4%. Therefore, the main attention is devoted to the bird strikes.

Table 21.1: Time of occurrence of wildlife strikes to civil aircraft USA 1990–2012.

Time of day	Birds		Terrestrial mammals	
	23 year total	% of total known	23 year total	% of total known
Dawn	3,038	4	51	3
Day	51,403	62	425	25
Dusk	3,805	5	142	8
Night	24,019	29	1066	63
Total known	**82,205**	**100**	**1084**	**100**
Unknown	44,947		1262	
Total	127,212		2.046	

Source: Weller (2014).

Most birds fly at less than 300 m above the ground. It is therefore inevitable that bird strikes are a continuing serious hazard for aircraft. At the very beginning of aviation, the low speed of flight made it possible for the birds to avoid a collision and the force of any impact was small. In most cases a collision resulted in minor damage to wind shields and leading edges of wings or the fuselage. The probability of collision was also small because of the small number of aircraft. Despite this, the test pilot Carl Rogers died after a collision between his aircraft and a bird in California in June 1912. His Wright Flyer hit a seagull, which jammed the steering mechanism and the aircraft and pilot fell into the sea.

Most of the birds quickly became used to the noise and speed of propeller aircraft and learnt to stay away from the dangerous airspace in the vicinity of the airport. The situation changed in the 1950s after turbo-prop aircraft and jet aircraft were introduced. This increased the speed of flight, the aircraft's acceleration and their dimensions. Therefore it was more difficult for the birds to avoid the flying aircraft and the impact force of a collision was greatly increased. The rate of collisions was also increased by the large quantity of air sucked into the jet engine and large dimensions of the engine intake. The jet engine also proved to be less resistant than piston engines to collision with the birds. Moreover, modern aircraft are quieter and therefore it is easier for them to escape the attention of the birds.

Figure 21.1: Typical damage of the wing leading edge. *Photo*: J. Šulc.

The cost to an airline as a consequence of damage to an aircraft after a bird strike can reach millions of dollars. It is not only the price for the repairs of damaged aircraft parts but also the indirect costs in connection with taking the aircraft out of service, redirecting passengers and freight to other flights and the costs for accommodation and board of the passengers and crew. Delay of the aircraft or its temporary withdrawal from service has a domino effect on the timetable of the airline and all connecting flights. Wildlife-related incidents can cause a substantial increase in operational costs and blemishes the good reputation of the airline. The consequences of bird strikes for new aircraft types with high composites contents may be even more serious. They are not only more 'fragile' than the traditional, aluminium-based constructions (Figure 21.1), but also damage detection and determination of the scope and method of repair is significantly more complex. On the other hand, the cost to airports in preventing collision of birds with aircraft is also not negligible. On New York's J.F. Kennedy airport they exceed half million dollars a year, which could be approximately the price of repairing two engines on a B 747 aircraft.

The size of the bird strike problem is different at each airport. Each airport operator has the responsibility to develop, implement and demonstrate an effective bird/wildlife strike and wildlife control programme at the airport. It should reflect the size and level of complexity at the airport. In order to prevent bird strikes effectively in the vicinity of airports it is necessary to have an in-depth understanding of the customs and behaviour of the birds. It is therefore important for the airport administration to co-operate with ornithologists.

Because of the importance of bird/wildlife control, each airport operator has the responsibility to develop, implement and demonstrate an effective bird/wildlife strike and wildlife control programme at the airport, and this should be tailored and commensurate to the size and level of complexity at the airport, Commission Regulation (EU) No 139/2014 (of 12 February 2014) also stipulates requirements for collection of information from aircraft operators, aerodrome personnel and other sources on the presence of wildlife constituting a potential hazard to aircraft operations and evaluation of the wildlife hazard by competent personnel.

21.2. BIRD STRIKE STATISTICS

ICAO began to monitor bird strikes in 1965, through the collection of bird strikes forms. In November 1979 ICAO requested all member states to report all collisions of aircraft with birds. An international system of collection and evaluation of data on collisions of aircraft with birds was set up, called the ICAO Bird Strike Information System (IBIS). A bird strike reporting form was sent to the member states. In order to maintain the continuity of data, this form is still used. Later a supplementary questionnaire was prepared for airlines, in which the costs in connection with the bird strike and detailed information on the damage to the engines should be entered.

To assist ICAO Member States in forwarding bird strike data to the ICAO Bird Strike Information System (IBIS) ICAO issued a Manual on the ICAO Bird Strike

Information System — Doc 9332. It describes IBIS, giving details on the data entry, data retrieval and the bird strike database layout. The analysis of bird strikes indicates that:

- the number of incidents with significant damage to aircraft represents approximately 5% of the total number of bird strikes,
- 62% of bird strikes take place in daylight, 29% take place at night and the rest of bird strikes take place at sunrise and sunset,
- 65% of damaged aircraft with a weight above 27,000 kg have engine intake damage,
- 29% of bird strikes take place during the approach and 25% during the take-off run,
- 51% of bird strikes take place below 30 m above the terrain and
- in 91% of cases the pilots were not warned of the appearance of the birds.

Collection of data on bird strikes by ICAO is only one part of the total collection and evaluation of data necessary for the functioning of an efficient airport system of bird management. In fact the data from the IBIS system are only of small significance for any specific airport. It is necessary to complement these data by long term and systematic observations of the appearance of birds, maintenance reports, observations of pilots and air controllers and ornithologists on the airport or in its vicinity. For example the pilot's report on the 'close approximation' of an aircraft with birds in the vicinity of the airport is just as important as the bird strike report, showing, as it does, that birds are a threat. The airport report should not be focussed only on whether there is a bird strike but also whether and where the birds appear in the vicinity of the airport. Only after all data had been processed and assessed is it possible to identify the scope of the problem and to propose and introduce an effective programme, the objectives of which are the decrease of the number of birds in the vicinity of the airport and the avoidance of bird strikes. The reporting procedure has to be co-ordinated by the responsible person and all airport employees must be acquainted with it.

The aircraft can be damaged by any kind or size of bird (Figure 21.2). The extent of the damage by the bird depends on the weight of the bird, speed of bird strike and on the strength of the aircraft structure. During the impact, the speed is changed within a very short period. The change of momentum of the body is proportional to the change of the speed and the change of momentum of the body equals the impulse force.

In mild climates there are indigenous, transient and migratory birds.

- Indigenous birds live all year in the same small area.
- The transient birds appear in a certain area only at nesting time. Then they move to other localities or roam over a vast area.

✈ During the year the migratory birds fly long distances. They have a specific direction and target, to which they fly by permanent 'flight routes'. Their nesting and hibernation places are very distant from each other.

Figure 21.2: Extent of aircraft damage depends also on the weight of the bird.
Photo: Courtesy of Brno-Tuřany Airport.

The birds have their natural daily cycle, just as humans do. The beginning of the activity, as the birds wake up, cannot be clearly determined. Some kinds of birds wake while it is still dark, others only after sunrise. The beginning of the activity is influenced by meteorological conditions wind direction, precipitation and temperature. If the sky is overcast, the birds wake later.

The total number of birds changes in the course of the year. Most kinds are represented in central Europe or North America during the spring and autumn migrations. In this period the birds are also most mobile. The spring passage is the shortest, while the autumn migration has been spread out over a longer period. The flight altitude is significantly affected by meteorological conditions. Smaller birds particularly do not want to lose visual contact with the ground. The average height of migratory flight is between 100 and 300 m.

The population of birds never spreads out evenly over a large area. The ecological conditions ensure that the birds are located in some 'islands'. As a rule, several species of birds appear in any specific environment, or habitat. The habitats can be:

✈ original and natural

✈ affected by human activities.

An airport's development normally means removing vegetation, trees and shrubs. This can cause the disturbance of the natural balance of habitats, and the eradication of existing plants and animals. Therefore it is necessary to consider the landscaping of the countryside so that it does not seriously disturb the ecological stability of the close vicinity of the airport, for example the ground water system. On the other hand the safety of air operations must not be endangered by an increase in birds, fog, dust etc.

21.3. Passive Management Techniques — Habitat Modification

The reduction of the number of birds in the vicinity of the airport can be achieved by changing the habitat so that it is unattractive for the birds. This regulatory harmony can be found through *compatible land-use planning*, a process that has resulted from the need to establish a co-operative environmental relationship between airports and the communities they serve. It is a system measure, which could be expensive but if the measures are considered during planning of the airport, the increase of costs is insignificant in comparison with the resulting benefits. While relatively simple, this concept has delivered, in practice, impressive results through the development of airport/community-system plans, as well as legislation for compatible land uses, easements and zoning.

However, an airport is always located amid fields, forests and other natural environments. Space with an area of several hundred hectares cannot be kept without fauna. Some airports even allocate sites, where appropriate, where animals can reside. However, animals are 'trained' not to enter critical areas like an air strip. To control airport-critical space effectively some airports use, for example cheetahs to be able to guard roe movements (e.g. Brno Airport).

System solutions connected with the change of habitat should represent perhaps 70% of the total bird control effort. The preconditions for birds to inhabit the airport are in particular:

✈ food

✈ water

✈ shelter

✈ en-route passage above the airport.

The *possibility of feeding* is probably the most important attraction. Waste dumps are very attractive for the birds. The construction of a new dump should not be situated closer than 13 km from the airport. An effective method which can prevent birds feeding in the waste dump is to cover the dump by nets or wiring. Landfills should also be covered daily with soil to reduce bird food sources. It often happens that visitors to the airport feed the birds. Earthworms are a delicacy for the birds, and there are many of them as well as other insects in the grass and on the paved surfaces after rain. Swarms of insects form over runways that have been heated by the sun, and they attract the birds to this worst possible location. Some species take advantage of a hard surface to break nuts open by dropping them. In the vicinity of the airport, birds of prey can catch mice, moles and other small rodents. Birds are also encouraged to find food on small ponds with an abundance of small fish, frogs, larvae and water plants. Significant foods for the birds are also various berries, seeds and agricultural plants. Because of the great variety of various kinds of bird food, it is practically impossible to remove the sources from the airport. For example during mowing of grass and hay collection, the birds collect the insects behind the machines. In addition to the elimination of natural and artificial sources of food it is possible to use chemicals to reduce the population of insects and rodents. Attention has also to be paid to the system of tilling soil which has to be agreed by the airport administration with the owners of surrounding land.

Surface and standing **water** attracts in particular waterfowl and shorebirds. Small water courses and swamps must be dried and backfilled, as in the vicinity of the Charles de Gaulle airport, or covered by nets, as at Bordeaux/Merignac or Marseilles/Marignane airports, in order to prevent the birds from accessing food. Using wire to cover bodies of water — such as lagoons — inhibits birds from landing. Small water areas could be also protected by nets. The banks surrounding water areas should be graded to a 4:1 slope to discourage birds from resting in the water — birds are unable to spot predators on top of the banks. Wetland areas of different types such as marshes, swamps, bogs or fens are sometimes protected as local systems of ecological stability and are not allowed to be filled or drained. In this case they should be protected by nets.

Birds often seek the *shelter* of airport buildings and long grass areas. Hangars and other premises on the airport can create safe nests. The birds nest directly on and in the buildings, forests, bushes and in other greenery, and sometimes also in the parked aircraft. On the aircraft themselves they nest most frequently in the gaps between the aerodynamic controls and the fixed surfaces, and in the engine nacelles. The nest can be built very quickly, within several hours. The nest can jam the controls, as can objects that are collected by the birds and stored there by magpies, jackdaws, crows etc.

548   Airport Design and Operation

In addition to modifying the construction of hangars and buildings in order to prevent penetration of birds into the buildings, it is necessary to eliminate other suitable places for nesting on the airport. It is advantageous to keep a forest a safe distance from the runway, as it will be an attractive place for birds to live.

The wide open area of an airport gives the birds a secure feeling, as they are always able to see a potential attacker. Seagulls feel so secure that sometimes they bring their food to the airport specifically to eat it there. They make the runway surface dirty, and if they are scared the whole flock rises and makes a wall which cannot be avoided by an aircraft.

Special attention has to be paid to the maintenance of the runway grass strip and other areas covered by grass. It is best if the grass height is kept around 20 cm or more. Only a few species of birds like pheasants and partridges prefer long grass, whereas most birds rest on short grass or they search for food in it. If the grass is higher than 20 cm, the view of the birds is limited and they cannot move easily in the grass. In order to avoid frequent grass cutting and also to make it less attractive, there are special grass mixes, which grow up to the height of around 20 cm and in addition to this, they contain particularly sharp and thorny blades.

Birds get used to attempts to disturb them, and take refuge in the areas adjacent to the airport. Their reactions are unpredictable and are also a serious danger to air operations. Indigenous birds, that is those that live in the vicinity of the airport the whole year, are familiar with the air traffic operation and can avoid the dangerous places, except when young or below average intelligence. A bigger problem is represented by 'strangers', that is those birds which fly into the airport from another locality.

If the airport is located on a migratory route, it may be used by the birds as a resting place. A great number of birds can flock together in this way. The only effective counter measure is not to build the airport in this locality. Information on the movement of birds in the vicinity of the airport is published in the air information publication (AIP).

21.4. ACTIVE MANAGEMENT USING DISPERSAL TECHNIQUES

It is not likely that the Passive Management Techniques, that is changes in habitat, will completely succeed in removing the birds from the vicinity of the airport. In fact, at any time a flock can fly onto the airport and sit on the runway. The sporadic appearance of birds at the airport must then be controlled by dispersal techniques to scare them off, particularly when there are air transport operations. The flock of birds must be scared and dispersed as soon as possible, because a flock of birds on the ground attracts other birds of the same kind and the flock gets bigger. Also, it is easier to disperse a small flock than a big one. The easiest way is to scare the birds before they settle down on the airport.

Bird scaring methods vary in their effectiveness depending on the situation. Some techniques can be used only sparingly, because the birds get used to them. It is often advantageous to combine the techniques or to change the intensity of use. Continual scaring can, nonetheless, significantly reduce the number of birds in the airport. Dispersal techniques discourage birds by scaring them with visual devices such as scarecrows, streamers, or with auditory devices such as cannons and pyrotechnics. The bird scaring schedule must be set in advance in order to avoid repetition. Only in this way is it possible to ensure that the birds do not get used to the scaring procedure. The birds react immediately to some methods and fly away, at other times they only fly away after some delay. Different species of birds react differently to various techniques.

The success of each technique is apparent immediately after it has been used. Dispersal techniques can be divided into:

- sound
- visual
- mechanical catching of birds
- falconry
- chemical.

Among the sound techniques are bio-acoustic techniques, artificial sounds, pyrotechnical means and gas guns.

The bio-acoustic technique is a very successful method of bird scaring. It is based on reproduction of the records of distress calls, warning calls, sounds of birds of prey or command sounds which are made by the flock of birds when taking-off. The most authentic sounds are records obtained directly in the natural conditions of birds reacting to danger. The reaction to warning calls differs between species of birds. Starlings fly away immediately, seagulls and crows rise, they can circle once or twice and then they fly to the loudspeaker and back several times until they gradually disperse, or they disperse only after the call has finished.

The advantage of warning calls is that the reaction of birds is inborn and this technique can therefore be used for a longer time than other techniques. The disadvantage is that the birds react only to the call of their own or closely related species. The warning call of some kinds of birds, for example of doves, is not known. The recording and its reproduction must be high quality, because the birds are capable of recognising that this is not the real call. It takes time to scare birds. Therefore it must not be used immediately before the take-off or landing of the aircraft, because the birds could just fly into its path. Fixed or mobile facilities can be used. The advantage of mobile facilities is that they are cheaper and more effective than fixed systems, and the birds do not get used to them so easily. It is advantageous to combine the bio-acoustic method, for example with shooting a 'very' pistol, but only *after* the call has stopped.

Some airports have successfully tested scaring birds by artificial sounds of various frequencies. In most cases this technique used stationary loudspeakers installed on the approach and along the runway.

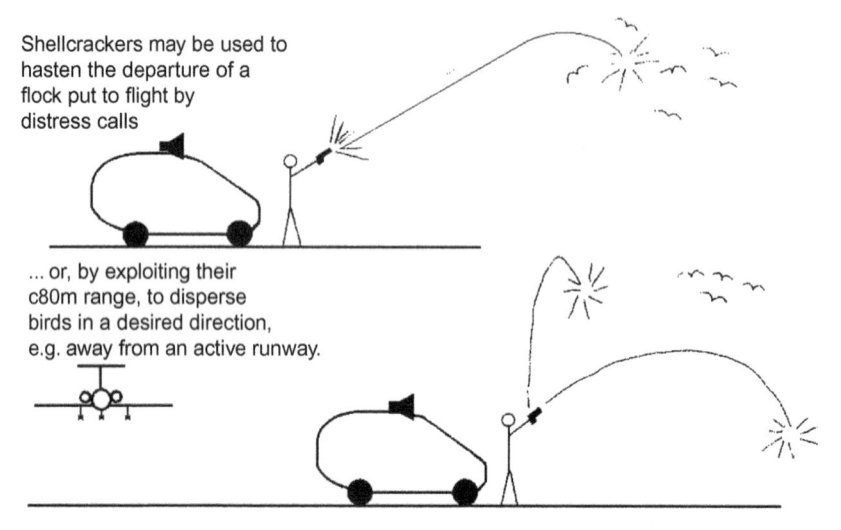

Figure 21.3: Scaring the birds by shell crackers.

After pyrotechnical methods have been used, the birds rise and usually fly away faster than after the use of bio-acoustic techniques. Special pistol shells are used for scaring the birds (Figure 21.3). The first effect is the blast from the gun, then during flight the shell glitters or smokes and at the end of the trajectory it explodes for a second time. The disadvantage is that the birds will be driven away only about 80 m and after a while they will return. It is better to work in two directions at the same time to herd the birds gradually from the danger area.

The gas gun automatic noise scaring device is a modified acetylene lamp with automatic fuse. The disadvantage is that explosions are periodically repeated. The birds find out very quickly that the gas gun is not a real danger for them and take little notice of it, unless this technique is sporadically complemented by, for example live ammunition. Gas guns are therefore less used these days.

Among visual techniques of scaring are:
+ driving off
+ scarecrows, ribbons or flags
+ lights

✈ radio-controlled model aircraft.

An effective way of dispersal is to scare the birds by people who face the birds and move their hands approximately 24 times a minute. They should be standing above the level of the birds' horizon. The birds consider them to be a large attacking predator. Visual techniques of scaring can be used in noise sensitive areas or when the use of pyrotechnics could cause a fire risk.

Static scarecrows or other mechanical means for scaring birds are not effective. Flashing lights, for example on the roof of a car, are used in combination with other methods, for example shooting. The birds gradually develop a conditioned reflex to this. A variation is to drive a car into the flock of birds while blowing the horn.

Figure 21.4: Falconry is highly professional work with high costs. *Photo*: Courtesy Bratislava M.R. Štefánik Airport.

Radio-controlled aircraft can be used to scare and disperse birds from airports. Aircraft models can be painted to resemble falcons, and are now available in the shape of these birds. Radio-controlled models have proven effective at some airports; however, well-trained operators are required.

Among the mechanical techniques of nets or thin wires, which prevent the birds gaining access to food or landing on water, have already been discussed. It also includes using plastic or metal soffits on buildings, which are smooth and thus make building nests more difficult.

Endangered species of birds can be caught and relocated into their new locality. Bird catching is highly qualified and time consuming work, seldom done by airport employees.

The use of falconry, scaring birds by their natural enemies, namely birds of prey, is preferred by some airports (Figure 21.4). The disadvantages of falconry are high acquisition and operational costs. In addition to the trained bird of prey, an all-terrain vehicle, space for locating the birds and training premises, aviaries, guns, hunting dogs etc. also have to be available. Some airports keep even horses to facilitate movement of airport personnel over the terrain. In some countries falconry has been forbidden, because the only suitable predators are protected as endangered species. There is also the issue of the legal responsibility of the airport administrator for the damage in the event of a collision between an aircraft and the trained predator. The bird of prey cannot be used in bad weather and it is not effective for all species of birds.

Use of chemicals is strictly limited in many countries. These are mostly used for reduce the population of insects or rodents, which could serve as feed for the birds. In order to prevent nesting or resting of birds in high inaccessible places like floodlighting masts, special pastes are used which repel the birds and mammals.

21.5. Mammals Control

Mammals could pose either direct or indirect strike hazards to aircraft. Large mammals such as roe deer, deer or wild boar are generally considered direct threats due to their size. Smaller mammals, including rodents, are considered as indirect threats because they attract direct-threat predators such as raptors and Canids. Mammals also pose specific airport-maintenance problems, such as chewing through electric cables, burrowing — which can damage mowing machinery as well as weaken aircraft movement areas — and tunnelling under fences.

Most mammal-control programs can include the following methods:

✈ Fencing to exclude large mammals, including electric fencing

✈ Poisoning through the use of lethal chemicals that repel or kill animals and

✈ Trapping and shooting to remove or disperse animals.

21.6. Ornithological Protection Zones

In some states ornithological protection zones are defined around airports. For example in line with a Transport Canada guideline contained in Land Use in the Vicinity of Airports (TP1247), no bird-attractant land use should be allowed within an 8 km radius of airport reference points. Some other states use a rectangular type of Ornithological Protection Zones in the direction of the runway as the highest bird strike risks are always during the take-off and landing phases. The objective of the declaration of ornithological protection zones is the limitation of bird strikes in the vicinity of airports, where the probability of bird strike is the highest.

In the inner ornithological protection zones, no dumps, stacks or silage must be located. The method of tilling agricultural soil will be agreed by the users of agricultural land with representatives of the airport administration. In some cases, areas belonging to the airport are leased out for agriculture. In these cases it is necessary to agree planting of suitable products.

Outer ornithological protection zones surround the inner ornithological protection zones. It is possible to establish agricultural facilities, such as chicken and pheasant farms, cowsheds, centres for waste collection and processing plants, lakes and other structures and facilities in any significant quantity in this zone only with the approval of the Aviation Authority.

REFERENCES

ACI (2005, December). *Survey on apron incidents/accidents 2004*. Geneva: ACI World Headquarters. Retrieved from http://www.aa2000.com.ar/boletin/downloads/APRON_2004_FINAL_PDF.pdf. Accessed on January 6, 2015.

ACI, DKMA: Ground Transportation. (2012). Airport Service Quality Best Practice. Benchmarking the Global Airport Industry. Report published by ACI.

Airbus Global Market Forecast 2014–2033. Airbus, Toulouse. Retrieved from http://www.airbus.com/company/market/forecast/. Accessed on January 3, 2015.

Airline profiler. *International low-cost airline market research*. Retrieved from http://www.airlineprofiler.eu/. Accessed on January 4, 2015.

Airbus Industrie. A380 Pavement Experimental Programme, Toulouse, France, October 25, 2001.

Airport CDM Operational Concept Document. (2006, September). Eurocontrol, EATMP Infocentre Reference: 05/04/05-1; Edition Number: 3.0. Retrieved from http://www.euro-cdm.org/library/cdm_ocd.pdf. Accessed on January 6, 2015.

Airports Commission. (2014, April). Appraisal framework. UK Airports Commission.

Airports International. (2006, April). Julietta: The New Taxiway.

Alaska Department of Environmental Conservation. *Technical review of leak detection technologies crude oil transmission pipelines*. Retrieved from http://dec.alaska.gov/spar/ipp/docs/ldetect1.pdf. Accessed on September 1, 2014.

Andrea Crisp: Asian airports battle for low-cost business. (2006, March 27). Retrieved from http://www.flightglobal.com/blogs/airline-business/2006/03/asian-airports-battle-for-lowc/#sthash.ukrakXIy.dpuf. Accessed on January 6, 2015.

Ashford, N. J., Saleh, M., & Paul, H. W. (2011). *Airport engineering: Planning, design and development of 21st century airports* (4th ed.). Hoboken, NJ: Wiley.

Ashford, N. J., Stanton, M. H. P., Moore, C. A., Coutu, P., & Beasley, J. R. (2013). *Airport operations* (3rd ed.). New York, NY: McGraw-Hill. ISBN: 978-0-07-177584-7.

Barringer, C. (2006, December 1). International airport review (Issue 4). Improving safety for ground handlers. Retrieved from http://www.internationalairportreview.com/1559/past-issues/improving-safety-for-ground-handlers/. Accessed on January 6, 2015.

Blow, C. (1996). *Airport terminals* (2nd ed.). Oxford: Architectural Press.

Boeing. (2014). *Current market outlook*. Seattle, WA: Boeing Commercial Airplane Group.

Boeing Current Market Outlook 2014–2033. Retrieved from http://www.boeing.com/boeing/commercial/cmo/. Accessed on January 3, 2015.

Brons, M., Pels, E., Nijkamp, P., & Rietveld, P. (2002). Price elasticities of demand for passenger air travel: A meta-analysis. *Journal of Air Transport Management*, 8, 165–175.

Burghouwt, G. (2007). *Airline network development in Europe and its implications for airport planning*. Aldershot: Ashgate.

CAP 1165, Managing Aviation Noise. (2014). UK Civil Aviation Authority.

Caves, R. E., & Gosling, G. D. (1999). *Strategic airport planning*. Oxford: Elsevier. ISBN: 0-08-042764-2.

Čihař, J. (1973). *Letiště a jejich zařízení I (Airports and its facilities I)*. Alfa Bratislava, Bratislava.

Commission Regulation (EU) No 965/2012 of 5 October 2012. Laying down technical requirements and administrative procedures related to air operations pursuant to Regulation (EC) No 216/2008 of the European Parliament and of the Council.

Commission Regulation (EU) No 139/2014 of 12 February 2014. Laying down requirements and administrative procedures related to aerodromes pursuant to Regulation (EC) No 216/2008 of the European Parliament and of the Council.

de Neufville, R., & Odoni, A. (2003). *Airport systems: Planning, design and management*. New York, NY: McGraw-Hill. ISBN: 0-07-138477-4.

de Neufville, R., & Odoni, A. (2013). *Airport systems: Planning, design, and management* (2nd ed.). ISBN-13: 978-0071770583. McGraw-Hill Professional.

de Wit, J. (1995). An urge to merge? – Competition and concentration in the European airline industry. Department of Civil Aviation, Ministry of Transport, Netherlands.

DIN 45631. Berechnung des Lautstärkepegels und der Lautheit aus dem Geräuschspektrum. Verfahren nach E. Zwicker.

DIN 45643:2011. Messung und Beurteilung von Fluggeräuschen [Measurement and assessment of aircraft sound].

Directive 2008/101/EC of the European Parliament and of the Council of 19 November 2008 amending Directive 2003/87/EC so as to include aviation activities in the scheme for greenhouse gas emission allowance trading within the Community. Retrieved from http://eur-lex.europa.eu/legal-content/EN/TXT/?uri=CELEX:32008L0101

EASA Certification Specifications and Acceptable Means of Compliance for Large Aeroplanes CS-25. (2013, December 19). Amendment 14.

ECAC. (2010, May). European Civil Aviation Conference Doc 30, Part II (restricted) ECAC Policy Statement in the Field of Civil Aviation Security (13th ed.).

Edwards, B. (2005). *The modern airport terminal* (2nd ed.). Oxford: Spon Press. ISBN: 978-0-415-24812-9.

FAA. (1988a, April 22). AC No: 150/5360-13: *Planning and design guidelines for airport terminal facilities*. Retrieved from http://www.faa.gov/documentLibrary/media/Advisory_Circular/150_5360_13.PDF. Accessed on June 1, 2015.

FAA. (1988b, April 4). AC 150/5360-9. P*lanning and design of airport terminal building facilities in nonhub locations*. Retrieved from http://www.faa.gov/documentLibrary/media/Advisory_Circular/150_5360_9.pdf. Accessed on January 6, 2015.

FAA. Change 2 – Design of aircraft de-icing facilities FAA. AC 150/5300-14. Washington, DC: US Department of Transportation, Federal Aviation Administration.

FAA. (2012). *Order 8000.94: Procedures for establishing airport low visibility operations and approval of low-visibility operations/surface movement guidance and control system operations* (21.8.12).

FAA. (2014). Aerospace forecasts, fiscal years 2014–2034. Washington, DC: US Federal Aviation Administration.

FAA Fact Sheet — Engineered Material Arresting System (EMAS). (2014, August 20).

Flyvbjerg, B., Mette, K., Holm, S., & Buhl, S. L. (2006). Inacccuracy in traffic forecasts. *Transport Reviews*, 26(1), 1–24.

Global Reporting Initiative (GRI) Sustainability Reporting Guidelines & Airport Operators Sector Supplement. 2000-2011 GRI Version 3.1/AOSS Final Version. Retrieved from https://www.globalreporting.org/resourcelibrary/AOSS-Complete.pdf. Accessed on December 21, 2012.

Horonjeff, R., & Mckelvey, F. X. (2010). *Planning and design of airports* (5th ed.). New York, NY: McGraw-Hill. ISBN: 13 978-0071446419.

Hromádka, M. (2014). *Optimalizácia procesov technického odbavenia lietadiel (Aircraft technical handling process optimization)*. Dissertation thesis, University of Žilina.

IATA. AHM, Airport Handling Manual.

IATA. (2004). *Airport development reference manual* (9th ed., p. 710). Montreal: International Air Transport Association (IATA). ISBN: 92-9195-086-6.

IATA. (2014). *Airport Handling Manual (AHM)* (35th ed.). Montreal: International Air Transport Association. ISBN: 978-92-9252-458-6

IATA Operational Safety Audit (IOSA). Retrieved from http://www.iata.org/whatwedo/safety/audit/iosa/Pages/index.aspx. Accessed on July 12, 2014.

ICAO Doc 9137. *Airport services manual*. Part 1 — Rescue and Fire Fighting and ICAO Fire Fighting Foam Testing.

ICAO Doc 9137. *Airport services manual*. Part 9 — Airport Maintenance Services.

ICAO Doc 9640-AN/940. *Manual of aircraft ground de-icing/anti-icing operations*.

ICAO. (1983a). Doc 9137-AN/898/2. *Airport services manual* (2nd ed.). Part 6 — Control of Obstacles.

ICAO. (1983b). Doc 9157-AN/901. *Aerodrome design manual* (1st ed.). Part 5 — Electrical Systems, Montreal.

ICAO. (1993). *Aerodrome design manual*. Part 3, Pavements, Doc 9157-AN/901, 1991, Corrigendum N 1, 1993, Montreal: International Civil Aviation Organisation.

ICAO. (2002a). Airport services manual (4th ed.). Part 2 — Pavement Surface Conditions (Chapter 7 snow removal and ice control), ICAO Doc 9137-AN/898, Montreal.

ICAO. (2002b). Doc 9137-AN//898, *Airport services manual* (4th ed.), Part 2 airport surface conditions.

ICAO. (2006a). *Aerodrome design manual* (3rd ed.). Part 1 — Runways, Doc 9157-AN/901.

ICAO. (2006b). *Aeronautical telecommunications*, Annex 10, Radio Navigation Aids (Vol. I, 6th ed.), International Civil Aviation Organisation, Montreal.

ICAO. (2006c). *Manual of air traffic forecasting* (3rd ed.). Doc. 8991. Montreal: International Civil Aviation Organisation.

ICAO. (2006d). *Procedures for Air Navigation Services. Construction of Visual and Instrument Flight Procedures* (Vol. II, 5th ed.), Aircraft operations. Doc 8168 OPS/611.

ICAO. (2007). *Review of noise abatement procedure research & development and implementation results; discussion of survey results; Montreal*. Retrieved from http://www.icao.int/environmental-protection/Documents/ReviewNADRD.pdf. Accessed on January 6, 2015.

ICAO. (2008, July). Annex 16. *Environmental protection*, Volume II Aircraft Engine Emissions; 3rd ed.

ICAO. (2009, July). Annex 14. *Heliports* (Vol. II, 3rd ed.).

ICAO. (2010, July). Annex 6. *Operation of aircraft*. Part I: International Commercial Air Transport — Aeroplanes (9th ed.).

ICAO. (2011). Doc 8973 (Restricted) *Aviation security manual* (8th ed.).

ICAO. (2012a). Doc 9137-AN/898 *Airport services manual*, Part 3 — Wildlife Control and Reduction, 4th ed.

ICAO. (2012b, September). EUR Doc 013 European guidance material on aerodrome operations under limited visibility conditions (4th ed.).

ICAO: (2012c). EUR Doc 013 European guidance material on all weather operations at aerodromes (4th ed.).
ICAO. (2013a). Annex 3. *Meteorological Service for International Air Navigation* (18th ed.).
ICAO. (2013b). Annex 14. *Aerodrome design and operations* (Vol. I, 6th ed.). Aerodromes. International Civil Aviation Organisation, Montreal.
ICAO. (2013c). Annex 14. *Heliports* (Vol. II, 4th ed.). Aerodromes. International Civil Aviation Organisation, Montreal.
ICAO. (2013d). Worldwide Air Transport Conference (ATCONF) (Sixth meeting). Night Flight Restrictions, Montréal, 18−22 March, ATConf/6 — WP/8, January 23.
ICAO. (2014a, July). Annex 16. *Environmental protection*. Aircraft noise (Vol. I, 7th ed.).
ICAO. (2014b, November 14). Annex 17. Safeguarding International Civil Aviation against Acts of Unlawful Interference, Amendment 14.
IEC 561. Electro-acoustical measuring equipment for aircraft noise certification.
IEC 616721-1. Electroacoustics — Sound level meters — Part 1: Specifications.
Index Mundi. (2011). *World jet fuel consumption by year*. Retrieved from http://www.indexmundi.com/energy.aspx. Accessed on January 6, 2015.
International Birdstrike Committee. (2006, October). Recommended Practices No. 1. Standards for Aerodrome. Bird/Wildlife Control. Issue 1.
ISO 20906-2009. Acoustics — Unattended monitoring of aircraft sound in the vicinity of airports.
Juan, C., Olmas, F., & Ashkeboussi, R. (2012). Flexible strategic planning of transport systems. *Transport Planning and Technology, 35*(6), 629−662.
Kahneman, D., & Tversky, A. (1979). Intuitive prediction and corrective procedures. *Management Science, 12,* 313−327.
Kazda, A. (1988). *Evaluation of the usability factor of aerodromes*. PhD thesis, University of Transport and Communications, Žilina.
Kirkland, I., & Caves, R. E. (1998, September). *Risk of overruns*. ACI Europe Airport Safety Symposium, Riga.
Kováč, M., Remišová, E., Čelko, J., Decký, M., & Ďurčanská, D. (2012). Diagnostika parametrov prevádzkovej spôsobilosti vozoviek. ISBN: 978-80-554-0568-1.
Krollova, S. (2005). *Dynamic* climatological data processing of occurrence of aviation weather hazards at Bratislava. Poprad-Tatry, Sliač and Košice airport. Dissertation work, University of Žilina.
Kurzweil Dr., L. (2014, October 31). Interní správa, Letište Praha.
Lythgoe, W. F., & Wordman, M. (2002). Demand for rail travel to and from airports. *Transportation, 29,* 125−143.
NAVAIR. (1992). *Aircraft refuelling handbook*. Naval Air Systems Command (NAVAIR). Retrieved from http://www.dtic.mil/dtic/tr/fulltext/u2/a261755.pdf. Accessed on December 17, 2013.
NAVAIR: Aircraft Refuelling Handbook. Naval Air Systems Command (NAVAIR). (1992). Retrieved from http://www.dtic.mil/dtic/tr/fulltext/u2/a261755.pdf. Accessed on December 17, 2013.
Regulation (EC) No 300/2008 of the European Parliament and of the Council of 11 March 2008. Common Rules in the field of civil aviation security and repealing Regulation (EC) No 2320/2002.
SITA: Air Transport Industry Insights; Airport IT Trends Survey. (2013).
SITA: Air Transport Industry Insights; The Passenger IT Trends Survey. (2014). Airports International, July 2006.

Šoltís, J. (1982). *Zborník prác Hydrometeorologického ústavu v Bratislave*. Zväzok 19. Prúdenie vzduchu na Slovensku. Alfa Bratislava.

The World Bank. (2015, January 6). Air transport, passengers carried. Retrieved from http://data.worldbank.org/indicator/IS.AIR.PSGR

Transport Canada. (1992). *Airports*. Retrieved from http://www.tc.gc.ca/eng/civilaviation/publications/tp13549-chapter6-406.htm. Accessed on January 6, 2015.

Turiak, M. (2015). *Airport typology – Impact of low cost carriers operations*. Dissertation thesis, FPEDAS ŽU, Žilina.

UK Civil Aviation Authority. (1991). Passengers at the London Airports, CAP 610, London.

Weller, J. R. (2014, March 24–26). Wildlife hazards at airports regulatory oversight. Program Evaluation and Strike Reporting Wildlife and Foreign Object Debris (FOD) Workshop, Cairo, Egypt.

INDEX

Aborted take-off, 82, 96, 114
Accelerate stop distance, 81, 83–86, 93
Accelerometer, 170, 179
Access mode, 23, 346
Access road, 4, 256, 310, 351, 453, 497
Acetate, 456
ACN-PCN, 149, 163, 167
Adhesion, 177–178, 197, 199
Aerodrome category, 474–478, 480, 482
Aerodrome reference code, 15, 73–75, 95
Aeronautical study, 66, 68, 399
Air Waybill, 250
Airbridge, 130, 138, 223, 231, 267
Air-conditioning, 186, 190, 195
Aircraft accident, 17, 325, 490, 494, 497, 506
Aircraft kerosene, 211, 214, 217, 229
Aircraft performance, 85, 90, 528
Aircraft refuelling, 189, 192, 211, 216, 221–223, 229, 232
Airport access, 45, 48, 344–345, 355, 359
Airport perimeter, 321, 323, 493
All Weather Operation Panel, 14, 374
ALPA ATA, 384, 388
Amsterdam, 7–8, 35, 134, 193, 279, 321, 362, 495, 539
Anchorage, 239–240
Anti-icing, 453, 456, 461–462, 468
Approach speed, 75, 95, 118, 374, 532–534
Approach surface, 63–64, 66
Apron accidents, 187
Arresting system, 107, 109

Arrestor bed, 109–111
A-SMGCS, 61, 202, 404, 421, 437, 494
Asphalt concrete, 148
Asphalt pavement, 97, 127, 150–151, 457
Atlanta, 8, 132–133, 269, 300–301
Automatic Ticket and Boarding (ATB), 288
Auxiliary Power Unit, 186, 195, 534

Background check, 239, 326
Baggage handling, 203, 270, 273, 288, 295–296, 490
Baggage reconciliation, 296, 329
Balanced field length, 75, 86
Bearing strength, 98, 113, 126, 145, 147–149, 151, 157, 160, 163, 494
Beijing, 4, 8, 10, 189–190, 224
Belly hold, 24, 234, 242, 254
Berlin, 2, 7, 203
Biodiversity, 231, 506
Biometric, 280, 290, 325–326, 328, 331
Bird
  catching, 552
  dispersal, 548
  food, 547
  indigenous, 544
  management, 544
  migratory, 545
  nest, 547
  of prey, 547, 549, 552
  scaring, 549
  transient, 544
Bird strike, 541, 543–544, 553

Bird Strike Information System, 22, 543
Birmingham, 300, 357, 499
Blast fence, 129, 142
Block paving, 127, 159
Bomb, 6, 307, 311, 319, 321, 339, 494
Brake-away thrust, 138, 140
Braking action, 48, 96, 145, 168, 176, 179–180, 453, 459
Braking distance, 83, 118, 196
Brussels, 30, 132, 202–203, 240, 258, 357
Bus, 134, 257, 274, 311, 345, 355–356, 361

Carbon dioxide, 456, 478–479, 504–505, 533
Carousel, 284–285, 296
Carrier
  combination, 237–239
  integrated, 42, 237, 241, 255, 257
  legacy, 235–236, 270, 275
  low-cost, xxiii, 5, 9–10, 24, 34, 37, 40, 43, 185, 193, 236, 270, 287, 344, 350, 356
  scheduled, 271
  traditional, 263, 344, 350
Catchment area, xxiii, 20, 47, 240, 298, 343, 347, 358, 362, 503
Category I, 51, 55, 57, 61, 63, 372, 379, 384, 387–388, 391, 401–402, 405, 411, 435
Category II, 51, 53, 57–58, 64, 379, 388, 391, 401, 406
Category III, 55, 57, 60, 259, 379, 391, 422, 430
CBR, 104, 149, 165, 167
CCTV, 250–251, 259, 315, 325
Cement concrete, 97, 148, 151
Charles de Gaulle, 10, 47, 132, 202, 219, 269, 354, 547
Check-in, 263, 267, 280
  common, 287
  counter, 263, 274, 288, 297
  curb, 289
  flight, 287
  internet, 278, 287, 304
  kiosk, 303
  mobile, 287, 304
China, 7, 10, 44, 240
Clearway, 86, 90, 93, 106
Coach, 290, 344, 349–350, 356–357
Coefficient of longitudinal friction, 174–176
Complementary agent, 475, 477, 480, 482
Composite, 148, 167, 189, 233, 332, 469, 472, 543
Compressed air, 189, 408, 498
Constant current regulator, 387, 391, 399, 401–403, 418, 420–422
Continuous Descent Approach, 213, 258, 530, 532
Convention
  Hague, 312
  Montreal, 312–313
  Tokyo, 312
Conveyor, 196, 242, 245–247, 255–256, 265, 276, 279, 284, 296, 336
Corrosion, 214, 220, 379, 391, 453, 461, 470, 475
Critical aircraft, 75, 78, 82, 110, 156, 167, 474
Critical and sensitive areas, 60, 122, 428
Critical area, 391, 428, 475, 479, 546
Critical engine, 84
Crosswind, 1, 3, 48, 54, 531
Cul-de-sacs, 134–135, 535
Customs, 24, 236, 247, 250, 253, 257, 259, 262, 283, 286, 288, 290, 327, 543
Cut-off angle, 58

Dallas-Fort Worth, 8, 270
De/anti-icing, 443, 468
Dead Man, 222, 229
Deceleration, 104–105, 111, 118–119, 279
Decision height, 55, 428
Decision speed, 82, 84, 106
De-icing, 453, 461

De-icing fluid, 405, 454, 456, 461–462, 467, 470, 537
Delay, 7, 24, 61, 70, 128, 142, 185, 187, 200, 243, 258, 272, 281, 298, 343, 364, 397, 435, 524, 549
Denver, 8, 269, 300
Deregulation, 5–6, 27, 41, 43
Diesel generator, 410, 412, 414–415
Disability, 274, 293
Disabled persons, 293–294
Dispenser, 220, 222–223, 230
Distribution system
  central, 195
  hydrant, 222
  parallel, 416
  serial, 387, 416–417
Dogs, 294, 332, 337–339, 552
Doorsill height, 192
Door-to-door, 30, 236, 343, 350, 353
Dowel bar, 154–155
Dowels, 154
Drainage, 99, 105, 147, 154, 169, 172, 175, 177, 226, 243
Drop off, 240, 275, 352, 354
Dubai, 4, 8, 215, 224, 240, 246
Dust, 215, 405, 473, 546
Dwell time, 244, 254, 273–274, 297, 349–350
Dynamometer, 175

EASA, 17, 55, 57, 79, 92, 315
East Midlands Airport, 111, 258, 357, 524
ECAC, 18, 200, 323, 503, 528, 530
Elasticity, 27–28, 29
EMAS, xxiii, 107, 109–110
Emergency landing, 320, 493–494, 496
Emissions, 189, 196, 229, 231, 337–338, 361, 504–505, 533–535
Engine failure, 82, 84, 92, 96, 107
Engineered Material Arresting System, 109
Environmental impact, 5, 19, 45, 343, 346, 361, 457, 504

Environmental protection, 468–469, 499, 503
Explosive Detection System, 336
Explosives, 6, 251, 307, 310, 319, 330–331, 335, 337
Extinguishing agent, 229, 472–473, 475, 477–478, 480, 482, 499
  complementary, 478
  principal, 475, 477, 480

Falconry, 549, 552
False alarms, 323–324
Feasibility
  economic, 21
  financial, 20, 23
  technical, 23, 457
Fence, 231, 324, 552
Fencing, 60, 231, 251, 322–323
  electric, 552
Filter
  blue, 390
  clay, 214, 219
  colour, 389, 391
  double-stage, 217
  fine, 217
  fuel, 215
  glass, 390
  micro, 214
  red, 394
  security, 318
  yellow, 390
Final approach and take-off area, 64, 397
Fixed ground systems, 186–187, 189
Flywheel, 413–414
Foam blanket, 494–496
Foam concentrate, 475, 480
Forecast, 23, 34–35, 37–38, 42
  air traffic, 38
  alternative, 459
  demand, 20, 39
  long-term, 24–25, 46
  market, 20
  meteorological, 444
  national, 27

rail, 34
regional, 44
short term, 24
traffic, 46
Fork lift, 245, 247, 259
Formate, 456–457
Frangibility, 373, 379, 384
Frankfurt, 8, 134, 136, 203, 233, 263, 344, 520
Free water, 215, 217
Freezing rain, 461, 465–466
Fuel consumption, 113, 211, 504
Fuel escapes, 226–227
Fuel farm, 4, 231–232, 340
Fuel hydrant, 186, 223
Fuel stores, 211, 217–218, 226, 536

Gas analyser, 331–332, 337–338
Gatwick, 7, 32, 34, 71, 218, 270, 300, 344, 357, 360, 364, 532
Gatwick Express, 358, 360
Gibb's phase rule, 454
Global warming, 232, 505, 533
Glycol, 456–457, 461–462, 469, 537
  collection, 537
  diethylene, 457
  ethylene, 337, 457
  pollution, 537
Glycol collection, 537
Gradient, 100
  climb, 84, 92
  gross, 91
  minimum, 91, 216
  net, 91
  temperature, 152
Grass carpet, 147–148
Gravel, 104–105, 147, 150–151, 173, 500
Grooving, 99, 169–170
Ground water, 468, 504–505, 535, 546

Habitat, 538, 546, 548
HAPI, 397–398
Heathrow, 4, 8, 10, 12, 120, 132, 219, 261, 269, 271, 282, 351, 493

Hijack, 305–308, 311–312, 333
Hold baggage, 295, 316–317, 330
Holdover time, 461, 463, 465–466
Hong Kong, 4, 8, 10, 120, 218, 240, 259, 276, 344, 364
Hub airports, 4, 298, 353, 356
Hub and spoke, 10, 72, 269, 344, 357
Hydrant system, 221–222, 224–225, 228
Hydrogen, 214, 232
Hygroscopic, 215, 444, 464

Immigration, 24, 247, 262, 274, 283, 289, 315
Impurities, 215, 217
Income, 23, 27, 29, 36–37, 331, 503
Infrared, 309, 324, 440
Instrument Flight Rules, 71, 423
Instrument Landing System, 60, 103, 424, 532
Investment, 36, 40, 43, 45, 57, 113, 147, 186, 197, 224, 281, 303, 409, 459
Isolation transformer, 390, 417–420

Jet A1, 211, 214–215, 218
Jet blast, 75, 125, 132, 137, 382, 391
Joints, 151–155, 159
  construction, 154
  contraction, 153–154
  expansion, 153
  isolation, 154
  longitudinal, 154

Kansai, 4, 113, 263
Kerb, 262, 268–269, 274–275, 291, 352–353, 355
Kerosene, 4, 127, 216, 226, 231–232, 499
Kuala Lumpur, 270, 300, 309, 358

Land use and compatibility planning, 504, 507, 520, 522
Leipzig, 240, 258, 279
Level of service, 69, 262, 272, 275, 280, 365

Liberalisation, 6, 9, 34
Light
  approach, 111, 378, 380, 385, 391, 397, 399, 537
  inset, 375–377, 381, 389, 404, 407, 421, 449
  LED, 376, 401, 535
  obstacle, 376, 399–400
Load factor, 24, 37–39, 44, 213, 271, 283, 295
Loading bridge, 132, 137–138, 140, 204, 223, 242, 267, 535
Localiser, 57, 108, 111, 425
Lockerbie, 6, 307, 328, 489
Long haul, 9, 62, 195, 221, 237, 269, 343, 349
Longitudinal friction, 146, 172, 174, 176
Longitudinal slope, 81, 99–101
LOS, 272–273, 276, 279, 283, 285
Los Angeles, 8, 131, 498
Loudness, 507, 509–510
Low Visibility
  conditions, 64
  operations, 60
  procedures, 56, 103
  take-off, 57
Luton, 34, 159, 286, 352

Macro-texture, 170, 172–174
Madrid, 10, 111, 203, 301, 357, 360, 443
Mammals, 541, 552
Manchester, 10, 30, 111, 227, 357, 536
Manoeuvring areas, 4, 70, 148, 447
Markings, 56, 113, 292, 367, 369–370, 374, 400
Marshaller, 137, 204
Master plan, 18–19, 46, 114, 243
Master planning, 16, 23, 42, 45, 243
Maximum take-off mass, 77, 81, 114, 165, 513
Meeters and greeters, 23, 348, 364, 489
Metal detector, 306, 308, 332
Method
  ACN-PCN, 163, 165

  bio-acoustic, 550
  dual mass, 168
  Durst, 50, 55
  dynamometer, 175
  judgemental, 35
  leak-detection, 228
  one-step, 468
  predictive, 516, 522, 530
  quantitative, 35
  sand, 175
  scaring, 549
  step-down, 27
  three segment, 119
  transparency, 49
Metro, 290, 360–361, 363
Micro-organisms, 214, 217
Micro-texture, 172, 174–175, 176
Middle East, 9–10, 44, 72, 307, 354
Milan Linate, 131, 133
Milan Malpensa, 10, 203
Missed approach, 55, 103, 533
Missile, 309
Mobile lounge, 131, 133, 144, 192, 300
Mobile stairs, 191–193
Model
  3D, 206
  aircraft, 551
  Airfield Capacity, 70
  behaviour, 30
  calibration, 34
  Collision Risk, 66
  disaggregate, 33
  distribution, 30
  econometric, 29, 34, 36, 43
  FAA, 272
  IATA, 272
  Integrated Noise, 526
  logit, 32
  macro, 37
  modal choice, 30
  simulation, 69
Modes of transport, 246, 315, 344, 346–347, 355
Monitoring system, 231–232, 402, 408, 420–421, 424, 459, 520

Munich, 4, 7, 9, 46, 131–132, 203, 261, 357, 412, 422, 469, 537

Nesting, 545, 548, 552
No-break, 412–413, 415
Noise
  abatement, 530–531
  aircraft, 512
  airport, 510
  attenuation, 527
  background, 521
  complains, 514
  continuous monitoring, 520
  contours, 514
  control, 508
  descriptors, 512
  emissions, 527
  energy, 511
  evaluation, 514–515
  external, 504
  impact, 45
  indices, 507
  insulation, 531
  intensity, 507
  limits, 518, 520–521
  long-term monitoring, 522
  maximum, 507
  mitigation, 530
  monitoring, 506, 513, 523–524
  night, 519
  nuisance, 364
  peaks, 515
  percieved, 510, 513
  prediction, 528
  protection wall, 531
  rating, 508
  reduction, 507
  strategic maps, 529
  transport, 512–513, 516
  zones, 519, 521
Nose-in, 129, 134, 136–137, 139, 197, 207
Nosewheel, 123, 206, 209

Obstacle clearance, 48, 59, 75, 90–91, 397, 399

Obstacle limitation surfaces, 62–65
One engine inoperative, 82–83, 85–86, 92
Operational towing, 197, 199
Oslo, 10, 203, 269, 357, 360
Outside air temperature, 81, 88, 91, 114
Overrun, 106–107, 109
Ozone layer, 478, 499, 505

PAPI, 393–395, 403, 421
Passenger bridge, 186, 190–193
Payload, 77–78, 235–236
Peak hour, 7, 24, 42, 44, 54, 72, 114, 117, 128, 142, 269, 273, 302, 331
People mover, 133, 268–269, 299–301
Performance, 110
  aircraft, xxiii, 74, 80, 463, 528
  approach, 431
  average, 80
  battery, 415
  characteristics, 82
  class A aircraft, 80
  class B aircraft, 81
  class C aircraft, 81
  extinguishing agent, 475
  flight, 441
  foam, 475
  gross, 80
  helicopter, 397
  ILS, 426
  landing, 89, 119
  level A, 476–477, 480
  level B, 476–477
  level C, 476–477, 480
  limitations, 88
  measured, 80
  navigation, 430
  net, 80
  radar, 219
  runway, 80
  snow blower, 449, 451
  take-off, 86
  wayfinding, 292

Planning horizon, 45–47
Plastic explosives, 313, 330, 337, 339
Precision approach runway, 55, 63, 96, 100, 103, 375, 382, 404–405
Propensity to fly, 26, 37–38
Protective fluids, 462, 464, 467
Protein foam, 475–476, 494
Push-back, 135, 137–138, 139

QR code, 296, 329

Radius
  of curvature, 101
  of relative stiffness, 157, 160
  relative stiffness, 157
  rolling, 178
  static, 178
  turning, 137–138, 243, 279
  turn-off, 116
Rail
  access, 258, 357, 359, 361
  air connection, 240
  conventional, 362
  dedicated, 357–358, 360
  deliveries, 218
  elevated, 362
  forecast, 34
  heavy, 290, 360–361
  high speed, 30, 358, 360–361
  integrated, 246, 257
  light, 357
  link, 360
  maintenance, 217
  municipal, 356
  performance, 34
  services, 26, 356, 358, 362
  share, 30
  station, 30, 300, 345, 361
  transport, 346, 357–359
Railway, 125, 215, 218, 345, 360, 363
Railway station, 301, 345, 360
Recycling, 469, 499, 537–539
Reference field length, 48, 75
Refinery, 214, 216–217
Relative slippage, 178

Reliability, 38, 134, 299, 301, 322, 330, 347, 349, 365, 374, 431, 439
RESA, xxiii, 107, 109, 112, 425
Response time, 473, 482, 487
Reverse thrust, 79, 89, 530–531
RFFS, 471, 473, 483, 486, 494
RFID, 279, 329
Rio Galleao, 131, 286
Rotation speed, 84, 463
Rubber deposits, 174, 405
Runway
  approach, 63
  extension, 77
  main, 3
  non-instrument, 63
  precision approach, 64
Runway end safety area, 107–108, 111
Runway End Safety Area, 425
Runway length, 2, 77–79, 87, 98, 106, 368, 471
Runway occupancy time, 70, 114
Runway reference code number, 103
Runway shoulders, 97, 105–106
Runway strip, 58, 93, 95, 103–104, 106, 109, 147, 378
Runway system, 3, 7, 10, 45, 48, 50, 69–70, 127, 149, 391, 487
Runway visual range, 54–55, 388, 402

Sabotage, 306–307, 311–312, 328
SAMs, 308–309
San Francisco, 201, 361
Schengen, 264, 267–268, 286, 326
Schiphol, 36, 193, 321, 359, 378, 539
Seattle, 246, 301
Security, xxiii, 52, 60, 135, 203, 231, 233, 239, 246, 251, 256, 262, 264, 266, 270, 275, 281–283, 288, 354, 360, 365
  centralised, 318
  decentralised, 318
Security management system, 315, 340
Security restricted areas, 316–317
Sedimentation, 216, 219, 536

Self-manoeuvring, 129, 132, 137, 139–140
Semtex, 335, 337
Seoul, 10, 239
Services
  aeronautical, 266, 297
  non-aeronautical, 3, 19, 297
Shanghai, 10, 240, 261, 364
Shannon, 218, 281
Shipper, 34, 233, 237–238, 244, 246–247, 259
Short-break, 410, 412–413, 415
Signs, 4, 105, 168, 292, 326, 353, 372, 375, 404, 437
Singapore, 8, 254, 261, 270, 318, 487
SITA, 249, 283, 295, 303–304
Slipform paver, 152, 155
Slippage, 178, 180
Slush, 87, 89, 442, 444, 461
Smart phone, 302, 355, 524
Snow blower, 111, 447, 450, 452
Snow removal, 111, 124, 442, 444, 450, 452
Soekarno-Hatta, 8
Sound exposure level, 511–513, 515, 517, 526, 528
Sound pressure, 508, 511–513, 529
Spilled fuel, 227, 472, 477, 495
Standard of living, 46, 346, 350, 503
Standards, 9, 14–15, 63, 80, 122, 168, 220, 266, 320, 536
Standards and Recommended Practices, 15, 17, 64, 314
Stansted, 34, 218, 258, 270, 276, 290, 532
Static electricity, 219, 229
Stockholm, 72, 205, 224, 357, 469, 493
Stopway, 79, 86, 90, 93, 95, 106
Storage tanks, 215–216, 218, 220–221, 223, 227, 230
Subgrade, 149–150, 152, 155, 157, 163, 167, 169
Suicide bomber, 311, 329
Surveillance, 57, 202, 315, 324, 424, 435–437

Tactile check, 468
Take-off and landing, 69, 96, 533, 553
Take-off climb, 63–64, 67–68, 84, 91
Take-off distance, 81, 83, 85–86, 90, 92–93, 106
Take-off length, 83, 87, 89
Take-off run, 86, 93, 368, 441, 544
Taxi, 353, 355
Taxi out, 127, 136, 138
Taxilane, 121, 129, 137
Taxiway width, 113, 123
Technical handling, 132, 185, 268
Teleconferencing, 25, 31
Terminal
  access, 203, 348, 353
  airport, 263
  automated, 254
  building, 4, 6, 127, 130, 132, 136, 139, 142, 204, 264–265, 350
  capacity, 69, 271, 286, 349
  cargo, 18, 24, 128, 243, 251–252
  central, 132, 135, 300
  centralised, 269–270
  charter, 273
  concept, 267, 269
  connection, 265
  decentralised, 269
  domestic, 273
  facilities, 3
  freight, 242–243
  frontage, 132–133, 300
  fuel, 217
  layout, 267
  life cycle, 271
  low-cost, 278, 304
  manual, 254
  midfield, 133
  multi-level, 257
  multiple, 268
  operation, 349
  passenger, 4, 9, 24, 243, 261, 487
  remote, 300
  satellite, 290
  simulation, 275
  single level, 275

traditional, 270
transportation, 299
two level, 268
wayfinding, 290
Terrorism, 5, 38, 251, 307, 312, 325
Thickened fluid, 462, 467
Threat, 37, 128, 251, 306–308, 312, 314, 317, 322, 336, 340, 494, 544
Time horizon, 23, 34, 46, 149, 271, 364
Tokyo, 8, 31, 239, 344, 498
Tokyo Convention, 311
Topsoil, 105, 147
Touch-down, 96, 116, 174, 377, 389, 496
Towbar, 197, 199
Towbarless, 139, 198–199
TPHP, 271–273, 275
Transport
  private, 347
  public, 345, 355, 357
  rail, 345, 357
  surface, 348, 353
Turf, 97, 105, 147
Turnround, 24, 128, 189, 193, 195, 202, 211–222, 236, 259, 299
Two-step procedure, 463, 467
Type I, 461, 468
Type II, 462, 467–468, 537

Type IV, 463–464, 537
Typical peak hour, 48, 271
Tyre pressure, 161–162, 165, 167

ULD, 244, 246, 254–255
Ultimate capacity, 18, 45, 243
Unlawful acts, 5, 305, 311–312
Upper deck, 192–193, 283
Urban areas, 45, 51, 109
Urea, 453–454, 456–457, 478
Usability factor, 48–49, 50, 54–55

V1, 82, 84, 86, 91, 107
Veering-off, 103–104
Vehicle free apron, 186, 190
Vienna, 7, 218, 268, 281, 307, 318, 357–358

Washington Dulles, 131, 133, 269, 301
Waste, 189, 281, 286, 455, 505, 537–539
Wayfinding, 263, 266, 290, 292, 294, 304
Wide body, 4, 39, 187, 215, 221, 233, 237–238, 285, 298

X-ray, 251, 280, 288, 307, 310, 329–330, 333–337